VOLUME ONE HUNDRED AND FIFTEEN

# ADVANCES IN HETEROCYCLIC CHEMISTRY

VOLUME ONE HUNDRED AND FIFTEEN

# ADVANCES IN HETEROCYCLIC CHEMISTRY

Editors

**ERIC F. V. SCRIVEN**
*Department of Chemistry,
University of Florida,
Gainesville, FL, USA*

**CHRISTOPHER A. RAMSDEN**
*Lennard-Jones Laboratories,
Keele University, Staffordshire,
United Kingdom*

ELSEVIER

AMSTERDAM • BOSTON • HEIDELBERG • LONDON
NEW YORK • OXFORD • PARIS • SAN DIEGO
SAN FRANCISCO • SINGAPORE • SYDNEY • TOKYO
Academic Press is an imprint of Elsevier

Academic Press is an imprint of Elsevier
225 Wyman Street, Waltham, MA 02451, USA
525 B Street, Suite 1800, San Diego, CA 92101-4495, USA
125 London Wall, London EC2Y 5AS, UK
The Boulevard, Langford Lane, Kidlington, Oxford OX5 1GB, UK

First edition 2015



ISBN: 978-0-12-802129-3
ISSN: 0065-2725

For information on all Academic Press publications visit our website at http://store.elsevier.com/

# CONTENTS

# CONTRIBUTORS

**R. Alan Aitken**
EaStCHEM School of Chemistry, University of St. Andrews, St. Andrews, Fife, UK

**Youcef Boubalouta**
EaStCHEM School of Chemistry, University of St. Andrews, St. Andrews, Fife, UK

**Jessica G. Greger**
Department of Chemistry, Hobart and William Smith Colleges, Geneva, NY, USA

**Ming Liu**
Institute of Organic Chemistry, Clausthal University of Technology, Clausthal-Zellerfeld, Germany

**Erin T. Pelkey**
Department of Chemistry, Hobart and William Smith Colleges, Geneva, NY, USA

**Sarah J. Pelkey**
Department of Chemistry, Hobart and William Smith Colleges, Geneva, NY, USA

**Andreas Schmidt**
Institute of Organic Chemistry, Clausthal University of Technology, Clausthal-Zellerfeld, Germany

**Viktor V. Zhdankin**
Department of Chemistry and Biochemistry, University of Minnesota Duluth, Minnesota, USA

# PREFACE

Volume 115 consists of 4 chapters; one on iodine heterocycles, one on a synthetic technique, and two on important nitrogen heterocycles.

Viktor Zhdankin of the University of Minnesota Duluth, USA has written a comprehensive survey of the chemistry of iodine heterocycles. This is the first time that these "hypervalent" rings have been systematically reviewed.

R. Alan Aitken and Youcef Boubalouta of the University of St Andrews, UK describe developments that have taken place since 2000 in the synthesis of heterocyclic compounds using flash vacuum pyrolysis.

Erin T. Pelkey and colleagues of Hobart and William Smith Colleges, USA provide a comprehensive and systematic account of 3-pyrrolin-2-ones, which occur widely in nature and are important synthetic building blocks.

Finally, Andreas Schmidt and Ming Liu of Clausthal University of Technology, Germany have reviewed advances in the chemistry of acridines that have been published in the last decade, including sections on biological activity, synthesis, and reactions.

**Chris Ramsden and Eric Scriven**
May, 2015

CHAPTER ONE

# Iodine Heterocycles

**Viktor V. Zhdankin**
Department of Chemistry and Biochemistry, University of Minnesota Duluth, Minnesota, USA
E-mail: vzhdanki@d.umn.edu

## Contents



*Advances in Heterocyclic Chemistry*, Volume 115
ISSN 0065-2725
http://dx.doi.org/10.1016/bs.aihch.2015.03.003

## Abstract

This chapter provides the first comprehensive review covering all main aspects of the chemistry of iodine heterocycles including preparation, structure, and synthetic applications. The most important and practically useful polyvalent iodine heterocycles have a five-membered ring, although several examples of three-membered, four-membered, and six-membered heterocycles with an iodine(III) or iodine(V) atom in the ring are also known. In general, iodine is not capable of forming conjugated cyclic systems with aromatic stabilization because of the large atom size and the semi-ionic nature of the hypervalent I—C, I—N, and I—O bonds. Despite the lack of aromatic conjugation, the five-membered heterocyclic iodine compounds (benziodoxoles and benziododazoles) have a considerably higher thermal stability compared to their acyclic analogs, which has made possible the preparation and isolation of iodine-substituted derivatives, such as bromobenziodoxoles, tosyloxybenziodoxoles, alkylperoxybenziodoxoles, azidobenziodoxoles, cyanobenziodoxoles, amidobenziodoxoles, alkynylbenziodoxoles, and trifluoromethylbenziodoxoles. Derivatives of benziodoxoles have found wide practical applications as oxidants and reagents for atom-transfer reactions in organic synthesis.

**Keywords:** 2-Iodoxybenzoic acid; Benziodoxoles; Cyclic iodonium; Dess—Martin periodinane; Hypervalent iodine; Iodine; Iodine heterocycles

## 1. INTRODUCTION

Since the end of the twentieth century, the organic chemistry of polycoordinated iodine compounds, which are commonly known as hypervalent iodine reagents, has experienced an unprecedented, explosive development. Several books (2014MI1, 2013MI2, 2003MI3) and numerous reviews (2014CAJ950, 2014CAJ972, 2013SL424, 2012COS247, 2011S517, 2011 NPR1722, 2011ARK370, 2009ARK1, 2008CR5299), some comprehensive, but most dealing with specific aspects of hypervalent organoiodine chemistry, have been published in recent years. This surging interest in iodine compounds is mainly due to the very useful oxidizing properties of hypervalent iodine reagents, combined with their benign environmental character and commercial availability. A particularly active area of hypervalent iodine chemistry involves the preparation and synthetic applications of hypervalent iodine heterocycles, mainly five-membered cyclic molecules with iodine(III) or iodine(V) in the ring, and known under the general name of benziodoxoles. Despite the widespread practical interest in heterocyclic hypervalent iodine reagents, particularly 2-iodoxybenzoic acid (IBX) and Dess—Martin periodinane (DMP) (2011JOC1185, 2010T7659, 2011AGE1524), the chemistry of iodine heterocycles has never been systematically reviewed.

Based on the available literature data, it can be stated that iodine is not capable of forming conjugated cyclic systems with aromatic stabilization because of the large atom size and the semi-ionic nature of the hypervalent I—C, I—N, and I—O bonds (2013MI2). Moreover, the high-level computational studies using an adaptive natural density partitioning bond modeling technique reveal that the double bond between iodine atoms and other elements does not exist (2014AGE9617). The bonds that are commonly shown in the literature, such as I=C, I=N, and I=O, in fact have a dative 2c—2e nature which precludes their involvement in a conjugated aromatic system.

Despite the lack of aromatic conjugation, five-membered heterocyclic iodine compounds have considerably higher thermal stability compared to their acyclic analogs. This stabilization is usually explained by the bridging of the apical and the equatorial positions at a hypervalent iodine center by a five-membered ring (1979JOC1779), and also by better overlap of the lone pair electrons on the iodine atom with the π-orbitals of the benzene ring (1983MI4). The greater stability of five-membered iodine—oxygen heterocycles (benziodoxoles) has enabled the preparation and isolation of otherwise unstable iodine(III) derivatives with I—OOR, I—$N_3$, I—$CF_3$, and other substituents. These various benziodoxole derivatives have found synthetic applications as "atom-transfer" reagents for functionalization of organic substrates.

# 2. GENERAL OVERVIEW OF ORGANIC POLYVALENT IODINE COMPOUNDS

## 2.1 Classification and Nomenclature

Iodine can form organic compounds with oxidation states of +3 and +5. The most common structural types of organic polyvalent iodine species are represented by structures **1–3**, shown in Figure 1. Structure **1**, the iodonium ion, formally does not belong to hypervalent species since it has only eight valence electrons on the iodine atom; however, in the modern literature iodonium salts are commonly treated as the 10-electron hypervalent

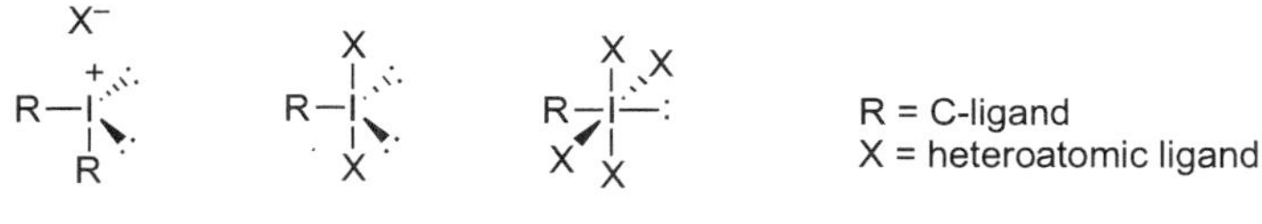


**1** **2** **3**

**Figure 1** Typical structural types of organic polyvalent iodine compounds.

compounds by including the closely associated anionic part of the molecule. The first two species, structures **1** and **2**, are conventionally considered as derivatives of trivalent iodine, while structure **3** represents the most typical structural type of pentavalent iodine.

In the older literature, the derivatives of iodine(III) were termed as iodinanes, while the compounds of pentavalent iodine were called periodinanes. According to the 1983 IUPAC recommendations (1984PAC769), these names were replaced by $\lambda^3$-iodanes for iodine(III) and by $\lambda^5$-iodanes for iodine(V) compounds. In the lambda nomenclature, the symbol $\lambda^n$ is used to indicate any heteroatom in a nonstandard valence state (n) in a formally neutral compound; for iodine the standard valence state is 1. The lambda terminology is broadly used in the modern literature to indicate the general type of hypervalent iodine compounds and to specify the number of primary bonds at the iodine atom.

It should be noted, however, that the lambda nomenclature is not used for naming common hypervalent iodine reagents, such as $PhICl_2$, $PhI(OAc)_2$, ArIO, $ArIO_2$, and heterocyclic derivatives of benziodoxole. According to the 1979 IUPAC rules (1979MI5), "compounds containing the group $-I(OH)_2$ or derivatives of this group are named by adding the prefixes 'dihydroxyiodo-', 'dichloroiodo-', 'diacetoxyiodo-', etc. to the name of the parent compound" (IUPAC Rule C−106.3). Likewise, "compounds containing the group −IO or $-IO_2$, are named by adding the prefix 'iodosyl-' or 'iodyl-'" (IUPAC Rule C−106.1), which replaces prefixes "iodoso-" and "iodoxy-" used in the older literature. According to the IUPAC Rule C−107.1 "cations of the type $R^1R^2I^+$ are given names derived from the iodonium ion $H_2I^+$ by substitution." In addition to the IUPAC recommended names, numerous common names and abbreviations are used for polyvalent iodine compounds; for example, 2-iodosylbenzoic acid (IBA) is the most commonly used name for 1-hydroxy-1*H*-1$\lambda^3$-benzo[*d*][1,2]iodoxol-3-one and IBX is the common, widely used name for 1-hydroxy-1-oxo-1*H*-1$\lambda^5$-benzo[*d*][1,2]iodoxol-3-one.

Organoiodine(III) compounds are generally classified by the type of ligands that are attached to the iodine atom. The following general classes of iodine(III) compounds have found broad applications as reagents in organic synthesis: (dichloroiodo)arenes **4**, iodosylarenes **5**, [bis(acyloxy) iodo]arenes **6**, aryliodine(III) organosulfonates **7**, iodonium salts **8**, iodonium ylides **9**, iodonium imides **10**, and five-membered iodine heterocycles (benziodoxoles **11** and **12**, and benziodazoles **13**) (Figure 2). The most important and commercially available representatives of aryliodine(III)

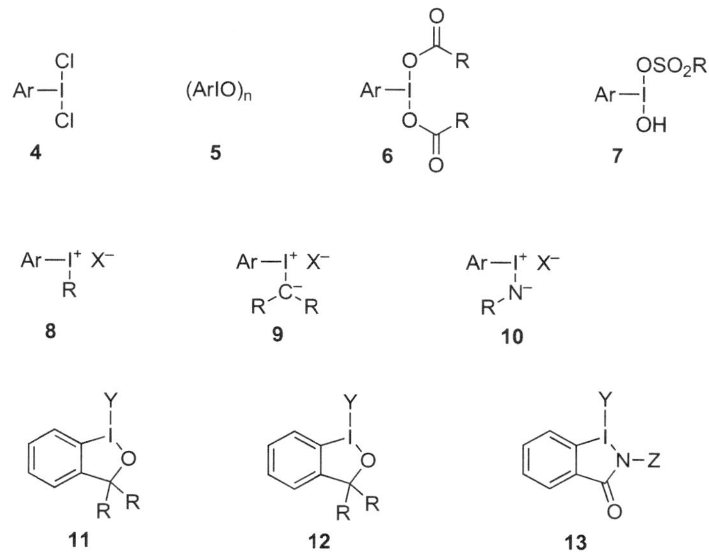


R = alkyl or aryl
X = Cl, Br, OTs, OTf, $BF_4$, etc.
Y = OH, OAc, $N_3$, CN, $CF_3$, etc.
Z = H, Ac, etc.

**Figure 2** Common classes of organoiodine(III) compounds.

carboxylates include (diacetoxyiodo)benzene $PhI(OAc)_2$, which has several commonly used abbreviations, such as DIB or PIDA (phenyliodine diacetate), and [bis(trifluoroacetoxy)iodo]benzene $PhI(OCOCF_3)_2$, abbreviated as BTI or PIFA [(phenyliodine bis(trifluoroacetate)]. The most important representative of aryliodine(III) organosulfonates, the commercially available [hydroxy(tosyloxy)iodo]benzene PhI(OH)OTs, is abbreviated as HTIB and is also known as Koser's reagent.

Organoiodine(V) compounds are represented by several typical classes **14**–**22** shown in Figure 3; all these compounds have found synthetic applications as efficient oxidizing reagents. Particularly important in organic synthesis are numerous five-membered heterocyclic benziodoxole derivatives **15** and **16**, pseudocyclic iodylarenes **17**, and 2-iodylbenzenesulfonic acid **20**.

## 2.2 Hypervalent Bonding and General Structural Features

The definition of "hypervalent" species as ions or molecules of the elements of Groups 15–18 bearing more than eight electrons within a valence shell was proposed by J. I. Musher in 1969 (1969AGE54). General aspects of bonding

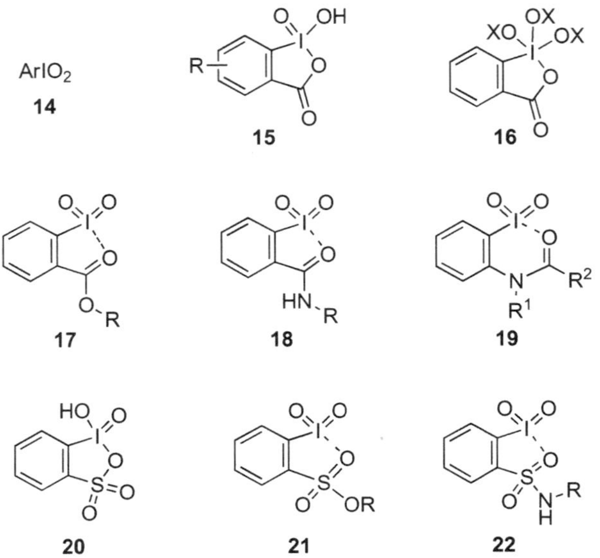


R = alkyl or aryl
X = Ac

**Figure 3** Common classes of organoiodine(V) compounds.

in hypervalent organic compounds were summarized by K.-y. Akiba in the book on Chemistry of Hypervalent Compounds (1999MI6). The special structural features and high reactivity of hypervalent iodine compounds are explained by the presence of hypervalent bonding involving a three-center—four-electron bond (1994CSR111, 1999MI6, 2014ARK109).

The molecular orbital description of a three-center—four-electron (3c—4e) bond was independently developed in 1951 by G. C. Pimentel (1951JCP446) and R. E. Rundle (1951JA4321). According to the fundamental description of the 3c—4e bond for L—X—L, one pair of bonding electrons is delocalized to the two ligands L, resulting in the charge distribution of almost −0.5 on each ligand and +1.0 on the central atom X. In molecules $RIL_2$, the interaction of the filled 5p orbital of the central iodine atom and the half-filled orbitals of the two ligands L *trans* to each other leads to formation of three molecular orbitals: bonding, nonbonding, and antibonding (Figure 4). Because the highest occupied molecular orbital contains a node at the central iodine, the hypervalent bonds show a highly polarized nature; hence, more electronegative atoms tend to occupy the axial positions formed by the interaction of the orbitals of three collinear atoms. The carbon substituent R is bound by a normal

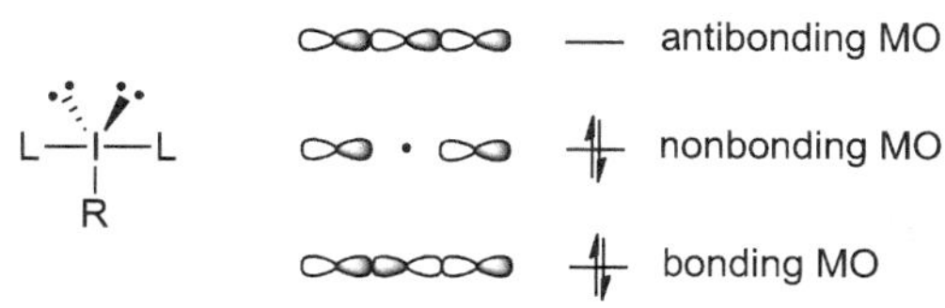


**Figure 4** Molecular orbital description of the 3c–4e bond in hypervalent iodine(III) molecules $RIL_2$.

covalent bond, and the overall geometry of molecule $RIL_2$ is a distorted trigonal bipyramid with two heteroatom ligands L occupying the apical positions, and the least electronegative carbon ligand R and both electron pairs residing in equatorial positions.

The bonding in iodine(V) compounds ($RIL_4$) with a square bipyramidal structure can be described as a normal covalent bond between iodine and the organic group R in an apical position, and two orthogonal, hypervalent 3c–4e bonds, accommodating four ligands L. The carbon substituent R and unshared electron pair in this case should occupy the apical positions with the electronegative ligands L residing at equatorial positions.

Bonding, structure, and reactivity of hypervalent iodine have some similarities with organometallic compounds. In fact, according to computational studies, hypervalent bonding commonly occurs in transition-metal complexes, and the 3c–4e bond is particularly important in the structure of transition-metal hydrides (1996JA6020, 1998JA12650, 1998JA2641, 1998JA1842, 1995JA1859). The importance of the effect of *trans* influence in transition-metal complexes is also typical of hypervalent iodine(III) compounds (2006ACE8203, 2005ARK19, 2012IC967). The reactions of hypervalent iodine reagents are commonly discussed in terms of oxidative addition, reductive elimination, ligand exchange, and ligand coupling, which are all typical of the transition-metal chemistry.

Structural aspects of polyvalent iodine compounds were previously summarized in several books and reviews (1983MI4, 2013MI2, 2008CR5299, 2003MI3). In general, molecular structure of $\lambda^3$- and $\lambda^5$-iodanes is predetermined by the nature of the hypervalent bonding. Key structural features of hypervalent organoiodine compounds can be summarized as follows:

1. $\lambda^3$-Iodanes $RIX_2$ (R = C-ligand, X = heteroatom ligands) have an approximately T-shaped structure with a collinear arrangement of the most electronegative ligands X. Including the nonbonding electron pairs, the geometry about iodine is a distorted trigonal bipyramid with the most electronegative groups occupying the apical positions, and the least

electronegative C—ligand R and both electron pairs residing in an equatorial position.

2. The I—C bond lengths in iodonium salts $R_2I^+ X^-$ and $\lambda^3$-iodanes $RIX_2$ are approximately equal to the sum of the covalent radii of iodine and carbon, ranging generally from 2.00 to 2.10 Å.
3. Iodonium salts $R_2I^+ X^-$ generally have a typical distance of 2.6—2.8 Å between iodine and the nearest anion $X^-$and in principle can be considered as ionic compounds with pseudotetrahedral geometry about the central iodine atom. However, with consideration of the anionic part of the molecule, the overall experimentally determined geometry is a distorted T-shaped structure similar to the $\lambda^3$-iodanes $RIX_2$.
4. For $\lambda^3$-iodanes RI(X)Y with two heteroatom ligands X and Y of the same electronegativity, both I—X and I—Y bonds are longer than the sum of the appropriate covalent radii, but shorter, than purely ionic bonds. For example, the I—Cl bond lengths in $PhICl_2$ are 2.45 Å (1953AC88), and the I—O bond lengths in $PhI(OAc)_2$ are 2.15—2.16 Å (1979JCS(D)851), while the sum of the covalent radii of I and O is 1.99 Å. In cases where the heteroatom ligands X and Y have different electronegativity, the *trans* influence of ligands has a strong effect on the structure, stability, and reactivity of $\lambda^3$-iodanes RI(X)Y (2006ACE8203, 2005ARK19, 2012IC967).
5. Various coordination types have been reported for the organoiodine(V) compounds. Depending on the ligands and taking into account secondary bonding, the overall observed geometry for the structural types **4** and **5** (Figure 1) can be pseudo-trigonal-bipyramidal, square bipyramidal, or pseudooctahedral.
6. Intramolecular positional isomerization (Berry pseudorotation) resulting in an exchange between the apical and the equatorial ligands occurs rapidly in both $\lambda^3$- and $\lambda^5$-iodanes. This process is important in explanation of the mechanisms of hypervalent iodine reactions.

Because of the highly polarized character of hypervalent bonds, noncovalent attractive interactions of a predominantly electrostatic nature are extremely important in the structural chemistry of hypervalent iodine compounds. Such attractive interactions are commonly termed as secondary bonds. As for hydrogen bonds, secondary bonds involving heavier atoms have strong electrostatic components and show directional preferences (1998NJC883, 1998ACE1887). Intermolecular secondary bonding in hypervalent iodine compounds is responsible for crystal packing in the solid state and for the self-assembly of individual molecules into complex supramolecular structures in the solid state and in the solution (2001JA4095, 2009IC4908). Intramolecular secondary bonding is commonly observed in the $\lambda^3$- and $\lambda^5$-aryliodanes, which have a sulfonyl or a carbonyl structural fragment in the *ortho*-position of the phenyl ring (1999JA7164, 2000ACE2007, 2005JOC6484, 2003ACE2194, 2009JOC8444, 2005ACE7127). The redirection of secondary bonding from the intermolecular to intramolecular mode due to the presence of an appropriate *ortho*-substituent leads to a partial disruption of polymeric network and enhancement of the solubility of a hypervalent iodine compound (2014CCR54).

# 3. PREPARATION, STRUCTURE, AND PROPERTIES OF IODINE HETEROCYCLES

## 3.1 Cyclic Iodonium Salts and Ylides

Three-membered cyclic iodonium salts are common intermediates in the reactions of alkenes with electrophilic iodine species. Usually these compounds are highly unstable; however, the three-membered cyclic iodonium salts derived from sterically hindered alkenes have been isolated and even characterized by single-crystal X-ray diffraction (1994JA2448). In particular, X-ray diffraction data on cyclic iodonium salt **23** derived from adamantylideneadamantane indicate that the halonium ion portion of salt **23** is essentially symmetrical with the following averaged structural parameters: I–C, 2.48 Å; C–C, 1.45 Å; I–C–C angle, 72°; C–I–C angle, 36°. The $^{13}C$ NMR spectra of ion **23** and the iodonium ion of bicyclo[3.3.l]nonylidenebicyclo[3.3.l]nonane **24** have been investigated in dichloromethane solution. The low-temperature $^{13}C$ NMR spectra of **23** and **24** indicate that the iodonium ion has two perpendicular planes of symmetry (Figure 5).

The most important representatives of stable cyclic iodonium salts, dibenziodolium or diphenyleneiodonium (DPI) salts, have been prepared as the iodide **26**, the hexafluorophosphate **28**, the tetrafluoroborate **29**, and the chloride **31**, and can be obtained by three different procedures

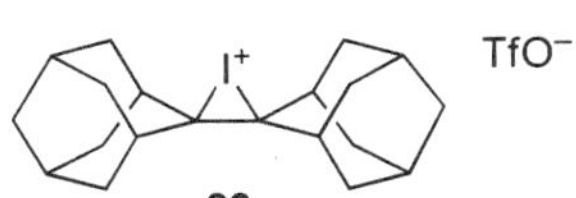


**23**

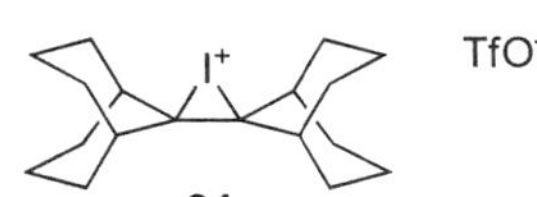


**24**

**Figure 5** Stable three-membered cyclic iodonium salts derived from sterically hindered alkenes.

summarized in Scheme 1. DPI iodide **26** was originally prepared by Mascarelli and Benati in 1909 (1909GCI619) by diazotization of 2,2′-diaminodiphenyl **25** with sodium nitrite in a hydrochloric acid solution followed by addition of potassium iodide. A similar reaction starting from 2-amino-2′-iododiphenyl **27** affords DPI as hexafluorophosphate **28** or tetrafluoroborate **29** in excellent yields (1968T3717). The third method involves the peracetic acid oxidation of 2-iodobiphenyl **30** to an iodine(III) intermediate which cyclizes to DPI in the acidic solution and is finally isolated as the chloride salt **31** (1956JA3819). More recently, these methods were used for the preparation of the tritium-labeled DPI and its 4-nitro derivative (2000JLCR515).

The structure of dibenziodolium tetrafluoroborate **29** was established by single-crystal X-ray analysis (1972JOC879). In particular, the dibenziodolium ion is planar with deviations from the mean molecular plane of less than 0.03 Å. The C—I—C bond angle of 83° in structure **29** is smaller than the corresponding bond angle in a non-cyclic iodonium salt, for example, DPI chloride (93°) (1956DAN71). The C—I bond lengths of 2.08 Å are close to the typical C—I bond lengths in hypervalent iodine compounds. A relatively long distance of 3.65 Å between the iodine center and the nearest tetrafluoroborate anion is consistent with the ionic character of this compound. Dibenziodolium salts have a relatively high thermal stability. The tetrafluoroborate salt **31** has a melting point of 239—240 °C; however,

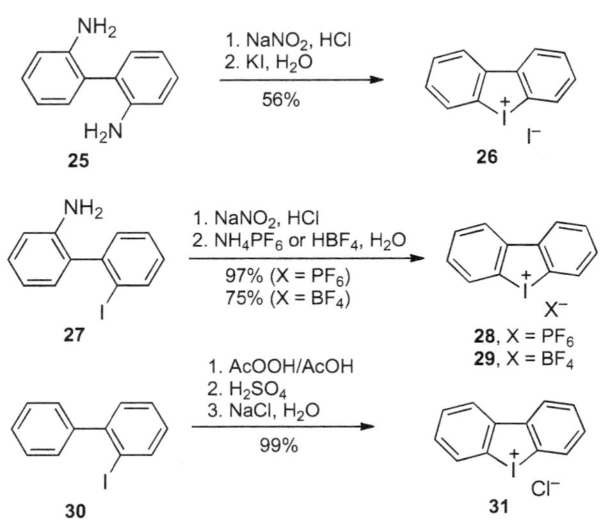


**Scheme 1**

the X-ray structural data do not support any aromatic character of the iodolium ring.

DPI chloride **31** has found some applications in biological studies (2013MI2). In particular, DPI is a potent hypoglycemic agent at a dose as low as 4 mg/kg body weight (1973JBC6050). It is assumed that DPI binds covalently to a 23.5 kDa protein within Complex I resulting in irreversible inhibition of NADH oxidation (1973JBC6050, 1976BJ307). DPI inhibits gluconeogenesis in isolated rat hepatocytes (1975BST333), causes swelling of rat liver mitochondria (1972BC39), induces cardiomyopathy (1986BST1209), induces mitochondrial myopathy (1988JNS335, 1988BP687), inhibits the superoxide production of neutrophils (1991BBRC143, 1988BJ887, 1986BJ111, 1987BP489), as well as nitric oxide synthase (1991FASEB98, 2007JFSN74), and NADPH oxidase (1993BJ41, 1992EJB61, 1991FRBM25). In modern biochemical and pharmacological research, DPI is often used as an NADPH oxidase inhibitor (2012JPET873, 2012BBRC329, 2012BP422, 2012PLoSe33817, 2012JCP1347, 2011WASJ67, 2011MGM241, 2011HMR619, 2010FS2437, 2010PNAS3030, 2009BP493, 2009ABP995, 2009TOL180, 2008AJPG99, 2008JPET50, 2007PP890, 2007CR663, 2007NPLG348, 2007BC1159, 2007JN1205, 2007CS1610, 2001FASEB2539, 1994NL63).

The chemistry of several other cyclic iodonium salts has been summarized in a review by Grushin (2000CSR315). Examples of known cyclic iodonium salts include 4,5-phenanthryleneiodonium salts **32** (1969T4339), 10*H*-dibenz[*b,e*]iodinonium salt **33** (1965JOC1141), 10,11-dihydrodibenz[*b,f*]iodeponium salt **34** (1965JOC1141), phenoxiodonium salt **35** (1965JOC1141), 10-acetylphenaziodonium salt **36** (1965JOC1141), 10-oxidophenothiiodonium salt **37** (1965JOC1141), the bicyclic bisiodonium salt **38** (1969JOC456), benziodolium chloride **39** (1972JOC879), and iodolium salt **40** (1981JOC4069; Figure 6).

Several macrocyclic iodonium salts, rhomboids **43**, a square **46**, and a pentagon **48**, have been prepared from the appropriate precursors **41, 42, 44, 45,** and **47** (Scheme 2; 1993JA9808, 2003JOC9209, 1993JA11626). The structures of these iodonium-containing charged macrocycles were established using elemental analysis, multinuclear NMR, and mass spectrometry. These iodonium-containing macromolecules may find potential application in nanotechnology (2003JOC9209).

The unusually stable cyclic iodonium ylides **50** can be synthesized via the intramolecular transylidation of a preformed acyclic ylide **49** (Scheme 3; 1992CC1487). X-ray structural analysis for cyclic ylide **50** shows a distorted

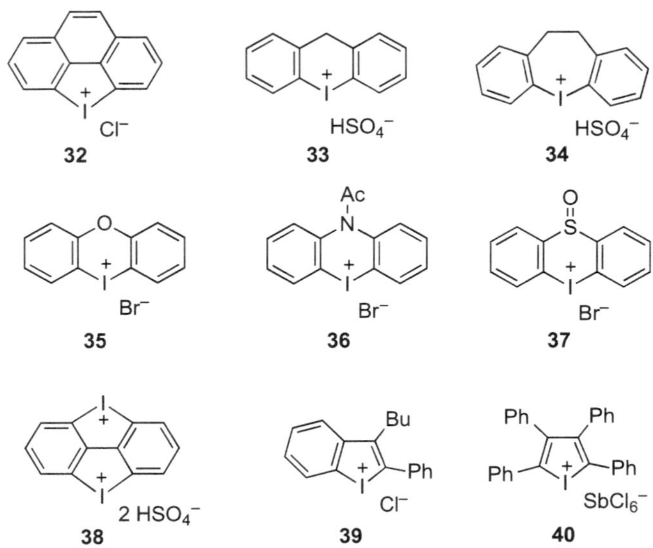

Figure 6 Examples of cyclic iodonium salts.

five-membered ring with an ylidic bond length of 2.1 Å and a C—I—C bond angle of 82°, which is smaller than the usual value (90°) for ylides (1992CC1487).

## 3.2 Trivalent Iodine Heterocycles

The most important and practically useful trivalent iodine heterocycles have a five-membered ring, although several examples of four-membered and six-membered heterocycles with iodine(III) atom in the ring are also known. The unsaturated heterocyclic systems with a hypervalent iodine atom in the ring generally do not possess any significant aromatic character because of the large iodine atom size precluding p-orbital overlap with the much smaller atoms of carbon, oxygen, or nitrogen, and also due to the electronic nature and the geometry of hypervalent bonding.

Typical examples of five-membered trivalent iodine heterocycles are represented by cyclic compounds **51—59**, which incorporate iodine, oxygen, nitrogen, and some other elements, in the heterocyclic ring (Figure 7). Iodoxolones **51** represent the only known examples of iodine heterocycles in which the hypervalent iodine atom is not connected to an aromatic ring. The collective name "benziodoxoles" is commonly used for the heterocycles **52** with iodine and oxygen atoms in a five-membered ring and various substituents X attached to iodine (1992MI7). The first

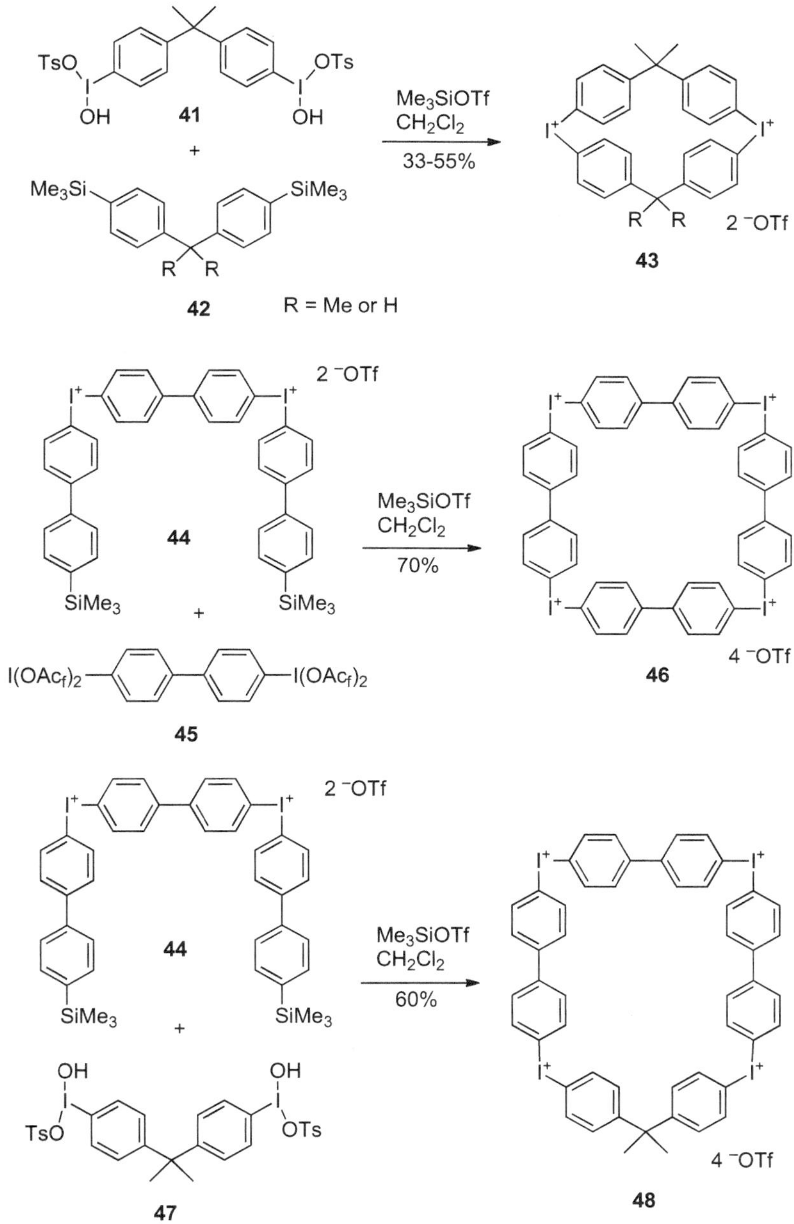
TsO
OTs
OH
41
OH
+
Me3SiOTf
CH2Cl2
33-55%
I+
I+
Me3Si
SiMe3
R R
2 -OTf
43
R R
42
R = Me or H
2 -OTf
I+
I+
44
Me3SiOTf
CH2Cl2
70%
SiMe3
SiMe3
+
I(OAcf)2
I(OAcf)2
45
4 -OTf
46
2 -OTf
44
Me3SiOTf
CH2Cl2
60%
SiMe3
SiMe3
+
OH
OH
TsO
OTs
47
4 -OTf
48

**Scheme 2**

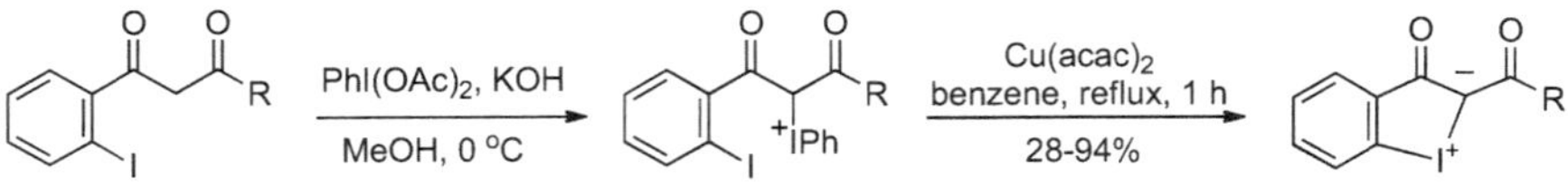


**49** **50**

R = Me, Ph, 4-MeC$_6$H$_4$, 4-MeOC$_6$H$_4$, 4-ClC$_6$H$_4$, 4-PhOC$_6$H$_4$, α-naphthyl

**Scheme 3**

representatives of benziodoxoles, 1-hydroxy-1,2-benziodoxol-3-(1*H*)-one and 1-chloro-1,2-benziodoxol-3-(1*H*)-one, were first prepared over 100 years ago by oxidation or chlorination of 2-iodobenzoic acid (1892B2632). In the mid-1980s, 1-hydroxybenziodoxoles attracted a significant research interest due to their excellent catalytic activity in the cleavage of toxic phosphates and reactive esters (2002CR2497). More recently, various new benziodoxole derivatives have been synthesized and their usefulness as reagents for organic synthesis, and particularly for atom–transfer reactions (2011CC102), has been demonstrated. Compared to benziodoxoles **52**, the analogous five-membered iodine–nitrogen heterocycles, benziodazoles **53**, have received much less attention and, in some cases, their

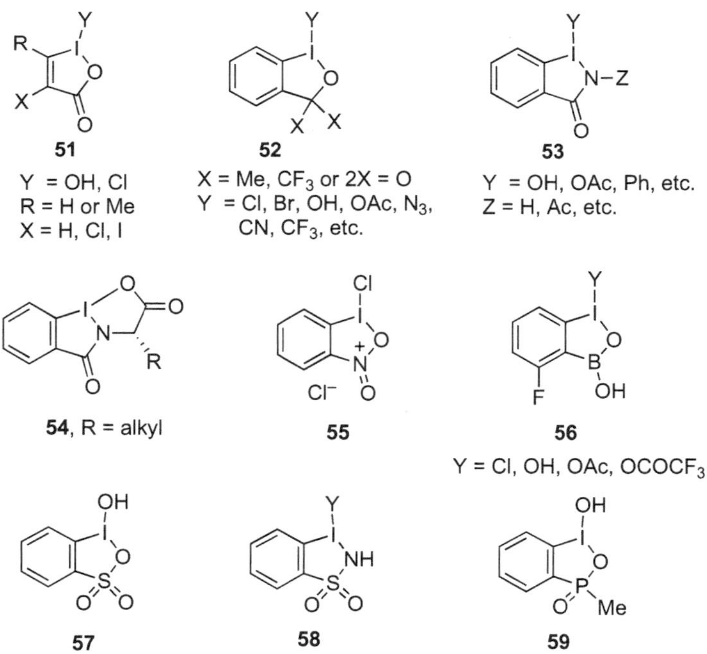

**51**: Y = OH, Cl; R = H or Me; X = H, Cl, I

**52**: X = Me, CF$_3$ or 2X = O; Y = Cl, Br, OH, OAc, N$_3$, CN, CF$_3$, etc.

**53**: Y = OH, OAc, Ph, etc.; Z = H, Ac, etc.

**54**, R = alkyl

**55**

**56**: Y = Cl, OH, OAc, OCOCF$_3$

**57**

**58**: Y = OH, OAc, Cl

**59**

**Figure 7** Typical examples of iodine(III) five-membered heterocyclic systems.

structural assignment was not reliable. The first readily available benziodazole derivative, acetoxybenziodazole (**53**, Y = OAc, Z = H), was synthesized in 1965 by the peracetic acid oxidation of 2-iodobenzamide (1965CC449).

Besides benziodoxoles and benziodazoles, the other known five-membered iodine(III) heterocyclic systems are represented by the less common compounds **54–59** (Figure 7), which include the fused benziodazoles **54** (2001JA4095), benziodoxazoles **55** (1965JOC617), benziodoxaboroles **56** (2011IC11263), benziodoxathioles **57** (1993JOC7310, 2006EJOC 4791), benziodathiazoles **58** (1975JOC797), and cyclic phosphonate **59** (1978 JOC4538). X-ray molecular structures have been reported for benziodoxole derivatives **52** (2006CEJ2579, 1986IC1415, 2002JA11614, 1996JA5192, 1979JOC1779, 1991JOC5511, 1965JPS104, 1976JA5326, 1974JSSC155, 1992JA6269, 1993CC218, 1996MC50, 1996H47, 2006CC1442, 2007 ACE754, 1995JCC857), benziodazoles **53** (1979JOC1447, 1998JOC6590, 2003OL1583, 1997JA7408), benziodoxaboroles **56** (2011IC11263), benziodoxathioles **57** (1993JOC7310, 2006EJOC4791), and cyclic phosphonate **59** (1978JOC4538). According to X-ray structural data, the five-membered rings in benziodoxoles **52** are highly distorted with almost linear alignment of the two electronegative ligands. The I–O bond length in benziodoxolones (**52**, 2X = O) varies in a wide range from 2.11 Å in carboxylates (**52**; Y = *m*-$ClC_6H_4CO_2$) to 2.48 Å in the phenyl derivative (**52**, Y = Ph) (1986IC1415), which indicates considerable changes in the ionic character of this bond. The endocyclic C–I–O bond angle is typically around 80°, which is a significant deviation from the expected angle of 90° for the normal T-shaped geometry of hypervalent iodine. The structural parameters of benziodazoles **53** (Y = OAc or Ph) in general are similar to those of benziodoxoles (1979JOC1447, 1998JOC6590, 2003OL1583, 1997JA7408).

Five-membered heterocyclic iodine(III) compounds are characterized by a generally higher thermal stability and lower reactivity compared to their acyclic analogs. This stabilization is explained by the bridging of the apical and the equatorial positions on hypervalent iodine by a five-membered ring (1979JOC1779), and also by better overlap of the lone pair electrons on the iodine atom with the π-orbitals of the benzene ring in compounds **52–59** (1983MI4). The greater stability of the benziodoxole system made possible the preparation and isolation of otherwise unstable trivalent iodine compounds with I–Br, I–OOR, I–$N_3$, I–CN, and I–$CF_3$ bonds. Some of these benziodoxole derivatives have found practical application as "atom-transfer" reagents for organic synthesis (2011CC102). The chemistry of benziodoxoles and benziodazoles has previously been summarized in several reviews (2011CC102, 2005COS121, 1997RHC133).

### 3.2.1 Iodoxolones

Iodoxolones represent the only known example of iodine heterocycles in which the hypervalent iodine atom is not connected to an aromatic ring. The first representatives of iodoxolones were prepared in the early 1900s by oxidative cyclization of iodofumaric acid derivatives (1910LA119). More recently, Moss and coworkers reported the synthesis of hydroxyiodoxolone **62** by the peracetic acid oxidation of iodofumaric acid **60** followed by decarboxylation of the initially formed 5-carboxylic acid derivative **61** (Scheme 4; 1989JA250). An alternative approach to iodoxolones (e.g., chloroiodoxolones **63** and **64**) involves the addition of $ICl_3$ to acetylenedicarboxylic acid followed by a spontaneous cyclization of the initially formed iododichlorides (1950IAN315).

Wirth and coworkers prepared iodoxolones **66**, **68**, **70**, and **71** by the oxidation of (*Z*)-3-iodo acrylic acid derivatives **65**, **67**, and **69** using peracetic acid or other common oxidants (Scheme 5; 2009OL3578). Structures of products **66** and **68** were established by X-ray analysis (2009OL3578). Bond angles and distances within the five-membered iodoxolone system are similar to the respective X-ray structural parameters of 1-hydroxybenziodoxole derivatives (see Section 3.2.2.2; 1964NAT512, 2007ACE6529). Iodoxolones **66**, **68**, **70**, and **71** have stability and reactivity similar to the noncyclic hypervalent iodine(III) compounds (2009OL3578).

A stable aliphatic iodoxolone **73**, the tautomer of *cis*-2-iodosylcyclopropanecarboxylic acid, has been prepared by two alternative methods: the

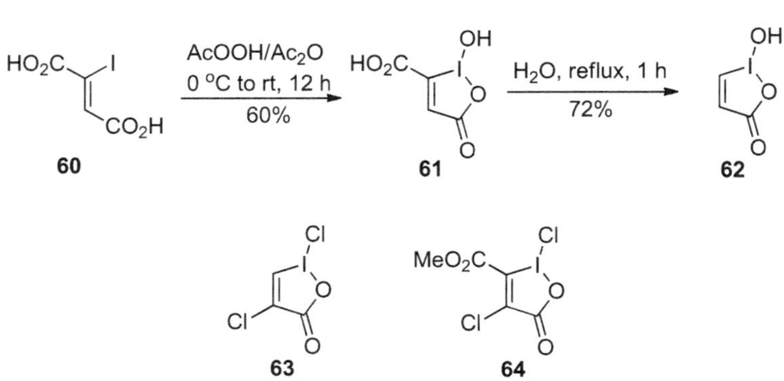


Scheme 4

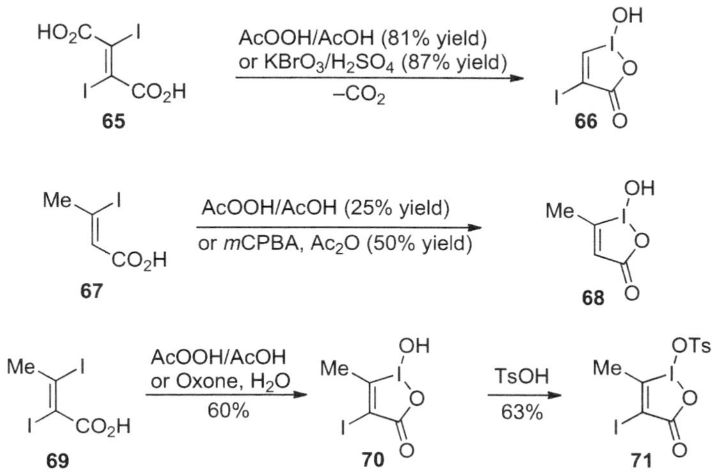

Scheme 5

peracetic acid oxidation of *cis*-2-iodocyclopropanecarboxylic acid **72** or the hydrolysis of the corresponding iododichloride **74** (Scheme 6; 1989JA6729).

Compound **73** exists in the hydroxyiodoxolone form, not in the open *cis*-2-iodosylcyclopropanecarboxylic acid form, as indicated by its low acidity (p$K_a$ 7.55) and by its ability to cleave phosphate esters in aqueous micellar solution (see Section 3.2.2.2; 1989JA6729).

### *3.2.2 Benziodoxoles*

#### 3.2.2.1 Halobenziodoxoles

1-Chloro-1,2-benziodoxol-3-(1*H*)-one (Figure 7, structure **52**; 2X = O, Y = Cl) is prepared by direct chlorination of 2-iodobenzoic acid (1894JPC466) or by oxidation of 2-iodobenzoic acid with sodium chlorite ($NaClO_2$) in aqueous hydrochloric acid media (2009S1163). An X-ray single-crystal structure of 1-chloro-1,2-benziodoxol-3-(1*H*)-one was originally published in 1976 (1976ZN687). More recently, Koser and coworkers reported a single-crystal X-ray analysis of a 1:1 complex of 1-chloro-1,2-benziodoxol-3-(1*H*)-one and tetra-*n*-butylammonium chloride (1995JCC857). The primary bond distances at iodine in this compound

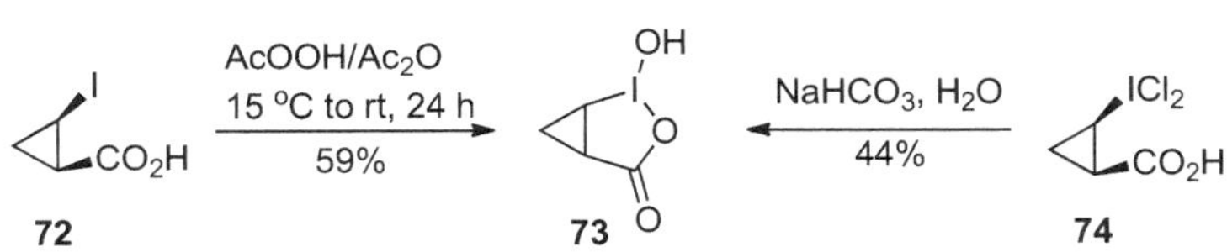

Scheme 6

are consistent with expectations for a trivalent iodine compound. In particular, both I—Cl and I—O bond distances of 2.454 and 2.145 Å are greater than the sums of the appropriate covalent radii, which is in agreement with the hypervalent nature of the apical iodine—heteroligand bond. A notable structural feature of this complex is the existence of a secondary bond I•••Cl of 2.943 Å between the chloride ion of $Bu_4NCl$ and the iodine atom of the chlorobenziodoxole moiety (1995JCC857). The analogous 1-fluoro- and 1-bromo-1,2-benziodoxol-3-(1*H*)-ones are unknown.

Amey and Martin reported the synthesis of halobenziodoxoles **76–78**, derivatives of the 3,3-bis(trifluoromethyl)-3-(1*H*)-1,2-benziodoxole system, by treatment of the benzylic alcohol **75** with the respective halogenating reagents (Scheme 7; 1979JOC1779). Bromobenziodoxole **77** represents an unusual, stable derivative of polyvalent iodine with an iodine—bromine bond.

More recently, Braddock and coworkers reported the preparation and X-ray structural analysis of two bromobenziodoxoles **80** and **81** (Scheme 8; 2006CC1442). Compounds **80** and **81** were synthesized in one step by the bromination of the corresponding alcohols **79** using stoichiometric *N*-bromosuccinimide in chloroform (Scheme 8).

Single-crystal X-ray diffraction analysis of compounds **80** and **81** confirmed the planar T-shaped geometry of the hypervalent iodine atoms. The five-membered benziodoxole rings have an envelope conformation, with the oxygen atom lying 0.23 Å (for **81**) and 0.44 Å (for **81**) out of the C—C—C—I plane. For both compounds, the closest intermolecular distance to the iodine center is from the oxygen of a centrosymmetrically related counterpart at 3.03 Å (for **80**) and 2.94 Å (for **81**), forming discrete dimer pairs. Most notably, while the I—C bond length is essentially constant [2.120(8) Å in **80** and 2.112(7) Å in **81**], the I—O and I—Br bond lengths vary considerably in **80** and **81**, presumably due to the effect of the electron—withdrawing *gem*-trifluoromethyl groups on the three-center—four-electron O—I—Br bond. The I—O bond length is *ca.* 2.12 Å in **80** and

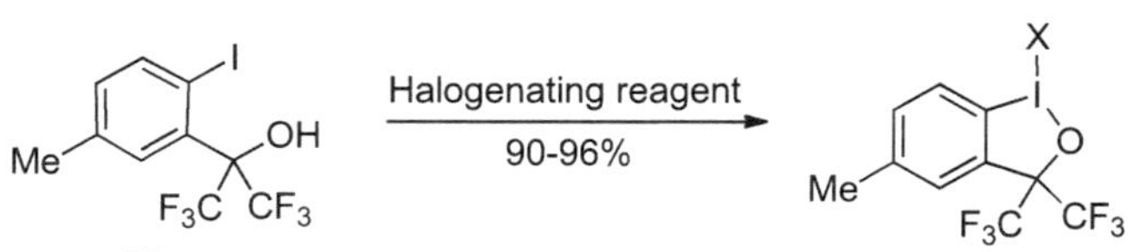


**75**

**76**, X = Cl, Halogenating reagent = $Bu^tOCl$ or $Cl_2$
**77**, X = Br, Halogenating reagent = $Br_2$
**78**, X = F, Halogenating reagent = $CF_3OF$

**Scheme 7**

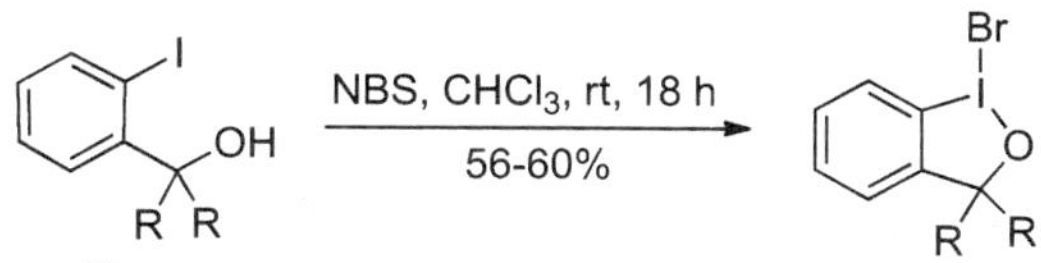

79

80, R = $CF_3$
81, R = $CH_3$

**Scheme 8**

*ca.* 2.05 Å in **81**. The I—Br bond lengths are *ca.* 2.59 and 2.69 Å, respectively (2006CC1442). The structure of compound **80** is isomorphous with that of its previously reported chloro analog (1996JCC371). The X-ray structure of the fluoro analog of **81**, 1-fluoro-3,3-dimethyl-1,3-dihydro-1$\lambda^3$-benzo[*d*][1,2]iodoxole, is characterized by a very short I—F distance of 2.045 Å, while the I—O and the I—C bond lengths (2.02 and 2.09 Å, respectively) are comparable with the analogous bond lengths in **81** (2012AX(E)o1238).

### 3.2.2.2 Hydroxybenziodoxoles

The most common and readily available benziodoxole derivative is 1-hydroxy-1,2-benziodoxole-3(1*H*)-one **83**, which is the cyclic tautomeric form of IBA **82** (Scheme 9). The tautomeric form **83** represents the actual structure of this compound, as confirmed by its unusually low acidity (p$K_a$ 7.25 against 2.85 for the hypothetical IBA **82**) (1983MI4), and unambiguously established by a single-crystal X-ray analysis of **83** in the solid state (1965JPS104, 1964NAT512). The I—O bond distance of 2.30 Å in the five-membered ring of **83** is significantly longer than the computed covalent I—O bond length of 1.99 Å, which is indicative of the hypervalent nature of this bond (1965JPS104). A report (1990JOC5202) of the existence of the non-cyclic tautomeric form of a ring-substituted IBA **82** has been shown to be wrong by more recent work (1998CC1559).

1-Hydroxy-1,2-benziodoxole-3(1*H*)-one **83** is commercially available or can be easily prepared by direct oxidation of 2-iodobenzoic acid (1986T155) or by basic hydrolysis of 2-(dichloroiodo)benzoic acid

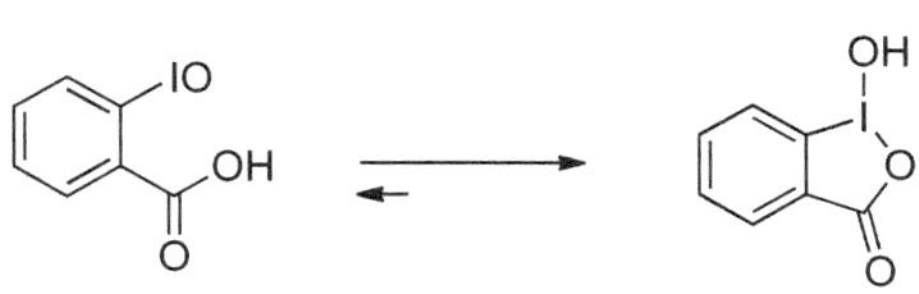

82 83

**Scheme 9**

(1894JPC466, 1892B2632). An alternative preparative procedure for benziodoxole **83** involves the oxidation of 2-iodobenzoic acid with acetyl nitrate in acetic anhydride at room temperature followed by aqueous work-up (1992SC1799).

During the 1980–1990s, benziodoxole **83** and other hydroxyiodoxoles attracted considerable research interest due to their excellent catalytic activity in the cleavage of toxic phosphates and reactive esters. This activity is explained by a pronounced O-nucleophilicity of the benziodoxole anion **84** due to the α-effect (2002CR2497, 1990JOC5202, 1995JOC7739). Spectroscopic and kinetic mechanistic studies indicate that highly unstable iodoxole derivatives, such as the phosphate **86**, are reactive intermediates in catalytic cleavage of phosphates, as shown for the catalytic hydrolysis of a typical substrate **85** (Scheme 10; 1987T251, 1994JA4471, 1986JA788). This mechanism was proved by the synthesis and reactivity studies of the phosphate intermediate **86**.

Hydroxybenziodoxole **83** can be readily converted to its acetoxy derivative, 1-acetoxy-1,2-benziodoxole-3(1*H*)-one (Figure 7, structure **52** with 2X = O, Y = OAc), by heating with acetic anhydride, and the acetoxy derivative can be further converted to alkoxy derivatives by treatment with an appropriate alcohol (1965JCS3721). The tetrabutylammonium salt of hydroxybenziodoxole **83** has been prepared by the reaction of hydroxybenziodoxole with tetrabutylammonium fluoride in THF; it is a mild oxidant useful for the preparation of epoxides from α,β-unsaturated carbonyl compounds (2000OL2923). 1-Hydroxy-1,2-benziodoxole-3(1*H*)-one and 1-acetoxy-1,2-benziodoxole-3(1*H*)-one have found wide application as

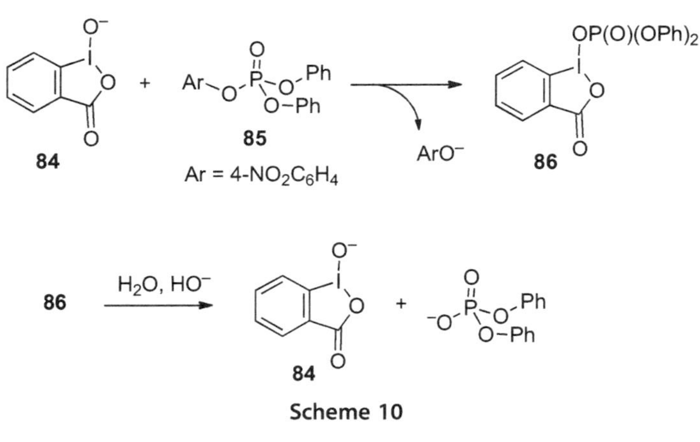


**Scheme 10**

starting compounds for the synthesis of various benziodoxole-based hypervalent iodine reagents by ligand exchange on iodine (2005COS121).

### 3.2.2.3 Organosulfonate Derivatives of Benziodoxoles

1-Organosulfonyloxy-1,2-benziodoxole-3(1*H*)-ones **88–90** and the analogous sulfonates **92–94** derived from the 3,3-bis(trifluoromethyl)-3-(1*H*)-1,2-benziodoxole system can be prepared by a simple, one-step procedure starting from hydroxybenziodoxoles **87** or **91** and the corresponding sulfonic acids or trimethylsilyltriflate (Scheme 11; 1994TL7323, 1996JOC6547). The organosulfonates **88–90** and **92–94** are isolated as moderately hygroscopic, but thermally stable, crystalline solids. Tosylates **90** and **94** have relatively high stability to moisture, while the mesylates and triflates are more hygroscopic and can only be isolated in the form of crystallohydrates; however, for further reactions they can be conveniently generated and used in situ (1996JOC6547). An alternative procedure for preparing benziodoxole triflate **98** in moderate yield involves the oxidation of 2-iodobenzoic acid with *m*-chloroperoxybenzoic acid in the presence of trifluoromethanesulfonic acid in dichloromethane at room temperature (2011EJOC3690).

Similarly to the noncyclic trivalent iodine derivatives of strong acids, sulfonates **88–90** and **92–94** are highly reactive toward unsaturated organic substrates and other carbon nucleophiles, and have found practical application for the preparation of C-substituted benziodoxoles.

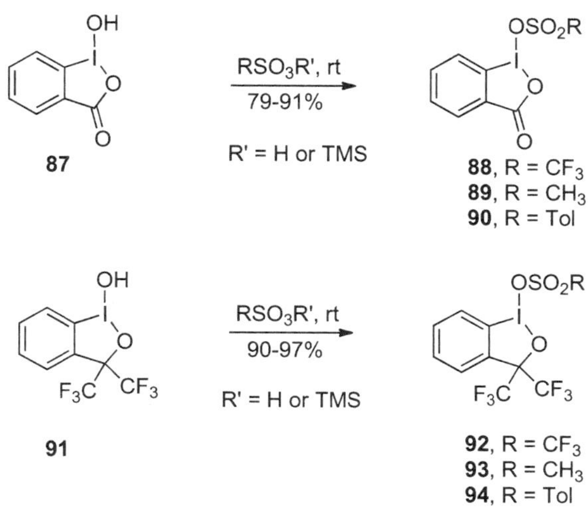


Scheme 11

### 3.2.2.4 Alkylperoxybenziodoxoles

The greater thermal stability of the benziodoxole system made possible the preparation and isolation of otherwise unstable trivalent iodine compounds with I—OOR bonds. Ochiai and coworkers first prepared 1-(*tert*-butylperoxy)benziodoxoles **96** and **97** by treatment of the corresponding benziodoxoles **87** and **95** with *tert*-butyl hydroperoxide in the presence of $BF_3 \cdot Et_2O$ (Scheme 12; 1992JA6269, 1996JA7716). The structure of 1-(*tert*-butylperoxy)benziodoxole **96** was established by single-crystal X-ray diffractometry. In particular, the *tert*-butylperoxy group occupies an apical position of distorted trigonal-bipyramidal geometry around the iodine. There are two primary hypervalent I—O bonds: a considerably ionic endocyclic bond [2.181 (5) Å] to the ring oxygen and a short exocyclic bond [2.039 (5) Å] to the peroxide oxygen. The peroxide O—O bond in structure **96** has a bond length of 1.459 (7) Å, which is comparable to that in anhydrous hydrogen peroxide (1992JA6269). A similar peroxybenziodoxole **98** was formed in an unusual reaction of hydroxybenziodoxole **87** with 1,3-bis(trimethylsilyl)-3-methylbut-1-yne in the presence of $BF_3$-etherate in dichloromethane; the structure **98** was determined by single-crystal X-ray analysis (1993CC218). Peroxides **96—98** are stable, crystalline products that can be safely stored at room temperature for an indefinite period of time.

Dolenc and Plesnicar developed an alternative procedure for the preparation of 1-(*tert*-butylperoxy)benziodoxoles **100** from the corresponding chlorobenziodoxoles **99** and *tert*-butyl hydroperoxide in the presence of potassium *tert*-butoxide in THF (Scheme 13; 1997JA2628).

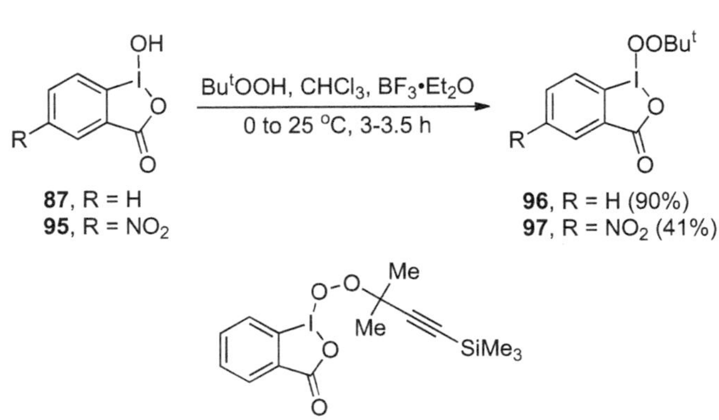


**98**

**Scheme 12**

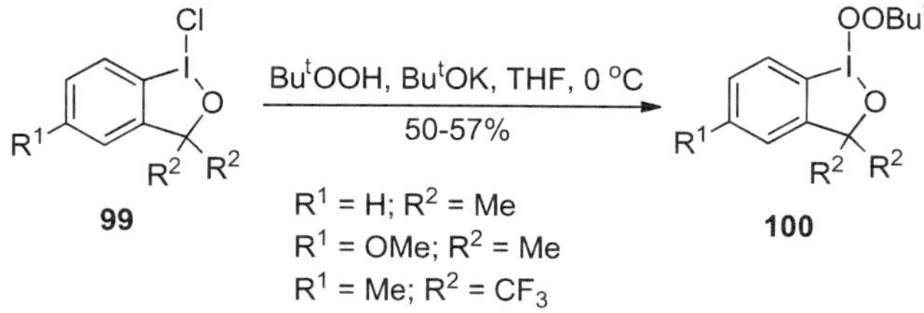


**Scheme 13**

Alkylperoxybenziodoxoles act as a strong oxidant toward a variety of organic substrates, such as ethers, organic sulfides, amides, and phenols (1996JA7716, 2001OL2387, 1997JOC4253, 1999TL5541, 1997H71, 1997 TL3927).

### 3.2.2.5 Azidobenziodoxoles

The noncyclic azido-substituted trivalent iodine compounds, for example, $PhI(N_3)OAc$ or $PhI(N_3)_2$, in general lack stability and rapidly decompose at −25 to 0 °C with the formation of iodobenzene and dinitrogen. The incorporation of a hypervalent iodine atom into a five-membered heterocycle results in significant stabilization of the azidoiodane. Stable azidobenziodoxoles **102–104** can be prepared by the reaction of hydroxybenziodoxoles **101** with azido(trimethyl)silane in acetonitrile (1996JA5192, 1994TL9677), or by treatment of acetoxybenziodoxoles **105** with azido(trimethyl)silane in dichloromethane in the presence of catalytic trimethylsilyl triflate (Scheme 14; 1996H47, 2013OL3246). All three azides **102–104** have been isolated as thermally stable, microcrystalline solids that can be stored indefinitely in a refrigerator.

The structure of azidobenziodoxole **103** was unambiguously established by single-crystal X-ray analysis (1996JA5192). The structural data revealed a typical hypervalent iodine T-shaped geometry with an N–I–O bond angle

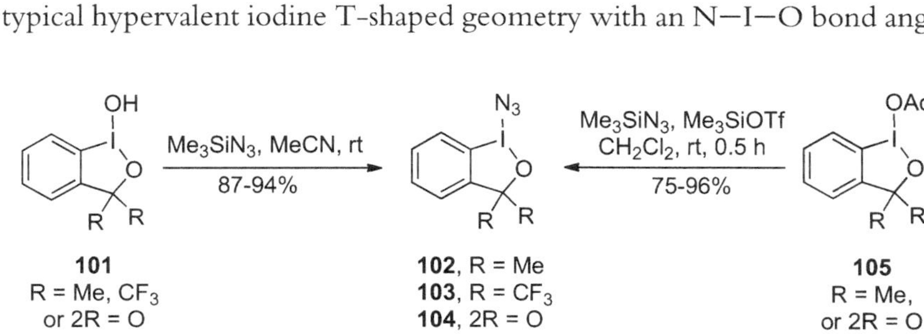


**Scheme 14**

of 169.5°. The lengths of the bonds to the iodine atom, I—N (2.18 Å), I—O (2.13 Å), and I—C (2.11 Å), are within the range of typical single covalent bonds in organic derivatives of polyvalent iodine (1996JA5192).

Azidobenziodoxoles can be used as efficient radical azidating reagents toward various organic substrates (1996JA5192, 1995SL1081).

Koser and Rabah reported the synthesis of the chiral, nonracemic 1-azido-1,3-dihydro-3-methyl-3-phenyl-1,2-benziodoxole **107** from the corresponding chlorobenziodoxole **106** (Scheme 15; 1996TL6453). This chiral azidobenziodoxole is a potentially useful reagent for asymmetric azidation reactions.

### 3.2.2.6 Amidobenziodoxoles

Amidobenziodoxoles **108—112** can be prepared in one step by the reaction of triflate **90** with an appropriate amide (Scheme 16; 1997TL21). Products **108—112** have been isolated as thermally stable and non-hygroscopic, microcrystalline solids. Amidobenziodoxoles can be used as C—H amidating reagents toward adamantane or *N,N*-dimethylanilines under radical conditions (1997TL21).

### 3.2.2.7 Cyanobenziodoxoles

Cyanobenziodoxoles **113—115** were originally synthesized in one step by treatment of hydroxybenziodoxoles **101** with cyanotrimethylsilane (Scheme 17; 1996MC50, 1995TL7975), or from acetoxybenziodoxole and cyanotrimethylsilane (1996H47).

More recently modified procedures involve the reactions of chlorobenziodoxole **116** or acetoxybenziodoxole **117** with cyanotrimethylsilane in the presence of fluoride anion (Scheme 18; 2015OBC365).

Products **113—115** are isolated as thermally stable, white, microcrystalline solids. Structures of two (**114** and **115**) were unambiguously established by single-crystal X-ray analysis (1996MC50, 1996H47). In particular, X-ray structural data for **114** revealed the expected hypervalent iodine distorted T-shaped geometry with an endocyclic C—I—O bond angle of 78.2° and

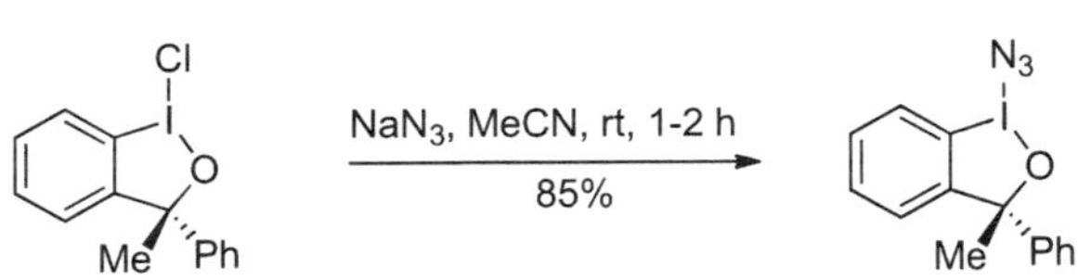


**106** **107**

**Scheme 15**

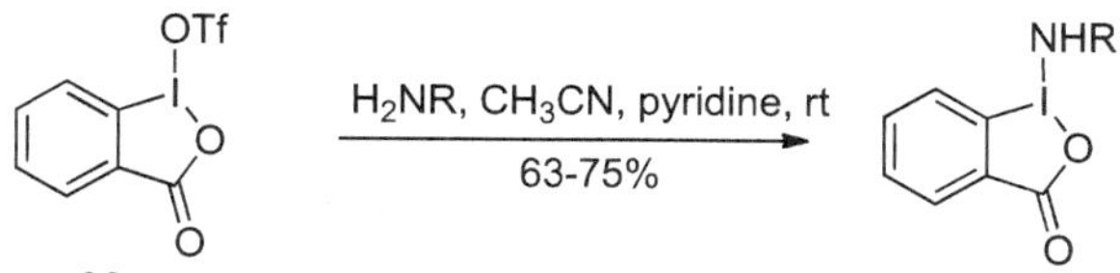

90

108, R = Ac
109, R = C(O)-4-ClC$_6$H$_4$
110, R = C(O)Et
111, R = C(O)NH$_2$
112, R = Ts

Scheme 16

an NC—I—O bond angle of 169.5° (1996MC50). The lengths of the bonds to the iodine atom, I—CN (2.167 Å), I—O (2.117 Å), and I—Ar (2.112 Å), are within the range of typical single covalent bond lengths in noncyclic organic derivatives of polyvalent iodine.

Chemical reactivity of cyanobenziodoxoles **113—115** is generally similar to that of azidobenziodoxoles, and they can be used as efficient cyanating reagents toward organic substrates (1996MC50, 2015OBC365).

### 3.2.2.8 Alkynylbenziodoxoles

Alkynylbenziodoxoles **118—120** were originally prepared by Ochiai and coworkers by a reaction of hydroxybenziodoxole **87** with alkynyltrimethylsilanes and $BF_3$•$Et_2O$ (Scheme 19; 1991JOC5511). The structure of product **119** was established by single-crystal X-ray diffraction analysis (1991 JOC5511). The structural data revealed the usual, distorted T-shaped geometry about the iodine atom. The I—O bond distance of 2.34 Å in **119** is significantly longer than the computed covalent single bond length of 1.99 Å, which is indicative of a highly ionic bond.

A modified method for the preparation of various alkynylbenziodoxoles **121—124** and **126—130** in high yields involves the reaction of triflates **90** or **125** with alkynyltrimethylsilanes according to Scheme 20 (1996 JOC6547). Further modification of this approach consists of the reaction

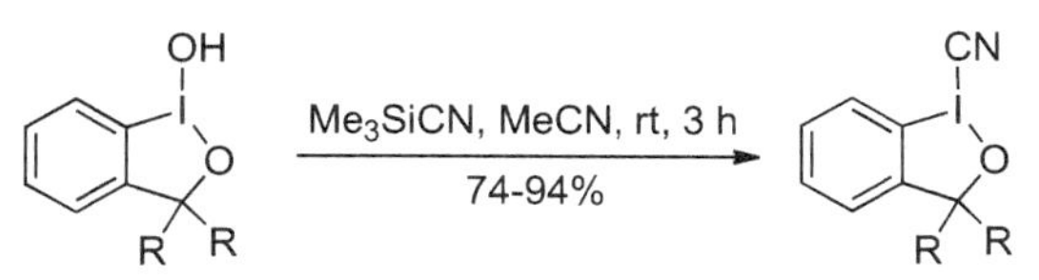

101
R = Me, CF$_3$
or 2R = O

113, R = Me
114, R = CF$_3$
115, 2R = O

Scheme 17

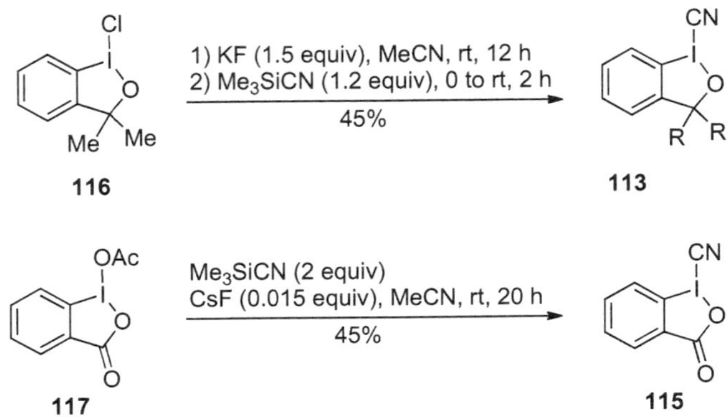


Scheme 18

of 1-acetoxy-1,2-benziodoxol-3(1*H*)-one with 1-alkynyl(diisopropyl)boronates in acetonitrile under reflux conditions (2000SL719).

An optimized procedure for the preparation of 1-[(triisopropylsilyl) ethynyl]-1,2-benziodoxol-3(1*H*)-one (TIPS−EBX) **123** on a 100-mmol (36 g) scale has been developed by Waser and Brand (Scheme 21; 2012S1155). In the larger scale procedure, removal of pyridine by treatment with acid and removal of 2-iodobenzoic acid by treatment with base are essential for efficient and reproducible purification. A single recrystallization from acetonitrile affords highly pure crystals of compound **123** which did not show any sign of decomposition in air after several months at room temperature.

TIPS−EBX **123** has found synthetic application as an efficient reagent for ethynylation of various organic substrates (2011CC102, 2009ACE9346, 2010CEJ9457, 2010OL384, 2011JOC565, 2010ACE4680, 2012CEJ5655).

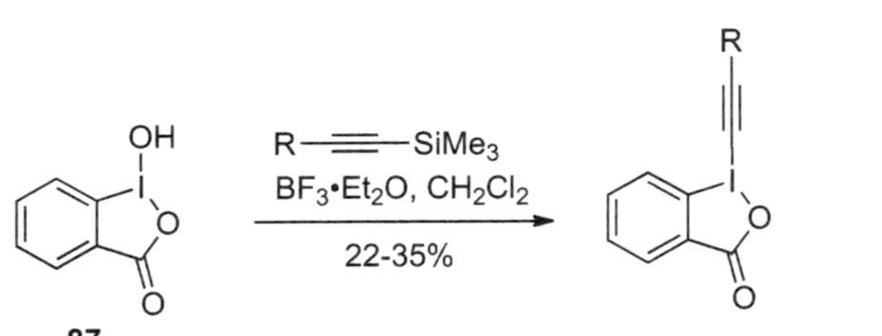


**118**, R = Bu$^t$
**119**, R = *cyclo*-$C_6H_{11}$
**120**, R = *n*-$C_8H_{17}$

Scheme 19

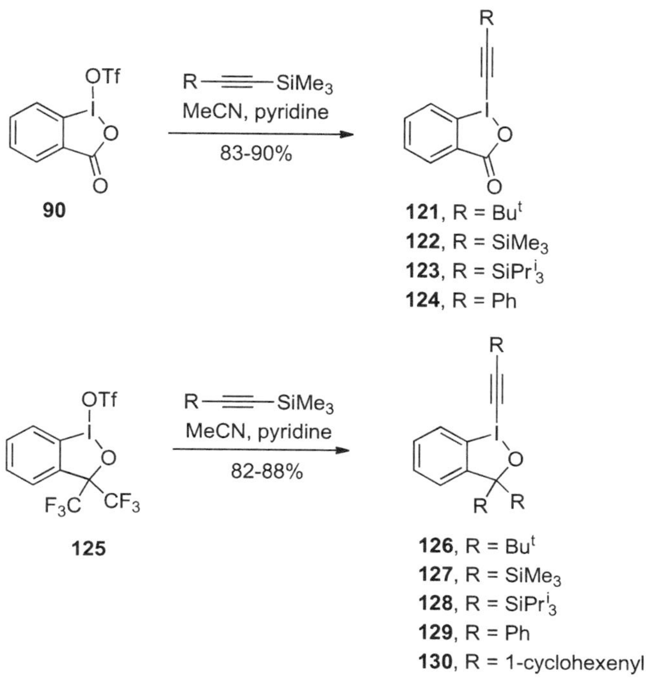


**Scheme 20**

### 3.2.2.9 Arylbenziodoxoles

1-Phenyl-1,2-benziodoxole-3(1*H*)-one **131**, also known as diphenyliodonium-2-carboxylate, is the most common commercially available and practically important representative of arylbenziodoxoles. Phenylbenziodoxole **131** can be prepared by oxidation of 2-iodobenzoic acid with potassium persulfate followed by an addition of benzene according to the optimized procedure published in *Organic Syntheses* (Scheme 22; 1966OS107). The original procedure was reported in 1960 by Beringer and Lillien (1960JA725), and has also been used for the synthesis of the substituted

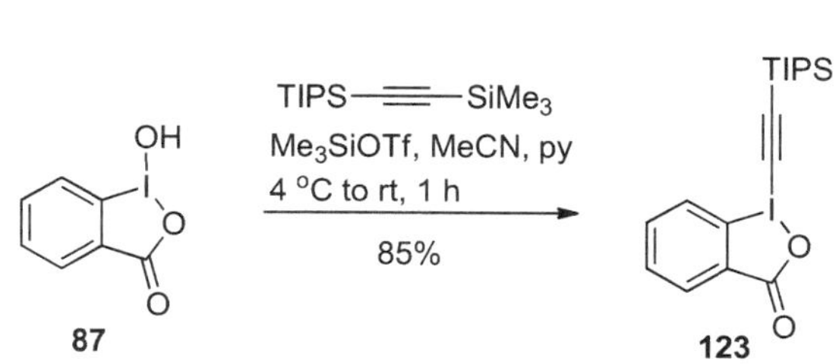


**Scheme 21**

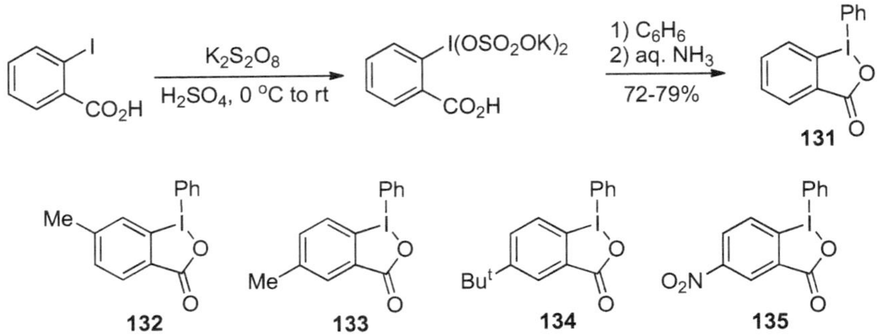


**Scheme 22**

phenylbenziodoxoles **132** (1978CB2510), **133** (1988JOC5799), and **134** (1985JOC3200). The nitro-substituted phenylbenziodoxole **135** was prepared by the treatment of 5-nitro-2-iodosybenzoic acid with benzene in concentrated sulfuric acid (1970JOC1196).

A recent modification of this procedure consists of the use of Oxone ($2KHSO_5{\bullet}KHSO_4{\bullet}K_2SO_4$) as an oxidant instead of $K_2S_2O_8$ and the use of $NaHCO_3$ as a base instead of $NH_3$. This modified method affords various arylbenziodoxoles **136** in generally high yields and is also applicable for the synthesis of the 7-methylbenziodoxolone ring system using 2-iodo-3-methylbenzoic acid as the starting compound (2013JOC3767, Scheme 23).

Merritt and Olofsson have reported a convenient modified procedure for the synthesis of phenylbenziodoxole **131** by a one-pot reaction of 2-iodobenzoic acid with *m*-chloroperoxybenzoic acid, trifluoromethanesulfonic acid, and benzene followed by an addition of aqueous ammonia (Scheme 24; 2011EJOC3690).

Batchelor, Birchall, and Sawyer reported X-ray crystal and molecular structure of phenylbenziodoxole **131** (1986IC1415). The geometry about

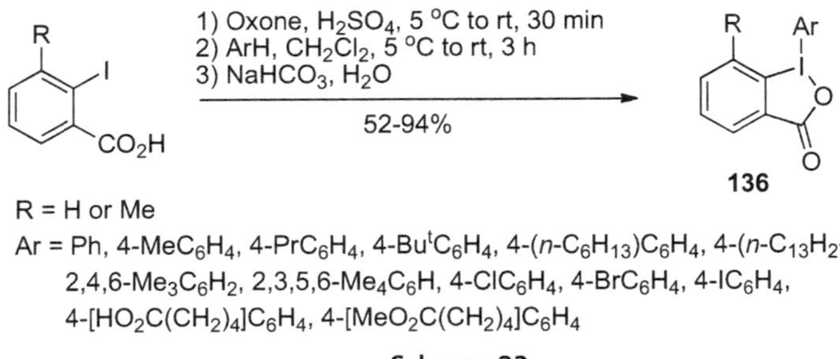


R = H or Me
Ar = Ph, 4-$MeC_6H_4$, 4-$PrC_6H_4$, 4-$Bu^tC_6H_4$, 4-($n$-$C_6H_{13}$)$C_6H_4$, 4-($n$-$C_{13}H_{27}$)$C_6H_4$, 2,4,6-$Me_3C_6H_2$, 2,3,5,6-$Me_4C_6H$, 4-$ClC_6H_4$, 4-$BrC_6H_4$, 4-$IC_6H_4$, 4-[$HO_2C(CH_2)_4$]$C_6H_4$, 4-[$MeO_2C(CH_2)_4$]$C_6H_4$

**Scheme 23**

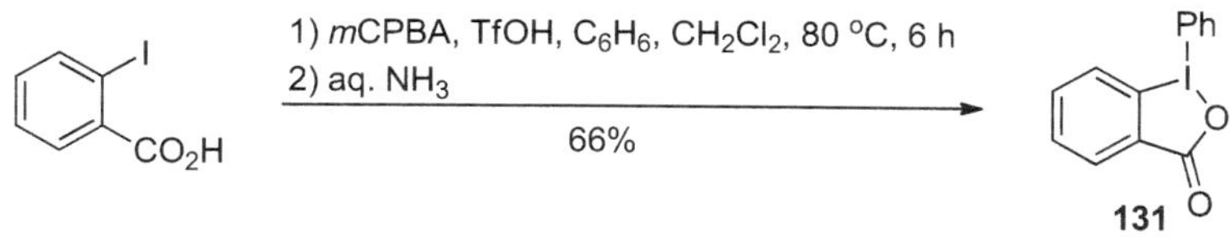


**Scheme 24**

iodine in **131** consists of primary bonds to the two phenyl rings (I—C 2.105 and 2.119 Å) and a strong intramolecular secondary I•••O bond of length 2.478 Å to the *ortho*-carboxylate group in an approximate trigonal-bipyramidal arrangement. The C—I—C angle (95.2°) is opened to accommodate the I•••O interaction and to reduce a C•••H interaction of 2.56 Å between the phenyl rings. Unexpectedly, no further I•••O secondary bonding occurs and the crystal packing is dictated by hydrogen bonds involving the water molecule in the lattice. The $^{127}$I Mössbauer spectral data for the structure of compound **131** are in agreement with that obtained from the crystallographic data (1986IC1415).

### 3.2.2.10 Phosphoranyl-Derived Benziodoxoles

Phosphoranyl-derived benziodoxoles **138** and **139** were prepared by the reaction of phosphoranes **137** with acetoxybenziodoxole **117** in the presence of trimethylsilyl triflate and pyridine (Scheme 25) and isolated as stable, microcrystalline solids (2002JA11614).

Structures of both benziodoxoles **138** and **139** were established by single crystal X-ray diffraction analysis (2002JA11614). According to X-ray crystallography, the benziodoxole ring system in structure **138** is essentially planar with an I—O bond length of 2.484 Å, which is similar to the structure of phenylbenziodoxole **131**. The solid-state packing of **138** demonstrates the significance of secondary bonding interactions, which link individual molecules of **138** into infinite chains via interactions between a carbonyl oxygen of one molecule and the I(III) center of its neighbor. This secondary

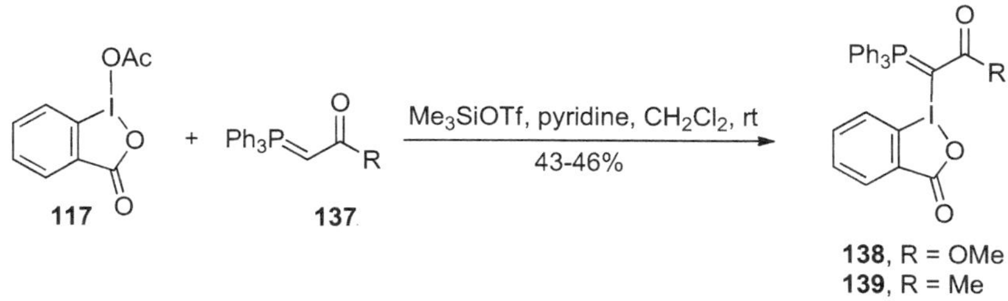


**Scheme 25**

bonding also affords an approximately square planar configuration about iodine, with a rather long I•••O bond length of 3.270 Å (2002JA11614).

### 3.2.2.11 Perfluoroalkylbenziodoxoles

The noncyclic $CF_3$-substituted trivalent iodine derivatives lack stability and cannot be isolated at room temperature; however, the incorporation of a hypervalent iodine atom into a five-membered heterocycle has a significant stabilization effect. The synthesis of stable trifluoromethylbenziodoxoles **141** and **143–145** by treatment of the corresponding methoxybenziodoxole **140** or acetoxybenziodoxole **142** with trimethyl(trifluoromethyl)silane was originally reported by Togni and coworkers in 2006 (Scheme 26; 2006CEJ2579).

Later, the synthesis of trifluoromethylbenziodoxole **141** was optimized to give the final product in high yield in three steps starting from IBX (2008JOC7678). Likewise, an improved one-pot synthesis of derivative **143** in 89% yield starting from chlorobenziodoxole **146** has been developed (Scheme 27; 2007ACE754).

Single-crystal X-ray structures have been reported for all four trifluoromethylbenziodoxoles **141, 143–145** (2007ACE754, 2006CEJ2579). X-ray structural data for the four products demonstrate typical trivalent iodine distorted T-shaped geometries. The I–$CF_3$ bond lengths increase from **141** (2.219 Å), through **144** (2.229 Å), to **145** (2.236 Å) and **143** (2.267 Å); simultaneously the I–O bond lengths decrease (**141**: 2.283 Å, **144**: 2.201 Å, **145**: 2.197 Å, **143**: 2.117 Å), whereas the I–C bond lengths

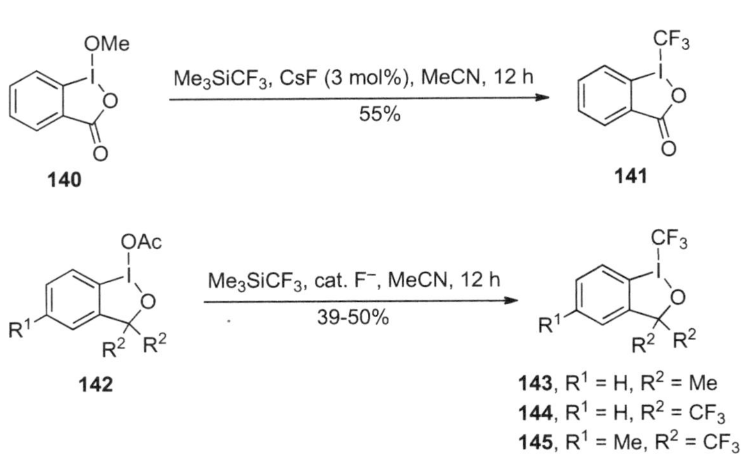


**Scheme 26**

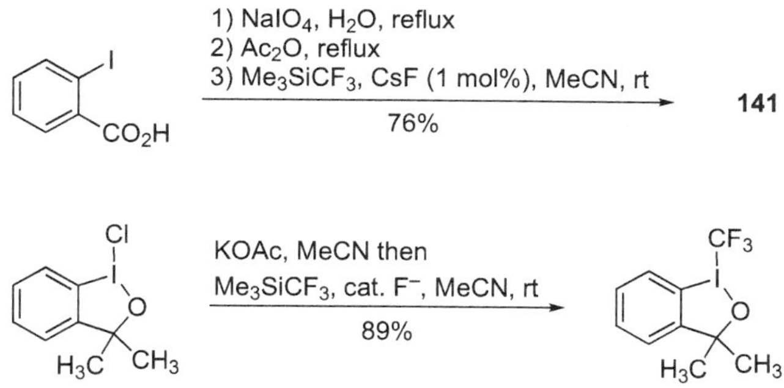


146 143

Scheme 27

remain constant within one standard deviation (2.114 Å). The $CF_3$—I—O bond angles in trifluoromethylbenziodoxoles are significantly smaller than 180° due to the repulsion of the two lone pairs at iodine, in agreement with the valence shell electron pair repulsion (VSEPR) theory, and are within the range of values typical of other benziodoxoles (around 170°) (2006CEJ2579).

Two additional examples of the fluoroalkyl-substituted benziodoxoles have been reported. 1-Pentafluoroethylbenziodoxolone has been prepared in a manner similar to that for 1-trifluoromethylbenziodoxolone **141** by reacting 1-acetoxybenziodoxolone **117** with $C_2F_5SiMe_3$ in the presence of CsF in acetonitrile (2012ACE8221). Likewise, benziodoxole **147** bearing a $PhSO_2CF_2$-substituent on the iodine atom has been synthesized from 1-chloro-1,3-dihydro-3,3-dimethyl-1,2-benziodoxole **146** and $PhSO_2CF_2SiMe_3$ using the procedure outlined in Scheme 28 (2008TL5006). Benziodoxole **147** can be used as an electrophilic (phenylsulfonyl)difluoromethylating reagent for a variety of S-nucleophiles under mild reaction conditions.

Trifluoromethylbenziodoxoles, especially 1-trifluoromethyl-1,3-dihydro-3,3-dimethyl-1,2-benziodoxole **143**, have found synthetic application as efficient reagents for trifluoromethylation of various organic substrates

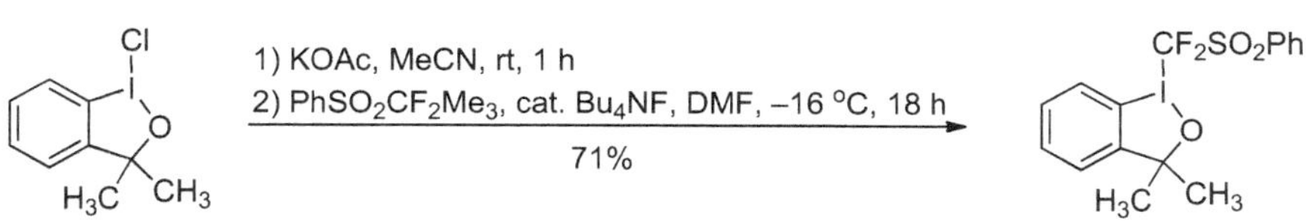


146 147

Scheme 28

(2007ACE754, 2006CEJ2579, 2008JOC7678, 2010OM1771, 2008HCA 2035, 2009S2818, 2008CC1575, 2010JOC1779, 2004HCA605, 2008C260, 2009CC5993, 2011ACE1059, 2010T5753, 2011JOC4189, 2010JFC951, 2012ACSC521, 2010JA4986).

#### 3.2.2.12 Isocyanate-Derived Benziodoxoles

The preparation and single-crystal X-ray structure of the isocyanatobenziodoxole **148** was reported by Buchwald and coworkers (2014ACE3125). Compound **148** has been isolated as the major product in the reaction of 1-chloro-1,3-dihydro-3,3-dimethyl-1,2-benziodoxole **146** and silver cyanate (Scheme 29). The formation of isomeric cyanatobenziodoxole as a minor product was also observed in this reaction. In the same paper (2014ACE3125), the previously reported structure of a trifluoromethylthiolate derivative of benziodoxole (2013ACE3457) was reinvestigated and proved to be wrongly assigned in the original paper.

### *3.2.3 Benziodazoles*

The preparation of a benziodazole heterocyclic system was originally reported by Wolf and Steinberg in 1965 (1965CC449). In this paper (1965CC449), the product isolated from the peracetic acid oxidation of 2-iodobenzamide **149** was incorrectly identified as *N*-acetyl-1-hydroxy-3-(1*H*)-1,2-benziodazole-3-one **151** based on infrared spectroscopy. The same structure **151** was also adopted in several other studies (1996H47, 1970JCS(C)862, 1986 JOC4303). In a more recent study, the product of peracetic acid oxidation of 2-iodobenzamide **149** (Scheme 30) was investigated by single-crystal X-ray analysis which showed the structure to be acetoxybenziodazole **150** and not the previously adopted structure **151** (1997JA7408).

Acetoxybenziodazole **150** (as a solvate with acetic acid) has the expected distorted T-shaped geometry at the iodine atom with an N—I—O bond angle of 162.1°. Bond lengths to the iodine center [I—N (2.101 Å), I—O (2.34 Å), and I—C (2.106 Å)] are all within the range of typical single covalent bonds in organic derivatives of polyvalent iodine and are in good

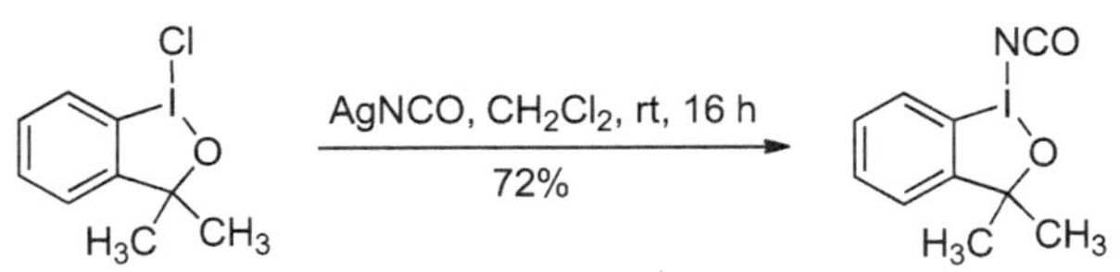


**146** **148**

**Scheme 29**

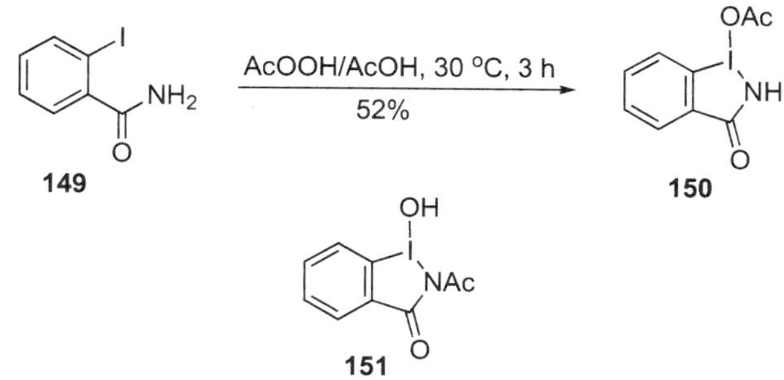


Scheme 30

agreement with the previously reported structures of chlorobenziodazoles (1979JOC1447, 1975JOC2129). According to molecular orbital calculations, the structure **150** is 6.31 kcal/mol more stable than **151** at the Hartree–Fock level of theory (1998JOC6590).

Acetoxybenziodazole **150** reacts at room temperature with carboxamides and alcohols in the presence of trimethylsilyl triflate leading to the rearranged products **152** and **153** (Scheme 31), the structures of which were established by X-ray analysis (1998JOC6590). The mechanism of this rearrangement includes ring opening and ring closure in the protonated benziodazole. Computational studies indicate that the driving force of this

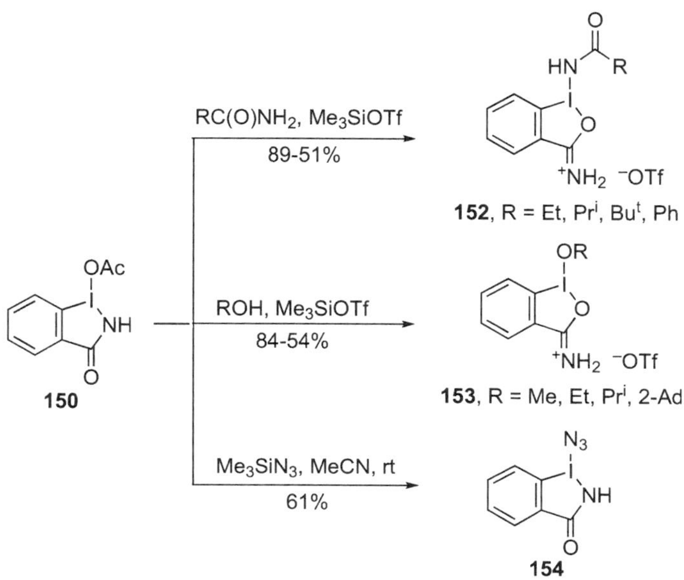


Scheme 31

rearrangement of benziodazoles to 3-iminiumbenziodoxoles is the greater thermodynamic stability of the N-protonated 3-iminobenziodoxoles **152** and **153** relative to the respective O-protonated benziodazole-3-ones by about 15 kcal/mol (1998JOC6590). The reaction of acetoxybenziodazole **150** with azidotrimethylsilane gives azide **154** in the form of a yellow, microcrystalline precipitate (1997JA7408). Azide **154** has reactivity similar to that of azidobenziodoxoles and can be used as an azidating reagent toward *N*,*N*-dimethylanilines.

The preparation of several N-functionalized benziodazoles derived from natural amino acids has been reported (2003OL1583). Acetoxybenziodazoles **156** and **157** were synthesized by peracetic acid oxidation of the corresponding 2-iodobenzamides **155** (Scheme 32) and isolated as stable, microcrystalline solids. Acetoxybenziodazole **156** can be further converted to the tosylate **158** by treatment with *p*-toluenesulfonic acid or to iodonium

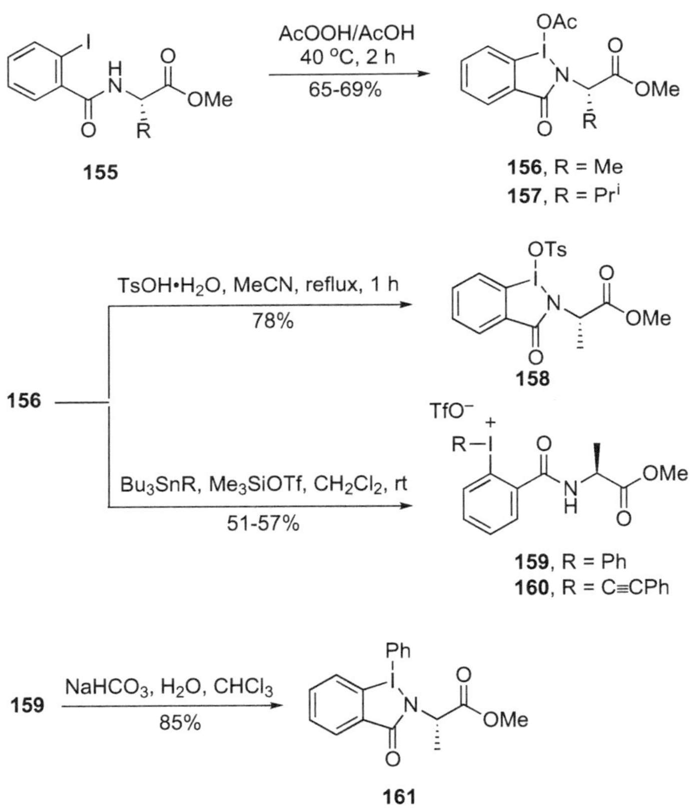


Scheme 32

salts **159** and **160** by reaction with tributylphenyltin or tributylphenylethynyltin and trimethylsilyl triflate.

The reaction of iodonium triflate **159** with sodium bicarbonate in water results in a restoration of the benziodazole ring with the formation of the phenylbenziodazole **161** (Scheme 32; 2003OL1583). An X-ray crystallographic study of **161** shows the essentially planar benziodoxole ring system with a relatively long I—N bond of 2.445 Å. This bond is longer than the analogous I—N bond in acetoxybenziodazole **155**, which is indicative of substantial ionic character. Overall, the geometry of **161** is similar to that observed for the previously reported structure of phenylbenziodoxole (2003OL1583).

Oxidation of *N*-(2-iodobenzoyl) amino acids **162** with dimethyldioxirane gives chiral and optically pure hypervalent iodine macrocycles **164—167** as the final isolated products (2001JA4095). Initial products in this oxidation are the monomeric amino acid-derived benziodazoles **163**, which undergo self-assembly leading to final products **164—167** (Scheme 33). This self-assembly is directed by secondary bonding between hypervalent iodine and oxygen atoms of the amino acid moiety (2001JA4095).

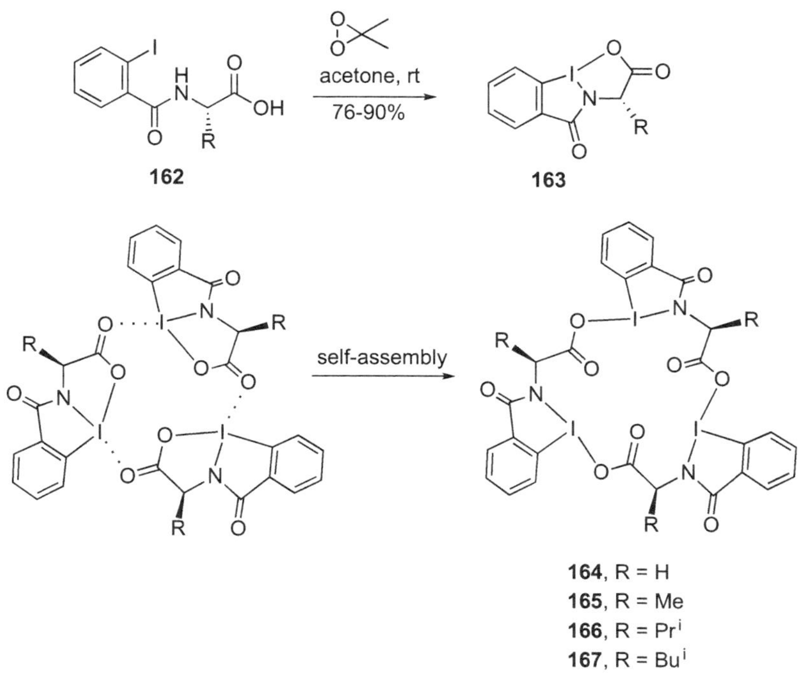


**Scheme 33**

Macrocycles **166** and **167** have been characterized by X-ray diffraction analysis (2001JA4095). Macrocyclic molecule **166** consists of a slightly distorted planar 15-atom cyclic system with three oxygens of the amino acid carboxylic function inside the ring and all three alkyl groups above the plane. Each iodine atom is covalently bonded to carbon (I—C 2.092 Å) and nitrogen (I—N 2.064 Å) and has three longer intramolecular contacts with oxygen atoms (I—O 2.368, 2.524, and 2.877 Å). With the consideration of primary and secondary bonds, the iodine atoms in **166** have a pentagonal-planar geometry, which is similar to that found in the solid state for $PhI(OAc)_2$. As a result of the central oxygen atoms, the electron-rich cavity of macrocycle **166** is suitable for complexation of metal cations. In particular, ESI—MS data indicate that macrocycles **164—167** can selectively form complexes with sodium cations in the presence of $K^+$, $Li^+$, $Ag^+$, or $Pb^{2+}$ (2001JA4095). The self-assembly of monomeric benziodazoles **163** into macrocyclic molecules **164—167** has been analyzed using molecular orbital calculations (2003ARK170). The driving force for this self—assembly is the formation of secondary bonding interactions between molecules and a rearrangement of primary and secondary bonding around iodine to place the least electronegative substituent in the equatorial position for each iodine atom in the trimer.

### 3.2.4 Benziodoxaboroles

Several heterocyclic compounds containing trivalent iodine, oxygen, and boron in a five-membered ring have been prepared by oxidative cyclization of the *ortho*-iodophenylboronic acids **168** and **171** (2011IC11263). 1-Chloro-4-fluoro-1*H*-1$\lambda^3$-benzo[*d*][1,2,3]iodoxaborol-3-ol **170** was formed in the reaction of 2-fluoro-6-iodophenylboronic acid **168** with chlorine followed by treatment of the intermediate iododichloride **169** with water. The 1-substituted acetoxybenziodoxaboroles **172** and **173** were synthesized by the hypochlorite oxidation of 2-fluoro-6-iodophenylboronic acid **168** or 2-iodophenylboronic acid **171** in acetic acid (Scheme 34).

Acetoxybenziodoxaboroles **172** and **173** were further converted to trifluoroacetates **174** and **175** by reaction with trifluoroacetic acid (Scheme 35). The 1-hydroxy derivative **176** can be prepared by basic hydrolysis of acetoxy- or trifluoroacetoxy benziodoxaboroles with aqueous $NaHCO_3$.

Single-crystal X-ray structural studies of 1-chloro- and 1-trifluoroacetoxy-substituted benziodoxaboroles (**170**, **174**, and **175**) demonstrated the presence of a planar five-membered heterocyclic ring with an unusually short endocyclic I—O bond distance of 2.04—2.09 Å (2011IC11263).

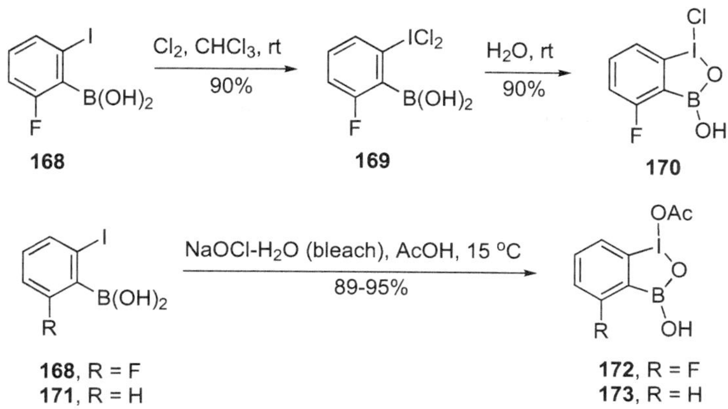


Scheme 34

Structural features of the five-membered iodoxaborol ring, such as the planar geometry and the short B—O and O—I bond lengths in the derivatives **170**, **174**, and **175**, may be explained by some conjugation present in this ring. The calculated NIST (0) and NIST (1) indexes for 1-chloro- and 1-trifluoroacetoxy-substituted benziodoxaboroles, however, are indicative of a significantly lower aromaticity compared to the classic aromatic systems (2011IC11263).

A tetrameric macrocyclic product **178**, resulting from self-assembly of the initially formed 4-fluoro-1,3-dimethoxy-1*H*-1$\lambda^3$-benzo[*d*][1,2,3] iodoxaborol **177**, was obtained by slow crystallization of trifluoroacetoxy-benziodoxaborole **174** from methanol (Scheme 36). The structure of macrocycle **178** was established by single-crystal X-ray analysis (2011IC11263).

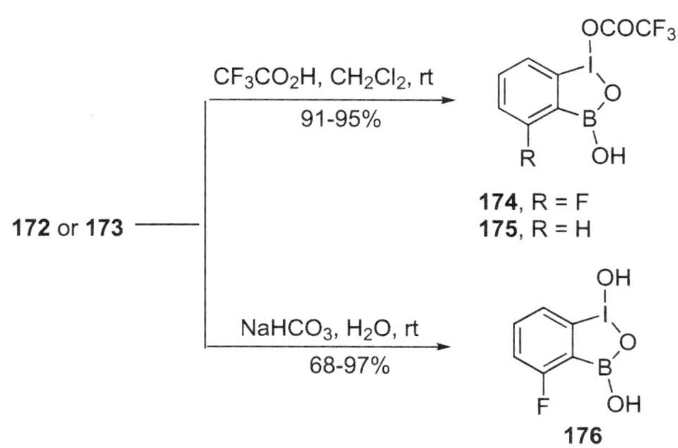


Scheme 35

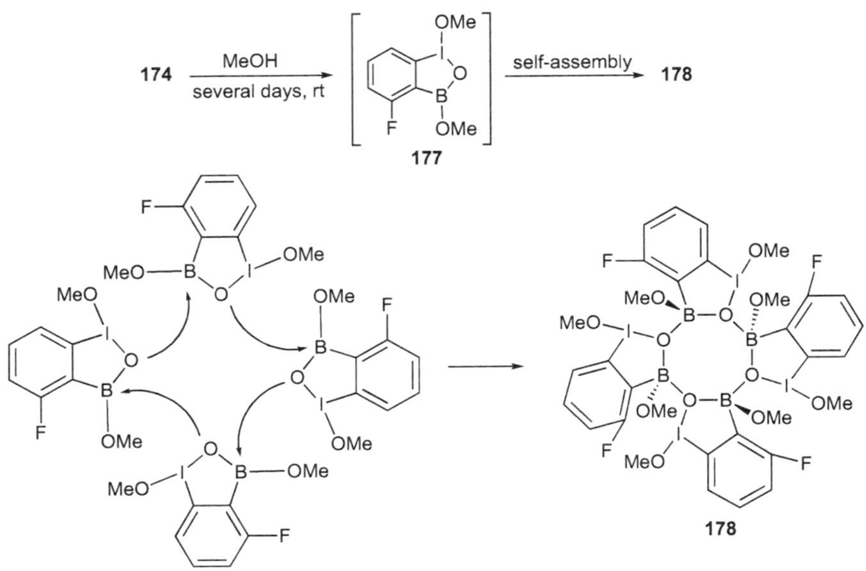


**Scheme 36**

The driving force for formation of the eight-membered cyclic system **178** is the transformation of initial trigonal-planar $sp^2$ hybridized boron atoms in **177** into tetrahedral $sp^3$ hybridized atoms. Each boron atom in tetramer **178** forms one covalent bond with carbon and three covalent bonds with oxygen atoms. The boron—oxygen bonds, which form the eight-membered cycle (1.48—1.50 Å), are significantly longer compared to the B—OMe bond distances (1.43—1.44 Å). The eight-membered cycle in **178** is non-planar with alternating larger B—O—B bond angles (126.2°, average) and smaller O—B—O bond angles (110.7°, average) with the latter being close to that expected for the $sp^3$ hybridized boron atoms. The change in hybridization of the boron atoms in tetramer **178** also leads to a significant elongation of boron—carbon and heterocyclic iodine—oxygen bonds compared to those observed in compounds **170**, **174**, and **175**. As for compounds **170**, **174**, and **175**, the iodine centers in **178** have the usual T-shape iodine(III) geometry with O—I—OMe bond angles varying between 167.2° and 170.5° (2011IC11263).

### 3.2.5 Benziodoxathioles

1*H*-1-Hydroxy-1,2,3-benziodoxathiole 3,3-dioxide (structure **57** in Figure 7) was originally described as a sodium salt in the 1914 book by Willgerodt (1914MI8). In 1993 Koser and coworkers reported the preparation of 1*H*-1-hydroxy-5-methyl-1,2,3-benziodoxathiole 3,3-dioxide **180** by oxidation of 2-iodo-5-methylbenzenesulfonic acid **179**. Benziodoxathiole **180** was further converted into a series of alkynylbenziodoxathioles **181** by reaction with the corresponding terminal alkynes in the presence of toluenesulfonic acid (Scheme 37; 1993JOC7310).

The structure of hydroxybenziodoxathiole **180** was established by single-crystal X-ray analysis (1993JOC7310). As expected for hypervalent iodine, the molecules **180** have a T-shaped geometry, although the O—I—O bond angle (168.6°) is less linear than the O—I—O angle in the noncyclic sulfonate PhI(OH)OTs (178.80°). The I—OH bond distances in **180** (1.933 Å) are nearly the same as in PhI(OH)OTs (1.940 Å) and slightly shorter than the computed covalent distance of 1.99 Å for an I—O single bond. The I—O bond to the sulfonate ligand in **180** (2.372 Å) is elongated; however, it is 0.1 Å shorter than it is in PhI(OH)OTs (2.473 Å).

A more recent paper reported an X-ray single-crystal structure of the unsubstituted 1*H*-1-hydroxy-1,2,3-benziodoxathiole 3,3-dioxide **57** (2006EJOC4791). In general, X-ray structural parameters of heterocycles **57** and **180** are very similar.

Justik, Protasiewicz, and Updegraff described the synthesis and X-ray structures of several arylbenziodoxathioles **182** (2009TL6072). Compounds **182** were prepared by reacting hydroxybenziodoxathiole **57** with appropriate arenes in a solvent mixture containing 2,2,2-trifluoroethanol (Scheme 38).

X-ray single-crystal structural analyses of compounds **182** (four structures for Ar = Ph, 4-$MeOC_6H_4$, 4-$ClC_6H_4$, and 2,4,6-$Me_3C_6H_2$) revealed that the I—O bonds are substantially longer (e.g., 2.676 Å in structure **182**, Ar = Ph) than the analogous bond in phenylbenziodoxole **131** (2.478 Å; see Section 3.2.2.9), suggesting a greater contribution of ionic bonding (2009TL6072). The I—O bonds in compounds **182** are also markedly longer than the analogous bond in hydroxybenziodoxathiole **57** (2.381 Å) (2006EJOC4791).

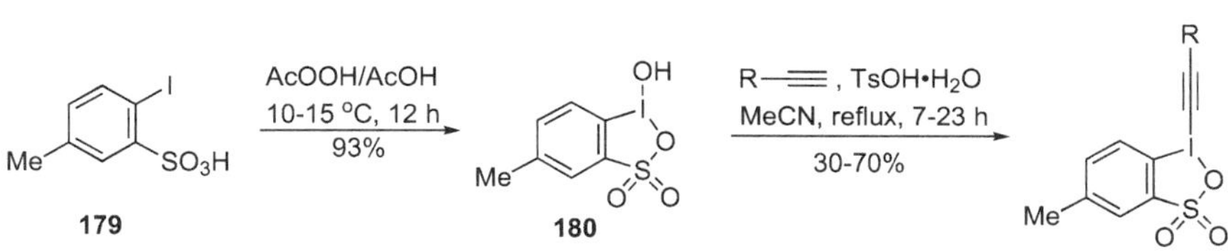


179 180 181

R = Pr, Bu, *n*-$C_5H_{11}$, *n*-$C_6H_{13}$, $Bu^i$, $Bu^s$, $Bu^t$, *c*-$C_6H_{11}$, Ph

**Scheme 37**

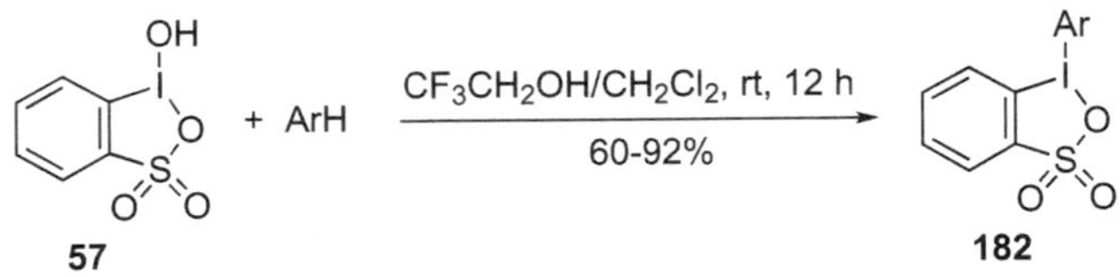

Ar = Ph, 4-$MeC_6H_4$, 4-$MeOC_6H_4$, 4-$ClC_6H_4$, 2,4,6-$Me_3C_6H_2$

**Scheme 38**

### *3.2.6 Benziodothiazoles*

Jaffe and Leffler in 1975 reported the synthesis of l-acetoxy-1,2-dihydro-1,3,2-benziodothiazole 3,3-dioxide **184**, which is the first representative of the benziodothiazole system (Scheme 39; 1975JOC797). The acetate **184** was prepared by the peracetic acid oxidation of 2-iodobenzenesulfonamide **183**. Hydrolysis of **184** affords l-hydroxy-1,2-dihydro-1,3,2-benziodothiazole 3,3-dioxide **185**. l-Chloro-1,2-dihydro-1,3,2-benziodothiazole 3,3-dioxide **186** can be obtained by treatment of a basic solution of acetate **184** with concentrated hydrochloric acid (1975JOC797).

### *3.2.7 Benziodoxaphospholes*

Balthazar and coworkers prepared 1,3-dihydro-1-hydroxy-3-methyl-1,2,3-benziodoxaphosphole 2-oxide **59** by oxidation of 2-iodophenylmethylphosphinic acid **187** (Scheme 40; 1978JOC4538). Benziodoxaphosphole **59** can be converted to the methoxy derivative **188** by treatment with hot methanol. Methoxybenziodoxaphosphole **188** is readily hydrolyzed in moist air to give the initial hydroxybenziodoxaphosphole **59** (1978 JOC4538). Freedman and DeMott reported a similar preparation of the benziodoxaphospholes **189** and **190** by peracetic acid oxidation of 2-iodophenylphosphonic acid or (2-iodophenyl)phenylphosphinic acid, respectively (1974PRGE277).

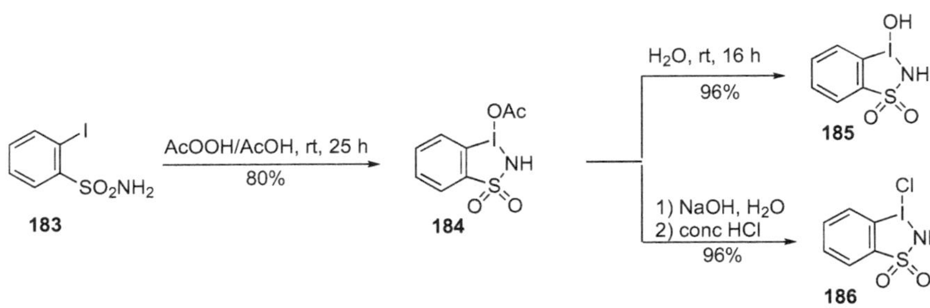

**Scheme 39**

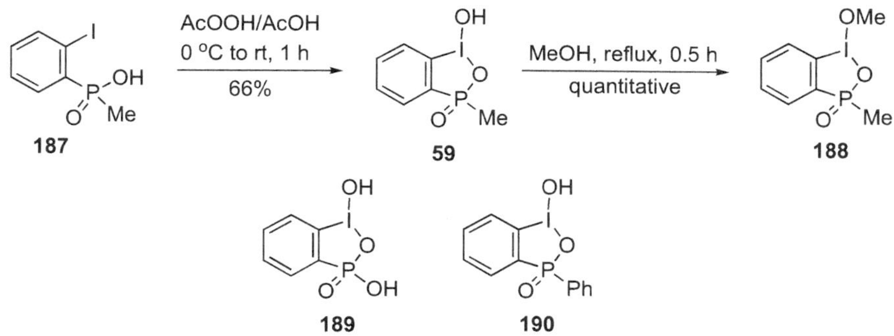

Scheme 40

The structure of hydroxybenziodoxaphosphole **59** was established by single-crystal X-ray analysis (1978JOC4538). Molecules of **59** are T-shaped about the iodine atom with the O—I—O bond angle of 171.3°, the I—OH bond distance of 1.952 Å, and the I—O bond to the phosphinate oxygen of 2.286 Å.

Moss and coworkers (1997JPOC27 ) investigated the catalytic activity of hydroxybenziodoxaphospholes in the hydrolysis of phosphate esters and found that the anion of **59** is 44 times less reactive compared to the hydroxybenziodoxole anion of **85** (see Scheme 10 in Section 3.2.2.2).

### *3.2.8 Six-membered Iodine(III) Heterocycles*

Because of the nature of hypervalent bonding and the T-shaped geometry of iodine(III) centers, the formation of six-membered iodine heterocycles is highly unfavorable. Several such compounds have been reported in the literature (1986JOC4303, 1962JA3443, 1993TL1729, 2009ACE4605, 2010 JOC8416, 1973JOC2719); however, their structures have not been confirmed by X-ray structural data. Moreover, based on the available X-ray data for several pseudocyclic six-membered iodine(V) derivatives (see Section 3.3.3), it can be expected that these compounds exist as their noncyclic tautomers (2010JOC8416).

Preparation of the first example of a six-membered iodine(III) heterocycle, the cyclic tautomer of 2-iodosylphenylacetic acid **192**, was reported in 1963 by Leffler and coauthors (1962JA3443). This compound was synthesized by chlorination of 2-iodophenylacetic acid **191** followed by hydrolysis of the initially formed, unstable 2-(dichloroiodo)phenylacetic acid (Scheme 41). Compound **192** is stable at room temperature but decomposes in solution at 80—100 °C; the proposed cyclic structure **192** is in agreement

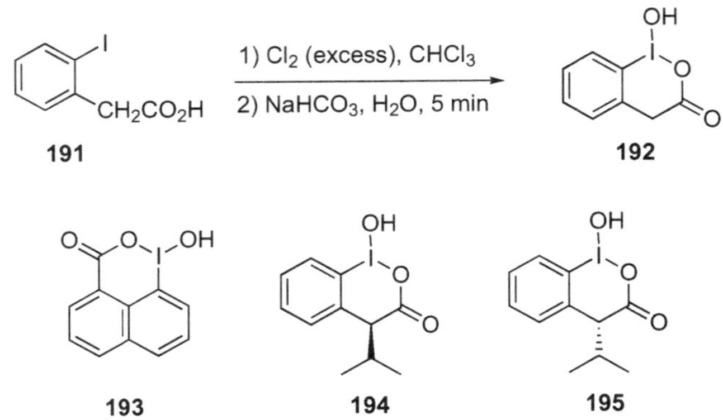


Scheme 41

with its relatively low acidity (p$K_a$ 7.45) (1962JA3443). 8-Iodosyl-1-naphthoic acid **193** has been prepared by the peracetic acid oxidation of 8-iodo-1-naphthoic acid (1993TL1729). Anions of 2-iodosylphenylacetic acid **192** and 8-iodosyl-1-naphthoic acid **193** (1993TL1729 ) have a moderate reactivity in the cleavage of phosphate esters in aqueous micellar solution. The chiral, enantiomerically pure substituted 2-iodosylphenylacetic acid derivatives **194** and **195** were synthesized from the corresponding aryl iodides by oxidation with dimethyldioxirane (2009ACE4605).

Leffler and Jaffe have also reported the synthesis of a presumably cyclic tautomer of 2-iodosylphenylphosphoric acid **197** and its methyl ester **198** starting from 2-iodophenylphosphoric acid **196** (Scheme 42; 1973 JOC2719). Products **197** and **198** were identified by NMR spectroscopy; an attempted X-ray diffraction analysis of **197** was unsuccessful.

### *3.2.9 Polycyclic Heterocycles with an Iodine(III) Atom at a Ring Junction*

The first representatives of fused hypervalent iodine heterocycles with an iodine atom at a ring junction were described by Agosta in 1965 (1965TL2681). Compounds **200** and **201** were originally prepared by oxidation of the corresponding iodides **199** and **201** with peracetic acid

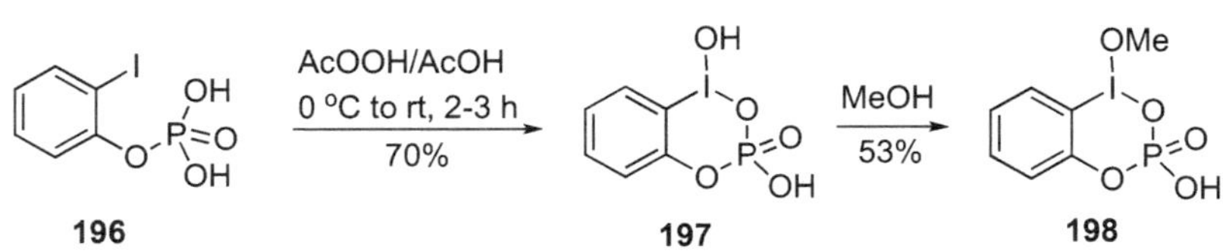


Scheme 42

(Scheme 43). A few years later, Martin and Chau reported an alternative way of obtaining product **200** by the thermal decomposition of the dibutyl ester of 2-iodoisophthalic acid (1974JA3319). The benziodoxolone derivative **200** has high thermal stability, with decomposition point above 260 °C; the formation of this heterocyclic system is particularly favorable due to the T-shaped geometry of hypervalent iodine(III). An X-ray crystallographic analysis of compound **200** showed an overall planar geometry with the expected T-configuration at the iodine(III) center (2012 OL3020). Several analogous iodine(III) heterocycles **203** have also been reported (1991JOC6565, 1982JOC1024).

An unusual hypervalent iodine heterocycle **205** with condensed four- and five-membered rings has been prepared by oxidative cyclization of 3-iodo-3-hexene-2,5-diol **204** using neat *tert*-butylhypochlorite (Scheme 44; 2001PSSS141, 2001JA1507).

The structures of products **203** (R = $Bu^t$) (1986JA3803) and **205** (2001JA1507) were determined by single-crystal X-ray analyses. Both structures have a distorted T-shape geometry around the central iodine atom. The apical O—I—O bond angles of **203** and **205** are 144.5(1)° and 158.2(2)°,

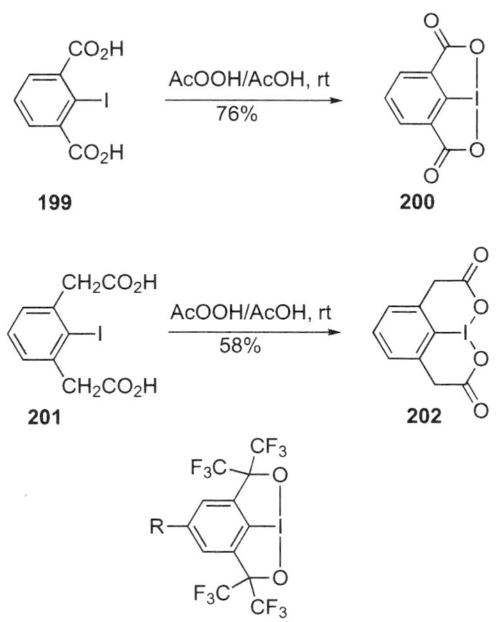


**203**, R = Me, $Bu^t$, $Me_2(FCH_2)C$

**Scheme 43**

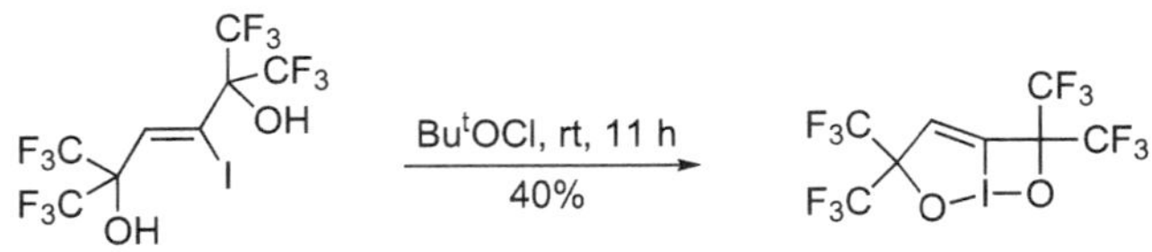

**204** **205**

Scheme 44

respectively, indicating that the deviation (35.5°) of the O—I—O bond angle of **205** from 180° is larger than that (21.8°) of **203**. The apical bond length (2.142(3) Å) of the four-membered ring of **205** is longer than that of the five-membered ring (2.070 Å). The four-membered ring is almost planar, judging from the I—C—C—O torsion angle of 0.0(3)° and the sum of bond angles of the four-membered ring of 359.9° (2001JA1507).

### *3.2.10 Pseudocyclic Iodine(III) Compounds*

Iodosylarenes (ArIO) and imidoiodanes (ArINR) bearing an appropriate substituent in the *ortho*-position to the iodine are characterized by the presence of a pseudocyclic structural moiety due to strong intramolecular secondary bonding between the hypervalent iodine center and the oxygen atom in the *ortho*-substituent. Compared to the non-cyclic aryliodyl derivatives, pseudocyclic iodine(III) compounds have much better solubility, which is explained by a partial disruption of their polymeric nature due to the redirection of secondary bonding (2014CCR54).

The *ortho*-substituted iodosylarene **208** bearing a *tert*-butylsulfonyl group was prepared as outlined in Scheme 45. Oxidation of aryliodide **206** with peracetic acid led to formation of the (diacetoxyiodo)arene **207**, which was further converted to iodosylarene **208** by treatment with aqueous base (1999JA7164).

Iodosylarene **208** has high solubility in organic solvents (up to 0.08 M in chloroform) and can be analyzed by NMR in solution (1999JA7164). A single-crystal X-ray analysis of compound **208** showed a pseudocyclic structure with an intramolecular distance of 2.707(5) Å between one of

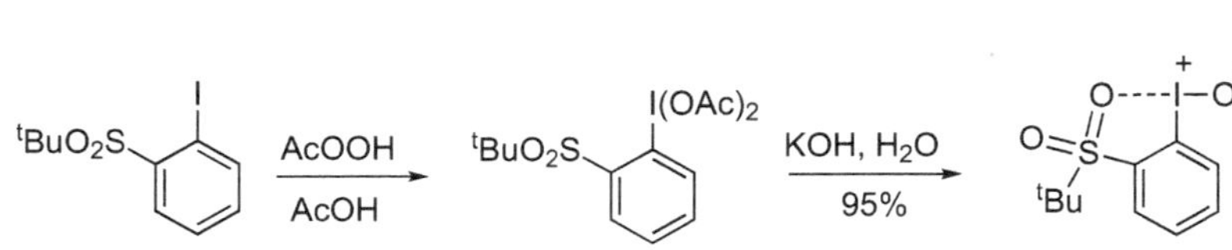

**206** **207** **208**

Scheme 45

the sulfone oxygen atoms and the hypervalent iodine center (2000ACE2007). The I—O bond length in the iodosyl group of **208** is 1.848(6) Å, and the intramolecular O—I—O bond angle is 167.3(2)°. The iodine centers in **208** have a pseudo square-planar geometry by the presence of an intermolecular I•••O secondary bond (2.665(6) Å) to a neighboring iodosyl oxygen atom (2000ACE2007, 2010T5768).

The *ortho*-substituted iminoiodane **209** bearing a *tert*-butylsulfonyl group has been prepared by treatment of the (diacetoxyiodo)arene **207** with tosylamide in basic methanol (1999JA7164), and the *ortho*-alkoxy iminoaryliodanes **211** were obtained using a similar procedure starting from the readily available diacetates **210** (Scheme 46; 2011CEJ10538).

Iminoiodane **209** has an excellent solubility in organic solvents (up to 0.14 M in chloroform at room temperature, at least a 50-fold increase over PhINTs). Compound **209** was analyzed by NMR in solution and by X-ray diffraction in the solid state. A single-crystal X-ray analysis of compound **209** demonstrated a pseudocyclic structure with an intramolecular distance of 2.677(3) Å between one of the sulfone oxygen atoms and the hypervalent iodine center (1999JA7164). Molecules of **209** are loosely associated into centrosymmetric dimers by long-range intermolecular I•••N and I•••O bonds (longer than 3.0 Å), quite unlike the infinite polymeric chains adopted in the solid state for PhINTs.

According to the single-crystal X-ray crystallographic analysis, (tosyliminoiodo)-*o*-methoxybenzene **211** (R = Me) has a polymeric, asymmetrically bridged structure similar to that observed for PhINTs (2011CEJ10538). However, in contrast to PhINTs, compound **211** has an additional weak

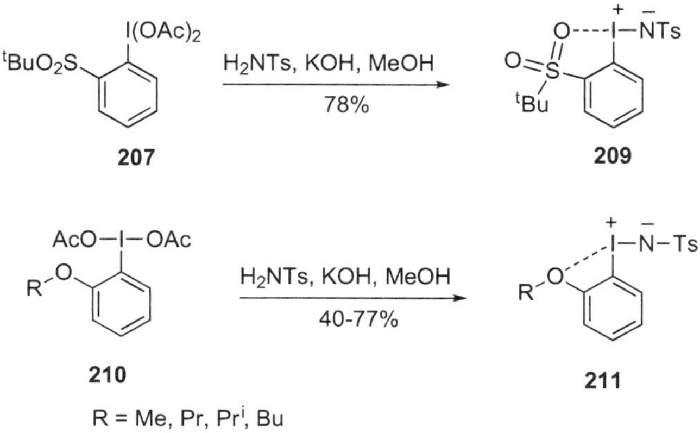


Scheme 46

intramolecular I•••O contact between the iodine center and the oxygen atom of the methoxy group. This additional interaction leads to the elongation of the I•••N intermolecular bond in **211** (2.735 Å) compared to that observed in PhINTs (2.482 Å). As a result, the polymeric structure of **211** is weakened and the solubility is significantly increased.

Tosyliminoiodoiodanes **209** and **211** have found application as readily available and soluble nitrene precursors that are useful for Cu-catalyzed aziridination of alkenes and metal-free amination of organic substrates under homogeneous conditions (2014CCR54).

## 3.3 Iodine(V) Heterocycles

The five-membered pentavalent iodine heterocycles represent a particularly important class of hypervalent iodine compounds. Cyclic iodine(V) compounds, such as IBX **212** and DMP **213**, have found broad practical application as mild and selective reagents for the oxidation of alcohols and some other useful oxidative transformations. Several comprehensive reviews of the chemistry and synthetic applications of IBX and DMP have been published (2011JOC1185, 2006ARK26, 2010T7659, 2011AGE1524, 2001ACE2812).

### *3.3.1 Five-Membered Iodine(V) Heterocycles with an Oxygen Atom in the Ring*

IBX **212** was first reported by Hartman and Mayer in 1893 (1893CB1727). Similar to the iodine(III) hydroxybenziodoxoles (see Section 3.2.2.3), the actual structure of IBX is best represented as a cyclic benziodoxole oxide (1-hydroxy-1-oxo-1*H*-1$\lambda^5$-benzo[*d*][1,2]iodoxol-3-one according to the IUPAC nomenclature), which is confirmed by X-ray structural analysis (1997CSC589, 1981CSC489, 1990JCS(P2)1657). The original procedure for the preparation of IBX involved oxidation of 2-iodobenzoic acid with potassium bromate in aqueous sulfuric acid (1893CB1727); an optimized procedure for the bromate oxidation has been published in *Organic Syntheses* (2000OS141). The samples of IBX prepared by the bromate oxidation were reported to be explosives under heating or impact (1990CEN3); this has been explained due to the presence of bromate impurities (1993JA2488). In a more recent work, Santagostino and coworkers described a convenient and safe procedure for the preparation of IBX **212** by the oxidation of 2-iodobenzoic acid using Oxone ($2KHSO_5$•$KHSO_4$•$K_2SO_4$) in water at 70 °C (1999JOC4537). This convenient procedure is now the most commonly used method for the large-scale preparation of IBX (Scheme 47).

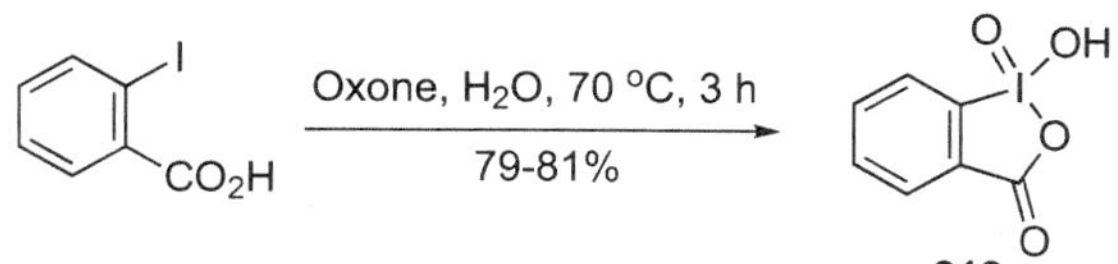


**Scheme 47**

In the solid-state, IBX has a three-dimensional polymeric structure formed by strong intermolecular secondary I•••O interactions and hydrogen bonding. An X-ray diffraction study of IBX samples, prepared by the oxidation of 2-iodobenzoic acid with potassium bromate in sulfuric acid, revealed the presence of both powder and macrocrystalline forms of IBX (1997CSC589). It was also reported that the powder form of IBX is more reactive than the macrocrystalline form in the reaction with acetic anhydride and thus is more useful as the precursor to DMP (see structure **213**). The macrocrystalline IBX can be converted to the more reactive powder form by treatment with aqueous sodium hydroxide and then with hydrochloric acid (1997CSC589).

A single-crystal X-ray analysis of the macrocrystalline form of IBX revealed that the pentavalent iodine center is chiral, and in the solid state IBX is a racemic mixture of two enantiomers (1997CSC589). The crystal structure of the racemic IBX consists of chains of molecules linked by I•••O secondary bonds (2.808 Å) and also by additional hydrogen bonding. Overall, molecules of IBX are linked together via one O—H•••O, four C—H•••O hydrogen bonds, and two I•••O secondary bonds to form a three-dimensional, hydrogen and iodine—oxygen bonded network in which the iodine atoms have distorted octahedral coordination. This three-dimensional network is further stabilized by $\pi$—$\pi$ stacking of the phenyl rings with C•••C contacts of 3.2 Å and H—ring center contacts of 2.4 Å (1997CSC589).

*IBX is a potentially explosive compound, and even the bromate-free samples of IBX are not safe.* Santagostino and coworkers reported that pure IBX typically explodes at 233 °C (1999JOC4537). The explosibility tests of analytically pure IBX samples (over 99% purity) confirmed the earlier observations by Plumb and Harper that IBX is explosive under impact or heating above 200 °C.

Quideau and coworkers have developed a nonexplosive formulation of IBX (SIBX), consisting of IBX, benzoic acid, and isophthalic acid (2003OL2903). SIBX has similar reactivity to IBX in the reactions of hydroxylative phenol dearomatization (2003ARK106, 2007JOC6280, 2007ACE1533), oxidation of sulfides into sulfoxides (2006TL5869), oxidative demethylation of phenolic methyl aryl ethers (2003OL2903), and other useful oxidative transformations (2003OL2903).

Theoretical and experimental studies of the p$K_a$ value and proton affinity of IBX solutions in aqueous media and dimethyl sulfoxide (DMSO) have been reported. In particular, the aqueous p$K_a$ value of 2.40 for IBX was obtained using standard potentiometric titration methods (2006ACE2929). The relatively high acidity of IBX should be taken into consideration when using this important reagent in the oxidation of complex organic molecules. The gas phase proton affinities of the anions of IBX (1300 ± 25 kJ/mol) and IBA (1390 ± 10 kJ/mol) using mass spectrometry-based experiments have been published (2008OBC2530). The experimental results were supported by theoretical calculations, which yielded proton affinities of 1336 and 1392 kJ/mol for the anions of IBX and IBA, respectively, at the B3LYP/aug-cc-PVDZ level of theory (2008OBC2530).

In 1983 Dess and Martin reported the first preparation of triacetoxybenziodoxolone **213** by heating IBX with acetic anhydride to 100 °C (1983JOC4155). Later, the triacetate **213** has emerged as the reagent of choice for the oxidation of alcohols to the respective carbonyl compounds (1993JA2488), and in the present literature it is commonly named as DMP. An improved procedure for the preparation of DMP **213** consists of the reaction of IBX with acetic anhydride in the presence of *p*-toluenesulfonic acid at 80 °C (Scheme 48; 1993JOC2899). DMP is also commercially available from several chemical companies. An optimized procedure for the preparation of DMP was published in *Organic Syntheses* (2000OS141). The synthetic applications of DMP have been highlighted in two short reviews (1996JPR588, 2000S278).

Freshly prepared, crystalline DMP has excellent solubility in dichloromethane and chloroform. It should be stored and handled under dry conditions; exposure to the atmosphere should be avoided. A careful hydrolysis of DMP **213** by slow addition of one equivalent of water in dichloromethane leads to formation of the monoacetate **214**, which can be isolated in 80% yield (Scheme 49; 1994JOC7549). The freshly prepared or generated in situ monoacetate **214** is a stronger oxidant compared to DMP.

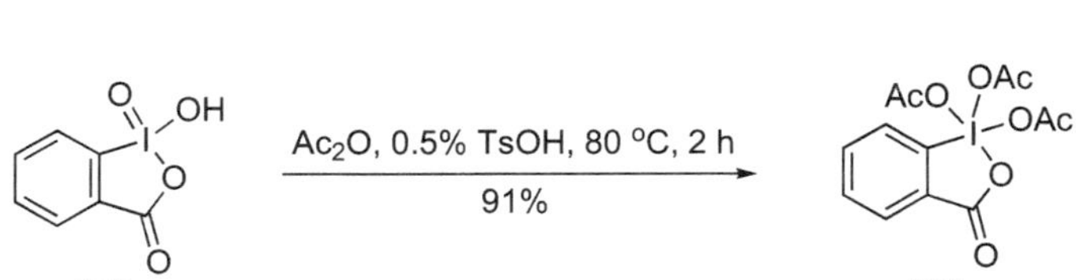


212 213

**Scheme 48**

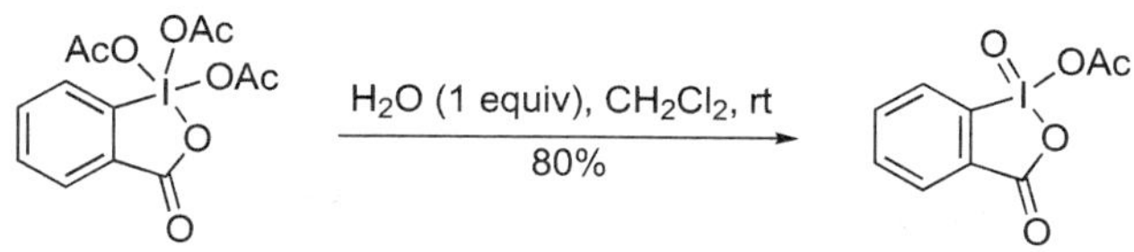



**213** **214**

**Scheme 49**

According to single-crystal X-ray structural data, DMP **213** crystallizes as a dimer (2012BJOC1523). The central iodine atom resides in a distorted octahedral environment that is in agreement with the VSEPR model. The acetoxy groups occupy equatorial positions, whereas the phenyl ring and the lone electron pair are in the apical positions. Due to the steric demands of the electron pair, the acetoxy substituents are pushed toward the phenyl ring. As a result, the iodine atom is located 0.315(1) Å below a plane formed by the oxygens of the three acetoxy groups. All three acetoxy groups are bound in a covalent fashion, showing typical iodine—oxygen bond lengths between 2.06 and 2.11 Å. The length of the iodine oxygen bond of the iodoxolone ring of 2.089 Å is also within the range of a covalent iodine oxygen bond. The iodine carbon bond has a length of 2.10 Å and forms a C—I—O bond angle of 79.6° within the iodoxolone ring. The unit cell is occupied by a centrosymmetric dimer that is held together, in part, by two intermolecular I•••O bonds between the iodine atom and a carbonyl group of the adjacent molecule. The intermolecular iodine—oxygen distance of 3.3 Å is well below the sum of the van der Waals radii (3.46 Å). In addition to I•••O bonds, the dimer is stabilized by two weak hydrogen bonds of the C—H•••O type.

Numerous derivatives and analogs of IBX have been described in the literature (Figure 8). In the early 1990s, Martin and coworkers reported the synthesis, structure, and properties of several cyclic $\lambda^5$-iodanes (**215**—**217**) (1991JOC6565, 1993JA2488, 1991JA7277). Martin and coworkers have also introduced bis(trifluoromethyl)benziodoxole oxides **218** and **219**, which are stable and non-explosive oxidizing reagents soluble in a wide range of organic solvents (1991JA7277, 1995TL9117). The chiral pentavalent benziodazole oxides **220** were prepared by the oxidation of the corresponding amino acid-derived 2-iodobenzamides with potassium bromate (2000 TL5299). Vinod and coworkers have developed water-soluble analogs of IBX; these are *m*-iodoxyphthalic acid (mIBX) **221** (2002TL569) and a similar derivative of terephthalic acid **222** (2008TL4378), which are useful oxidizing reagents in aqueous solutions. A zwitterionic, water-soluble IBX analog **223**

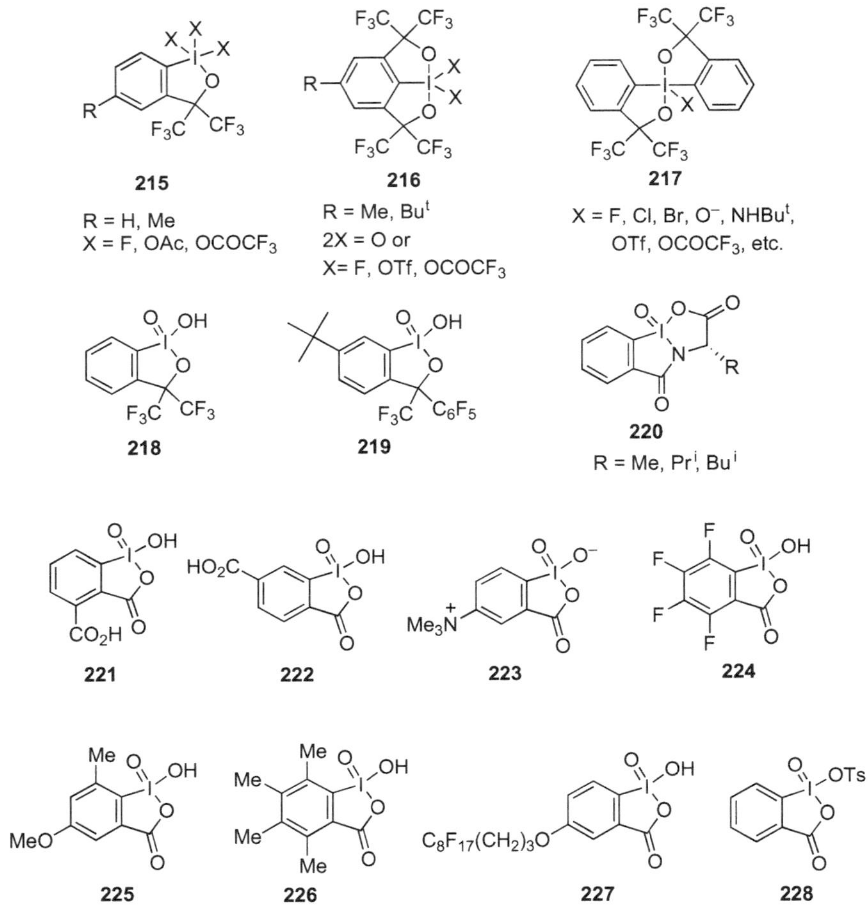


**Figure 8** IBX derivatives and analogs.

has been prepared by the oxidation of the corresponding iodide with dimethyldioxirane and characterized by single-crystal X-ray diffraction (2011OL6488). Wirth and coworkers have developed the tetrafluoro IBX derivative (FIBX **224**), which is more soluble and has higher reactivity than its nonfluorinated counterpart (2007ACE6529). Moorthy and coworkers have prepared *o*-methyl-substituted IBX (Me-IBX **225**) (2008TL80) and tetramethyl-IBX (TetMe-IBX **226**) (2011JOC9593) which are modified analogs of IBX that can oxidize alcohols in common organic solvents at room temperature due to the hypervalent twisting-promoted rate enhancement (2005JA14146). Fluorous IBX **227** is a recyclable oxidant that can also be used as a catalyst for oxidation of alcohols to the corresponding carbonyl compounds (2011CC1875). Tosylate derivative of 2-iodoxybenzoic acid (IBX-tosylate **228**), one of the most powerful hypervalent iodine(V) oxidants, can be prepared by the reaction of IBX with *p*-toluenesulfonic acid in acetic anhydride (2013CC11269). Several polymer-supported IBX derivatives have also been reported (2001ACE4393, 2001ACE4395, 2006RFP840, 2003TL1635, 2008CM2001).

A pyridinium salt of IBX in the form of stable, non-explosive, colorless crystals has been prepared by treatment of IBX with pyridine (2012MC129). Pyridinium 2-iodoxybenzoate has a reactivity pattern similar to IBX and can be used as a safe and soluble oxidant in organic solvents. It has also been demonstrated that IBX can form relatively stable complexes with *N*-oxides, sulfoxides (DMSO), and tetrahydrofuran (2002ACE993, 2002ACE996). Such ligand complexation allows fine-tuning of the reactivity profile of IBX (2002ACE993).

### 3.3.2 Five-Membered Iodine(V) Heterocycles with a Sulfur Atom in the Ring

2-Iodoxybenzenesulfonic acid, which is a thia-analog of IBX and a powerful oxidizing agent, exists in the cyclic tautomeric form of 1-hydroxy-1*H*-1,2,3-benziodoxothiole 1,3,3-trioxide **231** (Scheme 50). It can be prepared by two different pathways: hydrolysis of the methyl ester of 2-iodylbenzenesulfonic acid **229** or direct oxidation of 2-iodobenzenesulfonic acid **230** (2006EJOC4791). 2-Iodoxybenzenesulfonic acid **231** is thermally unstable and highly reactive toward organic solvents. The structure of its reductive decomposition product, 2-iodosylbenzenesulfonic acid in its cyclic tautomeric form **57**, has been established by single-crystal X-ray diffraction (Section 3.2.5; 2006EJOC4791). It has been demonstrated that thia-IBX

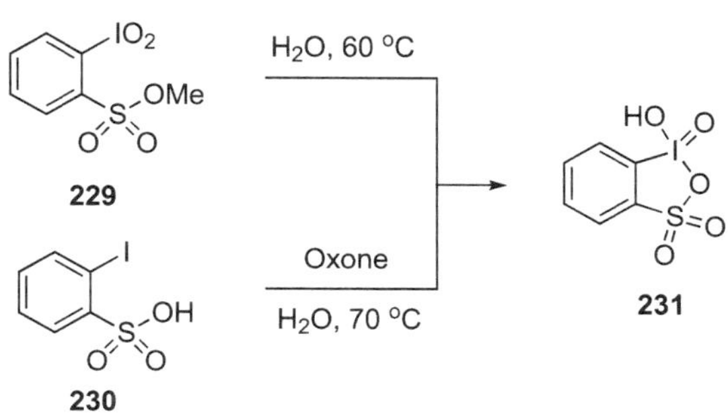


**Scheme 50**

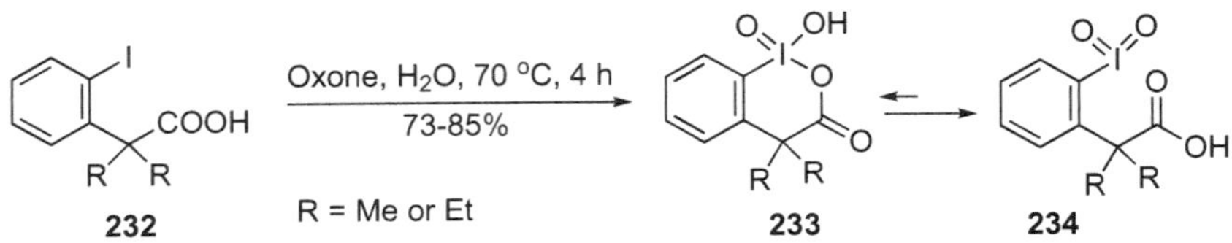

Scheme 51

**231** is the most powerful catalyst in the iodine(V)-catalyzed oxidation of alcohols using Oxone as a terminal oxidant (2009JA251).

### 3.3.3 Six-Membered Iodine(V) Heterocycles

The six-membered iodine heterocycles **233** have been synthesized by the Oxone oxidation of the corresponding 2-iodophenylacetic acids **232** in water (Scheme 51). Interestingly, the X-ray structural study demonstrated that the products **233** exist in the solid state in the open tautomeric form of the pseudocyclic acids **234** (2010JOC8416).

### 3.3.4 Polycyclic Heterocycles with an Iodine(V) Atom at a Ring Junction

A bicyclic iodoxole oxide **236**, the aliphatic analog of IBX, has been prepared by the fluorination of 1,2-iodoxetane **205** (see Section 3.2.9) with xenon difluoride followed by a hydrolysis of the intermediate, non-isolable, difluoride **235** (Scheme 52; 2004TL8173). Compound **236** has reactivity similar to IBX and can oxidize alcohols and sulfides to the corresponding carbonyl compounds and sulfoxides, respectively, in good yields under mild reaction conditions.

### 3.3.5 Pseudocyclic Iodine(V) Compounds

Iodylarenes bearing an appropriate substituent in the *ortho*-position to the iodyl group are characterized by the presence of a pseudocyclic structural moiety due to a strong intramolecular secondary bonding between the hypervalent iodine center and the oxygen atom in the *ortho*-substituent. Compared to the noncyclic iodylarenes, pseudocyclic pentavalent iodine compounds

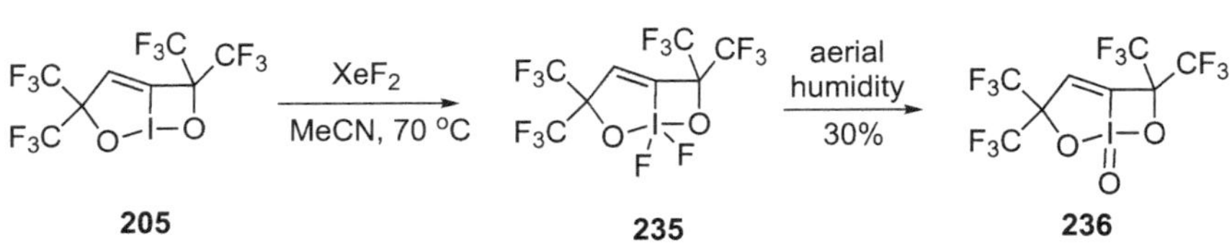

Scheme 52

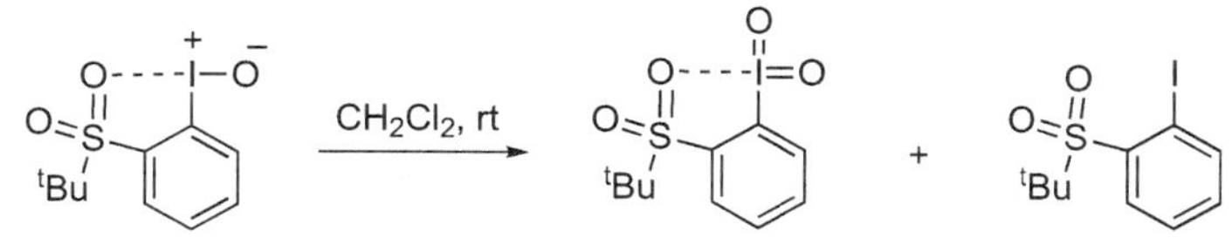

237 238

Scheme 53

have much better solubility, which is explained by a partial disruption of their polymeric nature owing to the redirection of secondary bonding. The first pseudocyclic iodylarene of this type, 1-(*tert*-butylsulfonyl)-2-iodylbenzene **238**, was prepared by Protasiewicz and coworkers by the disproportionation of iodosylarene **237** (Scheme 53; 2000ACE2007). A single-crystal X-ray structural study of product **238** revealed a pseudooctahedral geometry with the I—O bond lengths of 1.796 and 1.822 Å in the iodyl group and a close distance of 2.693 Å between one of the sulfone oxygen atoms and the hypervalent iodine center (2000ACE2007).

A similar, soluble *ortho*-phosphoryl-substituted iodylarene **240** was synthesized by the hypochlorite oxidation of the aryliodide **239** (Scheme 54; 2005TL5187). A single-crystal X-ray analysis of compound **240** demonstrated a strong interaction between the phosphoryl oxygen atom and the iodine(V) atom with an I•••O distance of 2.612 Å, which is significantly shorter than the I•••O distance of 3.291 Å determined for the unoxidized aryliodide **239** (2005TL5187).

Esters of 2-iodoxybenzoic acid (IBX-esters) **242** were obtained by the hypochlorite oxidation of appropriate 2-iodobenzoate esters **241** (Scheme 55) and isolated in the form of stable microcrystalline solids (2004CC106, 2005JOC6484). This procedure has been used for the synthesis of various IBX—esters **242** derived from primary, secondary, and tertiary alcohols, adamantanols, as well as from the optically active menthols and borneol. A single-crystal X-ray analysis of products **242** has revealed a pseudo-benziodoxole structure in which the intramolecular I•••O secondary bonds partially replace the intermolecular I•••O secondary bonds

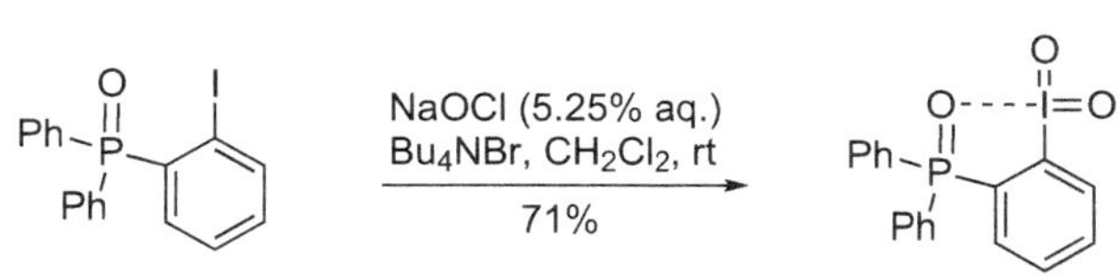


239 240

Scheme 54

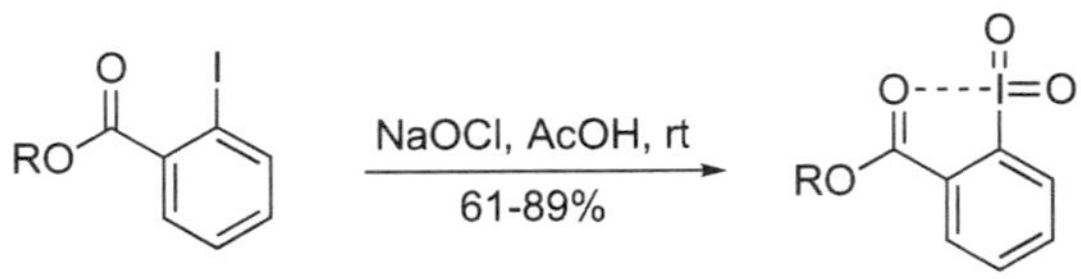

**241** **242**

R = Me, Et, $Pr^i$, (–)-menthyl, (+)-menthyl, (±)-menthyl, [(1*S*)-endo]-(–)-bornyl, 2-adamantyl, 1-adamantyl, $Bu^t$

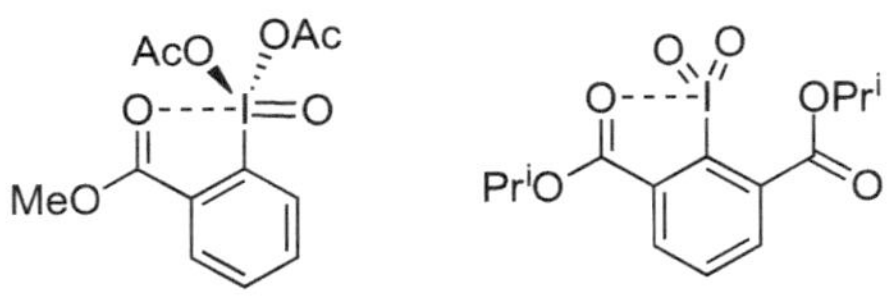

**243** **244**

**Scheme 55**

disrupting the polymeric structure characteristic of $PhIO_2$ and other previously reported iodylarenes (2005JOC6484). This structural feature is responsible for a substantially increased solubility of these compounds in comparison with other pentavalent iodine compounds.

Methyl-2-iodoxybenzoate can be further converted to the diacetate **243** or a similar bis(trifluoroacetate) by treatment with acetic anhydride or trifluoroacetic anhydride, respectively (2005JOC6484). A single-crystal X-ray diffraction analysis of methyl-2-[(diacetoxy)(oxo)iodo]benzoate **243** revealed a pseudo-benziodoxole structure with three relatively weak intramolecular I•••O interactions. Esters of 2-iodoxyisophthalic acid (e.g., **244**) were prepared by the oxidation of the respective iodoarenes with dimethyldioxirane. An X-ray structural analysis of diisopropyl 2-iodoxyisophthalate **244** has demonstrated strong intramolecular I•••O interaction with the carbonyl oxygen of only one of the two ester groups, while NMR spectra in solution indicated equivalence of both ester groups (2005JOC6484).

Amides of 2-iodoxybenzoic acid (IBX-amides) **246** can be prepared by the dioxirane oxidation of 2-iodobenzamides **245** (Scheme 56) in the form of stable, microcrystalline solids moderately soluble in dichloromethane and chloroform (2003ACE2194). This procedure has been used for the synthesis of the amides **246** derived from various types of amino compounds, such as esters of α-amino acids, esters of β-amino acids, and (*R*)-1-phenylethylamine. A single-crystal X-ray analysis of the phenylalanine derivative [**246**, R = (*S*)-$CH(CH_2Ph)CO_2Me$] shows a close intramolecular contact of 2.571 Å between the pentavalent iodine center and the oxygen atom of

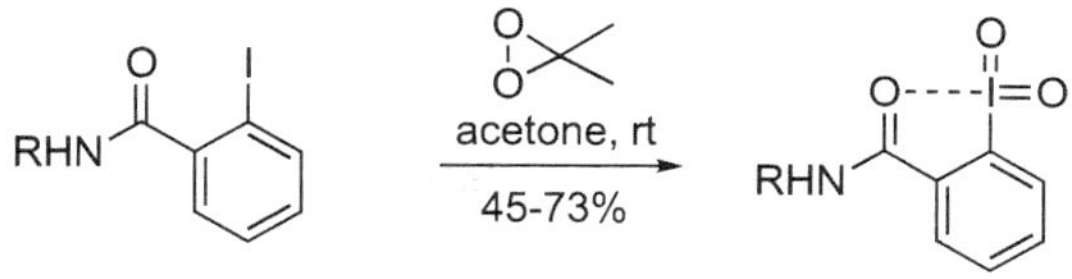

245 246

R = (*S*)-CH($CH_3$)$CO_2CH_3$, (*R*)-CH($CH_3$)$CO_2CH_3$, (*S*)-CH($CH_2$Ph)$CO_2CH_3$, (*S*)-CH($Bu^i$)$CO_2CH_3$, $CH_2CH_2CO_2H$, CH($CH_3$)$CH_2CO_2H$, (*R*)-CH(Ph)$CH_3$

**Scheme 56**

the amido group within each molecule enforcing a planar geometry of the resulting five-membered ring; a geometry that is analogous to that observed for IBX and other benziodoxoles (2003ACE2194). Polymer-supported IBX-amides and esters have also been reported (2003TL9251, 2005S279, 2005MD341).

Amides of 2-iodoxybenzenesulfonic acid **248** were prepared by a similar procedure from the corresponding 2-iodobenzenesulfamides **247** (Scheme 57; 2004TL2719). A single-crystal X-ray structural analysis of the alanine derivative **248** [R = (*S*)-CH($CH_3$)$CO_2$Me] shows a combination of intra- and intermolecular I•••O interactions resulting in a unique heptacoordinated iodine(V) center in this molecule (2005NJC998). The analogous esters **250** were synthesized similarly from the respective sulfonate esters **249** (2005ARK8).

The soluble and stable pentavalent iodine *N*-(2-iodylphenyl)acylamides **252** with a pseudo-benziodoxazine structure have been synthesized in good

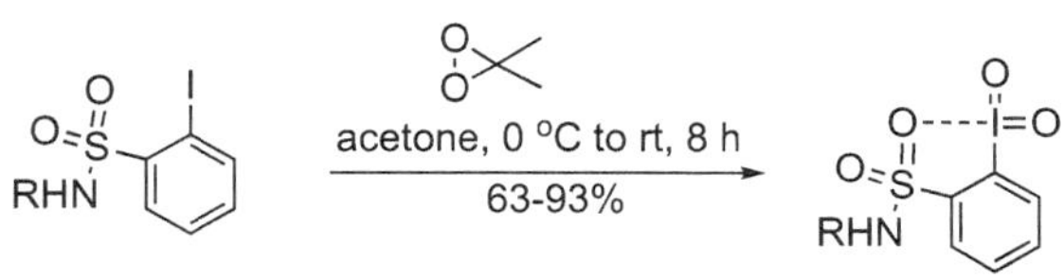

247 248

R = (*S*)-CH($CH_3$)$CO_2CH_3$, (*S*)-CH($CH_2$Ph)$CO_2CH_3$, (*S*)-CH($Pr^i$)$CO_2CH_3$, (*S*)-CH($Bu^i$)$CO_2CH_3$, (*R*)-CH(Ph)$CH_3$

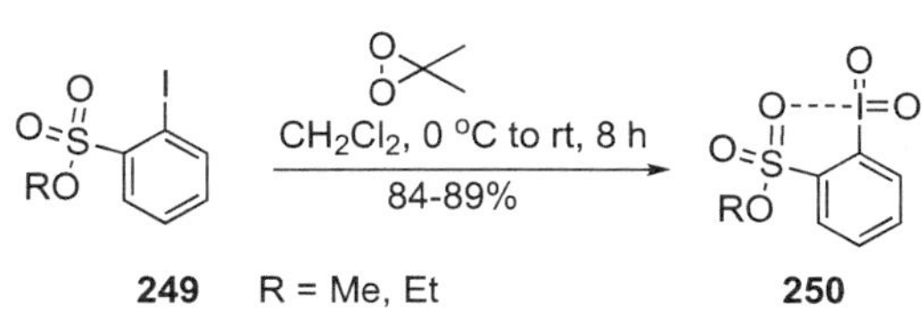

249 R = Me, Et 250

**Scheme 57**

yields by oxidation of the 2-iodoaniline derivatives **251** with 3,3-dimethyldioxirane under mild reaction conditions (Scheme 58; 2005ACE7127). X-ray data on compounds **252** revealed a unique pseudo-benziodoxazine structure with an intramolecular secondary I•••O bond distance of 2.647 Å. The synthesis of chiral pseudo-benziodoxazine reagents **253** and **254** has been achieved based on commercially available and inexpensive (*S*)-proline (2006TL6301). The synthesis of polymer-supported pseudo-benziodoxazine has also been reported (2006OL167, 2007S527).

The pseudocyclic tosyl derivatives of 2-iodylaniline **256** and 2-iodylphenol **258** were obtained by a similar dimethyldioxirane oxidation of the corresponding 2-iodophenyltosylamides **255** or 2-iodophenyl tosylate **257** (Scheme 59). Compounds **256** and **258** are isolated as stable, microcrystalline products (2009JOC8444). A single-crystal X-ray diffraction analysis of the tosylamide **256** (R = Me) revealed a pseudocyclic structure formed by intramolecular I•••O interactions between the hypervalent iodine center and the sulfonyl oxygens in the tosyl group (2009JOC8444). This tosylamide has an excellent solubility in organic solvents and is a potentially useful hypervalent iodine oxidant.

2-Iodylphenol ethers **260** were prepared by the dioxirane oxidation of the corresponding 2-iodophenol ethers **259** (Scheme 60) and isolated as chemically stable, microcrystalline products (2006JOC8452). A single-crystal X-ray diffraction analysis of 1-iodyl-2-isopropoxybenzene and 1-iodyl-2-butoxybenzene showed pseudo-polymeric arrangements in the solid state formed by intermolecular interactions between the $IO_2$ groups of different molecules.

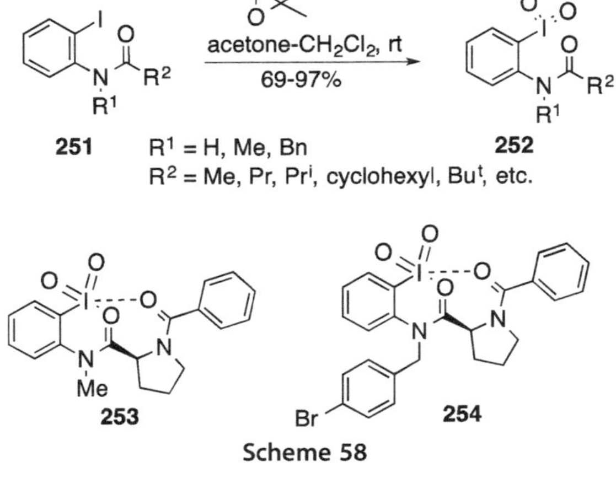


Scheme 58

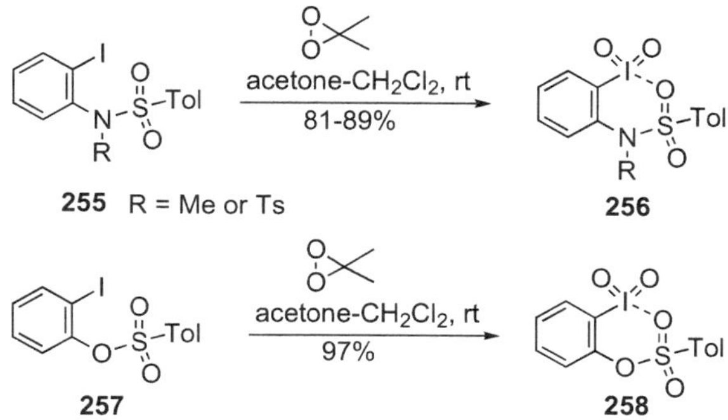


Scheme 59

Polymer-supported 2-iodophenol ethers have also been reported (2007JOC8149).

Several chiral, nonracemic, pseudocyclic 2-(*o*-iodylphenyl)-oxazolines **262** have been synthesized starting from chiral 2-amino alcohols **261** as shown in Scheme 61 (2009OL1221). Compounds **262** are obtained as white microcrystalline products soluble in most organic solvents.

# 4. IODINE HETEROCYCLES AS REAGENTS IN ORGANIC SYNTHESIS

Numerous stable iodine(III) five-membered heterocycles have found synthetic application as "atom-transfer" reagents, useful for oxidative functionalization of organic substrates (2011CC102). In particular, azidobenziodoxoles are efficient C—H azidating reagents, cyanobenziodoxoles are useful cyano group transfer reagents, alkynylbenziodoxoles have found application for direct electrophilic alkynylation of organic substrates, and trifluoromethylbenziodoxoles are important electrophilic trifluoromethylating reagents. The iodine(V) heterocycles, such as IBX **212** and DMP **213** (see Section 3.3.1), are extensively employed in organic synthesis as mild and highly selective reagents for the oxidation of alcohols to carbonyl compounds, as well as for a variety of other synthetically useful oxidative transformations.

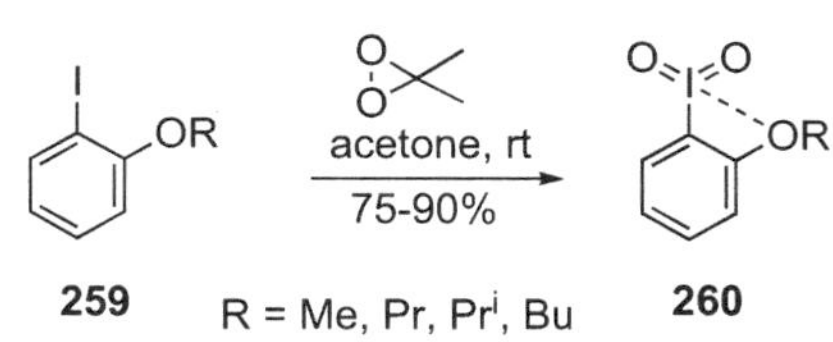


Scheme 60

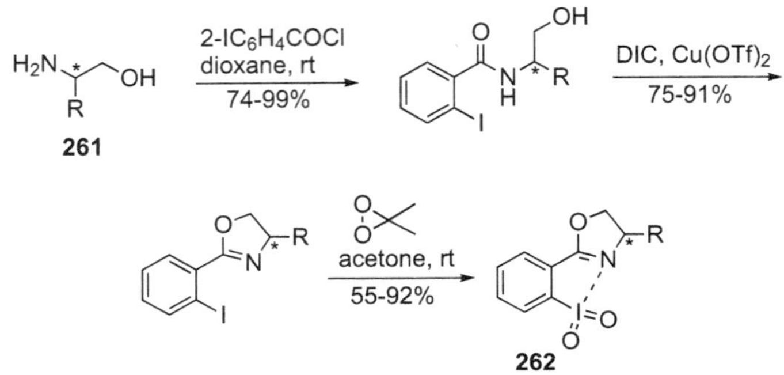


261

262

Scheme 61

## 4.1 Benziodoxoles as Atom-Transfer Reagents

The greater stability of the benziodoxole system made possible the preparation and isolation of otherwise unstable iodine(III) derivatives, such as bromobenziodoxoles, tosyloxybenziodoxoles, alkylperoxybenziodoxoles, azidobenziodoxoles, cyanobenziodoxoles, amidobenziodoxoles, alkynylbenziodoxoles, and trifluoromethylbenziodoxoles (Section 3.2.2). These benziodoxole derivatives have found practical applications as oxidants and reagents for "atom-transfer" reactions in organic synthesis, which have been summarized in several reviews (2011CC102, 2005COS121, 1997RHC133). In particular, bromobenziodoxoles (see Section 3.2.2.1) are active brominating agents. Bromobenziodoxole **80** can be used for selective allylic and benzylic brominations under radical conditions (1979JOC1779), and compound **81** has been utilized for electrophilic bromination of anisole or bromolactonization of 4-pentenoic acid (Scheme 62; 2006CC1442).

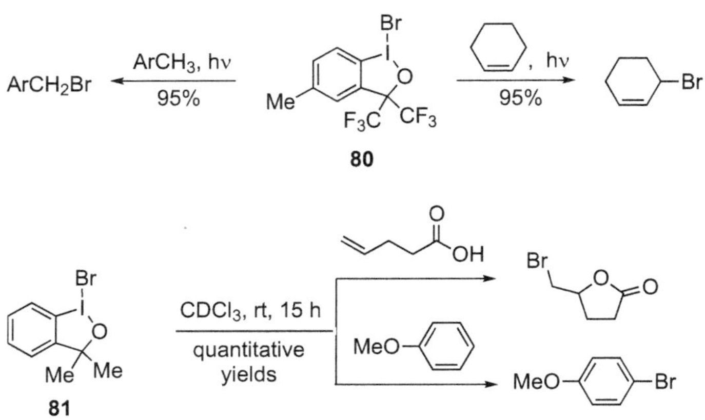


80

81

Scheme 62

Tosyloxybenziodoxole **90** (see Section 3.2.2.3) is an effective reagent for the oxidative iodination of aromatic compounds (1998SL286, 1999 JOC2883). Treatment of various aromatic compounds with reagent **90** and $I_2$ gives the corresponding iodinated compounds **263** in good yields (Scheme 63). Compared with other trivalent iodine reagents, the tosylate **90** shows the best reactivity as an oxidant for oxidative halogenation (1999JOC2883).

Furthermore, the **90**+ $I_2$ reagent system can be used for the iodotosyloxylation of alkynes to give the products of addition **264** in good yields (Scheme 64; 1999JOC2883). These reactions presumably proceed via intermediate formation of arenesulfonyl hypoiodites.

Alkylperoxybenziodoxoles (see Section 3.2.2.4) are useful oxidizing reagents toward organic sulfides, selenides, and phosphines (1992JA6269, 1997JOC4253, 1999TL5541). In particular, sulfides **265** are oxidized with peroxybenziodoxole **96** under mild reaction conditions affording sulfoxides **266** in high yields (Scheme 65; 1997JOC4253). A similar oxidation of dithioacetals **267** results in the regeneration of the parent carbonyl compound **268** and thus can be useful as a method for selective deprotection (1997JOC4253).

Azidobenziodoxoles (see Section 3.2.2.5) can be used as efficient azidating reagents toward various organic substrates (Scheme 66). For example, reagent **104** reacts with *N,N*-dimethylanilines in dichloromethane at reflux in 30 min to afford the respective *N*-azidomethyl-*N*-methylanilines **269** in excellent yields (1996JA5192). The main advantage of reagent **104** over the unstable $(PhIO)_n/TMSN_3$ reagent combination is high thermal stability allowing its use at higher temperatures. Azidobenziodoxole **104** can even be used for direct azidation of hydrocarbons at higher temperatures and in the presence of radical initiators (Scheme 66). Reagent **104** selectively reacts with isooctane upon reflux in 1,2-dichloroethane in the presence of catalytic amounts of benzoyl peroxide to afford the tertiary azide **273**. Under similar conditions,

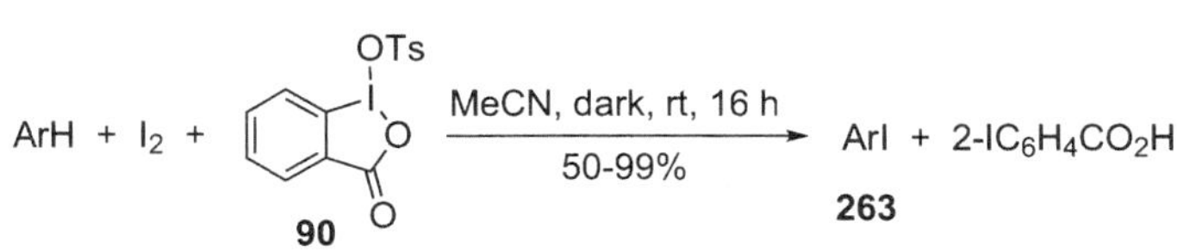


ArH = 1,3,5-$(MeO)_3C_6H_3$, 1,3,5-$(i\text{-}Pr)_3C_6H_3$, 1,3,5-$Me_3C_6H_3$, 1-MeO-4-$MeCO_2C_6H_4$, 1-MeO-4-$BrC_6H_4$, 1,4-$Me_2C_6H_4$, 1,3-$Me_2C_6H_4$, $MeOC_6H_5$, $Bu^tC_6H_5$, $AcOC_6H_5$, naphthalene, 2,3-benzothiophene, etc.

**Scheme 63**

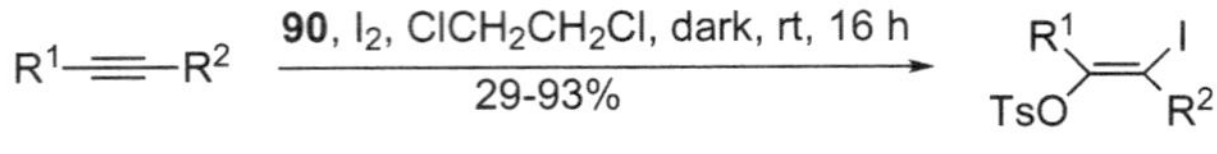

264

$R^1$ = Ph, Pr, Bu, H
$R^2$ = Ph, Pr, Me, H, $CO_2Et$

**Scheme 64**

reactions of azidobenziodoxole **104** with bicyclic and tricyclic hydrocarbons afford the respective alkyl azides **270–272**. Cyclohexene is azidated in the allylic position with the formation of 3-azidocyclohexene **274** (1996JA5192).

Azidobenziodoxole **102** (see Section 3.2.2.5) has been utilized as a reagent for efficient azidation of β-keto esters and silyl enol ethers. The azidation of cyclic β-keto esters can be achieved in excellent yields in the absence of any catalyst. In the case of less reactive linear β-keto esters and silyl enol ethers, complete conversion and good yields have been obtained using a zinc catalyst (2013OL3246).

Azidations with azidobenziodoxoles can be effectively catalyzed by iron salts, such as iron(II) propionate [$Fe(O_2CEt)_2$]. In the presence of chiral oxazoline ligands (boxmi ligands), the Fe-catalyzed azidations of β-keto esters and oxindoles afford the respective α-azido-β-keto esters and 3-azidooxindoles with high enantioselectivity (up to 94% ee) (2013JA5356).

An optimized procedure for the iron(II)-catalyzed azidations of tertiary C–H bonds using azidobenziodoxole **104** has been reported by Sharma and Hartwig (2015NAT600). This reaction tolerates aqueous environments and is suitable for the functionalization of complex structures in the late stages of a multistep synthesis.

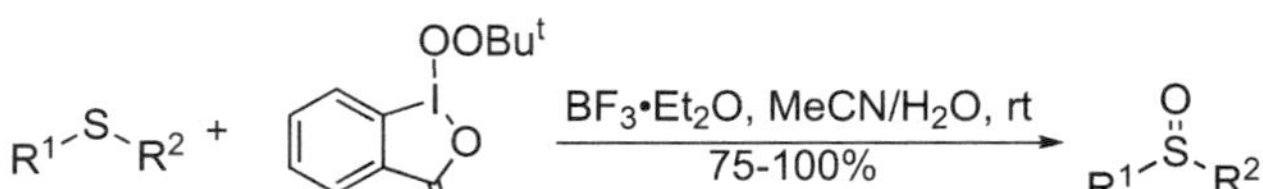

**265** **96** **266**

$R^1$ = Bu, $Bu^i$, $Bu^s$, $PhCH_2$, $Me(CH_2)_4$, $CH_2$=$CHCH_2$, 4-$MeC_6H_4$, 4-$MeOC_6H_4$, 4-$ClC_6H_4$, Ph
$R^2$ = $PhCH_2$, Ph, Me, $CH_2P(O)(OEt)_2$

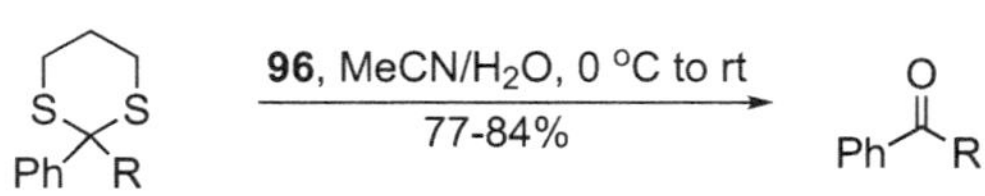

**267** R = Me, $C_5H_{11}$ **268**

**Scheme 65**

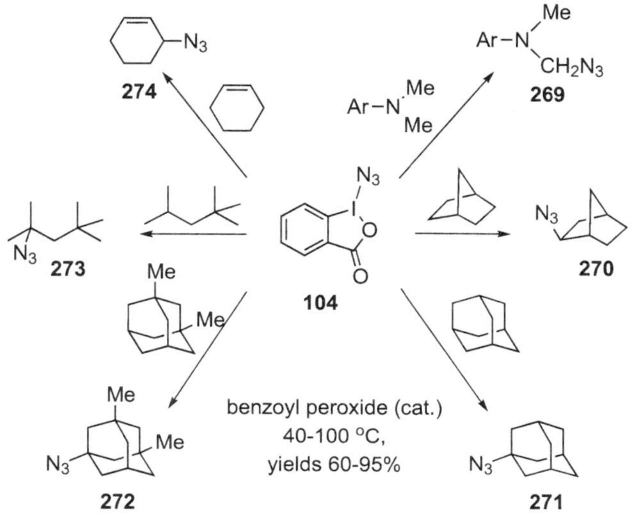


Scheme 66

Amidobenziodoxoles (see Section 3.2.2.6) can be used as amidating reagents toward polycyclic alkanes under radical conditions. For example, reagent **109** reacts with adamantane in chlorobenzene at 100−105 °C in the presence of a catalytic amount of benzoyl peroxide to afford 1-amidoadamantane **275** in a moderate yield (Scheme 67; 1997TL21).

C−Substituted benziodoxoles (e.g., cyanobenziodoxoles, trifluoromethylbenziodoxoles, and alkynylbenziodoxoles) are especially useful reagents for the electrophilic transfer of carbon ligands to nucleophilic substrates. Waser and coworkers have demonstrated that triisopropylsilyl- and trimethylsilyl-substituted ethynylbenziodoxolones (see Section 3.2.2.8) are excellent acetylene transfer reagents, both in metal-free and metal-catalyzed reaction conditions (2014C51, 2012CSR4165). For example, reagent **122** efficiently alkynylates β-dicarbonyl compounds and other C−H acidic substrates **276** in the presence of fluoride anion in THF under mild reaction conditions (Scheme 68; 2010CEJ9457). Indoles (e.g., **277**, Scheme 68),

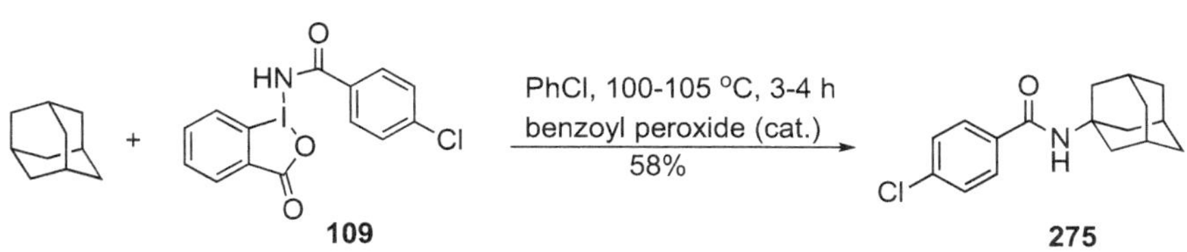


Scheme 67

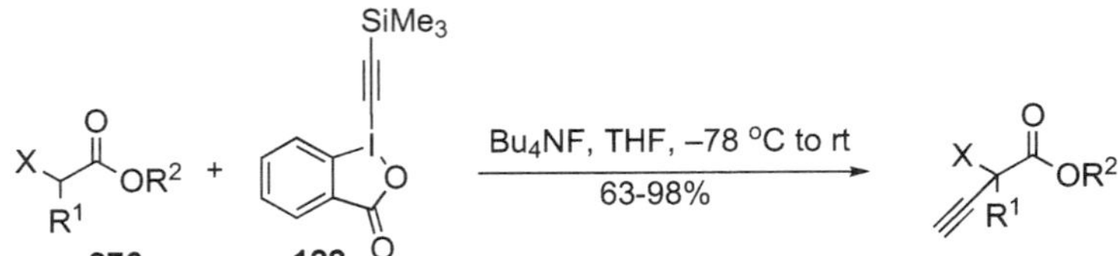


**276** **122**

X = C(O)R, CN, $NO_2$
$R^1$ = Me, Et, $PhCH_2$, 4-$BrC_6H_4CH_2$, Ph, allyl
$R^2$ = Me, Et, $Bu^t$

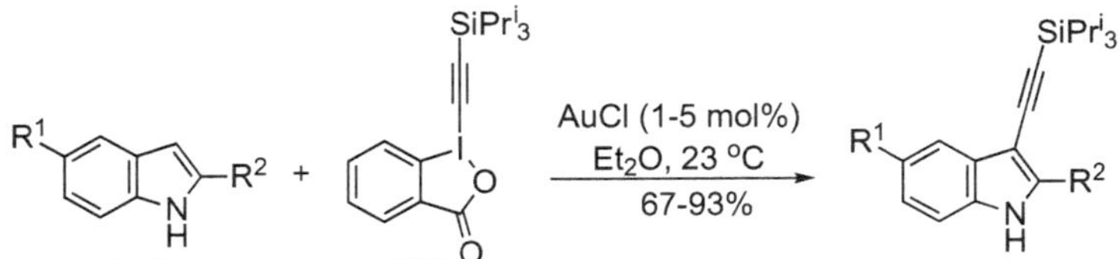


**277** **123**

$R^1$ = H, OH, $CO_2H$, Br, $NO_2$
$R^2$ = H, Ph, $HOCH_2$

**Scheme 68**

pyrroles, and thiophenes are directly alkynylated by reagent **123** in the presence of AuCl or Pd(II) catalysts (2013OL112, 2012OL744, 2012 S1155, 2009ACE9346, 2010CEJ9457, 2010OL384, 2011JOC565, 2010 ACE4680, 2012CEJ5655). Additional examples include the following: alkynylation of thiols (2013JA9620, 2014JA16563), alkynylation of H-phosphites and -phosphinates (2014CC12923), palladium-catalyzed intramolecular oxyalkynylation of non-activated alkenes using triisopropylsilylethynylbenziodoxolone (2010OL384), palladium-catalyzed alkynylations of *o*-allylphenols and carboxylic acids (2011CC102), *para*-selective gold-catalyzed direct alkynylation of anilines (2012OL744), and zinc-gold co-catalyzed direct alkynylation of benzofurans (2013BJOC1763). Aubineau and Cossy have found that trimethylsilyl-substituted ethynylbenziodoxolone is a useful reagent for direct chemoselective alkynylation of *N*-sulfonylamides (2013CC3303).

Togni and coworkers have found that 1-trifluoromethylbenziodoxoles (see Section 3.2.2.11) are useful reagents for electrophilic trifluoromethylation of nucleophilic substrates (2015CR650, 2014C624). For example, reagent **143** reacts with β-keto esters **278** under mild reaction conditions in the presence of potassium carbonate affording α-trifluoromethylated products **279** in good yields (Scheme 69; 2006CEJ2579, 2007ACE754). Likewise, this mild electrophilic trifluoromethylating reagent can be used to transfer a $CF_3$ group

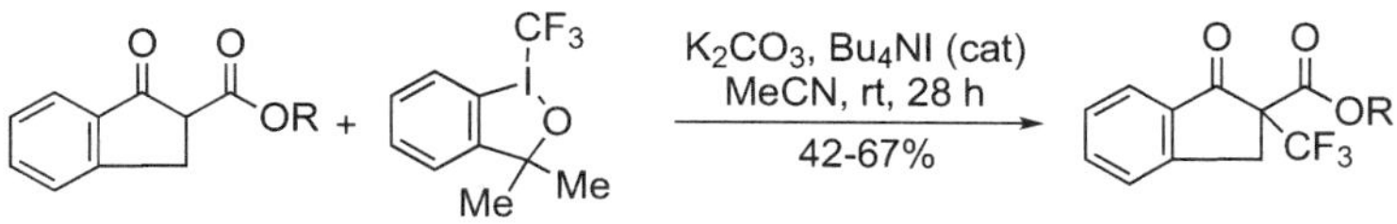

278 143 279

R = Me, Et, Bu[t]

**Scheme 69**

to other C-centered nucleophiles, such as α-nitro esters (2007ACE754), as well as to S-centered nucleophiles (thiols and S-hydrogen phosphorothioates) (2007ACE754, 2008HCA2035, 2011JOC4189), O-centered nucleophiles (alcohols, phenols, hydroxylamines, and triflate anions) (2014EJOC3087, 2008JOC7678, 2010JOC1779, 2009CC5993, 2009ACE4332), N-centered nucleophiles (azoles and nitriles) (2012ACE6511, 2011ACE1059, 2012 ACE6511), and P-centered nucleophiles (secondary or primary aryl- and alkylphosphines) (2008HCA2035, 2008CC1575, 2012JFC83).

Cyanobenziodoxole **115** (see Section 3.2.2.7) is an efficient cyanating reagent toward *N,N*-dialkylarylamines. In a typical example, reagent **115** reacts with *N,N*-dimethylanilines **280** in 1,2-dichloroethane at reflux to afford the respective *N*-cyanomethyl-*N*-methylanilines **281** in good yields (Scheme 70; 1995TL7975).

This procedure has been applied to the synthesis of *N*-cyanomethyl-*N*-cyclopropylamine, which is a possible metabolite of cyclopropylamine-derived drugs (2001JA349).

## 4.2 Oxidations with Iodine(V) Heterocycles

IBX **212** and DMP **213** (see Section 3.3.1) are extensively employed in organic synthesis as mild and highly selective reagents for the oxidation of alcohols to carbonyl compounds, as well as for a variety of other synthetically useful oxidative transformations. Synthetic applications of cyclic pentavalent iodine reagents have been summarized in several reviews (2011JOC1185, 2006ARK26, 2011AGE1524, 2010T7659, 2001ACE2812).

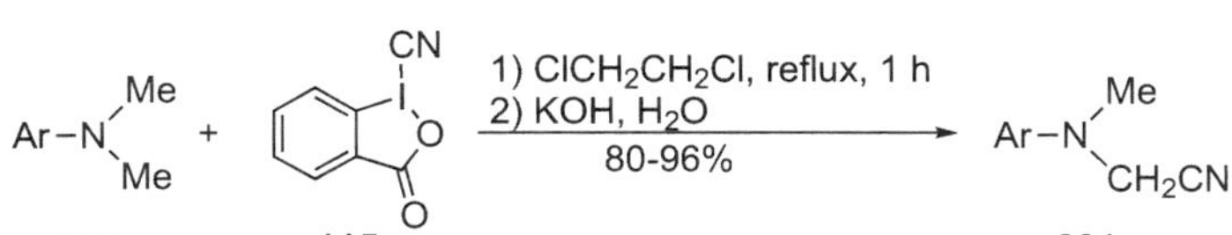

280 115 281

Ar = Ph, 4-$BrC_6H_4$, 4-$MeC_6H_4$, 1-naphthyl

**Scheme 70**

### 4.2.1 Oxidations with IBX

IBX is a particularly useful oxidant for the selective oxidation of alcohols to carbonyl compounds, even in complex molecules in the presence of other functional groups (2011AGE1524, 2010T7659). Primary alcohols are oxidized by IBX in DMSO to the corresponding aldehydes at room temperature without overoxidation to the acids. Chiral primary alcohols are oxidized without epimerization and various functional groups such as thioethers, amines, carboxylic acids, esters, carboxamides, and both conjugated and isolated double bonds are compatible with IBX (1994TL8019, 1995JOC7272). Representative examples of alcohol oxidations using IBX in DMSO are shown in Schemes 71–74.

The allylic alcohols **282** are selectively oxidized by IBX to the ketones **283** in high yields (Scheme 71; 1999TL6253). A similar IBX oxidation of the alcohols **284** selectively affords 5-monosubstituted 3-acyl-4-*O*-methyl tetronates **285**, which are structurally similar to the tetrodecamycin antibiotics (2001S2113).

The oxidation of diol **286** by IBX in DMSO solution was utilized in the synthesis of the functionalized hexahydroanthracene dione **287** (Scheme 72), which is a model for the D ring of taxoids (2001OL1021).

Total synthesis of the antifungal agent GM222712 was accomplished by a selective oxidation of diol **288** to hemiacetal **289** using IBX under similar reaction conditions (Scheme 73; 2000TL4379).

The IBX oxidation of carbohydrate **290** was employed in the synthesis of moenomycin A disaccharide analogs (Scheme 74; 1999T687).

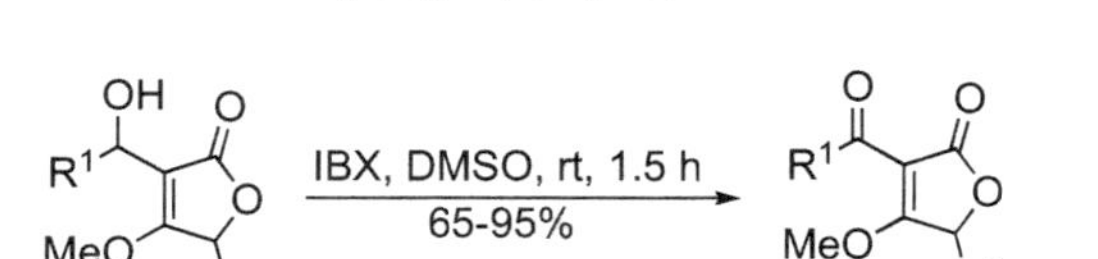


**282** $R^1$ = H, SPh; $R^2$ = H, $SO_2Ph$, $CO_2Me$, Ts **283**

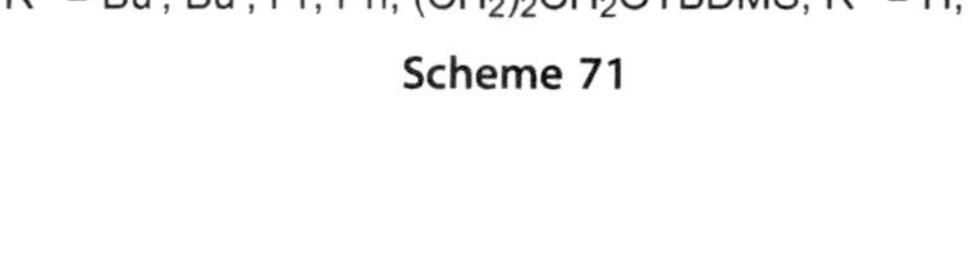


**284** **285**

$R^1$ = $Bu^i$, $Bu^t$, Pr, Ph, $(CH_2)_2CH_2OTBDMS$; $R^1$ = H, Me

**Scheme 71**

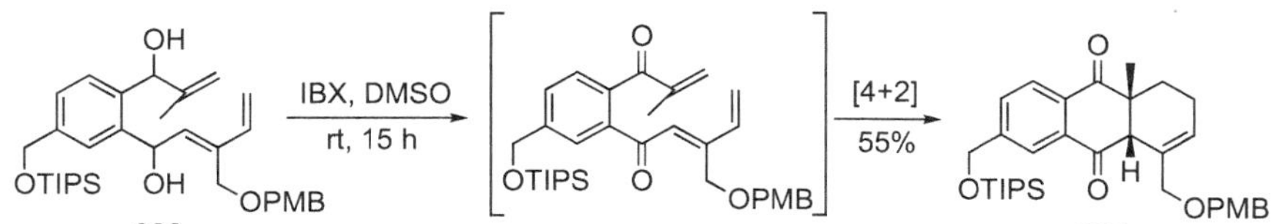


286 287

Scheme 72

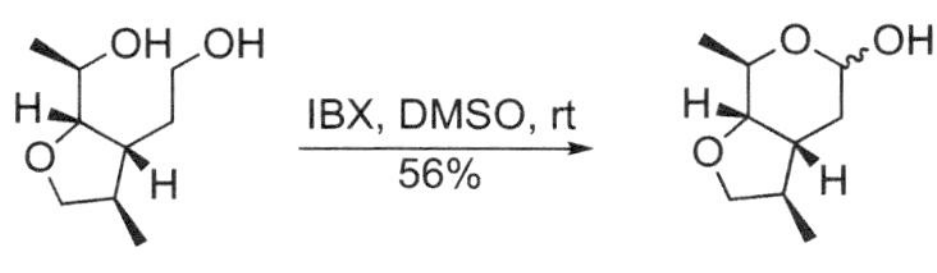


288 289

Scheme 73

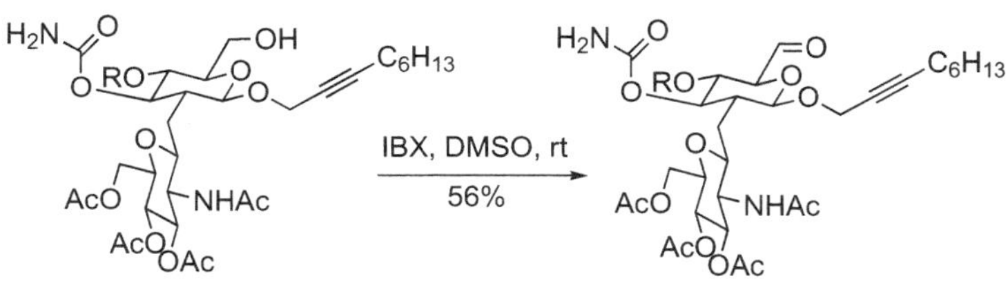


290 R = TBDMS

Scheme 74

Selective oxidation of alcohols using IBX has been utilized in numerous syntheses, such as: (1) the total synthesis of (−)-decarbamoyloxysaxitoxin (2007ACE8625); (2) the total synthesis of abyssomicin C and atrop−abyssomicin C (2006ACE3256); (3) the stereoselective synthesis of pachastrissamine (jaspine B) (2008TA209); (4) the syntheses of (±)-pterocarpans and isoflavones (2007TL8343); (5) the total synthesis of (±)-nitidanin (2007TL771); (6) the total synthesis of lagunamycin (2006TL6183); (7) the synthesis of (−)-agelastatin (2007OL2989); (8) the syntheses of heliannuols B and D (2007T6990); (9) the total syntheses of (−)-subincanadines A and B (2006OL4605); (10) the synthesis of marine sponge metabolite spiculoic acid A (2006CC2863); and (11) the total synthesis of a cyclic depsipeptide somamide A (2002TL8673). Likewise, the oxidation of alcohols with IBX in DMSO has also been used in the synthesis of optically pure highly functionalized tetrahydroisoquinolines (2008T2321); the kinetic study of organic reactions on polystyrene grafted microtubes (2000JCC224); the preparation of Fmoc-protected amino aldehydes from the corresponding alcohols (2007CC3493); expeditious and metal-free synthesis of quinolones, naphthyridones, and benzonaphthyridones (2014ASC3600); and a one-pot metal-free domino approach to functionalized pyridines (2013CC7926).

IBX is a useful reagent for the oxidation of 1,2-diols. In contrast to DMP and iodylarenes, IBX usually oxidizes 1,2-diols to α-ketols or α-diketones without cleaving the glycol C–C bond (1994TL8019, 2003ACE4685, 1995TL3485, 1995TL7945, 1996JOC9272). For example, in a key step of the total synthesis of the *streptomyces maritimus* metabolite, wailupemycin B, IBX oxidation of the diol precursor **291** leads to the desired hydroxyketone **292** without any cleavage of the glycol C–C bond (Scheme 75; 2003ACE4685).

A detailed investigation of IBX reactions with various vicinal diols revealed that, depending on the substrate and the reaction conditions, either oxidation to α-ketols or α-diketones, or oxidative cleavage of the C–C bond can occur (2007OBC767). In DMSO solutions, IBX oxidatively cleaves strained and sterically hindered *syn* 1,2-diols, while the non-hindered secondary glycols are oxidized to α-ketols or α-diketones. The use of trifluoroacetic acid as a solvent leads to an efficient oxidative fragmentation of 1,2-diols of all types (2007OBC767).

The synthetic usefulness of IBX in general is hampered by its low solubility in organic solvents except DMSO. However, in several publications it has been shown that IBX can be used as an effective oxidant in solvents other than DMSO (2002OL3001, 2003SL1307, 2003OL3321). More and Finney have reported that primary and secondary alcohols can be oxidized into the corresponding aldehydes or ketones in excellent yields (90–100%) by heating a mixture of alcohol and IBX in common organic solvents (2002OL3001). All reaction by-products can be completely removed by filtration. This method has been used for the efficient preparation of the ribosyl aldehyde **293** (Scheme 76), which is a key intermediate in the stereoselective synthesis of the polyoxin and nikkomycin antibiotics (2003SL1307).

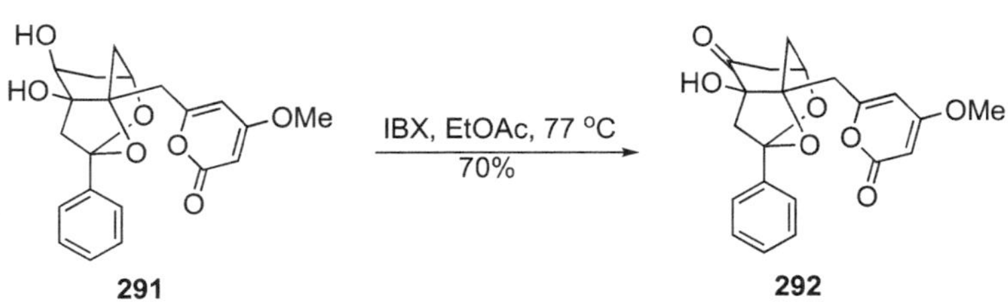


**Scheme 75**

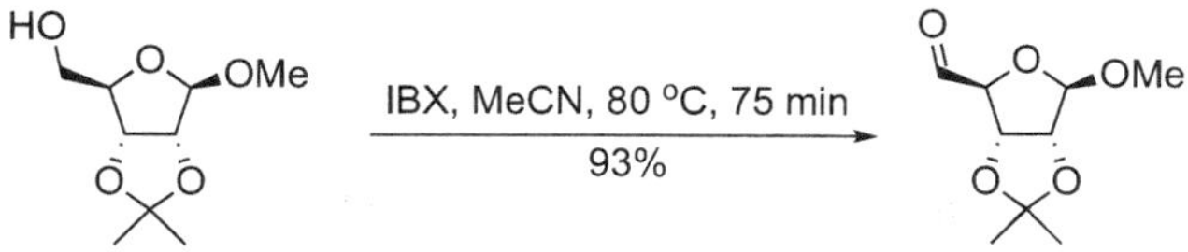

293

Scheme 76

IBX in DMSO or DMF has been shown to be an excellent reagent for the oxidation of various phenols to *o*-quinones (2002OL285). This procedure was used for the oxidation of phenol **294** to *o*-quinone **295** (Scheme 77), which is a key intermediate in the total synthesis of the novel cyclooxygenase inhibitor (±)-aiphanol (2003CL420). The same procedure was utilized in the synthesis of (±)-brazilin, a tinctorial compound found in the alcoholic extracts of trees collectively referred to as Brazil wood (2005OL5841).

The IBX-promoted oxygenative dearomatization of phenols leading to cyclohexadienone systems is an especially important synthetic transformation (2010T2235). Representative examples include the use of IBX in key oxidation steps in the total synthesis of the resveratrol-derived polyphenol natural products (−)-hopeanol and (−)-hopeahainol A (2010JA7540), the synthesis of carnosic acid and carnosol (2010CPB27), and the total synthesis of the bissesquiterpene (+)-aquaticol (2007ACE1533).

The practical value of IBX as a reagent has been extended to a variety of other synthetically useful oxidative transformations. In a series of papers, Nicolaou and coworkers have demonstrated the utility of IBX for the one-step synthesis of α,β-unsaturated carbonyl systems from saturated alcohols and carbonyl compounds (2000JA7596, 2002JA2245, 2000JA10246), for the selective oxidation of the benzylic carbon (2004JA5192), for the oxidative cyclization of anilides and related compounds (2000 JA622, 2002JA2233, 2001ACE202, 2000ACE2525), and for the synthesis of amino sugars and libraries thereof (2000ACE2525). Specifically, cycloalkanols **296** react with two equivalents of IBX in a mixture of either

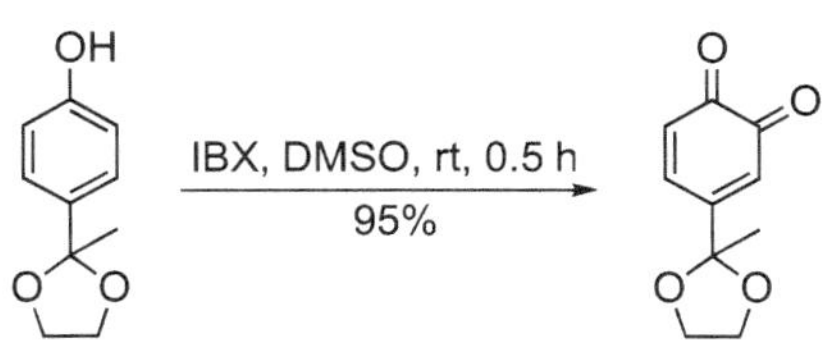

294 295

Scheme 77

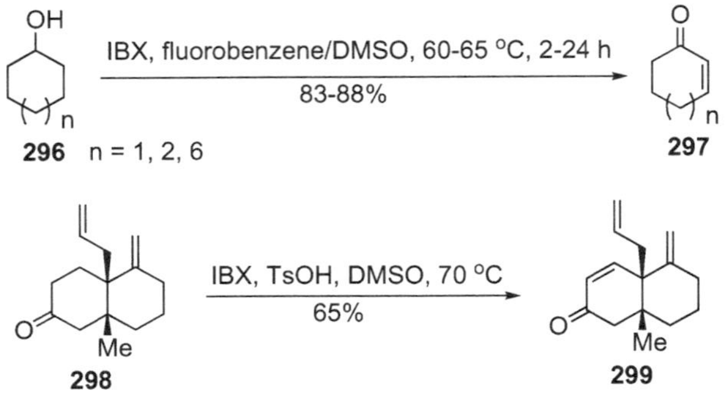

Scheme 78

fluorobenzene or toluene and DMSO with moderate heating affording the corresponding α,β-unsaturated ketones **297** in good yields (Scheme 78; 2002JA2245). The analogous oxidative dehydrogenation of a cyclohexanone derivative **298** to the respective enone **299** has been employed in the total synthesis of (−)-anominine (2010JA5966).

Various heterocycles **301** can be synthesized by the treatment of unsaturated aryl amides, carbamates, thiocarbamates, or ureas **300** with IBX (Scheme 79; 2000JA622, 2002JA2233). The mechanism of this reaction has been investigated in detail (2001ACE202). On the basis of solvent effects and D-labeling studies, it was proposed that the IBX-mediated cyclization of anilides in THF involves an initial single electron transfer to a THF−IBX complex followed by deprotonation, radical cyclization, and concluding termination by hydrogen abstraction from THF (2001ACE202). A similar IBX-mediated cyclization was used in the synthetic protocol for the stereoselective preparation of amino sugars (2000ACE2525).

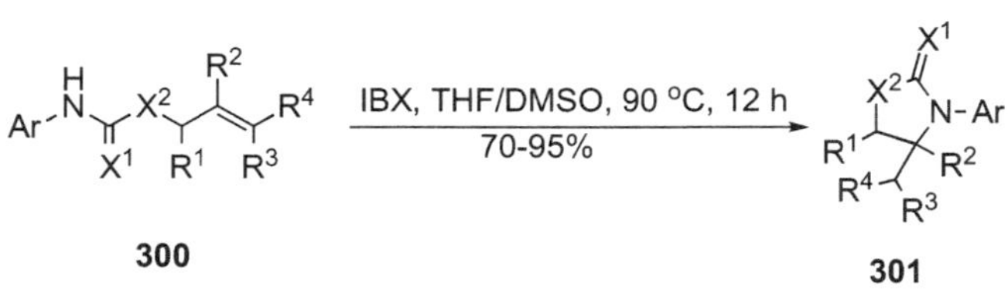

$X^1$ = O, S; $X^2$ = $CH_2$, O, N
Ar = Ph, 3-$EtC_6H_4$, 3-$BrC_6H_4$, 3-$FC_6H_4$, 4-$EtC_6H_4$, etc.
$R^1$ - $R^4$ = H, alkyl, cycloalkyl, etc.

Scheme 79

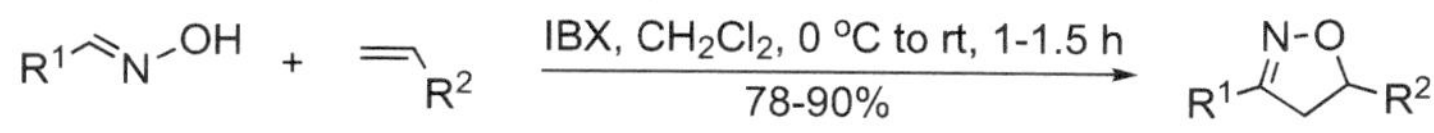


302 303 304

$R^1$ = aryl, heteroaryl; $R^2$ = $CO_2Me$, CN, OAc

**Scheme 80**

IBX has been used for the preparation of the 3,5-disubstituted isoxazolines **304**. The mechanism of this reaction includes initial oxidation of aldoximes **302** with IBX giving the respective nitrile oxides which then undergo 1,3-dipolar cycloaddition with an alkene component **302** to afford final products **304** (Scheme 80; 2004TL7347).

The IBX-mediated oxidative Ugi-type multicomponent reaction of tetrahydroisoquinoline with isocyanides and carboxylic acids affords the nitrogen- and carbon-functionalized tetrahydroisoquinolines **305** in good-to-excellent yields (2006ACE3495). Likewise, the three-component Passerini reaction of an alcohol, carboxylic acid, and an isonitrile in the presence of IBX affords the corresponding α-acyloxy carboxamides **306** in generally high yields (Scheme 81; 2007ACE5775).

IBX is an efficient and selective reagent for the oxidation of alkylarenes at the benzylic positions to give the respective ketones (2001JA3183). This reaction is quite general and can tolerate a variety of substituents on the aromatic ring. Overoxidation to the corresponding carboxylic acids is not observed even in the presence of electron-rich substituents. Secondary amines can be oxidized with IBX in DMSO to yield the corresponding imines in good-to-excellent yields (2004JA5192).

Additional representative examples of synthetic applications of IBX include the following oxidative transformations: (1) aromatization of

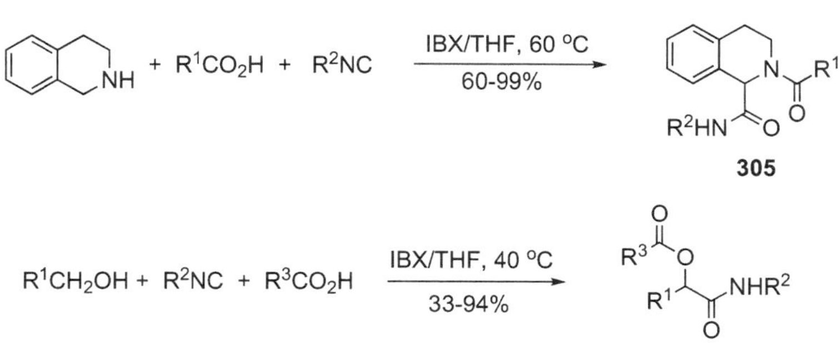


305

306

$R^1$, $R^2$, $R^3$ = alkyl, aryl, etc.

**Scheme 81**

tetrahydro-β-carbolines under mild reaction conditions applied in a total synthesis of the marine indole alkaloid eudistomin U (2010OL4086); (2) oxidation of glycosides to the respective 6-carbaldehydes used as precursors in the synthesis of amino-bridged oligosaccharides (2010EJOC900); (3) oxidation of amidoximes to carboxamides or nitriles with IBX or IBX/tetraethylammonium bromide (2010ARK118); (4) aromatic hydroxylations of flavonoids (2010T6047); (5) hydroxylation of resveratrol diacyl derivatives (2009M4669); (6) synthesis of DOPA and DOPA peptides by oxidation of tyrosine residue (2009TL6519); (7) oxidative preparation of γ-hydroxy-α-nitroolefins from α,β-epoxyketoximes (2009TL7395); (8) aromatization of 1,4-dihydropyridines using IBX in water/acetone in the presence of β-cyclodextrin (2009SC3521); (9) iodohydroxylation of alkenes and iodination of aromatics using IBX/$I_2$ in aqueous acetone (2009JOC6287); (10) conversion of alkenes and alkynes into α-iodo ketones using IBX/$I_2$ in water (2008TL5880); (11) oxidation of primary amines to nitriles (2009S1370, 2008S2045); (12) oxidative cleavage of acetals using IBX/tetraethylammonium bromide in water (2009S929); (13) one-pot synthesis of trifluoromethyl-containing pyrazoles via sequential $Yb(PFO)_3$-catalyzed three-component reaction and IBX-mediated oxidation (2008SL3058); (14) oxidative thiocyanation of indoles, pyrrole, and arylamines (20083779); (15) oxidative functionalization of Baylis–Hillman adducts (2009TL715); (16) construction of multisubstituted 2-acyl furans by the IBX-mediated cascade oxidation/cyclization of *cis*-2-en-4-yn-1-ols (2008CEJ9495); (17) one-pot synthesis of substituted salicylnitriles via oxidation of the corresponding imines with IBX (2008TL4443); (18) conversion of indoles into isatins using indium(III) chloride/IBX (2007S693); (19) synthesis of iminoquinones from anilines (2007S412); and (20) oxidative transformation of primary carboxamides to one-carbon dehomologated nitriles (2007JOC662).

### 4.2.2 Oxidations with DMP

In modern organic synthesis DMP (structure **213**) has emerged as the reagent of choice for oxidation of primary and secondary alcohols to the respective carbonyl compounds. DMP is commercially available or can be conveniently prepared by the reaction of IBX with acetic anhydride (see Section 3.3.1; 2000OS141). Applications of DMP in organic synthesis have been summarized in several overviews (2011JOC1185, 2006ARK26, 1996JPR588, 2000S278).

The DMP-promoted oxidations of alcohols proceed with high chemoselectivity under mild reaction conditions (room temperature, absence of acidic or basic additives). DMP is especially useful for the oxidation of alcohols containing sensitive functional groups, and in the case of epimerization-sensitive substrates DMP allows clean oxidation with no loss of enantiomeric excess. Oxidations with DMP are accelerated by the addition of water to the reaction mixture immediately before or during the reaction (1994JOC7549).

In numerous synthetic studies, it has been demonstrated that DMP can be used for the selective oxidation of alcohols containing sensitive functional groups, such asunsaturated alcohols (2001OL1649, 2001T9689, 2001 SL789, 2001TL3939, 2001JFC127), carbohydrates and polyhydroxy derivatives (2001SL789, 2001JA4974, 2001SL597, 2000JOC9172), silyl ethers (2001EJOC2623, 1999T12059), amines and amides (2000TL1359, 2001 TL1875, 2000OL2037, 2001TL5167), various nucleoside derivatives (1997 JOC1754, 1997JA3637, 1997JMC1608, 1997T447), selenides (2001 TL3331), tellurides (1997OM651), phosphine oxides (1998JCS(P1)3405), homoallylic and homopropargylic alcohols (2002S326), fluoroalcohols (2012JFC99, 2001CC1710, 2000EJOC2621, 1998TL8857), and boronate esters (2012ACE11092). Several representative examples of these oxidations are shown in Schemes 82—85. Specifically, the functionalized allylic alcohols **307**, which are the Baylis—Hillman adducts of aryl aldehydes and alkyl acrylates, are selectively oxidized by DMP to the corresponding α-methylene-β-keto esters **308** (Scheme 82; 2001TL3939). The attempted Swern oxidation of the same adducts **307** resulted in substitution of the allylic hydroxyl group by chloride.

Cyclic enecarbamates **310** have been prepared in high yields by oxidation of ω-hydroxycarbamates **309** with DMP followed by cyclocondensation—dehydration of the intermediate aminoaldehydes (Scheme 83; 2001TL5167).

α-Hydroxyboronates **311** can be selectively oxidized to acylboronates **312** by Dess—Martin oxidation (Scheme 84; 2012ACE11092).

DMP is particularly useful for the oxidation of optically active, epimerization-sensitive substrates without loss of enantiomeric purity (2000 TL1359, 1999JOC6931, 1998TA1929). In a representative example, DMP has been used as the oxidant for efficient, epimerization-free synthesis of the optically active N-protected α-amino aldehydes **314** from the corresponding N-protected β-amino alcohols **313** (Scheme 85; 2000TL1359). In contrast, the Swern oxidation of amino alcohols **313** affords products **314** of only 50—68% ee.

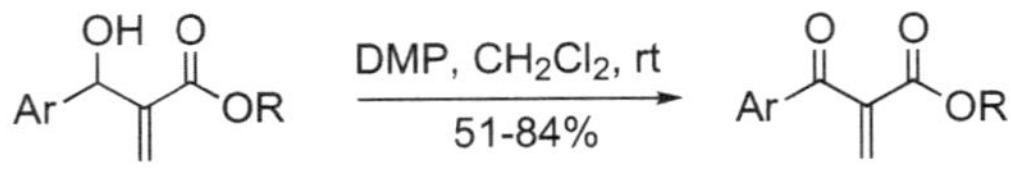

**307** **308**

Ar = Ph, 2-Cl$C_6H_4$, 4-Me$C_6H_4$, 3,4-$F_2C_6H_3$, 4-F$C_6H_4$, 4-$CF_3C_6H_4$
R = Me, Et

**Scheme 82**

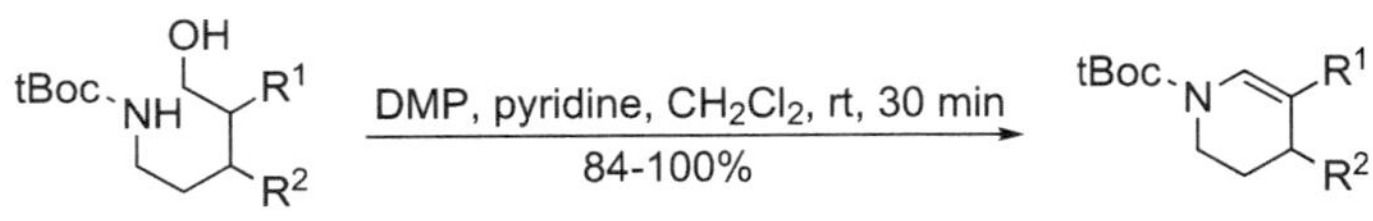

**309** **310**

$R^1$ = H, Me, $CH_2OSi(Ph)_2Bu^t$
$R^2$ = H, $OCH_2OMe$, $OCH_2OCH_2CH_2SiMe_3$

**Scheme 83**

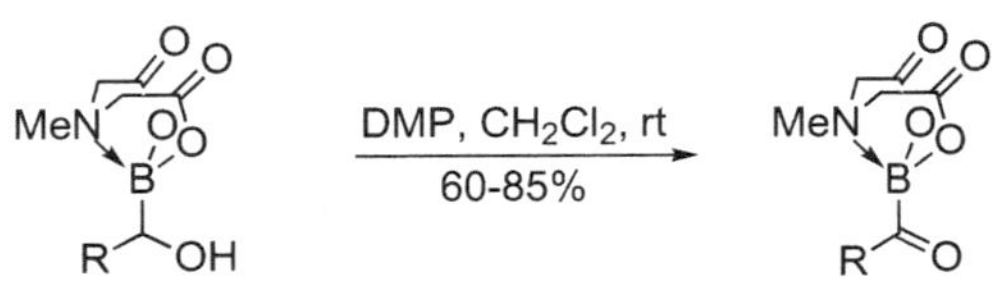

**311** R = alkyl, aryl **312**

**Scheme 84**

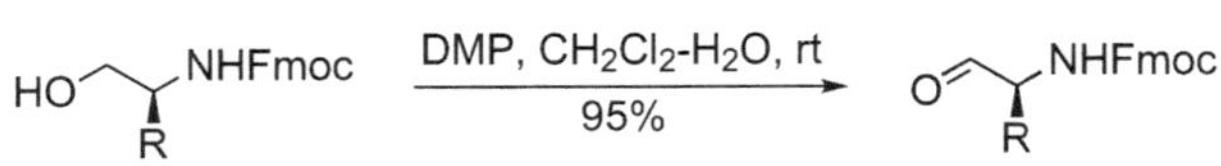

**313**, 99% ee **314**, 99% ee

R = Me, Ph
Fmoc = fluorenylmethoxycarbonyl

**Scheme 85**

The DMP oxidation of 1,2-diols usually cleaves the glycol C—C bond as illustrated by the synthesis of tricyclic enol ether **316** from diol **315** via tandem 1,2-diol cleavage-intramolecular cycloaddition (Scheme 86; 2001SL597).

Owing to its unique oxidizing properties and convenience of use, DMP has been widely employed in the synthesis of biologically important natural products. Representative examples include the use of DMP in key oxidation steps of the following total syntheses: (±)-deoxypreussomerin A (1998JOC3530), racemic brevioxime (1999CC2251), erythromycin B

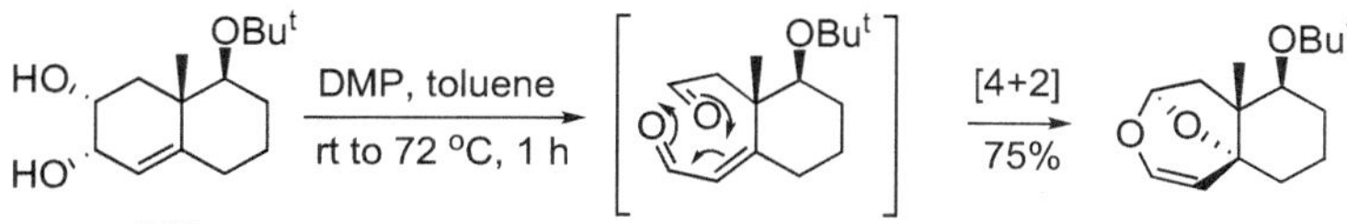


315 316

Scheme 86

(2001JA9535), (+)-cephalostatin 7 (1999JA2071), (+)-cephalostatin 12 (1999JA2071), (+)-ritterazine K (1999JA2071), fredericamycin A (2001JA 3214), angucytcline antibiotics (1996JOC5681), tricyclic β-lactam antibiotics (1996JOC1014), ent-hyperforin (2010T6569), (−)-spirotryprostatin B (2010T6514), (+)-peloruside A (2010ACE6147), (+)-ambruticin S (2010JOC5601), (±)-platensimycin (2007CC1922), (−)-pseudolaric acid B (2007JA14556), azadirachtin (2007ACE7633), salvinorin A (2007JA 8968), amphidinol 3 (2006ACE7258), FD-891 16-membered macrolide (2012JOC11109), (+)-bretonin B (2012CC11629), (±)-maoecrystal V (2012JA18860), resolvin D1 (2012TL6990), (−)-tirandamycin C (2012JOC9628), gambieric acid A (2012JA11984), (+)-sieboldine A (2013 JOC9), ripostatin B (2012ACE3405), 15-deoxyripostatin A (2012ACE 3401), spirastrellolide A methyl ester (2012ACE2749), (−)-fusarisetin A (2012JA920), halichondrin C (2012JA893), (−)-calyciphylline N (2014 JA870), (±)-yezo'otogirin C (2014OL496), (+)- and (−)-frondosins A and B (2014JA4309), (+)-salvileucalin B (2014JOC3410), mandelalide A (2014ACE6533), kendomycin (2014JOC9922), sceptrin and massadine (2014SCI219), (−)-maoecrystal V (2014JA17738), 3-epi-(+)-lycoricidine (2015TL146), and numerous other synthetic works.

The unique oxidizing properties of DMP can be illustrated by its application in the total synthesis of the CP-molecules, lead structures for cardiovascular and anticancer drugs (2002JA2183, 2002JA2190, 2002JA2202). For example, in the course of this synthetic study, a hindered secondary alcohol **317** was oxidized by DMP to give diol **319** via intermediate formation of the hemiketal **318** (Scheme 87; 2002JA2190).

The use of DMP as a reagent has been extended to numerous other synthetically useful oxidative transformations, such as (1) dehydration of primary alcohols under extraordinarily mild reaction conditions (2010 TL1863); (2) synthesis of various polycyclic heterocycles via the oxidative cascade cyclization of anilides with pendant double bonds (2002 JA2212); (3) one-pot oxidative allylation of Morita–Baylis–Hillman adducts with allyltrimethylsilane promoted by DMP/$BF_3$•$OEt_2$ (2008S469);

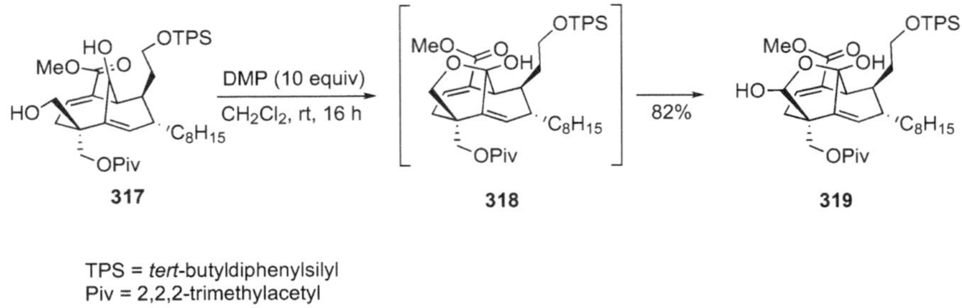


**Scheme 87**

(4) synthesis of 2-amino-1,4-benzoquinone-4-phenylimides from anilines via DMP oxidation (2007S1679); (5) α-tosyloxylation of ketones using DMP and *p*-toluenesulfonic acid (2008SL987); and (6) DMP-mediated oxidative aromatization of 1,3,5-trisubstituted pyrazolines (2012SC2617).

DMP is a particularly useful reagent for the oxidation of nitrogen substrates. Several specific examples are shown in Schemes 88–91. The oxidation of *N*-acyl hydroxylamines **320** with DMP generates the highly reactive acyl nitroso compounds **321**, which can be trapped by conjugated dienes to give the corresponding cycloadducts **322** (Scheme 88; 2000SC947).

2-Hydroxyporphyrins and 2-aminoporphyrins **323**, as well as 2,3-aminoporphyrins, are oxidized by DMP to porphyrin-α-diones **324** (Scheme 89; 1998JCS(P1)2847, 1999JCS(P1)583, 2001JCS(P1)14). This reaction has been applied to the preparation of *meso*-functionalized porphyrin-α-diones, which are the basic building blocks for bis-porphyrin arrays (1999 JCS(P1)583).

Anilides with pendant double bonds **325** undergo DMP-induced stereoselective oxidative cyclization to give complex and diverse natural product-like polycycles **326** (Scheme 90; 2000JA622, 2002JA2212). This oxidative cyclization is proposed to occur by the initial *ortho*-directed oxidation of the anilide **325** to give an *ortho*-hydroxylated benzene ring which is further

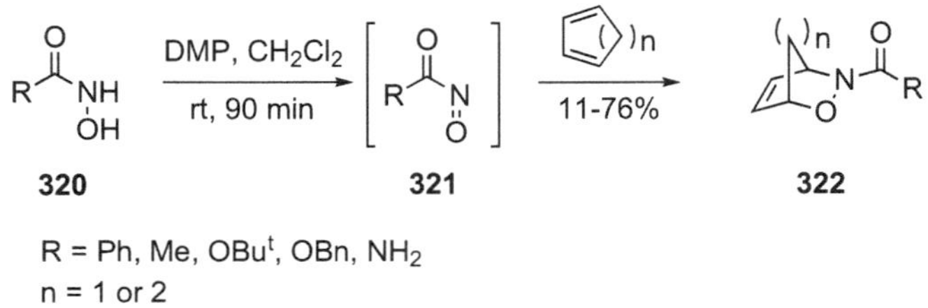


**Scheme 88**

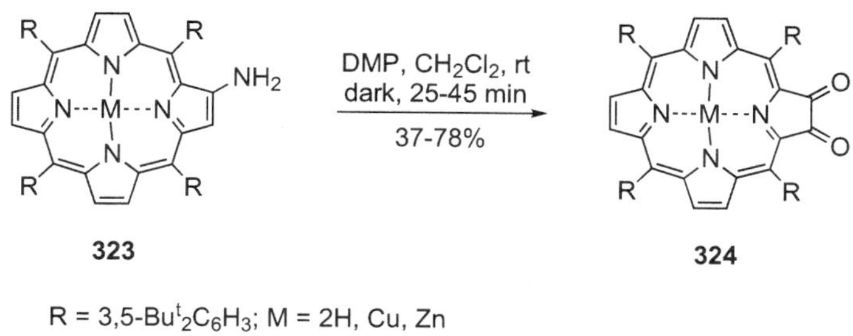


Scheme 89

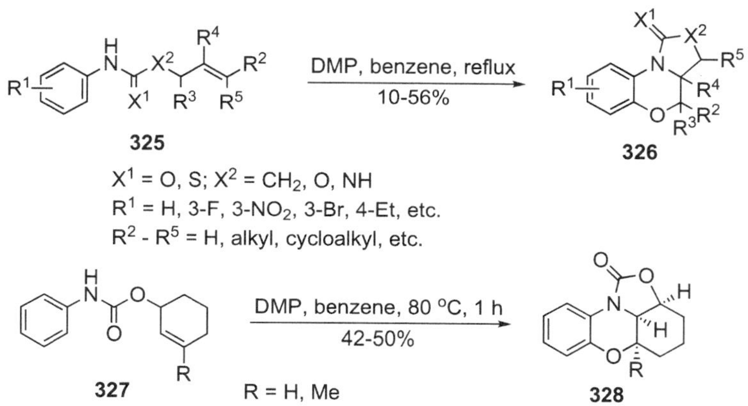


Scheme 90

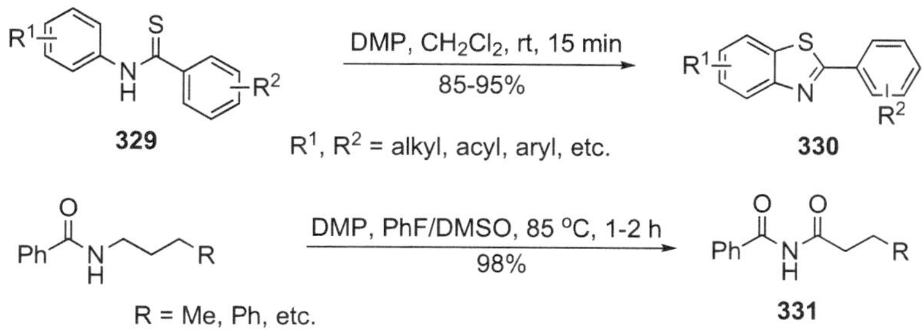


Scheme 91

oxidized to the quinone imine; the intramolecular Diels–Alder cyclization of the quinone imine with the pendant alkene gives the final product **326** (2000JA622). A specific example is the oxidation of carbamates **327** leading to the benzomorpholine derivatives **328**.

Additional examples of the DMP-mediated oxidations of nitrogen substrates include the synthesis of 2-substituted benzothiazoles **330** via oxidative cyclization of thioanilides **329** (2006JOC8261), and the synthesis of imides (e.g., **331**), *N*-acyl vinylogous carbamates and ureas, and nitriles by the oxidation of amides and amines with DMP (Scheme 91; 2005ACE5592).

### *4.2.3 Oxidations with Pseudocyclic Iodine(V) Compounds*

Iodylaryl derivatives bearing an appropriate substituent in the *ortho*-position to the iodine are characterized by the presence of a pseudocyclic structural moiety due to a strong intramolecular secondary bonding between the hypervalent iodine center and the oxygen atom in the *ortho*-substituent. Compared to the noncyclic aryliodyl derivatives, pseudocyclic iodine(V) compounds have much better solubility, which is explained by a partial disruption of their polymeric nature due to the redirection of secondary bonding (2007SL527, 2000ACE2007, 2005JOC6484, 2003ACE2194, 2005ACE7127). Particularly useful reagents of this type are IBX-esters (see Section 3.3.5). IBX-esters can oxidize alcohols to the respective aldehydes or ketones in the presence of trifluoroacetic acid or boron trifluoride etherate (2005JOC6484). Isopropyl-2-iodoxybenzoate is a useful reagent for clean and selective oxidation of organic sulfides to sulfoxides (2005 S22). This reaction proceeds without overoxidation to sulfones and is compatible with the presence of the hydroxy group, double bond, phenol ether, benzylic carbon, and various substituted phenyl rings in organic sulfides. Duschek and Kirsch have reported that isopropyl-2-iodoxybenzoate in the presence of trifluoroacetic anhydride can be used for α-hydroxylation of β-keto esters at room temperature in THF (2009CEJ10713).

IBX-esters can serve as stable and efficient sources of oxygen in metalloporphyrin-catalyzed oxidations of hydrocarbons, and the reactivity of isopropyl 2-iodoxybenzoate as an oxygenating reagent is similar to the reactivity of commonly used iodosylbenzene, which is a thermally unstable and potentially explosive compound (2009ASC733, 2008TL7410, 2012ACE5006).

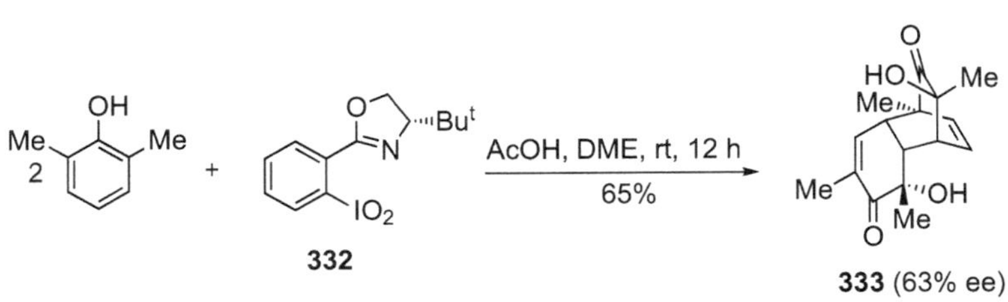


**Scheme 92**

The chiral, pseudocyclic 2-(*o*-iodoxyphenyl)-oxazolines **332** have been found to transform *ortho*-alkylphenols into *ortho*-quinol Diels–Alder dimers (e.g., **333**) with significant levels of asymmetric induction (Scheme 92; 2009OL1221).

## REFERENCES


1892B2632 V. Meyer and W. Wachter, *Ber*, **25**, 2632 (1892).
1893CB1727 C. Hartman and V. Mayer, *Chem. Ber.*, **26**, 1727 (1893).
1894JPC466 C. Willgerodt, *J. Prakt. Chem.*, **49**, 466 (1894).
1909GCI619 L. Mascarelli and G. Benati, *Gazz. Chim. Ital.*, **38**, 619 (1909).
1910LA119 J. Thiele and W. Peter, *Justus Liebigs Annalen der Chemie*, **369**, 119 (1910).
1914MI8 C. Willgerodt, *Die Organischen Verbindungen mit Mehrwertigen Jod*, Ferdinand Enke Verlag: Stuttgart (1914).
1950IAN315 E.M. Brainina and R.K. Freidlina, *Izv. Akad. Nauk SSSR, Ser. Khim.*, 315 (1950).
1951JA4321 R.J. Hach and R.E. Rundle, *J. Am. Chem. Soc.*, **73**, 4321 (1951).
1951JCP446 G.C. Pimentel, *J. Chem. Phys.*, **19**, 446 (1951).
1953AC88 E.M. Archer and T.G.D. van Schalkwyk, *Acta Cryst.*, **6**, 88 (1953).
1956DAN71 T.L. Khotsyanova, *Dokl. Akad. Nauk SSSR*, **110**, 71 (1956).
1956JA3819 J. Collette, D. McGreer, R. Crawford, F. Chubb, and R.B. Sandin, *J. Am. Chem. Soc.*, **78**, 3819 (1956).
1960JA725 F.M. Beringer and I. Lillien, *J. Am. Chem. Soc.*, **82**, 725 (1960).
1962JA3443 J.E. Leffler, L.K. Dyall, and P.W. Inward, *J. Am. Chem. Soc.*, **85**, 3443 (1963).
1964NAT512 E. Shefter and W. Wolf, *Nature*, **203**, 512 (1964).
1965CC449 W. Wolf and L. Steinberg, *J. Chem. Soc., Chem. Commun.*, 449 (1965).
1965JCS3721 G.P. Baker, F.G. Mann, N. Sheppard, and A.J. Tetlow, *J. Chem. Soc.*, 3721 (1965).
1965JOC1141 F.M. Beringer, L. Kravetz, and G.B. Topliss, *J. Org. Chem.*, **30**, 1141 (1965).
1965JOC617 E.A. Jeffery, L.J. Andrews, and R.M. Keefer, *J. Org. Chem.*, **30**, 617 (1965).
1965JPS104 E. Shefter and W. Wolf, *J. Pharm. Sci.*, **54**, 104 (1965).
1965TL2681 W.C. Agosta, *Tetrahedron Lett.*, 2681 (1965).
1966OS107 L.F. Fieser and M.J. Haddadin, *Org. Synth.*, **46**, 107 (1966).
1968T3717 H. Heaney and P. Lees, *Tetrahedron*, **24**, 3717 (1968).
1969AGE54 J.I. Musher, *Angew. Chem., Int. Ed. Engl.*, **8**, 54 (1969).
1969JOC456 R.B. Sandin, *J. Org. Chem.*, **34**, 456 (1969).
1969T4339 F.M. Beringer, L.L. Chang, A.N. Fenster, and R.R. Rossi, *Tetrahedron*, **25**, 4339 (1969).
1970JCS(C)862 H.J. Barber and M.A. Henderson, *J. Chem. Soc., C*, 862 (1970).
1970JOC1196 G.F. Morrison and J. Hooz, *J. Org. Chem.*, **35**, 1196 (1970).
1972BC39 P.C. Holland and H.S.A. Sherratt, *Biochem. J.*, **129**, 39 (1972).
1972JOC879 F.M. Beringer, P. Ganis, G. Avitabile, and H. Jaffe, *J. Org. Chem.*, **37**, 879 (1972).
1973JBC6050 P.C. Holland, M.C. Clark, D.P. Bloxham, and H.A. Lardy, *J. Biol. Chem.*, **248**, 6050 (1973).
1973JOC2719 J.E. Leffler and H. Jeffe, *J. Org. Chem.*, **38**, 2719 (1973).
1974JA3319 J.C. Martin and M.M. Chau, *J. Am. Chem. Soc.*, **96**, 3319 (1974).
1974JSSC155 J.Z. Gougoutas and L. Lessinger, *J. Solid State Chem.*, **9**, 155 (1974).

1974PRGE277 L.D. Freedman and R.P. DeMott, *Phosphorus Relat. Group V Elem.*, **3**, 277 (1974).
1975BST333 S.J. Gatley, S.S. Al-Bassam, J.R. Taylor, and H.S.A. Sherratt, *Biochem. Soc. Trans.*, **3**, 333 (1975).
1975JOC2129 D.G. Naae and J.Z. Gougoutas, *J. Org. Chem.*, **40**, 2129 (1975).
1975JOC797 H. Jaffe and J.E. Leffler, *J. Org. Chem.*, **40**, 797 (1975).
1976BJ307 S.J. Gatley and H.S.A. Sherratt, *Biochem. J.*, **158**, 307 (1976).
1976JA5326 M.C. Etter, *J. Am. Chem. Soc.*, **98**, 5326 (1976).
1976ZN687 K. Prout, N.M. Stevens, A. Coda, V. Tazzoli, R.A. Shaw, T. Demir, and Z. Naturforsch, *B: Anorg. Chem., Org. Chem.*, **31B**, 687 (1976).
1978CB2510 J.B.S. Bonilha, N. Petragnani, and V.G. Toscano, *Chem. Ber.*, **111**, 2510 (1978).
1978JOC4538 T.M. Balthazor, J.A. Miles, and B.R. Stults, *J. Org. Chem.*, **43**, 4538 (1978).
1979JCS(D)851 N.W. Alcock and R.M. Countryman, *J. Chem. Soc., Dalton Trans.*, 851 (1979).
1979JOC1447 T.M. Balthazor, D.E. Godar, and B.R. Stults, *J. Org. Chem.*, **44**, 1447 (1979).
1979JOC1779 R.L. Amey and J.C. Martin, *J. Org. Chem.*, **44**, 1779 (1979).
1979MI5 *Nomenclature of Organic Chemistry, Sections A, B, C, D, E, F, and H*, Pergamon Press: Oxford (1979).
1981CSC489 J.Z. Gougoutas, *Cryst. Struct. Commun.*, **10**, 489 (1981).
1981JOC4069 V.R. Sandel, G.R. Buske, S.G. Maroldo, D.K. Bates, D. Whitman, and G. Sypniewski, *J. Org. Chem.*, **46**, 4069 (1981).
1982JOC1024 T.T. Nguyen, R.L. Amey, and J.C. Martin, *J. Org. Chem.*, **47**, 1024 (1982).
1983JOC4155 D.B. Dess and J.C. Martin, *J. Org. Chem.*, **48**, 4155 (1983).
1983MI4 G.F. Koser, In S. Patai and Z. Rappoport, editors: *Chem. Halides, Pseudo-Halides Azides*, Wiley-Interscience: Chichester (1983). Chapters 18 and 25.
1984PAC769 W.H. Powell, *Pure Appl. Chem.*, **56**, 769 (1984).
1985JOC3200 R.D. Miller, L. Franz, and G.N. Fickes, *J. Org. Chem.*, **50**, 3200 (1985).
1986BJ111 A.R. Cross and O.T.G. Jones, *Biochem. J.*, **237**, 111 (1986).
1986BST1209 M.J. Brosnan, D.J. Hayes, R.A.J. Challiss, and G.K. Radda, *Biochem. Soc. Trans.*, **14**, 1209 (1986).
1986IC1415 R.J. Batchelor, T. Birchall, and J.F. Sawyer, *Inorg. Chem.*, **25**, 1415 (1986).
1986JA3803 T.T. Nguyen, S.R. Wilson, and J.C. Martin, *J. Am. Chem. Soc.*, **108**, 3803 (1986).
1986JA788 R.A. Moss, K.Y. Kim, and S. Swarup, *J. Am. Chem. Soc.*, **108**, 788 (1986).
1986JOC4303 R.A. Moss, S. Chatterjee, and B. Wilk, *J. Org. Chem.*, **51**, 4303 (1986).
1986T155 F. Gavina, S.V. Luis, A.M. Costero, and P. Gil, *Tetrahedron*, **42**, 155 (1986).
1987BP489 A.R. Cross, *Biochem. Pharmacol.*, **36**, 489 (1987).
1987T251 R.A. Moss, P. Scrimin, and R.T. Rosen, *Tetrahedron Lett.*, **28**, 251 (1987).
1988BJ887 J.A. Ellis, S.J. Mayer, and O.T.G. Jones, *Biochem. J.*, **251**, 887 (1988).
1988BP687 J.M. Cooper, R.K.H. Petty, D.J. Hayes, J.A. Morgan-Hughes, and J.B. Clark, *Biochem. Pharmacol.*, **37**, 687 (1988).
1988JNS335 J.M. Cooper, R.K.H. Petty, D.J. Hayes, R.A.J. Challiss, M.J. Brosnan, E.A. Shoubridge, G.K. Radda, J.A. Morgan-Hughes, and J.B. Clark, *J. Neurol. Sci.*, **83**, 335 (1988).

1988JOC5799 D. Del Mazza and M.G. Reinecke, *J. Org. Chem.*, **53**, 5799 (1988).
1989JA250 R.A. Moss, B. Wilk, K. Krogh-Jespersen, J.T. Blair, and J.D. Westbrook, *J. Am. Chem. Soc.*, **111**, 250 (1989).
1989JA6729 R.A. Moss, B. Wilk, K. Krogh-Jespersen, and J.D. Westbrook, *J. Am. Chem. Soc.*, **111**, 6729 (1989).
1990CEN3 J.B. Plumb and D.J. Harper, *Chem. Eng. News*, **68**, 3 (1990).
1990JCS(P2)1657 A.R. Katritzky, G.P. Savage, G.J. Palenik, K. Qian, Z. Zhang, and H.D. Durst, *J. Chem. Soc., Perkin Trans. 2*, 1657 (1990).
1990JOC5202 C.A. Panetta, S.M. Garlick, H.D. Durst, F.R. Longo, and J.R. Ward, *J. Org. Chem.*, **55**, 5202 (1990).
1991BBRC143 J. Doussiere and P.V. Vignais, *Biochem. Biophys. Res. Commun.*, **175**, 143 (1991).
1991FASEB98 D.J. Stuehr, O.A. Fasehun, N.S. Kwon, S.S. Gross, J.A. Gonzalez, R. Levi, and C.F. Nathan, *FASEB J.*, **5**, 98 (1991).
1991FRBM25 J.T. Hancock, J.I. White, O.T.G. Jones, and I.A. Silver, *Free Radical Biol. Med.*, **11**, 25 (1991).
1991JA7277 D.B. Dess and J.C. Martin, *J. Am. Chem. Soc.*, **113**, 7277 (1991).
1991JOC5511 M. Ochiai, Y. Masaki, and M. Shiro, *J. Org. Chem.*, **56**, 5511 (1991).
1991JOC6565 L. Weclas-Henderson, T.T. Nguyen, R.A. Hayes, and J.C. Martin, *J. Org. Chem.*, **56**, 6565 (1991).
1992CC1487 R. Yang, L. Dai, and C. Chen, *J. Chem. Soc., Chem. Commun.*, 1487 (1992).
1992EJB61 J. Doussiere and P.V. Vignais, *Eur. J. Biochem.*, **208**, 61 (1992).
1992JA6269 M. Ochiai, T. Ito, Y. Masaki, and M. Shiro, *J. Am. Chem. Soc.*, **114**, 6269 (1992).
1992MI7 A. Varvoglis, *The Organic Chemistry of Polycoordinated Iodine*, VCH Publishers, Inc.: New York (1992).
1992SC1799 H.E. Folsom and J. Castrillon, *Synth. Commun.*, **22**, 1799 (1992).
1993BJ41 V.B. O'Donnell, D.G. Tew, O.T.G. Jones, and P.J. England, *Biochem. J.*, **290**, 41 (1993).
1993CC218 M. Ochiai, T. Ito, and M. Shiro, *J. Chem. Soc., Chem. Commun.*, 218 (1993).
1993JA11626 R.J. Hinkle, G.T. Poulter, and P.J. Stang, *J. Am. Chem. Soc.*, **115**, 11626 (1993).
1993JA2488 D.B. Dess, S.R. Wilson, and J.C. Martin, *J. Am. Chem. Soc.*, **115**, 2488 (1993).
1993JA9808 P.J. Stang and V.V. Zhdankin, *J. Am. Chem. Soc.*, **115**, 9808 (1993).
1993JOC2899 R.E. Ireland and L. Liu, *J. Org. Chem.*, **58**, 2899 (1993).
1993JOC7310 G.F. Koser, G. Sun, C.W. Porter, and W.J. Youngs, *J. Org. Chem.*, **58**, 7310 (1993).
1993TL1729 R.A. Moss, H. Zhang, S. Chatterjee, and K. Krogh-Jespersen, *Tetrahedron Lett.*, **34**, 1729 (1993).
1994CSR111 C.A. Ramsden, *Chem. Soc. Rev.*, **23**, 111 (1994).
1994JA2448 R.S. Brown, R.W. Nagorski, A.J. Bennet, R.E.D. McClung, G.H.M. Aarts, M. Klobukowski, R. McDonald, and B.D. Santarsiero, *J. Am. Chem. Soc.*, **116**, 2448 (1994).
1994JA4471 R.A. Moss and H. Zhang, *J. Am. Chem. Soc.*, **116**, 4471 (1994).
1994JOC7549 S.D. Meyer and S.L. Schreiber, *J. Org. Chem.*, **59**, 7549 (1994).
1994NL63 C.N. Wyatt, E.K. Weir, and C. Peers, *Neurosci. Lett.*, **172**, 63 (1994).
1994TL7323 V.V. Zhdankin, C.J. Kuehl, J.T. Bolz, M.S. Formaneck, and A.J. Simonsen, *Tetrahedron Lett.*, **35**, 7323 (1994).
1994TL8019 M. Frigerio and M. Santagostino, *Tetrahedron Lett.*, **35**, 8019 (1994).

1994TL9677 V.V. Zhdankin, C.J. Kuehl, A.P. Krasutsky, M.S. Formaneck, and J.T. Bolz, *Tetrahedron Lett.*, **35**, 9677 (1994).
1995JA1859 C.R. Landis, T. Cleveland, and T.K. Firman, *J. Am. Chem. Soc.*, **117**, 1859 (1995).
1995JCC857 G.F. Koser, D.B. McConville, G.A. Rabah, and W.J. Youngs, *J. Chem. Cryst.*, **25**, 857 (1995).
1995JOC7272 M. Frigerio, M. Santagostino, S. Sputore, and G. Palmisano, *J. Org. Chem.*, **60**, 7272 (1995).
1995JOC7739 R.A. Moss, K. Bracken, and T.J. Emge, *J. Org. Chem.*, **60**, 7739 (1995).
1995SL1081 A.P. Krasutsky, C.J. Kuehl, and V.V. Zhdankin, *Synlett*, 1081 (1995).
1995TL3485 E.J. Corey and A. Palani, *Tetrahedron Lett.*, **36**, 3485 (1995).
1995TL7945 E.J. Corey and A. Palani, *Tetrahedron Lett.*, **36**, 7945 (1995).
1995TL7975 V.V. Zhdankin, C.J. Kuehl, A.P. Krasutsky, J.T. Bolz, B. Mismash, J.K. Woodward, and A.J. Simonsen, *Tetrahedron Lett.*, **36**, 7975 (1995).
1995TL9117 S.H. Stickley and J.C. Martin, *Tetrahedron Lett.*, **36**, 9117 (1995).
1996H47 S. Akai, T. Okuno, T. Takada, H. Tohma, and Y. Kita, *Heterocycles*, **42**, 47 (1996).
1996JA5192 V.V. Zhdankin, A.P. Krasutsky, C.J. Kuehl, A.J. Simonsen, J.K. Woodward, B. Mismash, and J.T. Bolz, *J. Am. Chem. Soc.*, **118**, 5192 (1996).
1996JA6020 T. Cleveland and C.R. Landis, *J. Am. Chem. Soc.*, **118**, 6020 (1996).
1998JA12650 T.K. Firman and C.R. Landis, *J. Am. Chem. Soc.*, **120**, 12650 (1998).
1998JA2641 C.R. Landis, T. Cleveland, and T.K. Firman, *J. Am. Chem. Soc.*, **120**, 2641 (1998).
1996JA7716 M. Ochiai, T. Ito, H. Takahashi, A. Nakanishi, M. Toyonari, T. Sueda, S. Goto, and M. Shiro, *J. Am. Chem. Soc.*, **118**, 7716 (1996).
1996JCC371 M. Takahashi, H. Nanba, T. Kitazawa, M. Takeda, and Y. Ito, *J. Coord. Chem.*, **37**, 371 (1996).
1996JOC1014 C. Niu, T. Pettersson, and M.J. Miller, *J. Org. Chem.*, **61**, 1014 (1996).
1996JOC5681 D.S. Larsen and M.D. O'Shea, *J. Org. Chem.*, **61**, 5681 (1996).
1996JOC6547 V.V. Zhdankin, C.J. Kuehl, A.P. Krasutsky, J.T. Bolz, and A.J. Simonsen, *J. Org. Chem.*, **61**, 6547 (1996).
1996JOC9272 S. De Munari, M. Frigerio, and M. Santagostino, *J. Org. Chem.*, **61**, 9272 (1996).
1996JPR588 A. Speicher, V. Bomm, and T. Eicher, *J. Prakt. Chem.*, **338**, 588 (1996).
1996MC50 V.V. Zhdankin, C.J. Kuehl, A.M. Arif, and P.J. Stang, *Mendeleev Commun.*, 50 (1996).
1996TL6453 G.A. Rabah and G.F. Koser, *Tetrahedron Lett.*, **37**, 6453 (1996).
1997CSC589 P.J. Stevenson, A.B. Treacy, and M. Nieuwenhuyzen, *J. Chem. Soc., Perkin Trans. 2*, 589 (1997).
1997H71 M. Ochiai, D. Kajishima, and T. Sueda, *Heterocycles*, **46**, 71 (1997).
1997JA2628 D. Dolenc and B. Plesnicar, *J. Am. Chem. Soc.*, **119**, 2628 (1997).
1997JA3637 M.J. Robins, Z. Guo, and S.F. Wnuk, *J. Am. Chem. Soc.*, **119**, 3637 (1997).
1997JA7408 V.V. Zhdankin, R.M. Arbit, M. McSherry, B. Mismash, and V.G. Young, *J. Am. Chem. Soc.*, **119**, 7408 (1997).
1997JMC1608 S.F. Wnuk, C.-S. Yuan, R.T. Borchardt, J. Balzarini, E. De Clercq, and M.J. Robins, *J. Med. Chem.*, **40**, 1608 (1997).
1997JOC1754 R.E. Harry-O'kuru, J.M. Smith, and M.S. Wolfe, *J. Org. Chem.*, **62**, 1754 (1997).
1997JOC4253 M. Ochiai, A. Nakanishi, and T. Ito, *J. Org. Chem.*, **62**, 4253 (1997).

1997JPOC27 R.A. Moss, S. Bose, and K. Krogh-Jespersen, *J. Phys. Org. Chem.*, **10**, 27 (1997).

1997OM651 L.H.S. Rahmeier and J.V. Comasseto, *Organometallics*, **16**, 651 (1997).

1997RHC133 V.V. Zhdankin, *Rev Heteroatom Chem.*, **17**, 133 (1997).

1997T447 M.J. Robins, S. Sarker, V. Samano, and S.F. Wnuk, *Tetrahedron*, **53**, 447 (1997).

1997TL21 V.V. Zhdankin, M. McSherry, B. Mismash, J.T. Bolz, J.K. Woodward, R.M. Arbit, and S. Erickson, *Tetrahedron Lett.*, **38**, 21 (1997).

1997TL3927 M. Ochiai, A. Nakanishi, and A. Yamada, *Tetrahedron Lett.*, **38**, 3927 (1997).

1998ACE1887 G.A. Landrum and R. Hoffmann, *Angew. Chem., Int. Ed*, **37**, 1887 (1998).

1998CC1559 R.A. Moss and T.J. Emge, *Chem. Commun.*, 1559 (1998).

1998JA1842 C.R. Landis, T.K. Firman, D.M. Root, and T. Cleveland, *J. Am. Chem. Soc.*, **120**, 1842 (1998).

1998JCS(P1)2847 R. Beavington, P.A. Rees, and P.L. Burn, *J. Chem. Soc., Perkin Trans. 1*, 2847 (1998).

1998JCS(P1)3405 C. Gueguen, P. O'Brien, H.R. Powell, P.R. Raithby, and S. Warren, *J. Chem. Soc., Perkin Trans. 1*, 3405 (1998).

1998JOC3530 P. Wipf and J.-K. Jung, *J. Org. Chem.*, **63**, 3530 (1998).

1998JOC6590 V.V. Zhdankin, R.M. Arbit, B.J. Lynch, P. Kiprof, and V.G. Young, *J. Org. Chem.*, **63**, 6590 (1998).

1998NJC883 G.A. Landrum, N. Goldberg, R. Hoffmann, and R.M. Minyaev, *New J. Chem.*, **22**, 883 (1998).

1998SL286 T. Muraki, H. Togo, and M. Yokoyama, *Synlett*, 286 (1998).

1998TA1929 C. Botuha, M. Haddad, and M. Larcheveque, *Tetrahedron Asymmetry*, **9**, 1929 (1998).

1998TL8857 L. Leveque, M. Le Blanc, and R. Pastor, *Tetrahedron Lett.*, **39**, 8857 (1998).

1999CC2251 D.L.J. Clive and S. Hisaindee, *J. Chem. Soc., Chem. Commun.*, 2251 (1999).

1999JA2071 J.U. Jeong, C. Guo, and P.L. Fuchs, *J. Am. Chem. Soc.*, **121**, 2071 (1999).

1999JA7164 D. Macikenas, E. Skrzypczak-Jankun, and J.D. Protasiewicz, *J. Am. Chem. Soc.*, **121**, 7164 (1999).

1999JCS(P1)583 R. Beavington and P.L. Burn, *J. Chem. Soc., Perkin Trans. 1*, 583 (1999).

1999JOC2883 T. Muraki, H. Togo, and M. Yokoyama, *J. Org. Chem.*, **64**, 2883 (1999).

1999JOC4537 M. Frigerio, M. Santagostino, and S. Sputore, *J. Org. Chem.*, **64**, 4537 (1999).

1999JOC6931 F.A. Davis, V. Srirajan, and D.D. Titus, *J. Org. Chem.*, **64**, 6931 (1999).

1999MI6 K.y. Akiba, *Chemistry of Hypervalent Compounds*, Wiley-VCH: New York (1999).

1999T12059 D.L.J. Clive and J. Zhang, *Tetrahedron*, **55**, 12059 (1999).

1999T687 D. Weigelt, R. Krahmer, K. Bruschke, L. Hennig, M. Findeisen, D. Muller, and P. Welzel, *Tetrahedron*, **55**, 687 (1999).

1999TL5541 M. Ochiai, D. Kajishima, and T. Sueda, *Tetrahedron Lett.*, **40**, 5541 (1999).

1999TL6253 T. Zoller, P. Breuilles, D. Uguen, A. De Cian, and J. Fischer, *Tetrahedron Lett.*, **40**, 6253 (1999).

2000ACE2007 D. Macikenas, E. Skrzypczak-Jankun, and J.D. Protasiewicz, *Angew. Chem., Int. Ed.*, **39**, 2007 (2000).
2000ACE2525 K.C. Nicolaou, P.S. Baran, Y.-L. Zhong, and J.A. Vega, *Angew. Chem., Int. Ed.*, **39**, 2525 (2000).
2000CSR315 V.V. Grushin, *Chem. Soc. Rev.*, **29**, 315 (2000).
2000EJOC2621 C. Rocaboy, W. Bauer, and J.A. Gladysz, *Eur. J. Org. Chem.*, 2621 (2000).
2000JA10246 K.C. Nicolaou, P.S. Baran, and Y.-L. Zhong, *J. Am. Chem. Soc.*, **122**, 10246 (2000).
2000JA622 K.C. Nicolaou, Y.-L. Zhong, and P.S. Baran, *Angew. Chem., Int. Ed.*, **39**, 622 (2000).
2000JA7596 K.C. Nicolaou, Y.L. Zhong, and P.S. Baran, *J. Am. Chem. Soc.*, **122**, 7596 (2000).
2000JCC224 W. Li, A.W. Czarnik, J. Lillig, and X.-Y. Xiao, *J. Comb. Chem.*, **2**, 224 (2000).
2000JLCR515 J.-L. Berthon, M. Dias, R. Mornet, and J.-M. Camadro, *J. Labelled Compd. Radiopharm.*, **43**, 515 (2000).
2000JOC9172 L.A. Paquette, J. Tae, B.M. Branan, D.G. Bolin, and S.W.E. Eisenberg, *J. Org. Chem.*, **65**, 9172 (2000).
2000OL2037 A.B. Smith 3rd, H. Liu, and R. Hirschmann, *Org. Lett.*, **2**, 2037 (2000).
2000OL2923 M. Ochiai, A. Nakanishi, and T. Suefuji, *Org Lett.*, **2**, 2923 (2000).
2000OS141 R.K. Boeckman, P. Shao, and J.J. Mullins, *Org. Synth.*, **77**, 141 (2000).
2000S278 S.S. Chaudhari, *Synlett*, 278 (2000).
2000SC947 N.E. Jenkins, R.W. Ware Jr., R.N. Atkinson, and S.B. King, *Synth. Commun.*, **30**, 947 (2000).
2000SL719 V.V. Zhdankin, P.J. Persichini III, R. Cui, and Y. Jin, *Synlett*, 719 (2000).
2000TL1359 A.G. Myers, B. Zhong, M. Movassaghi, D.W. Kung, B.A. Lanman, and S. Kwon, *Tetrahedron Lett.*, **41**, 1359 (2000).
2000TL4379 J.M. Bueno, J.M. Coteron, J.L. Chiara, A. Fernandez-Mayoralas, J.M. Fiandor, and N. Valle, *Tetrahedron Lett.*, **41**, 4379 (2000).
2000TL5299 V.V. Zhdankin, J.T. Smart, P. Zhao, and P. Kiprof, *Tetrahedron Lett.*, **41**, 5299 (2000).
2001ACE202 K.C. Nicolaou, P.S. Baran, R. Kranich, Y.-L. Zhong, K. Sugita, and N. Zou, *Angew. Chem., Int. Ed.*, **40**, 202 (2001).
2001ACE2812 T. Wirth, *Angew. Chem., Int. Ed.*, **40**, 2812 (2001).
2001ACE4393 M. Muelbaier and A. Giannis, *Angew. Chem., Int. Ed.*, **40**, 4393 (2001).
2001ACE4395 G. Sorg, A. Mengei, G. Jung, and J. Rademann, *Angew. Chem., Int. Ed.*, **40**, 4395 (2001).
2001CC1710 R.J. Cox, A.T. Hadfield, and M.B. Mayo-Martin, *Chem. Commun.*, 1710 (2001).
2001EJOC2623 R. Heck, A.P. Henderson, B. Kohler, J. Retey, and B.T. Golding, *Eur. J. Org. Chem.*, 2623 (2001).
2001FASEB2539 H. Sauer, B. Klimm, J. Hescheler, and M. Wartenberg, *FASEB J.*, **15**, 2539 (2001).
2001JA1507 T. Kawashima, K. Hoshiba, and N. Kano, *J. Am. Chem. Soc.*, **123**, 1507 (2001).
2001JA3183 K.C. Nicolaou, P.S. Baran, and Y.-L. Zhong, *J. Am. Chem. Soc.*, **123**, 3183 (2001).
2001JA3214 Y. Kita, K. Higuchi, Y. Yoshida, K. Iio, S. Kitagaki, K. Ueda, S. Akai, and H. Fujioka, *J. Am. Chem. Soc.*, **123**, 3214 (2001).

2001JA349 C.L. Shaffer, M.D. Morton, and R.P. Hanzlik, *J. Am. Chem. Soc.*, **123**, 349 (2001).

2001JA4095 V.V. Zhdankin, A.E. Koposov, J.T. Smart, R.R. Tykwinski, R. McDonald, and A. Morales-Izquierdo, *J. Am. Chem. Soc.*, **123**, 4095 (2001).

2001JA4974 L.A. Paquette and J. Tae, *J. Am. Chem. Soc.*, **123**, 4974 (2001).

2001JA9535 I. Paterson, G.J. Florence, K. Gerlach, J.P. Scott, and N. Sereinig, *J. Am. Chem. Soc.*, **123**, 9535 (2001).

2001JCS(P1)14 V. Promarak and P.L. Burn, *J. Chem. Soc., Perkin Trans. 1*, 14 (2001).

2001JFC127 Z. Wang, Y. Gu, A.J. Zapata, and G.B. Hammond, *J. Fluorine Chem.*, **107**, 127 (2001).

2001OL1021 C. Martin, N. Macintosh, N. Lamb, and A.G. Fallis, *Org. Lett.*, **3**, 1021 (2001).

2001OL1649 Y. Hu, C. Li, B.A. Kulkarni, G. Strobel, E. Lobkovsky, R.M. Torczynski, and J.A. Porco Jr., *Org. Lett.*, **3**, 1649 (2001).

2001OL2387 T. Sueda, S. Fukuda, and M. Ochiai, *Org. Lett.*, **3**, 2387 (2001).

2001PSSS141 T. Kawashima, K. Hoshiba, and N. Kano, *Phosphorus, Sulfur, Silicon Relat. Elem*, **168-169**, 141 (2001).

2001S2113 F.F. Paintner, L. Allmendinger, and G. Bauschke, *Synthesis*, 2113 (2001).

2001SL597 J.I. Candela Lena, J.I. Martin Hernando, M.d. R. Rico Ferreira, E. Altinel, and S. Arseniyadis, *Synlett*, 597 (2001).

2001SL789 J. Roels and P. Metz, *Synlett*, 789 (2001).

2001T9689 T. Satoh, A. Nakamura, A. Iriuchijima, Y. Hayashi, and K-i. Kubota, *Tetrahedron*, **57**, 9689 (2001).

2001TL1875 D. Bonnet, P. Joly, H. Gras-Masse, and O. Melnyk, *Tetrahedron Lett.*, **42**, 1875 (2001).

2001TL3331 H. Kumamoto, J. Ogamino, H. Tanaka, H. Suzuki, K. Haraguchi, T. Miyasaka, T. Yokomatsu, and S. Shibuya, *Tetrahedron*, **57**, 3331 (2001).

2001TL3939 N.J. Lawrence, J.P. Crump, A.T. McGown, and J.A. Hadfield, *Tetrahedron Lett.*, **42**, 3939 (2001).

2001TL5167 C. Yu and L. Hu, *Tetrahedron Lett.*, **42**, 5167 (2001).

2002ACE993 K.C. Nicolaou, T. Montagnon, and P.S. Baran, *Angew. Chem., Int. Ed.*, **41**, 993 (2002).

2002ACE996 K.C. Nicolaou, D.L.F. Gray, T. Montagnon, and S.T. Harrison, *Angew. Chem., Int. Ed.*, **41**, 996 (2002).

2002CR2497 H. Morales-Rojas and R.A. Moss, *Chem. Rev.*, **102**, 2497 (2002).

2002JA11614 V.V. Zhdankin, O. Maydanovych, J. Herschbach, R. McDonald, and R.R. Tykwinski, *J. Am. Chem. Soc.*, **124**, 11614 (2002).

2002JA2183 K.C. Nicolaou, J. Jung, W.H. Yoon, K.C. Fong, H.S. Choi, Y. He, Y.L. Zhong, and P.S. Baran, *J. Am. Chem. Soc.*, **124**, 2183 (2002).

2002JA2190 K.C. Nicolaou, P.S. Baran, Y.L. Zhong, K.C. Fong, and H.S. Choi, *J. Am. Chem. Soc.*, **124**, 2190 (2002).

2002JA2202 K.C. Nicolaou, Y.L. Zhong, P.S. Baran, J. Jung, H.S. Choi, and W.H. Yoon, *J. Am. Chem. Soc.*, **124**, 2202 (2002).

2002JA2212 K.C. Nicolaou, P.S. Baran, Y.L. Zhong, and K. Sugita, *J. Am. Chem. Soc.*, **124**, 2212 (2002).

2002JA2233 K.C. Nicolaou, P.S. Baran, Y.L. Zhong, S. Barluenga, K.W. Hunt, R. Kranich, and J.A. Vega, *J. Am. Chem. Soc.*, **124**, 2233 (2002).

2002JA2245 K.C. Nicolaou, T. Montagnon, P.S. Baran, and Y.L. Zhong, *J. Am. Chem. Soc.*, **124**, 2245 (2002).

2002OL285 D. Magdziak, A.A. Rodriguez, R.W. Van De Water, and T.R.R. Pettus, *Org. Lett.*, **4**, 285 (2002).

2002OL3001 J.D. More and N.S. Finney, *Org. Lett.*, **4**, 3001 (2002).
2002S326 L. Wavrin and J. Viala, *Synthesis*, 326 (2002).
2002TL569 A.P. Thottumkara and T.K. Vinod, *Tetrahedron Lett.*, **43**, 569 (2002).
2002TL8673 F. Yokokawa and T. Shioiri, *Tetrahedron Lett.*, **43**, 8673 (2002).
2003ACE2194 V.V. Zhdankin, A.Y. Koposov, B.C. Netzel, N.V. Yashin, B.P. Rempel, M.J. Ferguson, and R.R. Tykwinski, *Angew. Chem., Int. Ed.*, **42**, 2194 (2003).
2003ACE4685 S. Kirsch and T. Bach, *Angew. Chem., Int. Ed.*, **42**, 4685 (2003).
2003ARK106 S. Quideau, L. Pouysegu, D. Deffieux, A. Ozanne, J. Gagnepain, I. Fabre, and M. Oxoby, *Arkivoc*, **vi**, 106 (2003).
2003ARK170 P. Kiprof and V. Zhdankin, *Arkivoc*, **vi**, 170 (2003).
2003CL420 A. Kuboki, T. Yamamoto, and S. Ohira, *Chem. Lett.*, **32**, 420 (2003).
2003JOC9209 U. Radhakrishnan and P.J. Stang, *J. Org. Chem.*, **68**, 9209 (2003).
2003MI3 T. Wirth, editor: *Hypervalent Iodine Chemistry: Modern Developments in Organic Synthesis*, Top. Curr. Chem., **224**, (2003).
2003OL1583 V.V. Zhdankin, A.Y. Koposov, L.S. Su, V.V. Boyarskikh, B.C. Netzel, and V.G. Young, *Org. Lett.*, **5**, 1583 (2003).
2003OL2903 A. Ozanne, L. Pouysegu, D. Depernet, B. Francois, and S. Quideau, *Org. Lett.*, **5**, 2903 (2003).
2003OL3321 Z. Liu, Z.-C. Chen, and Q.-G. Zheng, *Org. Lett.*, **5**, 3321 (2003).
2003SL1307 J.D. More and N.S. Finney, *Synlett*, 1307 (2003).
2003TL1635 Z. Lei, C. Denecker, S. Jegasothy, D.C. Sherrington, N.K.H. Slater, and A.J. Sutherland, *Tetrahedron Lett.*, **44**, 1635 (2003).
2003TL9251 W.-J. Chung, D.-K. Kim, and Y.-S. Lee, *Tetrahedron Lett.*, **44**, 9251 (2003).
2004CC106 V.V. Zhdankin, D.N. Litvinov, A.Y. Koposov, T. Luu, M.J. Ferguson, R. McDonald, and R.R. Tykwinski, *Chem. Commun.*, 106 (2004).
2004HCA605 H. Ibrahim, F. Kleinbeck, and A. Togni, *Helv. Chim. Acta*, **87**, 605 (2004).
2004JA5192 K.C. Nicolaou, C.J.N. Mathison, and T. Montagnon, *J. Am. Chem. Soc.*, **126**, 5192 (2004).
2004TL2719 A.Y. Koposov, D.N. Litvinov, and V.V. Zhdankin, *Tetrahedron Lett.*, **45**, 2719 (2004).
2004TL7347 B. Das, H. Holla, G. Mahender, J. Banerjee, and M.R. Reddy, *Tetrahedron Lett.*, **45**, 7347 (2004).
2004TL8173 N. Kano, M. Ohashi, K. Hoshiba, and T. Kawashima, *Tetrahedron Lett.*, **45**, 8173 (2004).
2005ACE5592 K.C. Nicolaou and C.J.N. Mathison, *Angew. Chem., Int. Ed.*, **44**, 5992 (2005).
2005ACE7127 U. Ladziata, A.Y. Koposov, K.Y. Lo, J. Willging, V.N. Nemykin, and V.V. Zhdankin, *Angew. Chem., Int. Ed.*, **44**, 7127 (2005).
2005ARK19 P. Kiprof, *Arkivoc*, **iv**, 19 (2005).
2005ARK8 V. Zhdankin, R.N. Goncharenko, D.N. Litvinov, and A.Y. Koposov, *Arkivoc*, **iv**, 8 (2005).
2005COS121 V.V. Zhdankin, *Curr. Org. Synth.*, **2**, 121 (2005).
2005JA14146 J.T. Su and W.A. Goddard III, *J. Am. Chem. Soc.*, **127**, 14146 (2005).
2005JOC6484 V.V. Zhdankin, A.Y. Koposov, D.N. Litvinov, M.J. Ferguson, R. McDonald, T. Luu, and R.R. Tykwinski, *J. Org. Chem.*, **70**, 6484 (2005).
2005MD341 P. Lecarpentier, S. Crosignani, and B. Linclau, *Mol. Diversity*, **9**, 341 (2005).
2005NJC998 A.Y. Koposov, V.N. Nemykin, and V.V. Zhdankin, *New J. Chem.*, **29**, 998 (2005).

2005OL5841 Y. Huang, J. Zhang, and T.R.R. Pettus, *Org. Lett.*, **7**, 5841 (2005).
2005S22 A.Y. Koposov and V.V. Zhdankin, *Synthesis*, 22 (2005).
2005S279 D.-K. Kim, W.-J. Chung, and Y.-S. Lee, *Synlett*, 279 (2005).
2005TL5187 B.V. Meprathu, M.W. Justik, and J.D. Protasiewicz, *Tetrahedron Lett*, **46**, 5187 (2005).
2006ACE2929 M.J. Gallen, R. Goumont, T. Clark, F. Terrier, and C.M. Williams, *Angew. Chem., Int. Ed.*, **45**, 2929 (2006).
2006ACE3256 K.C. Nicolaou and S.T. Harrison, *Angew. Chem., Int. Ed.*, **45**, 3256 (2006).
2006ACE3495 T. Ngouansavanh and J. Zhu, *Angew. Chem., Int. Ed.*, **45**, 3495 (2006).
2006ACE7258 J. de Vicente, J.R. Huckins, and S.D. Rychnovsky, *Angew. Chem., Int. Ed.*, **45**, 7258 (2006).
2006ACE8203 M. Ochiai, T. Sueda, K. Miyamoto, P. Kiprof, and V.V. Zhdankin, *Angew. Chem., Int. Ed.*, **45**, 8203 (2006).
2006ARK26 U. Ladziata and V.V. Zhdankin, *Arkivoc*, **ix**, 26 (2006).
2006CC1442 D.C. Braddock, G. Cansell, S.A. Hermitage, and A.J.P. White, *Chem. Commun.*, 1442 (2006).
2006CC2863 J.E.D. Kirkham, V. Lee, and J.E. Baldwin, *Chem. Commun.*, 2863 (2006).
2006CEJ2579 P. Eisenberger, S. Gischig, and A. Togni, *Chem. Eur. J.*, **12**, 2579 (2006).
2006EJOC4791 A.Y. Koposov, D.N. Litvinov, V.V. Zhdankin, M.J. Ferguson, R. McDonald, and R.R. Tykwinski, *Eur. J. Org. Chem.*, 4791 (2006).
2006JOC8261 D.S. Bose and M. Idrees, *J. Org. Chem.*, **71**, 8261 (2006).
2006JOC8452 A.Y. Koposov, R.R. Karimov, I.M. Geraskin, V.N. Nemykin, and V.V. Zhdankin, *J. Org. Chem.*, **71**, 8452 (2006).
2006OL167 U. Ladziata, J. Willging, and V.V. Zhdankin, *Org. Lett.*, **8**, 167 (2006).
2006OL4605 K. Suzuki and H. Takayama, *Org. Lett.*, **8**, 4605 (2006).
2006RFP840 Z.Q. Lei, H.C. Ma, Z. Zhang, and Y.X. Yang, *React. Funct. Polym.*, **66**, 840 (2006).
2006TL5869 A. Ozanne-Beaudenon and S. Quideau, *Tetrahedron Lett.*, **47**, 5869 (2006).
2006TL6183 S. Hosokawa, S. Kuroda, K. Imamura, and K. Tatsuta, *Tetrahedron Lett.*, **47**, 6183 (2006).
2006TL6301 U. Ladziata, J. Carlson, and V.V. Zhdankin, *Tetrahedron Lett.*, **47**, 6301 (2006).
2007ACE1533 J. Gagnepain, F. Castet, and S. Quideau, *Ang. Chem. Int. Ed.*, **46**, 1533 (2007).
2007ACE5775 T. Ngouansavanh and J. Zhu, *Angew. Chem., Int. Ed.*, **46**, 5775 (2007).
2007ACE6529 R.D. Richardson, J.M. Zayed, S. Altermann, D. Smith, and T. Wirth, *Angew. Chem., Int. Ed.*, **46**, 6529 (2007).
2007ACE754 I. Kieltsch, P. Eisenberger, and A. Togni, *Angew. Chem., Int. Ed.*, **46**, 754 (2007).
2007ACE7633 G.E. Veitch, E. Beckmann, B.J. Burke, A. Boyer, C. Ayats, and S.V. Ley, *Angew. Chem., Int. Ed.*, **46**, 7633 (2007).
2007ACE8625 O. Iwamoto, H. Koshino, D. Hashizume, and K. Nagasawa, *Angew. Chem., Int. Ed.*, **46**, 8625 (2007).
2007BC1159 L. Serrander, V. Jaquet, K. Bedard, O. Plastre, O. Hartley, S. Arnaudeau, N. Demaurex, W. Schlegel, and K.-H. Krause, *Biochimie*, **89**, 1159 (2007).
2007CC1922 K.C. Nicolaou, Y. Tang, and J. Wang, *Chem. Commun.*, 1922 (2007).
2007CC3493 J.J. Chen and V. Aduda, *Synth. Commun.*, **37**, 3493 (2007).
2007CR663 F.J. Miller, M. Filali, G.J. Huss, B. Stanic, A. Chamseddine, T.J. Barna, and F.S. Lamb, *Circ. Res.*, **101**, 663 (2007).

2007CS1610 D.S. Hutchinson, R.I. Csikasz, D.L. Yamamoto, I.G. Shabalina, P. Wikstroem, M. Wilcke, and T. Bengtsson, *Cell. Signal.*, **19**, 1610 (2007).
2007JA14556 B.M. Trost, J. Waser, and A. Meyer, *J. Am. Chem. Soc.*, **129**, 14556 (2007).
2007JA8968 J.R. Scheerer, J.F. Lawrence, G.C. Wang, and D.A. Evans, *J. Am. Chem. Soc.*, **129**, 8968 (2007).
2007JFSN74 H.-R. Lee, H. Do, S.-R. Lee, E.-S. Sohn, S. Pyo, and E. Son, *J. Food Sci. Nutr.*, **12**, 74 (2007).
2007JN1205 S.W. Barger, M.E. Goodwin, M.M. Porter, and M.L. Beggs, *J. Neurochem.*, **101**, 1205 (2007).
2007JOC6280 N. Lebrasseur, J. Gagnepain, A. Ozanne-Beaudenon, J.-M. Leger, and S. Quideau, *J. Org. Chem.*, **72**, 6280 (2007).
2007JOC662 D.S. Bhalerao, U.S. Mahajan, K.H. Chaudhari, and K.G. Akamanchi, *J. Org. Chem.*, **72**, 662 (2007).
2007JOC8149 R.R. Karimov, Z.-G.M. Kazhkenov, M.J. Modjewski, E.M. Peterson, and V.V. Zhdankin, *J. Org. Chem.*, **72**, 8149 (2007).
2007NPLG348 C. Ho, P.-H. Lee, W.-J. Huang, Y.-C. Hsu, C.-L. Lin, and J.-Y. Wang, *Nephrology*, **12**, 348 (2007).
2007OBC767 J.N. Moorthy, N. Singhal, and K. Senapati, *Org. Biomol. Chem.*, **5**, 767 (2007).
2007OL2989 Y. Ichikawa, T. Yamaoka, K. Nakano, and H. Kotsuki, *Org. Lett.*, **9**, 2989 (2007).
2007PP890 H.W. Choi, Y.J. Kim, S.C. Lee, J.K. Hong, and B.K. Hwang, *Plant Physiol.*, **145**, 890 (2007).
2007S1679 H.C. Ma and X.Z. Jiang, *Synlett*, 1679 (2007).
2007S412 H.C. Ma and X.Z. Jiang, *Synthesis*, 412 (2007).
2007S527 U. Ladziata and V.V. Zhdankin, *Synlett*, 527 (2007).
2007S693 J.S. Yadav, B.V.S. Reddy, C.S. Reddy, and A.D. Krishna, *Synthesis*, 693 (2007).
2007SL527 U. Ladziata and V.V. Zhdankin, *Synlett*, 527 (2007).
2007T6990 J. Zhang, X. Wang, W. Wang, W. Quan, X. She, and X. Pan, *Tetrahedron*, **63**, 6990 (2007).
2007TL771 A. Kuboki, T. Yamamoto, M. Taira, T. Arishige, and S. Ohira, *Tetrahedron Lett.*, **48**, 771 (2007).
2007TL8343 R. Skouta and C.-J. Li, *Tetrahedron Lett.*, **48**, 8343 (2007).
20083779 J.S. Yadav, B.V.S. Reddy, and B.B.M. Krishna, *Synthesis*, 3779 (2008).
2008AJPG99 A. Masamune, T. Watanabe, K. Kikuta, K. Satoh, and T. Shimosegawa, *Am. J. Physiol.*, **294**, G99 (2008).
2008C260 I. Kieltsch, P. Eisenberger, K. Stanek, and A. Togni, *Chimia*, **62**, 260 (2008).
2008CC1575 P. Eisenberger, I. Kieltsch, N. Armanino, and A. Togni, *Chem. Commun.*, 1575 (2008).
2008CEJ9495 X. Du, H. Chen, and Y. Liu, *Chem. Eur. J.*, **14**, 9495 (2008).
2008CM2001 L. Bromberg, H. Zhang, and T.A. Hatton, *Chem. Mater.*, **20**, 2001 (2008).
2008CR5299 V.V. Zhdankin and P.J. Stang, *Chem. Rev.*, **108**, 5299 (2008).
2008HCA2035 S. Capone, I. Kieltsch, O. Flogel, G. Lelais, A. Togni, and D. Seebach, *Helv. Chim. Acta*, **91**, 2035 (2008).
2008JOC7678 K. Stanek, R. Koller, and A. Togni, *J. Org. Chem.*, **73**, 7678 (2008).
2008JPET50 J.-R. Chen, K. Shankar, S. Nagarajan, T.M. Badger, and M.J.J. Ronis, *J. Pharmacol. Exp. Ther.*, **324**, 50 (2008).

| | |
|---|---|
| 2008OBC2530 | T. Waters, J. Boulton, T. Clark, M.J. Gallen, C.M. Williams, and R.A.J. O'Hair, *Org. Biomol. Chem.*, **6**, 2530 (2008). |
| 2008S2045 | S. Chiampanichayakul, M. Pohmakotr, V. Reutrakul, T. Jaipetch, and C. Kuhakarn, *Synthesis*, 2045 (2008). |
| 2008S469 | J.S. Yadav, B.V.S. Reddy, A.P. Singh, and A.K. Basak, *Synthesis*, 469 (2008). |
| 2008SL3058 | L. Shen, J. Zhang, S. Cao, J. Yu, N. Liu, J. Wu, and X. Qian, *Synlett*, 3058 (2008). |
| 2008SL987 | U.S. Mahajan and K.G. Akamanchi, *Synlett*, 987 (2008). |
| 2008T2321 | Z. Kaluza, D. Mostowicz, G. Dolega, and R. Wojcik, *Tetrahedron*, **64**, 2321 (2008). |
| 2008TA209 | K. Venkatesan and K.V. Srinivasan, *Tetrahedron: Asymmetry*, **19**, 209 (2008). |
| 2008TL4378 | A. Kommreddy, M.S. Bowsher, M.R. Gunna, K. Botha, and T.K. Vinod, *Tetrahedron Lett.*, **49**, 4378 (2008). |
| 2008TL4443 | H.F. Anwar and T.V. Hansen, *Tetrahedron Lett.*, **49**, 4443 (2008). |
| 2008TL5006 | W. Zhang, J. Zhu, and J. Hu, *Tetrahedron Lett.*, **49**, 5006 (2008). |
| 2008TL5880 | J.S. Yadav, B.V. Subba Reddy, A.P. Singh, and A.K. Basak, *Tetrahedron Lett.*, **49**, 5880 (2008). |
| 2008TL7410 | I.M. Geraskin, M.W. Luedtke, H.M. Neu, V.N. Nemykin, and V.V. Zhdankin, *Tetrahedron Lett.*, **49**, 7410 (2008). |
| 2008TL80 | J.N. Moorthy, N. Singhal, and K. Senapati, *Tetrahedron Lett.*, **49**, 80 (2008). |
| 2009ABP995 | Z.-H. Hu, Y.-B. Shen, F.-Y. Shen, and X.-H. Su, *Acta Physiol. Plant.*, **31**, 995 (2009). |
| 2009ACE4332 | R. Koller, K. Stanek, D. Stolz, R. Aardoom, K. Niedermann, and A. Togni, *Angew. Chem., Int. Ed.*, **48**, 4332 (2009). |
| 2009ACE4605 | S. Quideau, G. Lyvinec, M. Marguerit, K. Bathany, A. Ozanne- Beaudenon, T. Buffeteau, D. Cavagnat, and A. Chenede, *Angew. Chem., Int. Ed.*, **48**, 4605 (2009). |
| 2009ACE9346 | J.P. Brand, J. Charpentier, and J. Waser, *Angew. Chem., Int. Ed.*, **48**, 9346 (2009). |
| 2009ARK1 | V.V. Zhdankin, *Arkivoc*, **i**, 1 (2009). |
| 2009ASC733 | I.M. Geraskin, O. Pavlova, H.M. Neu, M.S. Yusubov, V.N. Nemykin, and V.V. Zhdankin, *Adv. Synth. Catal.*, **351**, 733 (2009). |
| 2009BP493 | R.K. Sairam, D. Kumutha, K. Ezhilmathi, V. Chinnusamy, and R.C. Meena, *Biol. Plant.*, **53**, 493 (2009). |
| 2009CC5993 | R. Koller, Q. Huchet, P. Battaglia, J.M. Welch, and A. Togni, *Chem. Commun.*, 5993 (2009). |
| 2009CEJ10713 | A. Duschek and S.F. Kirsch, *Chem. Eur. J.*, **15**, 10713 (2009). |
| 2009IC4908 | V.N. Nemykin, A.Y. Koposov, B.C. Netzel, M.S. Yusubov, and V.V. Zhdankin, *Inorg. Chem.*, **48**, 4908 (2009). |
| 2009JA251 | M. Uyanik, M. Akakura, and K. Ishihara, *J. Am. Chem. Soc.*, **131**, 251 (2009). |
| 2009JOC6287 | J.N. Moorthy, K. Senapati, and S. Kumar, *J. Org. Chem.*, **74**, 6287 (2009). |
| 2009JOC8444 | A.K. Mailyan, I.M. Geraskin, V.N. Nemykin, and V.V. Zhdankin, *J. Org. Chem.*, **74**, 8444 (2009). |
| 2009M4669 | R. Bernini, M. Barontini, and C. Spatafora, *Molecules*, **14**, 4669 (2009). |
| 2009OL1221 | J.K. Boppisetti and V.B. Birman, *Org. Lett.*, **11**, 1221 (2009). |
| 2009OL3578 | A-u-H.A. Shah, Z.A. Khan, N. Choudhary, C. Loholter, S. Schafer, G.P.L. Marie, U. Farooq, B. Witulski, and T. Wirth, *Org. Lett.*, **11**, 3578 (2009). |

2009S1163 X.-Q. Li and C. Zhang, *Synthesis*, 1163 (2009).
2009S1370 F. Drouet, P. Fontaine, G. Masson, and J. Zhu, *Synthesis*, 1370 (2009).
2009S2818 J. Cvengros, D. Stolz, and A. Togni, *Synthesis*, 2818 (2009).
2009S929 C. Kuhakarn, W. Panchan, S. Chiampanichayakul, N. Samakkanad, M. Pohmakotr, V. Reutrakul, and T. Jaipetch, *Synthesis*, 929 (2009).
2009SC3521 J.-M. Chen and X.-M. Zeng, *Synth. Commun.*, **39**, 3521 (2009).
2009TL6072 M.W. Justik, J.D. Protasiewicz, and J.B. Updegraff, *Tetrahedron Lett.*, **50**, 6072 (2009).
2009TL6519 R. Bernini, M. Barontini, F. Crisante, M.C. Ginnasi, and R. Saladino, *Tetrahedron Lett.*, **50**, 6519 (2009).
2009TL715 L.D.S. Yadav and C. Awasthi, *Tetrahedron Lett.*, **50**, 715 (2009).
2009TL7395 A. Souto, J. Rodriguez, and C. Jimenez, *Tetrahedron Lett.*, **50**, 7395 (2009).
2009TOL180 V. Souza, M.d. C. Escobar, L. Bucio, E. Hernandez, L.E. Gomez-Quiroz, and M.C. Gutierrez Ruiz, *Toxicol. Lett.*, **187**, 180 (2009).
2010ACE4680 S. Nicolai, C. Piemontesi, and J. Waser, *Angew. Chem., Int. Ed.*, **50**, 4680 (2011).
2010ACE6147 M.A. McGowan, C.P. Stevenson, M.A. Schiffler, and E.N. Jacobsen, *Angew. Chem., Int. Ed.*, **49**, 6147 (2010).
2010ARK118 S.S. Deshmukh, S.N. Huddar, D.S. Bhalerao, and K.G. Akamanchi, *Arkivoc*, **ii**, 118 (2010).
2010CEJ9457 D. Fernandez Gonzalez, J.P. Brand, and J. Waser, *Chem. Eur. J.*, **16**, 9457 (2010).
2010CPB27 M. Tada, T. Ohkanda, and J. Kurabe, *Chem. Pharm. Bull.*, **58**, 27 (2010).
2010EJOC900 J. Neumann and J. Thiem, *Eur. J. Org. Chem.*, 900 (2010).
2010FS2437 E. Miraglia, C. Lussiana, D. Viarisio, C. Racca, A. Cipriani, E. Gazzano, A. Bosia, A. Revelli, and D. Ghigo, *Fert. Ster.*, **93**, 2437 (2010).
2010JA4986 A.E. Allen and D.W.C. MacMillan, *J. Am. Chem. Soc.*, **132**, 4986 (2010).
2010JA5966 B. Bradshaw, G. Etxebarria-Jardi, and J. Bonjoch, *J. Am. Chem. Soc.*, **132**, 5966 (2010).
2010JA7540 K.C. Nicolaou, Q. Kang, T.R. Wu, C.S. Lim, and D.Y.K. Chen, *J. Am. Chem. Soc.*, **132**, 7540 (2010).
2010JFC951 M.S. Wiehn, E.V. Vinogradova, and A. Togni, *J. Fluorine Chem.*, **131**, 951 (2010).
2010JOC1779 S. Fantasia, J.M. Welch, and A. Togni, *J. Org. Chem.*, **75**, 1779 (2010).
2010JOC5601 S. Hanessian, T. Focken, X. Mi, R. Oza, B. Chen, D. Ritson, and R. Beaudegnies, *J. Org. Chem.*, **75**, 5601 (2010).
2010JOC8416 J.N. Moorthy, K. Senapati, and K.N. Parida, *J. Org. Chem.*, **75**, 8416 (2010).
2010OL384 S. Nicolai, S. Erard, D.F. Gonzalez, and J. Waser, *Org. Lett.*, **12**, 384 (2010).
2010OL4086 J.D. Panarese and S.P. Waters, *Org. Lett.*, **12**, 4086 (2010).
2010OM1771 N. Armanino, R. Koller, and A. Togni, *Organometallics*, **29**, 1771 (2010).
2010PNAS3030 F.L. Van De Veerdonk, S.P. Smeekens, L.A.B. Joosten, B.J. Kullberg, C.A. Dinarello, J.W.M. Van Der Meer, and M.G. Netea, *Proc. Natl. Acad. Sci. U.S.A.*, **107**, 3030 (2010).
2010T2235 L. Pouysegu, D. Deffieux, and S. Quideau, *Tetrahedron*, **66**, 2235 (2010).
2010T5753 K. Niedermann, J.M. Welch, R. Koller, J. Cvengros, N. Santschi, P. Battaglia, and A. Togni, *Tetrahedron*, **66**, 5753 (2010).
2010T5768 B.V. Meprathu and J.D. Protasiewicz, *Tetrahedron*, **66**, 5768 (2010).
2010T6047 M. Barontini, R. Bernini, F. Crisante, and G. Fabrizi, *Tetrahedron*, **66**, 6047 (2010).

2010T6514 L.E. Overman and M.D. Rosen, *Tetrahedron*, **66**, 6514 (2010).
2010T6569 Y. Shimizu, S.-L. Shi, H. Usuda, M. Kanai, and M. Shibasaki, *Tetrahedron*, **66**, 6569 (2010).
2010T7659 V. Satam, A. Harad, R. Rajule, and H. Pati, *Tetrahedron*, **66**, 7659 (2010).
2010TL1863 T. Andreou, J. Bures, and J. Vilarrasa, *Tetrahedron Lett.*, **51**, 1863 (2010).
2011ACE1059 K. Niedermann, N. Frueh, E. Vinogradova, M.S. Wiehn, A. Moreno, and A. Togni, *Angew. Chem., Int. Ed.*, **50**, 1059 (2011).
2011AGE1524 A. Duschek and S.F. Kirsch, *Angew. Chem. Int. Ed.*, **50**, 1524 (2011).
2011ARK370 M.S. Yusubov, A.V. Maskaev, and V.V. Zhdankin, *Arkivoc*, **i**, 370 (2011).
2011CC102 J.P. Brand, D.F. Gonzalez, S. Nicolai, and J. Waser, *Chem. Commun.*, **47**, 102 (2011).
2011CC1875 T. Miura, K. Nakashima, N. Tada, and A. Itoh, *Chem. Commun.*, **47**, 1875 (2011).
2011CEJ10538 A. Yoshimura, V.N. Nemykin, and V.V. Zhdankin, *Chem. Eur. J.*, **17**, 10538 (2011).
2011EJOC3690 E.A. Merritt and B. Olofsson, *Eur. J. Org. Chem.*, 3690 (2011).
2011HMR619 Y. Ishibashi, T. Matsui, M. Takeuchi, and S. Yamagishi, *Horm. Metab. Res.*, **43**, 619 (2011).
2011IC11263 V.N. Nemykin, A.V. Maskaev, M.R. Geraskina, M.S. Yusubov, and V.V. Zhdankin, *Inorg. Chem.*, **50**, 11263 (2011).
2011JOC1185 V.V. Zhdankin, *J. Org. Chem.*, **76**, 1185 (2011).
2011JOC4189 N. Santschi and A. Togni, *J. Org. Chem.*, **76**, 4189 (2011).
2011JOC565 J.P. Brand, C. Chevalley, and J. Waser, *Beilstein J. Org. Chem.*, **7**, 565 (2011).
2011JOC9593 J.N. Moorthy, K. Senapati, K.N. Parida, S. Jhulki, K. Sooraj, and N.N. Nair, *J. Org. Chem.*, **76**, 9593 (2011).
2011MGM241 L. Lu, X. Gu, D. Li, L. Liang, Z. Zhao, and J. Gao, *Mol. Genet. Metab.*, **104**, 241 (2011).
2011NPR1722 J.L.F. Silva and B. Olofsson, *Nat. Prod. Rep.*, **28**, 1722 (2011).
2011OL6488 L.-Q. Cui, Z.-L. Dong, K. Liu, and C. Zhang, *Org. Lett.*, **13**, 6488 (2011).
2011S517 E.A. Merritt and B. Olofsson, *Synthesis*, 517 (2011).
2011WASJ67 M. Pakizeh, A.A. Kuliev, Z.M. Mammadov, M. Ardjmand, and M. Hasani, *World Appl. Sci. J.*, **14**, 67 (2011).
2012ACE11092 Z. He, P. Trinchera, S. Adachi, J.D. St Denis, and A.K. Yudin, *Angew. Chem., Int. Ed.*, **51**, 11092 (2012).
2012ACE2749 I. Paterson, P. Maltas, S.M. Dalby, J.H. Lim, and E.A. Anderson, *Angew. Chem., Int. Ed.*, **51**, 2749 (2012).
2012ACE3401 W. Tang and E.V. Prusov, *Angew. Chem., Int. Ed.*, **51**, 3401 (2012).
2012ACE3405 F. Glaus and K.-H. Altmann, *Angew. Chem., Int. Ed.*, **51**, 3405 (2012).
2012ACE5006 W. Ye, D.M. Ho, S. Friedle, T.D. Palluccio, and E.V. Rybak- Akimova, *Inorg. Chem.*, **51**, 5006 (2012).
2012ACE6511 K. Niedermann, N. Frueh, R. Senn, B. Czarniecki, R. Verel, and A. Togni, *Angew. Chem., Int. Ed.*, **51**, 6511 (2012).
2012ACE8221 Y. Li and A. Studer, *Angew. Chem., Int. Ed.*, **51**, 8221 (2012).
2012ACSC521 E. Mejia and A. Togni, *ACS Catal.*, **2**, 521 (2012).
2012AX(E)o1238 C.Y. Legault and J. Prevost, *Acta Crystallogr., Sect. E: Struct. Rep.*, **E68**, o1238 (2012).
2012BBRC329 S. Hino, A. Kito, R. Yokoshima, R. Sugino, K. Oshima, T. Morita, T. Okajima, D. Nadano, K. Uchida, and T. Matsuda, *Biochem. Biophys. Res. Commun.*, **421**, 329 (2012).

2012BJOC1523 A. Schroeckeneder, D. Stichnoth, P. Mayer, and D. Trauner, *Beilstein J. Org. Chem.*, **8**, 1523 (2012).

2012BP422 H. Gong, G. Chen, F. Li, X. Wang, Y. Hu, and Y. Bi, *Biol. Plant.*, **56**, 422 (2012).

2012CC11629 T. Neubauer, C. Kammerer-Pentier, and T. Bach, *Chem. Commun.*, **48**, 11629 (2012).

2012CEJ5655 J.P. Brand, C. Chevalley, R. Scopelliti, and J. Waser, *Chem. Eur. J.*, **18**, 5655 (2012).

2012COS247 M.S. Yusubov and V.V. Zhdankin, *Curr. Org. Synth.*, **9**, 247 (2012).

2012CSR4165 J.P. Brand and J. Waser, *Chem. Soc. Rev.*, **41**, 4165 (2012).

2012IC967 P.K. Sajith and C.H. Suresh, *Inorg. Chem.*, **51**, 967 (2012).

2012JA11984 H. Fuwa, K. Ishigai, K. Hashizume, and M. Sasaki, *J. Am. Chem. Soc.*, **134**, 11984 (2012).

2012JA18860 F. Peng and S.J. Danishefsky, *J. Am. Chem. Soc.*, **134**, 18860 (2012).

2012JA893 A. Yamamoto, A. Ueda, P. Bremond, P.S. Tiseni, and Y. Kishi, *J. Am. Chem. Soc.*, **134**, 893 (2012).

2012JA920 J. Deng, B. Zhu, Z. Lu, H. Yu, and A. Li, *J. Am. Chem. Soc.*, **134**, 920 (2012).

2012JCP1347 K.-H. Tsai, W.-J. Wang, C.-W. Lin, P. Pai, T.-Y. Lai, C.-Y. Tsai, and W.-W. Kuo, *J. Cell. Physiol.*, **227**, 1347 (2012).

2012JFC83 N. Santschi, P. Geissbuehler, and A. Togni, *J. Fluorine Chem.*, **135**, 83 (2012).

2012JFC99 Y. Tanaka, T. Ishihara, and T. Konno, *J. Fluorine Chem.*, **137**, 99 (2012).

2012JOC11109 J.S. Yadav, S.K. Das, and G. Sabitha, *J. Org. Chem.*, **77**, 11109 (2012).

2012JOC9628 J.S. Yadav, S. Dhara, S.S. Hossain, and D.K. Mohapatra, *J. Org. Chem.*, **77**, 9628 (2012).

2012JPET873 T.W. Moody, N. Osefo, B. Nuche-Berenguer, L. Ridnour, D. Wink, and R.T. Jensen, *J. Pharmacol. Exp. Ther.*, **341**, 873 (2012).

2012MC129 I.M. Kumanyaev, M.A. Lapitskaya, L.L. Vasiljeva, and K.K. Pivnitsky, *Mendeleev Commun.*, **22**, 129 (2012).

2012OL3020 J. Tian, W.-C. Gao, D.-M. Zhou, and C. Zhang, *Org. Lett.*, **14**, 3020 (2012).

2012OL744 J.P. Brand and J. Waser, *Org. Lett.*, **14**, 744 (2012).

2012PLoSe33817 J. Liu, J. Zhou, and D. Xing, *PLoS One*, **7**, e33817 (2012).

2012S1155 J.P. Brand and J. Waser, *Synthesis*, **44**, 1155 (2012).

2012SC2617 S.V. Gamapwar, N.P. Tale, and N.N. Karade, *Synth. Commun.*, **42**, 2617 (2012).

2012TL6990 A.R. Rodriguez and B.W. Spur, *Tetrahedron Lett.*, **53**, 6990 (2012).

2013ACE3457 X. Shao, X. Wang, T. Yang, L. Lu, and Q. Shen, *Angew. Chem., Int. Ed.*, **52**, 3457 (2013).

2013BJOC1763 Y. Li and J. Waser, *Beilstein J. Org. Chem.*, **9**, 1763 (2013).

2013CC11269 M.S. Yusubov, D.Y. Svitich, A. Yoshimura, V.N. Nemykin, and V.V. Zhdankin, *Chem. Commun.*, **49**, 11269 (2013).

2013CC3303 T. Aubineau and J. Cossy, *Chem. Commun.*, **49**, 3303 (2013).

2013CC7926 N.R. Gade, V. Devendram, M. Pal, and J. Iqbal, *Chem. Commun.*, **49**, 7926 (2013).

2013JA5356 Q.-H. Deng, T. Bleith, H. Wadepohl, and L.H. Gade, *J. Am. Chem. Soc.*, **135**, 5356 (2013).

2013JA9620 R. Frei and J. Waser, *J. Am. Chem. Soc.*, **135**, 9620 (2013).

2013JOC3767 M.S. Yusubov, R.Y. Yusubova, V.N. Nemykin, and V.V. Zhdankin, *J. Org. Chem.*, **78**, 3767 (2013).

2013JOC9 S.M. Canham, D.J. France, and L.E. Overman, *J. Org. Chem.*, **78**, 9 (2013).

2013MI2 V.V. Zhdankin, *Hypervalent Iodine Chemistry: Preparation, Structure, and Synthetic Applications of Polyvalent Iodine Compounds*, Wiley: Chichester (UK) (2013).

2013OL112 G.L. Tolnai, S. Ganss, J.P. Brand, and J. Waser, *Org. Lett.*, **15**, 112 (2013).

2013OL3246 M.V. Vita and J. Waser, *Org. Lett.*, **15**, 3246 (2013).

2013SL424 M. Brown, U. Farid, and T. Wirth, *Synlett*, **24**, 424 (2013).

2014ACE3125 E.V. Vinogradova, P. Mueller, and S.L. Buchwald, *Angew. Chem., Int. Ed.*, **53**, 3125 (2014).

2014ACE6533 H. Lei, J. Yan, J. Yu, Y. Liu, Z. Wang, Z. Xu, and T. Ye, *Angew. Chem., Int. Ed.*, **53**, 6533 (2014).

2014AGE9617 A.S. Ivanov, I.A. Popov, A.I. Boldyrev, and V.V. Zhdankin, *Angew. Chem., Int. Ed.*, **53**, 9617 (2014).

2014ARK109 C.A. Ramsden, *Arkivoc*, **i**, 109 (2014).

2014ASC3600 N.J. Victor and K.M. Muraleedharan, *Adv. Synth. Catal.*, **356**, 3600 (2014).

2014C516 F. de Nanteuil, Y. Li, M.V. Vita, R. Frei, E. Serrano, S. Racine, and J. Waser, *Chimia*, **68**, 516 (2014).

2014C624 H.P. Luthi and A. Togni, *Chimia*, **68**, 624 (2014).

2014CAJ950 F.V. Singh and T. Wirth, *Chem. Asian J.*, **9**, 950 (2014).

2014CAJ972 R.M. Romero, T.H. Woeste, and K. Muniz, *Chem. Asian J.*, **9**, 972 (2014).

2014CC12923 C.C. Chen and J. Waser, *Chem. Commun.*, **50**, 12923 (2014).

2014CCR54 V.V. Zhdankin and J.D. Protasiewicz, *Coord. Chem. Rev*, **275**, 54 (2014).

2014EJOC3087 V. Matousek, E. Pietrasiak, L. Sigrist, B. Czarniecki, and A. Togni, *Eur. J. Org. Chem.*, **2014**, 3087 (2014).

2014JA16563 R. Frei, M.D. Wodrich, D.P. Hari, P.-A. Borin, C. Chauvier, and J. Waser, *J. Am. Chem. Soc.*, **136**, 16563 (2014).

2014JA17738 P. Lu, A. Mailyan, Z. Gu, D.M. Guptill, H. Wang, H.M.L. Davies, and A. Zakarian, *J. Am. Chem. Soc.*, **136**, 17738 (2014).

2014JA4309 E.Z. Oblak, M.D. VanHeyst, J. Li, A.J. Wiemer, and D.L. Wright, *J. Am. Chem. Soc.*, **136**, 4309 (2014).

2014JA870 A. Shvartsbart and A.B. Smith III, *J. Am. Chem. Soc.*, **136**, 870 (2014).

2014JOC3410 D.F. Taber and C.M. Paquette, *J. Org. Chem.*, **79**, 3410 (2014).

2014JOC9922 K. Tanaka, H. Matsuyama, M. Watanabe, Y. Fujimori, K. Ishibashi, T. Ozawa, T. Sato, Y. Saikawa, and M. Nakata, *J. Org. Chem.*, **79**, 9922 (2014).

2014MI1 T. Kaiho, editor: *Iodine Chemistry and Applications*, Wiley: Chichester (UK) (2014).

2014OL496 S. He, W. Yang, L. Zhu, G. Du, and C.-S. Lee, *Org. Lett.*, **16**, 496 (2014).

2014SCI219 Z. Ma, X. Wang, X. Wang, R.A. Rodriguez, C.E. Moore, S. Gao, X. Tan, Y. Ma, A.L. Rheingold, P.S. Baran, and C. Chen, *Science*, **346**, 219 (2014).

2015CR650 J. Charpentier, N. Fruh, and A. Togni, *Chem. Rev.*, **115**, 650 (2015).

2015NAT600 A. Sharma and J.F. Hartwig, *Nature*, **517**, 600 (2015).

2015OBC365 Y.-F. Wang, J. Qiu, D. Kong, Y. Gao, F. Lu, P.G. Karmakera, and F.-X. Chen, *Org. Biomol. Chem.*, **13**, 365 (2015).

2015TL146 S. Bera, S.K. Das, T. Saha, and G. Panda, *Tetrahedron Lett.*, **56**, 146 (2015).

CHAPTER TWO

# Recent Advances in the Synthesis of Heterocyclic Compounds Using Flash Vacuum Pyrolysis[a]

**R. Alan Aitken*, Youcef Boubalouta**
EaStCHEM School of Chemistry, University of St. Andrews, St. Andrews, Fife, UK
*Corresponding author: E-mail: raa@st-andrews.ac.uk

## Contents



[a] Dedicated to the memory of two major figures in this field of research: Gloria Inés Yranzo (1957–2008) and Hamish McNab (1949–2010).

*Advances in Heterocyclic Chemistry*, Volume 115
ISSN 0065-2725
http://dx.doi.org/10.1016/bs.aihch.2015.03.002

## Abstract

Applications of flash vacuum pyrolysis to the synthesis of heterocyclic compounds published between 2000 and 2015 are described in six major sections with 150 literature references. The emphasis is on methods suitable for preparation of stable heterocyclic products in multi-gram quantities. Consideration of the relatively few reports of synthesis of three- and four-membered rings is followed by the largest single section on the synthesis of five-membered rings, mainly those containing one or two nitrogen atoms. A further major section discusses synthesis of six-membered ring products, again predominantly containing nitrogen, before the final shorter sections on seven-membered, eight-membered, and larger rings.

**Keywords:** Azepines; Benzimidazoles; Benzofurans; Benzopyranones; Benzothiazoles; Benzothiophenes; Benzoxazoles; Benzoxocanes; Dioxepanes; Dioxocanes; Dioxolanes; Dithiiranes; Dithioles; Furans; Imidazoles; Indazoles; Indoles; Indolines; Indolizines; Isoquinolines; Isoxazoles; Isoxazolines; Oxazines; Oxazoles; Oxepanes; Oxiranes; Pyrazines; Pyrazoles; Pyridazines; Pyridines; Pyrimidines; Pyrroles; Pyrrolidines; Pyrrolines; Pyrrolizines; Quinolines; Thietanes; Thiophenes; Triazoles

## 1. INTRODUCTION

The technique of flash vacuum pyrolysis (FVP) is now well established as a clean and selective method for bringing about unimolecular thermal reactions. The apparatus required is relatively simple and inexpensive to construct (2014AJC1150), and its use is now spreading beyond groups specializing in the technique and finding occasional routine use in target-directed synthesis. In this review, we have aimed to describe all studies involving the use of FVP in the synthesis of heterocyclic compounds published since 2000. Coverage is restricted to work in which the heterocyclic products are stable enough to be isolated and characterized, and studies involving matrix isolation and spectroscopic detection of heterocyclic reactive intermediates are excluded. The material is arranged according to the heterocyclic ring system present in the product in order of increasing ring size, increasing number of heteroatoms, and decreasing seniority of ring systems.

There have been recent reviews on pyrolysis of isoxazoles and pyrazoles (2004COC1071), and use of FVP in general including heterocyclic synthesis (2004MI19). There are several reviews of earlier work on FVP including many heterocyclic synthesis methods (1980MI1, 1982RTC317, 1982RTC 365, 1984MI31, 1996COS373, 1997MI1). More specific reviews are also available on intramolecular reactions of carbenes and nitrenes in heterocyclic chemistry (1981AHC(28)231), extrusion of $SO_2$ from cyclic systems (1992PHC1, 1993PHC1), pyrolysis involving compounds with C=C, C=N, and C=O double bonds (1997MI473), retro- Diels–Alder reactions including their use in the synthesis of heterocycles (1999CR1163), and applications of pyrolysis of Meldrum's acid derivatives (2001S2059).

## 2. THREE-MEMBERED RINGS

### 2.1 Azirines

The two isomeric spiro azirines **1** and **2** were synthesized and separated chromatographically but FVP of either compound at 400 °C produced a 1:1 mixture of the two (Scheme 1) with the interconversion postulated to take place by means of a vinylnitrene intermediate (2011CEJ10071).

### 2.2 Oxiranes

An unusual example where an oxirane ring is actually formed under FVP conditions involves the $\lambda^5$-antimony compound **3** which undergoes reductive elimination upon FVP at 500 °C to give the bis(trifluoromethyl)oxirane

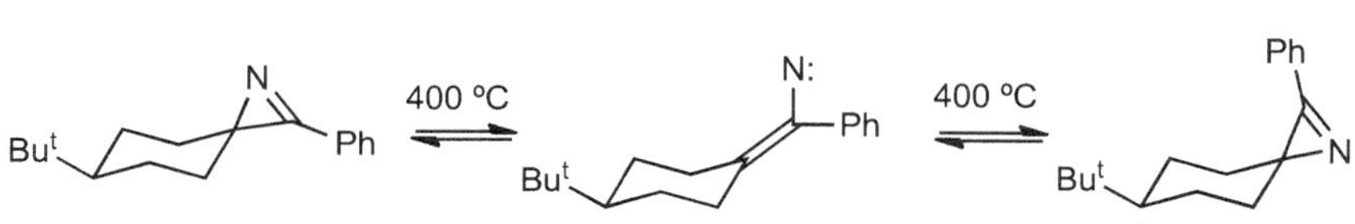


(1) (2)

Scheme 1

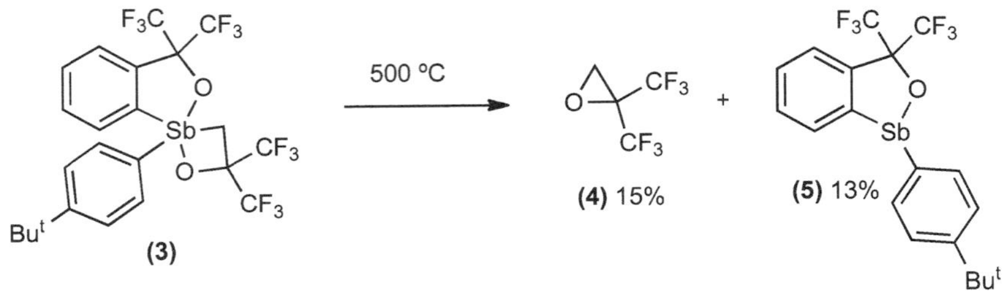


Scheme 2

**4** and the $\lambda^3$-benzoxastibole **5** as the main products (Scheme 2; 2006JOC659). This is in contrast to an analogous phosphorane which gives the alkene and cyclic phosphinate.

Several other studies in which oxiranes are synthesized involve a retro-Diels–Alder reaction conducted by FVP as the final deprotection step in routes to otherwise difficultly accessible multi-functional compounds (Scheme 3). The FVP of compound **6** at 400 °C gives the cyclopentadienone monoepoxide **7** by loss of cyclopentadiene but only as a minor product accompanied by a larger amount of the ring-expanded isomer, (pyranone **8**), which is the sole product at 500 °C (2000TL9189). A sequence of transformations starting from the cyclopentadiene–benzoquinone Diels–Alder adduct gives epoxide **9** which is deprotected using FVP at 600 °C to give the racemic epoxycyclohexenol **10** in a good yield (2004JOC4477). By starting from the enantiomerically pure diene **11**, differentiation between the two diastereotopic double bonds of 4-allyl-4-fluorocyclohexa-1,5-dien-4-one was achieved giving, after final deprotection of the adduct **12**, the chiral epoxycyclohexenone **13** (2004ARK(ii)4).

## 2.3 Diazirinone

The FVP of carbonyl azide **14** at 400 °C gave a yellow solid shown by IR (infrared) spectroscopy, including isotope shifts, to be the diazirinone **15** (Scheme 4). Despite its high energy, this was unexpectedly stable with a half-life of over an hour in the gas phase at room temperature, decomposing to $N_2$ and CO (2011AGE1720). For safety reasons this preparation was only conducted on a very small scale.

## 2.4 Dithiiranes

Pyrolytic decomposition of 1,2,4-trithiolanes results in cycloreversion to give a thiocarbonyl compound and a thiocarbonyl *S*-sulfide or thiosulfine

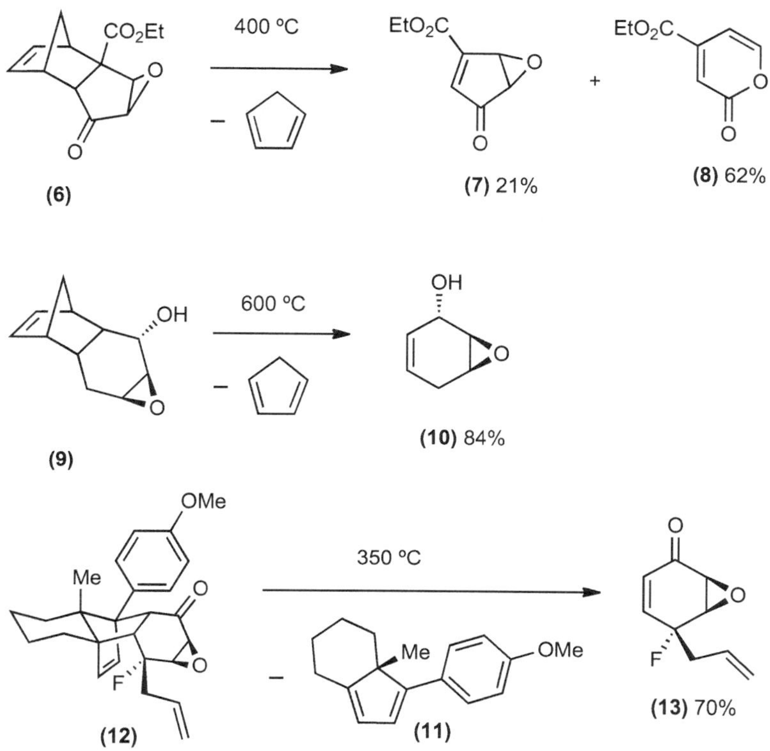

Scheme 3

(Scheme 5). The latter may undergo electrocyclic ring closure to give a dithiirane with the ease of this process depending on the substituents present. In the case of the parent compound **16**, fragmentation with formation of thioformaldehyde and thiosulfine **17** only occurs at high temperatures such as 950 °C, and under these conditions a third of the thiosulfine **17** cyclizes to the dithiirane **18**, detected by matrix IR spectroscopy (2001AGE393). The methyl and dimethyl thiosulfines are similarly formed by FVP of the corresponding trithiolanes at 700 °C but cyclization to the

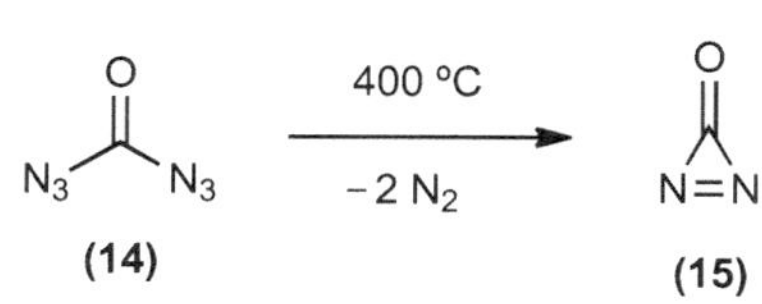

Scheme 4

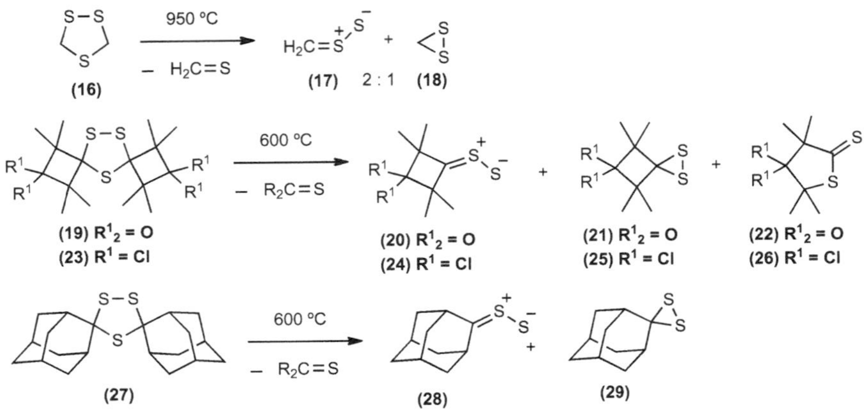


Scheme 5

dithiirane is insignificant, except upon photolysis and other isomerization processes intervene (2006EJO3721). In more sterically hindered cases, the dithiiranes are more readily formed and might possibly be isolated, although studies to date have relied on argon matrix IR spectroscopy to prove their presence. Thus, compound **19** loses the thione at 600 °C to give mainly the dithiirane **21**, formed from **20**, along with its rearrangement product, the dithiolactone **22** (2008EJO2998). For the chlorinated analog **23**, the main product is dithiirane **25** which only isomerizes to the dithiolactone **26** photochemically, together with a little of the thiosulfine **24**. In the adamantane series, **27** reacts to give a little **28** and mainly the dithiirane **29**, which does not isomerize even on photolysis (2008EJO2998).

## 3. FOUR-MEMBERED RINGS

### 3.1 Thietes

A range of 4*H*-benzo[*d*][1,3]oxathiin-2-ones **30** lose $CO_2$ to afford the corresponding benzothietes **31** in good yields upon FVP at 350 °C (Scheme 6; 2008CJC119). Pyridothietone **33** is formed as a stable liquid in 41% yield by FVP of 2-mercaptonicotinic acid **32** at 550 °C along with the cyclic trimer **34** (2002TL5285). Perhaps unexpectedly the benzo[*d*][1,3]oxathiin-4-thione **35** undergoes initial isomerization to **36** followed by loss of thioformaldehyde to generate benzothietone **37**, which upon addition of methanol is trapped as methyl 2-mercaptobenzoate (2006HCA991, 2008HCA766).

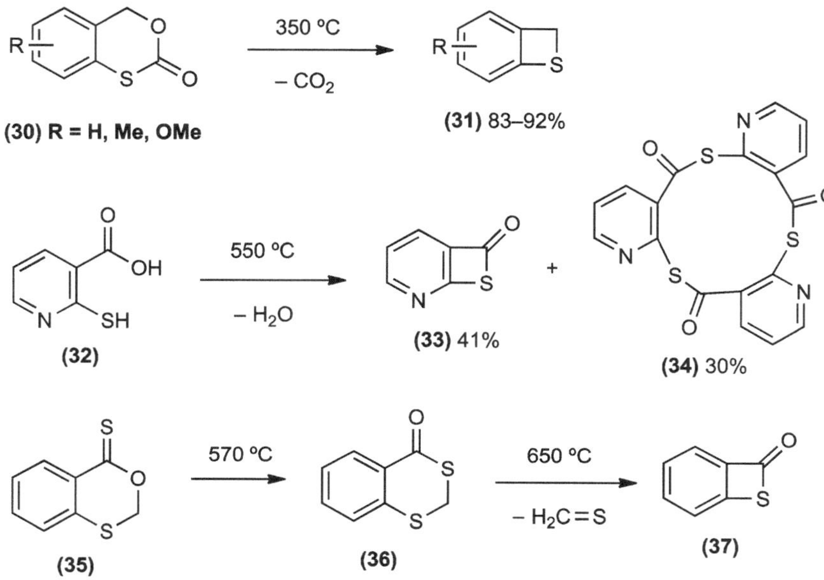


Scheme 6

## 3.2 Dithietes

Both the 1,3-benzodithiin-4-thione **38** and the analogous dihydrobenzothiazinethione **39** generate the same intermediate **40** upon FVP at, respectively, 1000 and 850 °C (Scheme 7). This disproportionates by a complex mechanism to afford the phenanthrodithiete **41** and the 1,2-benzodithiole-3-thione **42** (2009PS(184)1269, 2006HCA991). Compound **41** is a stable, high-melting solid isolated in 67% yield from **38** and in 70% yield from **39**.

# 4. FIVE-MEMBERED RINGS

## 4.1 Pyrroles, Pyrrolidines, and Ring-Fused Derivatives

The convenient formation of nitriles by dehydration of oximes using FVP over molecular sieves has been applied to the synthesis of pyrrole-3-carbonitrile **43** and indole-3-carbonitrile **44** (Scheme 8; 2015UP9).

Further examples in the well-established use of isoxazolones as precursors for substituted pyrrole synthesis include pyrolysis of the esters **45** to give the pyrroles **46** (Scheme 9; 2001ARK(vii)88).

Use of the retro-Diels–Alder reaction in the synthesis of chiral pyrrolidine and pyrrolinone products has been described by several groups. The

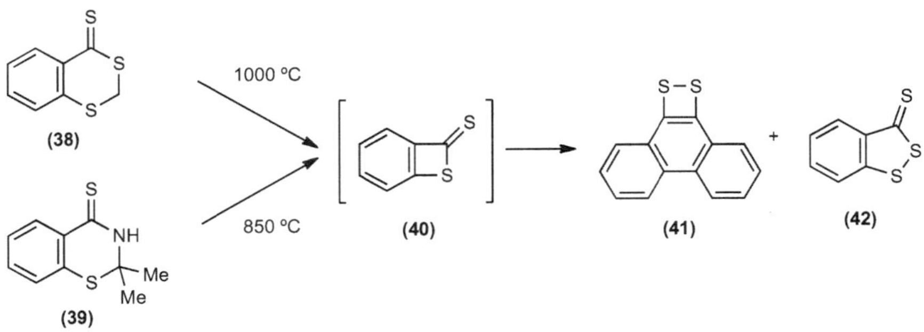


Scheme 7

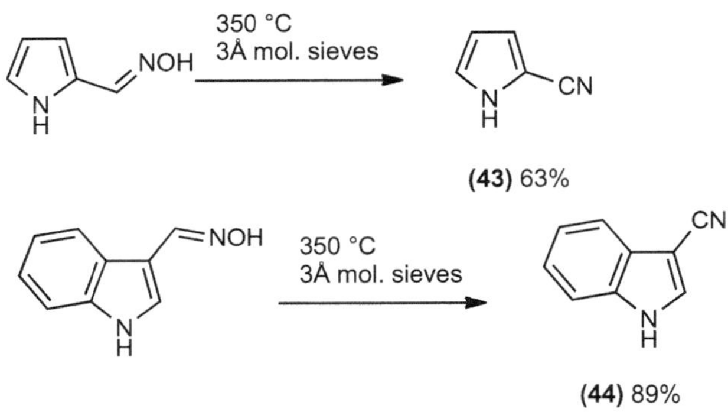


Scheme 8

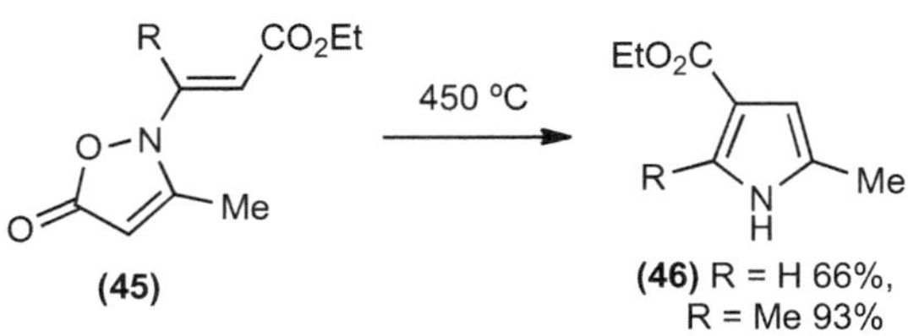


Scheme 9

tricyclic lactams **48** are obtained in good yields and without racemization by pyrolytic deprotection of the cyclopentadiene adducts **47** (Scheme 10; 2002SL593).

The cyclopentadiene **11** has been used as a chiral template to allow multi-step construction of the fluoropyrrolidine function in **49** with the final deprotection by FVP to give the hexahydroindolone product **50** in a quantitative yield (Scheme 11; 2004ARK(ii)4).

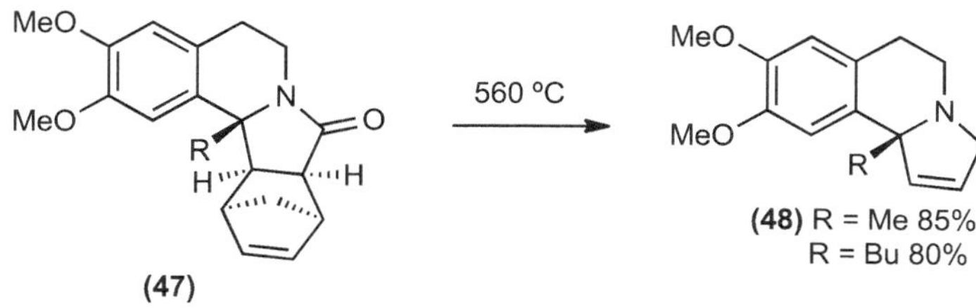

Scheme 10

Various chiral anthracenes have been used as templates to manipulate maleimides with a final retro-Diels—Alder step using FVP giving chiral pyrrolinone products, and this approach has been reviewed (2005BMC5299). Typical examples are provided by the synthesis of products **51** and **52** using dimethoxyethylathracene (2005OL31) and product **53** using α-methylbenzylaminoathracene (2005TL2475; Scheme 12).

More recently the α-methoxyethylanthracene system has been used to prepare chiral pyrrolidinone esters **54** (2010TA2719) and pyrrolinones **55** (2011OBC4353; Scheme 13).

The chemistry of azafulvenium methylides generated by extrusion of $SO_2$ from cyclic sulfone precursors has been extensively investigated. Depending upon the substituents present and the FVP temperature, the products can be either polysubstituted pyrroles such as **56** (2001JCS(P1)1795), **58** (2013T3646), **59** (2014EJO2933), and **60** (2001JCS(P1)1795), or cyclization products such as the pyrrolizinones **57** and **61** (2004TL3889, 2005JOC6629, 2007T1833, 2008T9745) or the cycloheptatrienone-fused pyrrole **62** (2004TL3889, 2005JOC6629; Scheme 14).

A range of substituted cinnamic acid amides are found to undergo cyclization under high-temperature FVP conditions to afford the corresponding pyrrolidinones. The reaction was initially described for *N,N*-dimethyl amides

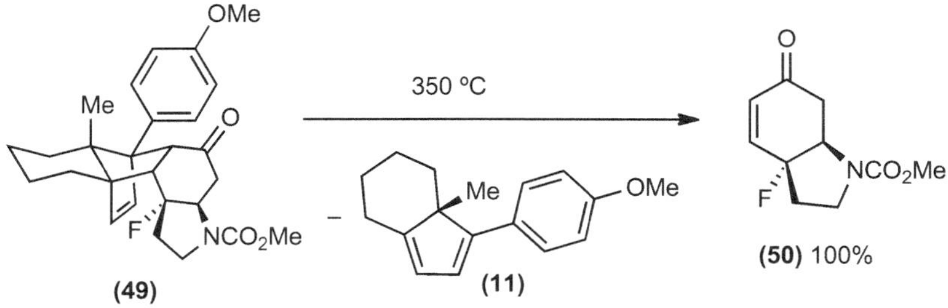

Scheme 11

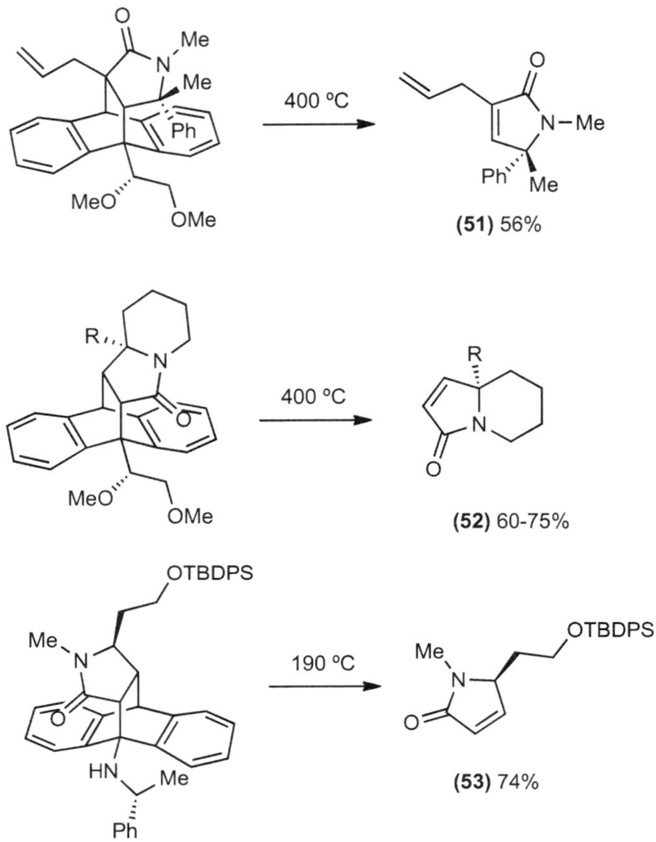

Scheme 12

**63** to give **64** with the corresponding *N*-ethyl and *N*-benzyl amides giving products **65** (R = H or Et or Ph, i.e., **65** is formed not only from NHEt but also from $NEt_2$ and NPhEt amides) and **66**, respectively, as 2:1 mixtures of *trans*- and *cis*-isomers (Scheme 15; 2005TL3093). This reaction was later extended to the cinnamic amides **67** derived from piperidine, morpholine, and thiomorpholine which gave bicyclic lactams **68** (2009T6364).

Pyrolysis of carbonyl-stabilized phosphonium ylides has been used to gain access to a variety of pyrrolidinone and carbazole products. The proline-derived ylides **69** undergo a loss of $Ph_3PO$ to give the pyrrolidinyl acetylenic esters **70** without racemization (Scheme 16; 2002JCS(P1)533). On the other hand, the amino acid-derived ylides **71** with a free amino group undergo cyclization with loss of ethanol to give the stable cyclic ylide

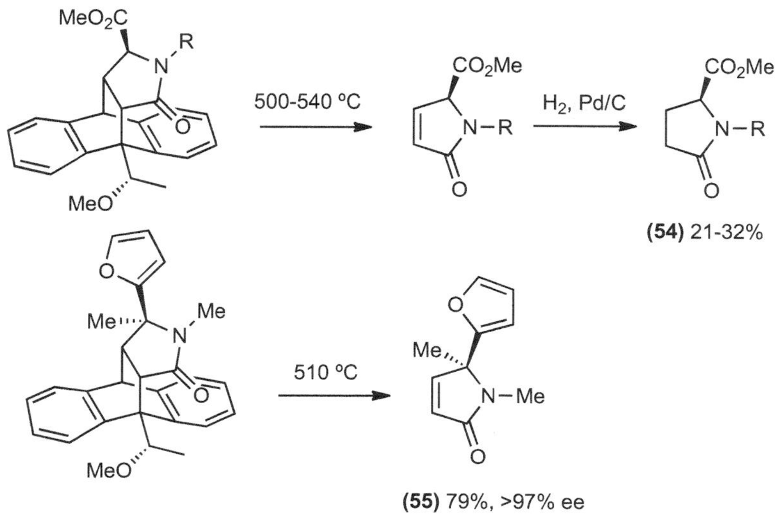


**Scheme 13**

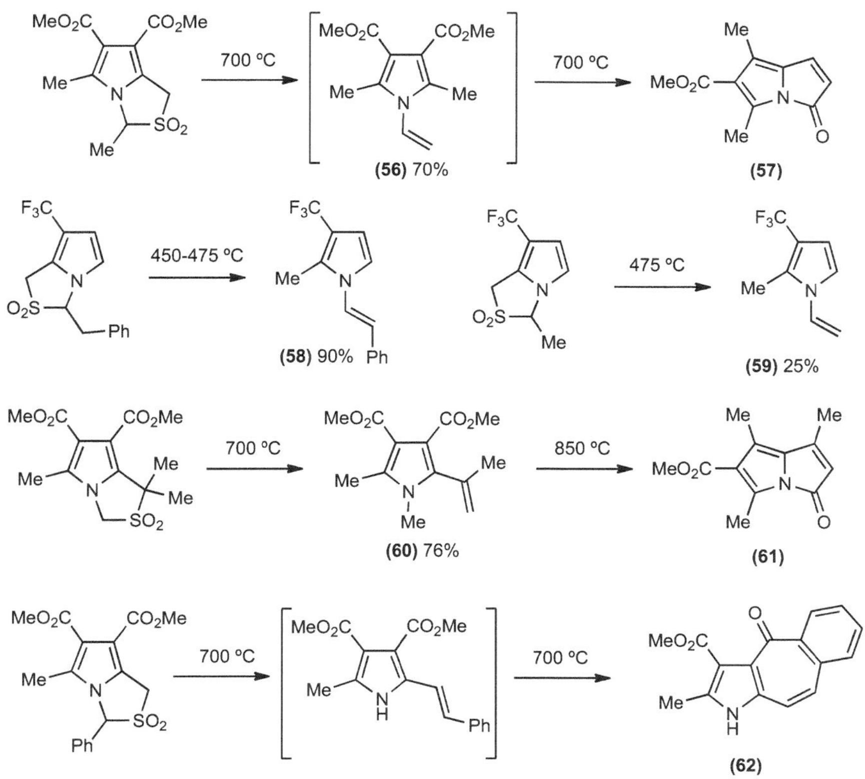


**Scheme 14**

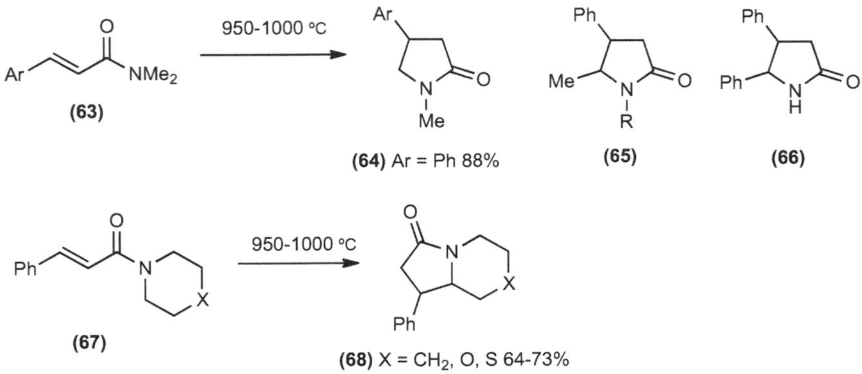


Scheme 15

products **72**, and in this case the ylide derived from proline gives the bicyclic product **73** (2001TL141). Tandem cyclization of suitably designed ylides **74** proceeds by loss of $Ph_3PO$ to give the alkyne, loss of benzyl radical to give an aminyl radical which then undergoes a 5-*endo-dig* cyclization to form the

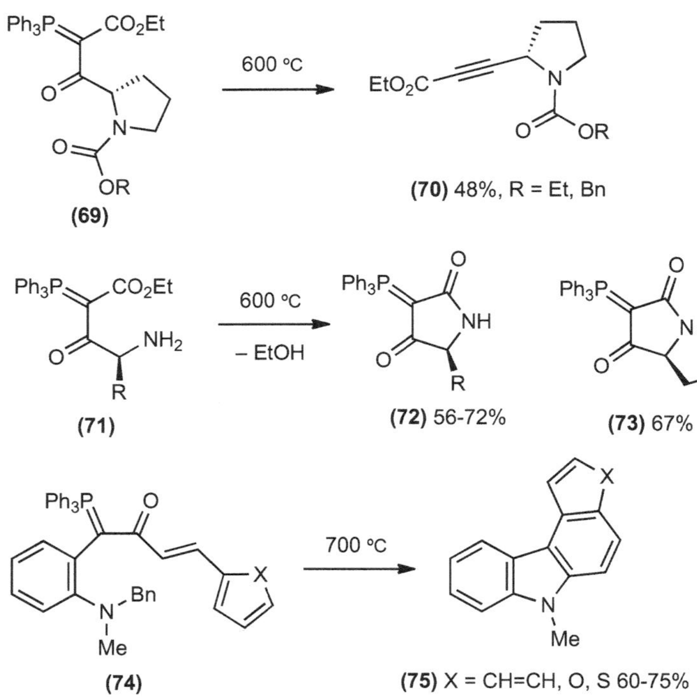


Scheme 16

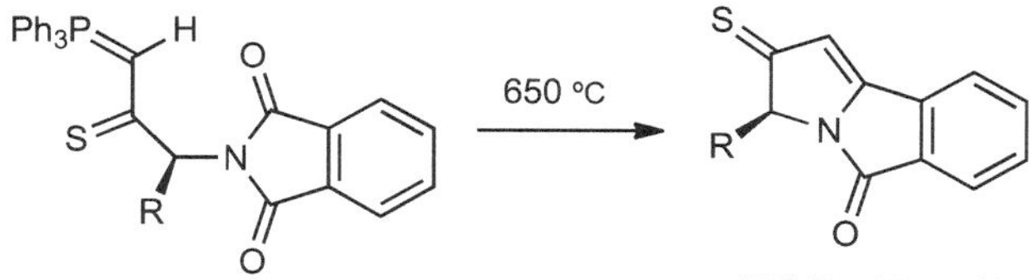


**(76)**

**(77)** R = Me 55%
R = Bn 52%

**Scheme 17**

3-indolyl radical which is set up for a further intramolecular homolytic aromatic substitution to give the ring-fused carbazoles **75** (Scheme 16; 2008 JOC9781).

The thiocarbonyl-stabilized ylides **76**, derived in a few steps from alanine and phenylalanine, undergo an intramolecular Wittig reaction involving one carbonyl group of the phthalimide function to give products **77**, which are derivatives of the previously unknown pyrrolo[2,1-*a*]isoindol-5-one-2-thione ring system (Scheme 17; 2014H(88)1135).

A versatile new approach to ring closure involves generation of a radical either on nitrogen of a 2-*o*-tolylazole or more commonly at the *ortho* position of the aryl group in an *N*-aryl-2-methylazole. The former approach is illustrated by conversion of the *N*-aminobenzimidazole **78** into the benzimidazoisoindole **79** (Scheme 18; 2015UP8).

In the latter approach, FVP of a range of *N*-(2-nitrophenyl)-2-methylazoles at 850 °C results in a loss of the nitro group to give the aryl radical, hydrogen atom abstraction from methyl, and finally cyclization to give ring-fused indoles. The process is illustrated by the conversion of the imidazole **80** into **81**, the pyrazole **82** into **83**, and the 1,2,4-triazole **84** into **85** (Scheme 19; 2015UP6). The reaction was later extended to the benzimidazole derivative **86**, which gives the indolobenzimidazole **87**, and analogs of both **84** and **86** with ring substituents that react similarly (2015UP7).

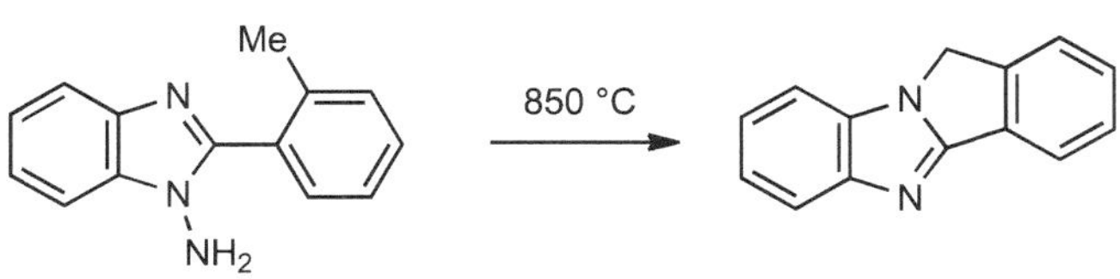


**(79)** 56%

**(78)**

**Scheme 18**

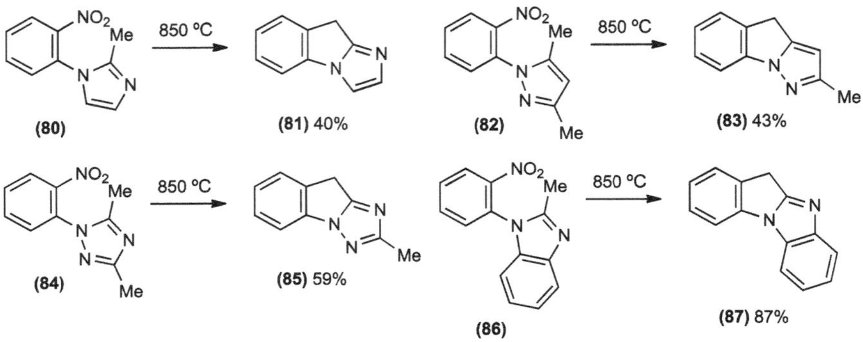


Scheme 19

The corresponding processes for *N*-(3-nitro-2-pyridyl)azoles are also successful although generally in lower yields, as illustrated by the conversion of the azoles **88** into **89**, **90** into **91**, **92** into **93**, and **94** into **95** (Scheme 20; 2015UP6, 2015UP7).

The dibenzazepinyl radical, generated by high-temperature pyrolysis of either *N*-benzyl or *N*-allyl derivatives **96**, undergoes a rearrangement to give the pyrrolocarbazole **97**; here again, an *ortho* nitrophenyl derivative **98** provides convenient alternative access (Scheme 21; 2008JOC6642). The analogous nitrophenylcarbazole **99** similarly allows convenient preparation of the indolocarbazole **100** (2009CEJ5482), and this synthesis has been extended to substituted derivatives (2011JPC(A)5435). In both these systems, the use of nitropyridyl starting materials gives aza analogs of general structures **101** and **102**, where nitrogen can be at any of the positions marked X (2010S923).

Allyl esters are also suitable aryl radical precursors, and the triazole **103** gives the product **104**, an isomer of product **85**, in a moderate yield at 850 °C (Scheme 22; 2015UP7).

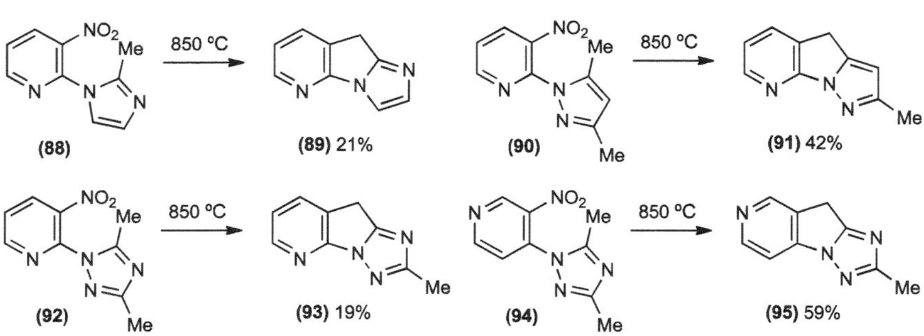


Scheme 20

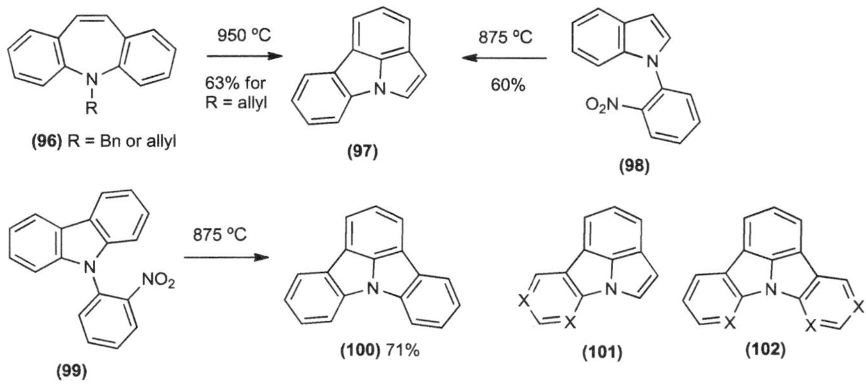


**Scheme 21**

The imine **105** acts as a convenient source of a benzylic radical, which undergoes cyclization at 750 °C to give the pyrroloindole **106** in a good yield (Scheme 22). The corresponding thiophene precursor **107** reacts similarly but with an additional 1,5-shift leading to the product **108**, an isomer of the expected product, in a low yield (2015UP6).

Radical processes are also responsible for the pyrolytic behavior of the *tert*-butyl imines derived from benzaldehyde and the three isomeric pyridinecarbaldehydes. FVP of **109–112** at 800 °C gives mainly the cyclization products shown in Scheme 23, although the reactions are complex and several other products are formed in each case (2013CEJ14983, 2014 EJO3020).

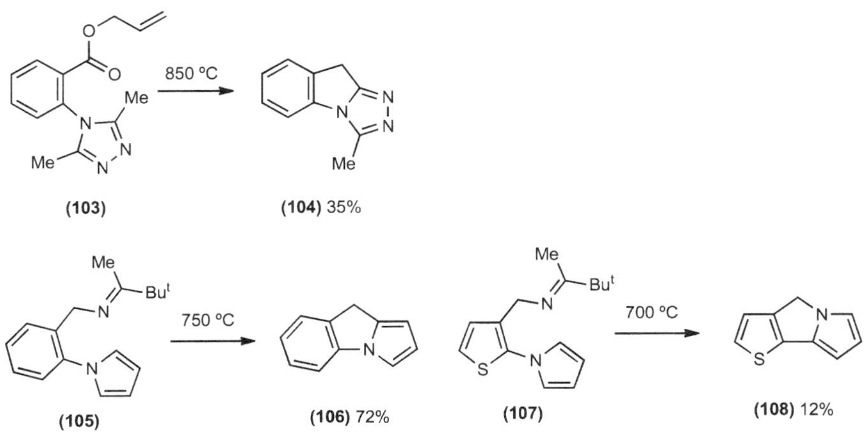


**Scheme 22**

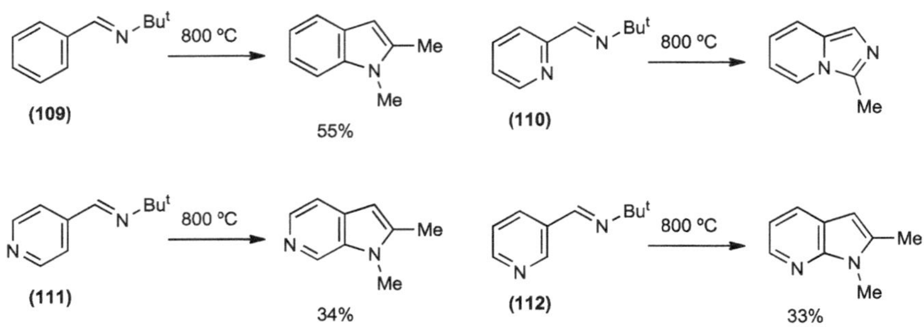

Scheme 23

The indolin-3-one **115** is conveniently prepared by FVP of the benzisoxazole **113** at 700 °C (Scheme 24). Electrocyclic formation of the nitrene **114**, which then inserts intramolecularly into CH, is the most likely mechanism (2015UP3). There has been considerable progress in generating heterocycle-fused analogs of such indolones, so-called heteroindoxyls.

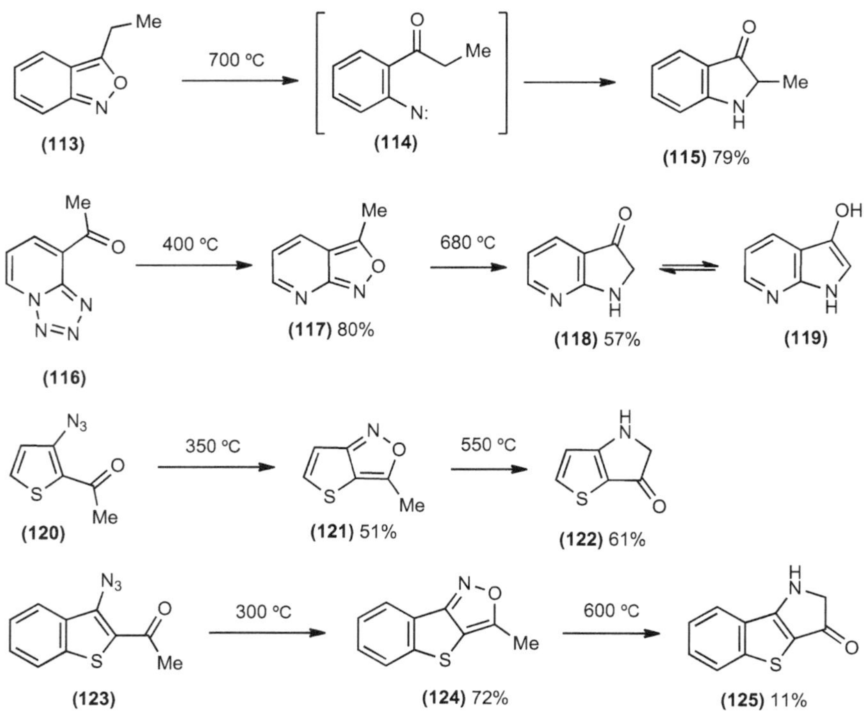

Scheme 24

In the pyridine series, the tetrazolopyridine **116** gives the isoxazole **117** at low temperature, and at higher temperatures this rearranges to the product **118**, which exists partly as the hydroxypyrrole tautomer **119** (2010OBC5166). The same pattern is repeated for both the thiophene-fused system **120** which gives **122** by way of **121** (2009JOC4278), and the benzothiophene system **123** which gives **124** and then **125** in lower yield (Scheme 24; 2015UP2).

Convenient access to indolin-2-ones **127** is provided by FVP of *o*-aminodiazo ketones **126** at 300 °C resulting in a Wolff rearrangement to give the ketene, which then undergoes intramolecular addition of NH (Scheme 25). The quinoline-fused example **129** is similarly formed by FVP of the quinoline derivative **128** at 450 °C (2015UP3).

Pyrolysis of aminomethylene Meldrum's acid derivatives with loss of acetone and $CO_2$ has yielded a range of further interesting products. Simple derivatives **130** give good-to-excellent yields of pyrrolinones **131** which, depending upon the substituents, may exist as the hydroxypyrrole tautomer (Scheme 26; 2009S2531). This observation led to the first synthesis of the parent 3-hydroxypyrrole **133** from the *tert*-butoxycarbonyl derivative **132** (2009S2535). By using lactam-functionalized Meldrum's acids, the bi- and tricyclic products **134** and **135** were formed (2009OBC4936), and the isomeric system **136** was also accessible (2010S1361).

The oxazolidine-containing Meldrum's acids **137** and **138** lose formaldehyde, as well as $CO_2$ and acetone, to give novel *N*-alkenylpyrrolinones together with a rearranged isomer in the first case (Scheme 27; 2002JCS(P1)548). The competition between cyclization onto two different

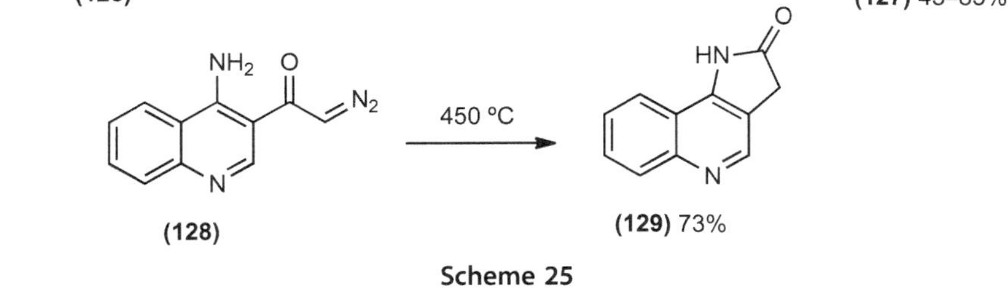


Scheme 25

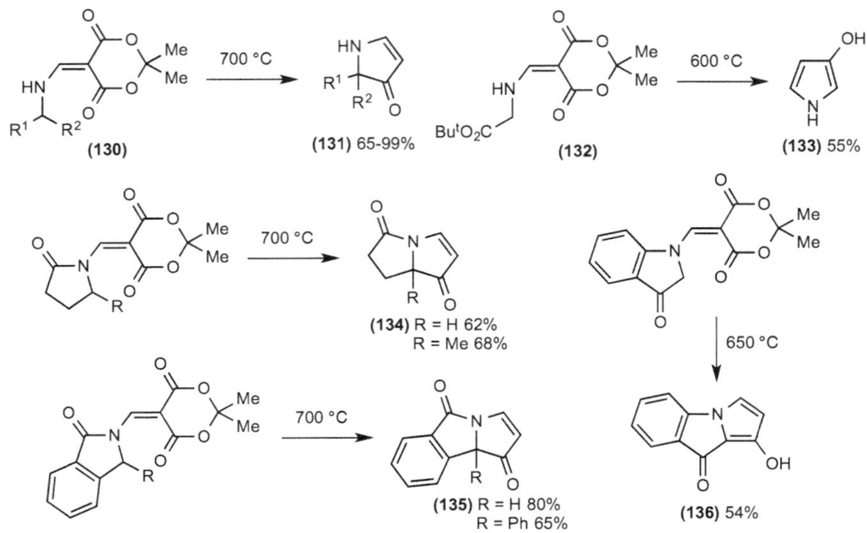


Scheme 26

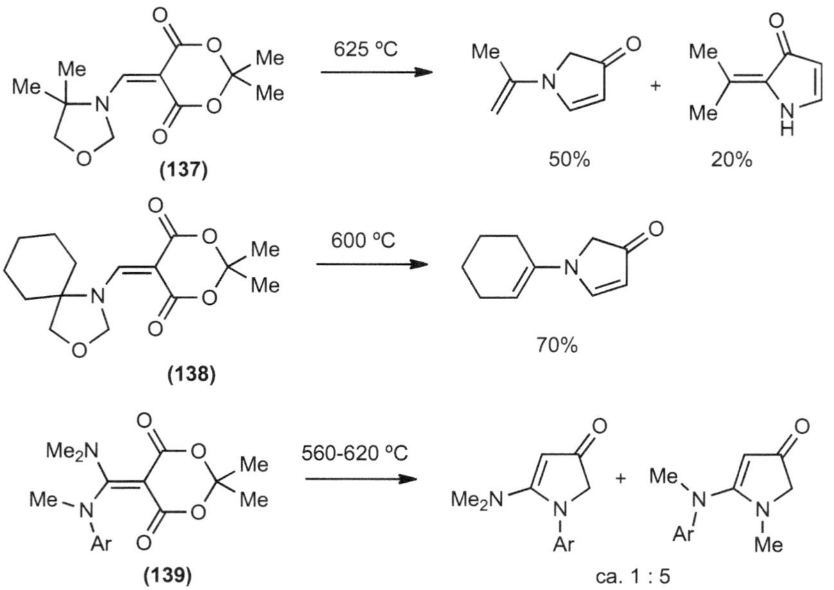


Scheme 27

methylamino groups was examined in the case of compounds **139** (Ar = Ph, 4-Me$OC_6H_4$, 4-Cl$C_6H_4$) (2002ARK(viii)125).

The *N*-propargyl Meldrum's acid compounds **140** and **141** lose acetone, $CO_2$, and CO to give dipyrrolylethanes (Scheme 28) but the mechanism of this process is unclear (2000MOL1130).

The FVP of pyrrolylacrylates proceeds with loss of an alcohol and cyclization of the resulting ketene to form pyrrolizinones, and various new examples have been reported in this area. Thus, a range of *N*-substituted 2-pyrrolylacrylates **142** give pyrrolizinones **143** (Scheme 29; 2007ARK(xi)85, 2008T9745) and both 1- and 3-indolylacrylates **144** and **146** afford the corresponding benzopyrrolizinones **145** and **147** (2007JOC8760). The pyrrolyl diester **148** gives the rather unstable product **149** which undergoes [2 + 2]-dimerization (2009OBC2187). Finally in this section, the indoloindolone **151** is conveniently formed from precursor **150** by high-temperature pyrolysis with loss of methanol (2008OBC2334).

## 4.2 Furans and Ring-Fused Derivatives

Cyclization of a range of epoxide-containing α,β-unsaturated esters and amides **152** gives the dihydrofuran products **153** with high *cis*-selectivity (Scheme 30; 2006OBC2912).

FVP of the lactones **154** and **156** results in decarbonylation and cyclization of the resulting acylallene to give furans **155** and **157**, respectively (Scheme 31; 2014JOC65).

Manipulation of the Diels–Alder adducts of cyclic anhydrides followed by the retro-Diels–Alder reaction using FVP has been used to prepare substituted

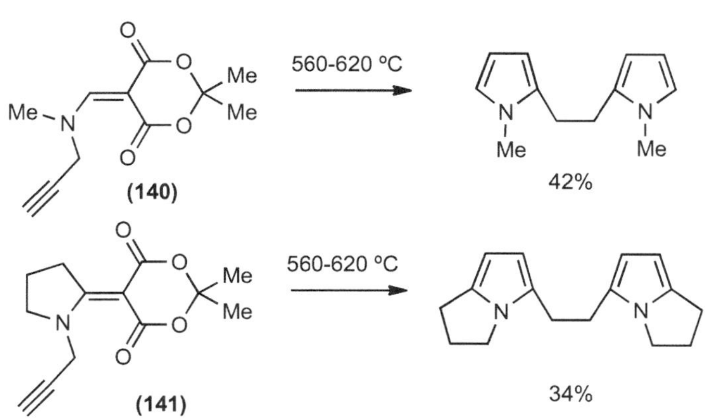


Scheme 28

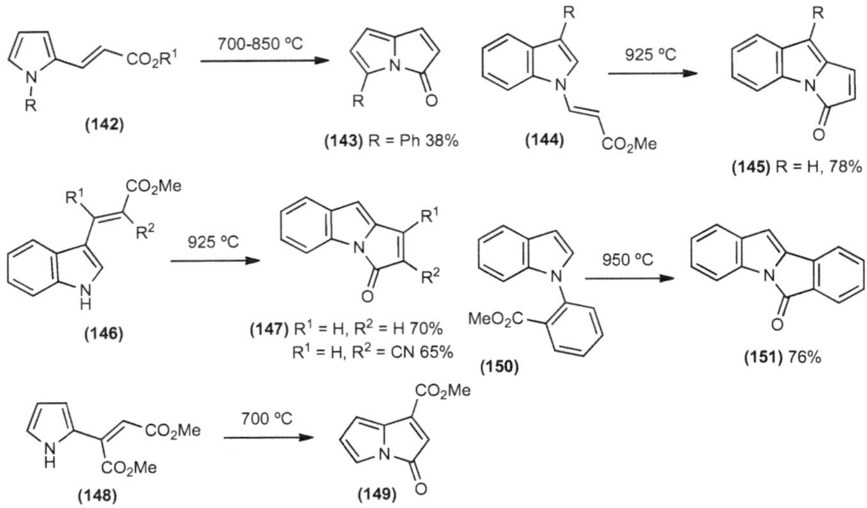


Scheme 29

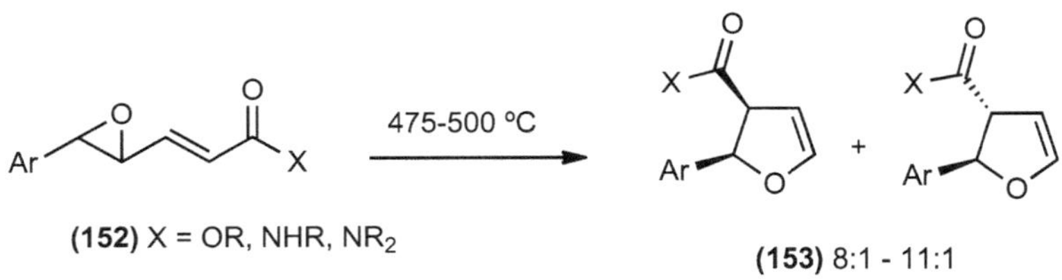


Scheme 30

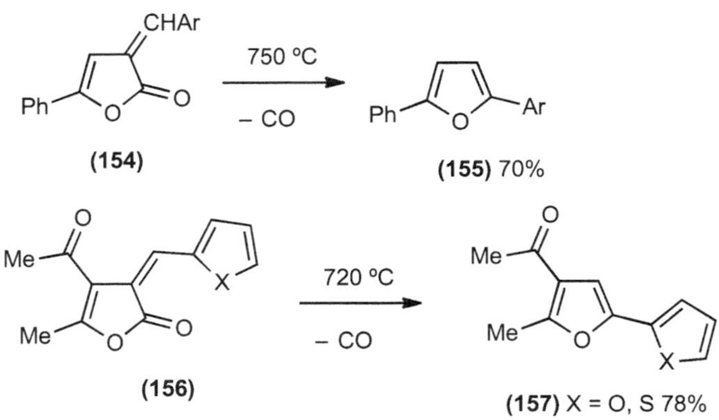


Scheme 31

butenolides such as **158** (2013OBC6650) and **159** (2006TL4377; Scheme 32). Use of chiral anthracene templates allows this approach to be applied to asymmetric synthesis as illustrated by formation of **161** from **160** (2005TL2475) and formation of the natural product analogs **163** from **162** (2001TA1913), with further work toward controlling the stereochemistry reported in the latter case (2009T6382). Compound **6**, which undergoes a retro-Diels–Alder reaction upon simple FVP (Scheme 3; Section 2.2), isomerizes upon FVP over a solid acid catalyst to the polycyclic tetrahydrofuran derivative **164** (2000TL9189).

The unusual benzodifuran quinone **166** is formed by the retro-Diels–Alder elimination of ethane from **165**, itself obtained by hydrogenation of a bis(aryne-furan) cycloadduct (Scheme 33); a similar approach gives the analogs **167–169** (2007T12621).

The pyrolysis of aryl propiolates **170** as a route to cycloheptafuranones **171** has been extended to highly substituted examples with three to five methyl or ethyl substituents (Scheme 34; 2000HCA1022). The formation of benzofurans upon FVP of stabilized phosphonium ylides carrying an *ortho*-methoxyphenyl group, previously reported for ylides with C=$PPh_3$ and C=O functions interchanged, also occurs with the isomeric compounds

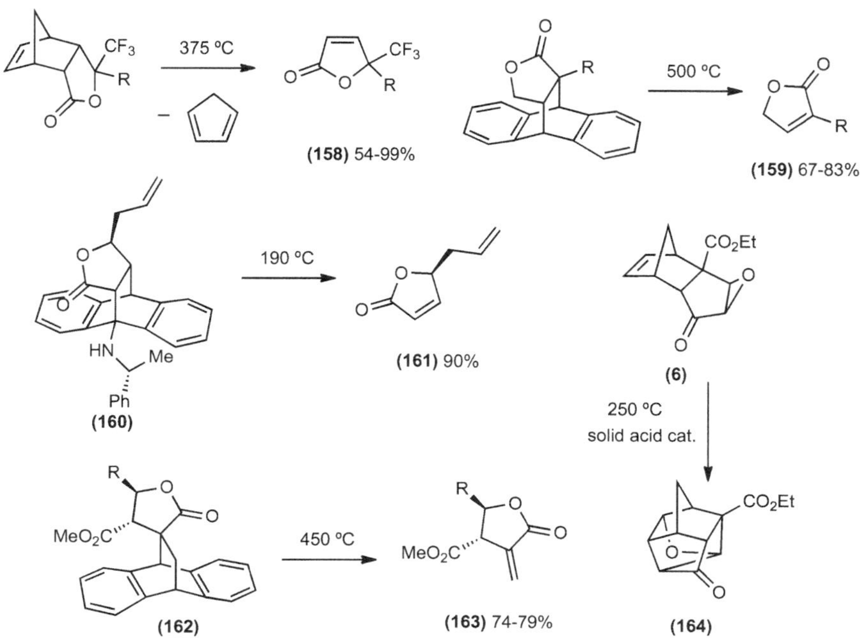


Scheme 32

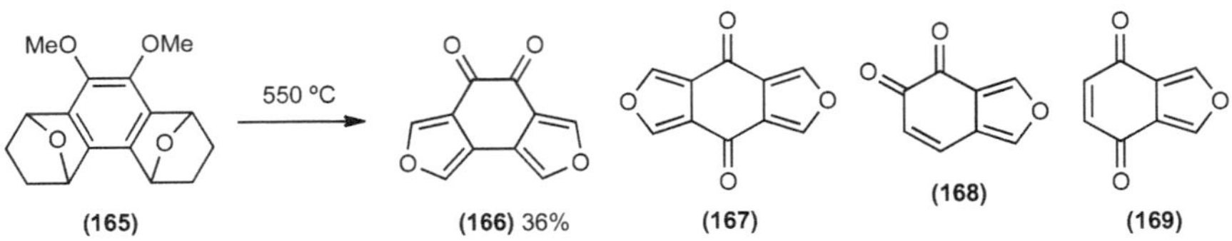


Scheme 33

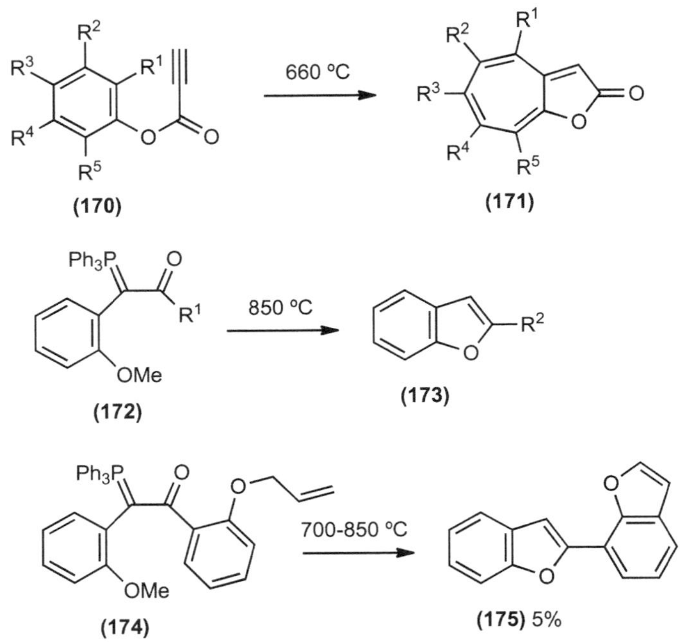


Scheme 34

**172** to give **173** (2000ARK(v)798). This approach has been extended to the methoxy allyloxy ylide **174** which undergoes a multi-step cascade reaction to afford a low yield of compound **175**, which is the first unsymmetrical bibenzofuryl (2001SL228).

The FVP of aryl 2-allyloxybenzoates **176** proceeds with decarboxylation to provide a high-yielding synthesis of the dibenzofurans **177** (Scheme 35) and the naphthyl ester **178** similarly gives benzonaphthofuran isomers **179** and **180** (2010OBC2961).

FVP of the ester **181** at 600 °C results in a loss of methanol to produce the heterocyclic phenanthrol analog **182** directly in a low yield (Scheme 36; 2015UP10).

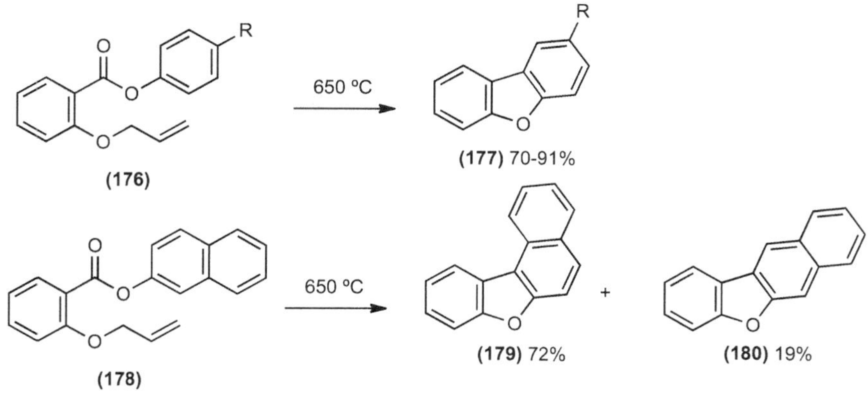


Scheme 35

## 4.3 Thiophenes and Ring-Fused Derivatives

The efficient formation of nitriles by dehydration of oximes using FVP over molecular sieves has been applied to the synthesis of thiophene-3-carbonitrile **183** (Scheme 37; 2015UP9). FVP of the acid chloride **184** at 850 °C results in a loss of HCl to produce the heterocyclic phenanthrol analog **185** in a moderate yield (2015UP10). The benzofuranthiones **186** undergo quantitative isomerization to the benzothiophenones **187** on pyrolysis at 850 °C (2006HCA991).

Alkylthiomethylene Meldrum's acid derivatives **188** provide convenient access to 2-alkyl-3-hydroxythiophene products **189**, and the bifunctional analog **190** gives the dithienylbutane **191** (Scheme 38; 2000MOL1130).

The thienothiapyran *S,S*-dioxides **192** undergo a clean loss of SO at 700 °C to give the alkenyl ketone products **193** (Scheme 39). The *tert*-butyl analog **194** gives mainly the corresponding alkenyl aldehyde **195** but also a low yield of the ethylidenecyclobutathiophene **196** formed in a multi-step process involving a novel radical rearrangement and extrusion of acetone (2014AJC1288).

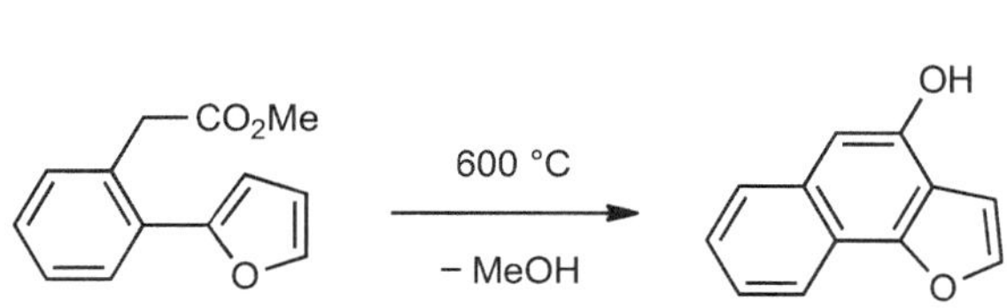


(181) (182) 20%

Scheme 36

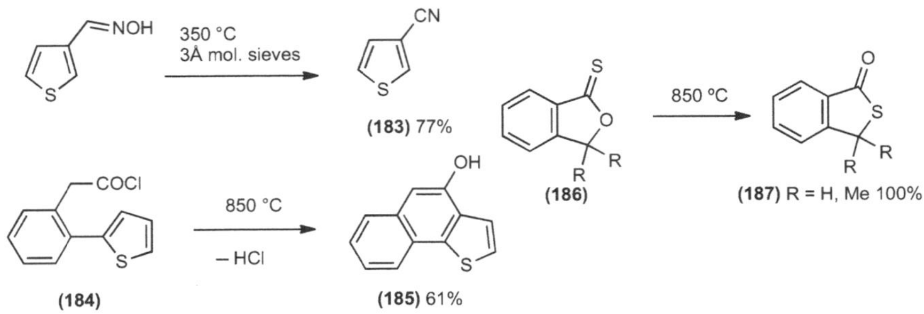


Scheme 37

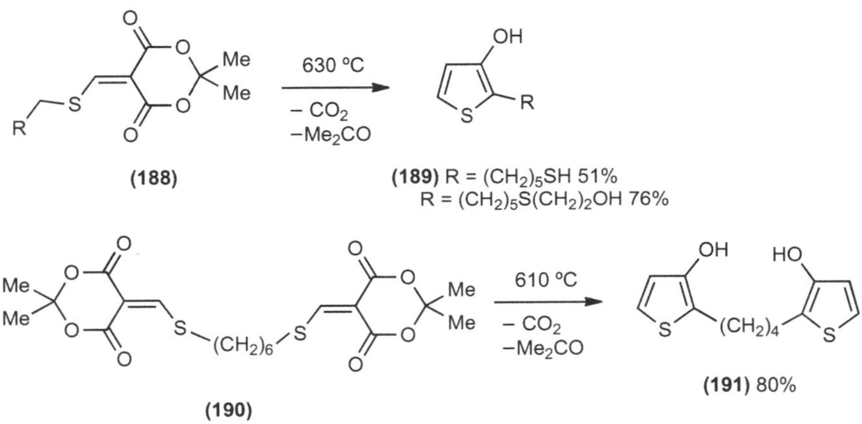


Scheme 38

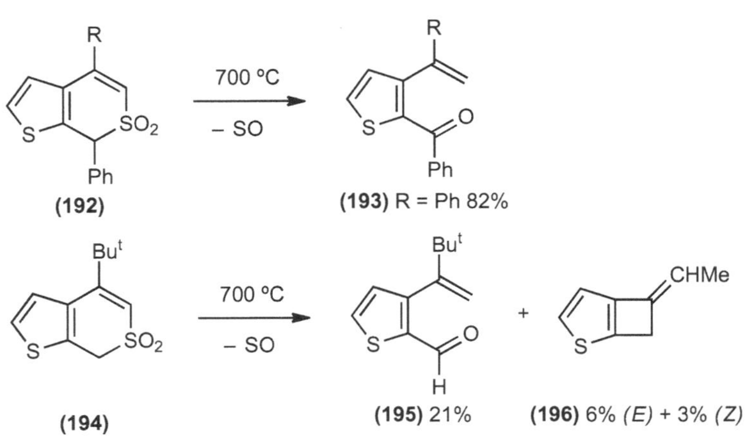


Scheme 39

FVP of aryl 2-allylthiobenzoates **197** results in radical cyclization and subsequent decarboxylation to give dibenzothiophenes **198** (Scheme 40; 2010OBC2961). Extrusion of $Ph_3PO$ from the *S*-methyl *O*-allyl ylide **199** is accompanied by a cascade process including a Claisen rearrangement to give the first benzothienylbenzofuran **200** (2001SL228).

Stabilized ylides containing a 2-(methylthio)pyridine group provide access to novel fused thienopyridine ring systems, as illustrated by conversion of ylide **201** into **202** and ylide **203** into **204** (Scheme 41; 2009NJC2402).

Dehalogenative coupling is observed when dichloromethylthiophenes **205** are pyrolyzed through a finely divided metal such as magnesium or calcium to give the dithienylethenes **206** (Scheme 42). When this method is applied to substrates bearing an extra adjacent halogen atom, such as **207**, the expected benzodithiophene **208** is accompanied by the isomers **209** and **210** (2015RSCA19379). Their formation is likely to involve equilibration via cyclopropenylmethyl radicals, which is supported by the formation of the same products from alkene **211**.

Similar isomerization processes occur upon high-temperature FVP of the brominated dithienylhexadienynes **212** and **213**, which gives mixtures of isomeric naphthodithiophenes **214–218** (Scheme 43; 2009CRC378).

## 4.4 Pyrazoles, Imidazoles, and Ring-Fused Derivatives

The highly substituted vinylpyrazole **220** is formed via a diazafulvenium methylide when $SO_2$ is eliminated from the cyclic sulfone **219** (Scheme 44; 2006TL791). Similar processes occur for the benzo-fused analogs **221** and **223**, which give the indazoles **222** and **224**, respectively (2013JOC628).

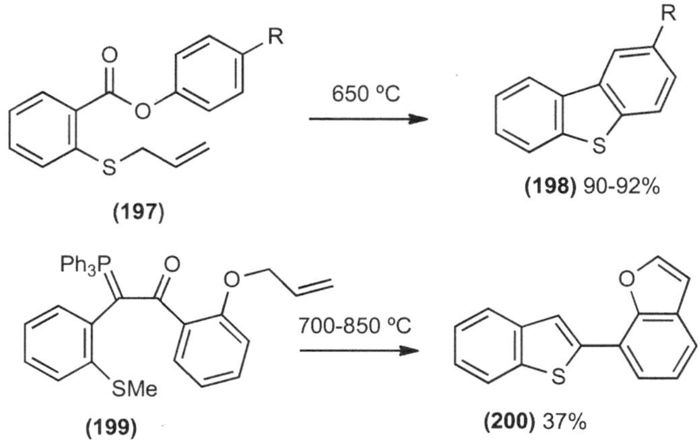


Scheme 40

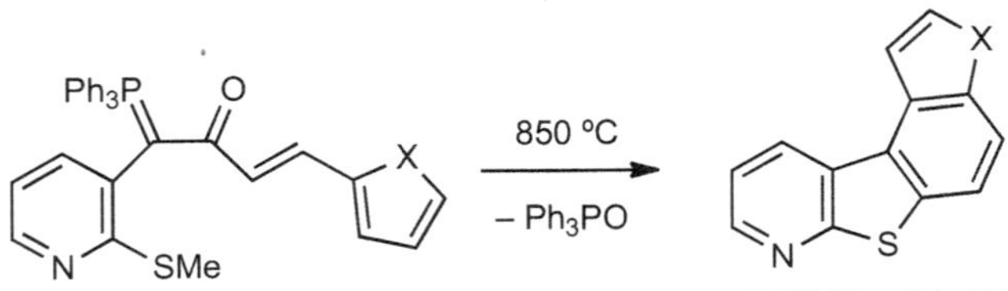


**(201)**

**(202)** X = CH=CH 52%
X = S 52%
X = O 42%

**(203)**

**(204)** X = S 95%
X = O 31%

**Scheme 41**

Aryltetrazolylacroleins **225** undego loss of nitrogen and efficient conversion into the indazolylacrolein products **226** (Scheme 45; 2012T1299).

The surprising formation of 2-methylimidazole **228** in a good yield from pyrolysis of the azadiene **227** (Scheme 46; 2013T722) bears an obvious similarity to the formation of the imidazopyridine from **110** mentioned in Section 4.1 (Scheme 23). The pyridylbenzotriazole **229** is a good precursor for synthesis of the benzimidazopyridine **230** and the corresponding

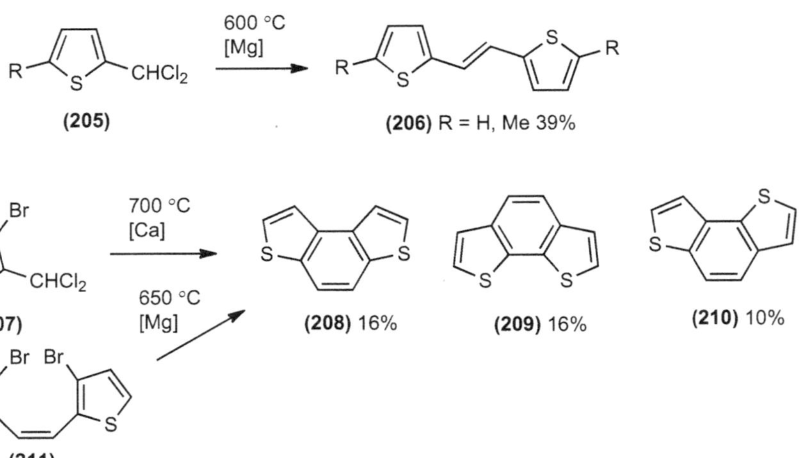


**(205)**

**(206)** R = H, Me 39%

**(207)**

**(208)** 16%

**(209)** 16%

**(210)** 10%

**(211)**

**Scheme 42**

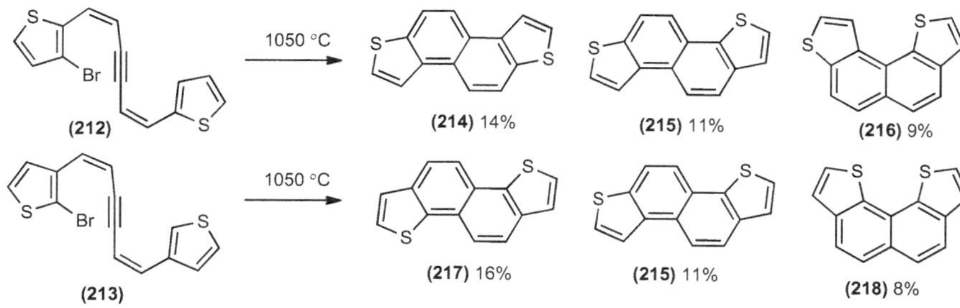


Scheme 43

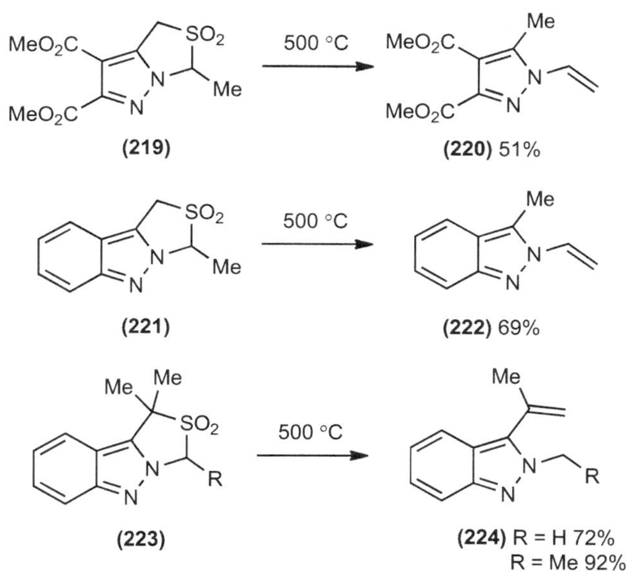


Scheme 44

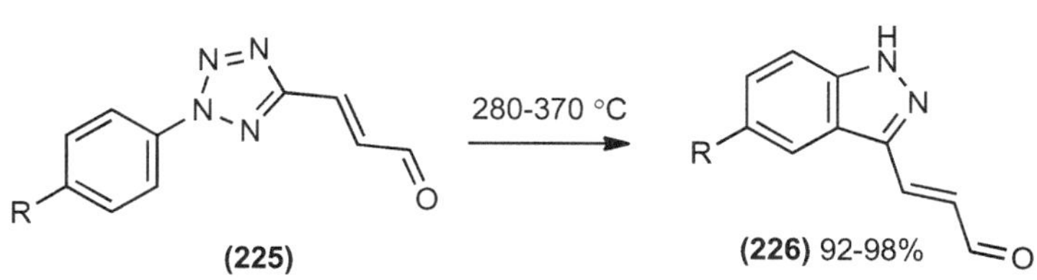


Scheme 45

isoquinolyl compound **231** affords the product **232** in an even higher yield (2000JCS(P1)3212).

Further results have appeared upon pyrolysis of isoxazolones bearing a pyridyl group on nitrogen as a route to imidazopyridine products (Scheme 47).

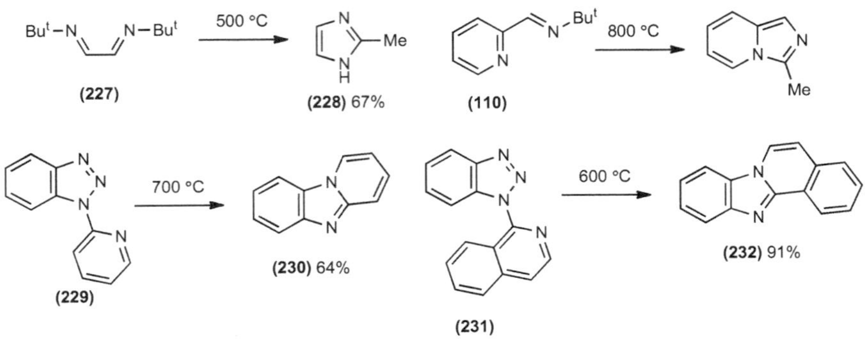


Scheme 46

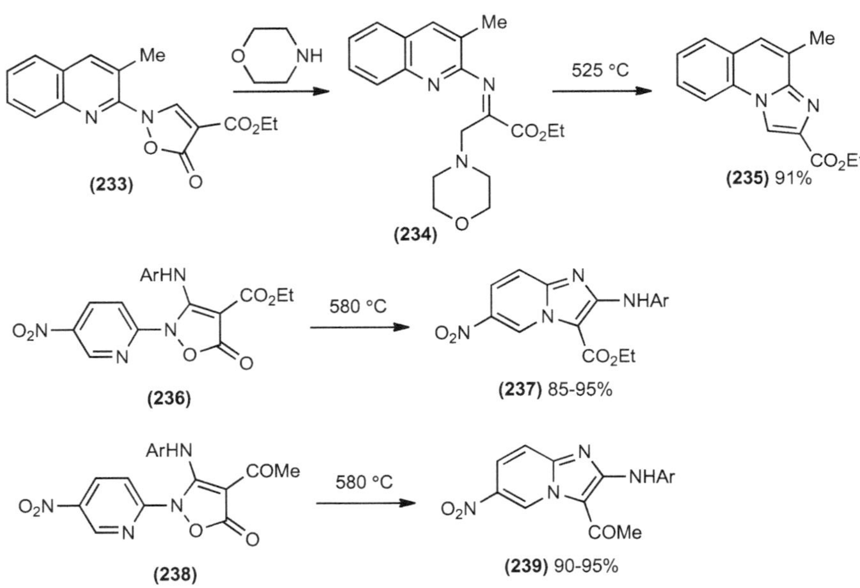


Scheme 47

Thus, the morpholinyl compound **234**, obtained from **233**, gives a good yield of the rearrangement product **235**, in contrast to pyrolysis of **233** directly which affords the isomer of **235** with the ester group on the other imidazole carbon (2002MI389). Loss of $CO_2$ and cyclization to give good yields of imidazopyridines occurs readily for isoxazolones **236** (2012ASJC2884) and **238** (2013OJC1443, 2014AJC1228) giving products **237** and **239**, respectively.

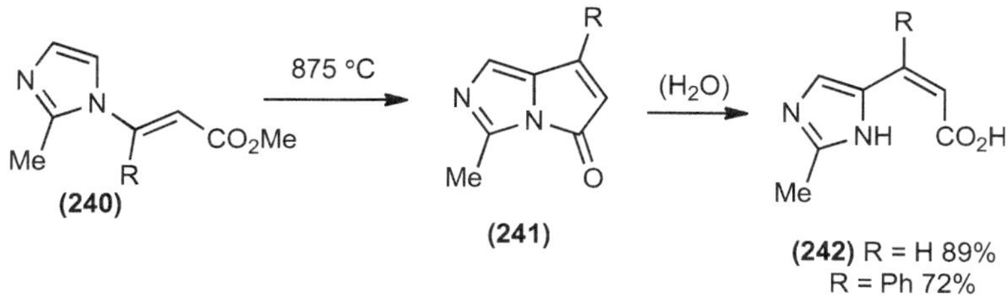


Scheme 48

Analogs **242** of (*Z*)-urocanic acid have been prepared by FVP of imidazolylacrylates **240** followed by hydrolysis of the resulting azapyrrolizinones **241** (Scheme 48; 2008S1676).

## 4.5 Isoxazoles, Oxazoles, and Ring-Fused Derivatives

As described in Section 4.1, a variety of azides bearing an adjacent acetyl group initially form ring-fused isoxazoles upon pyrolysis at lower temperatures but are then converted at higher temperatures to the pyrrolinones (Scheme 24). Studies on the pyrolysis of 1,2,5-oxadiazoles (furazans) **243** (2009ARK(xiv)200) and their *N*-oxides (furoxans) **244** (2010ARK(ix)34) used trapping with hex-1-ene to form isoxazolines **245** in good yields as a means of demonstrating the generation of nitrile oxides (Scheme 49).

FVP of the imines **246**, formed from aromatic aldehydes and *o*-anisidine, provides a convenient synthesis of 2-arylbenzoxazoles **247** (2011JCCS301).

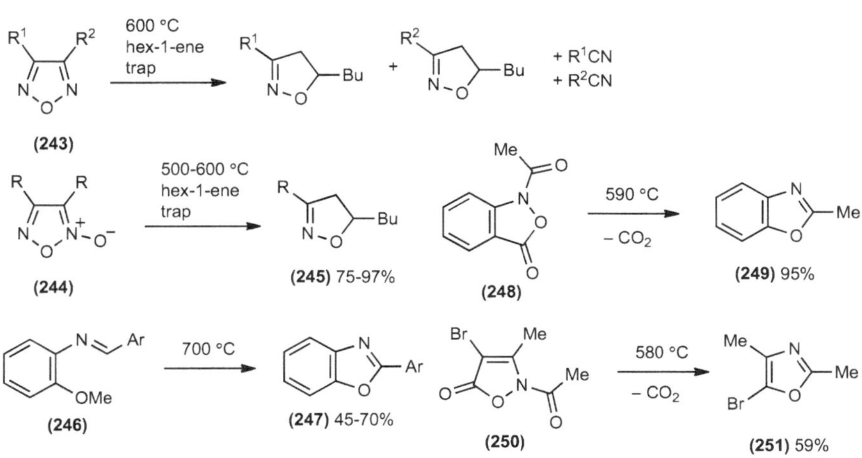


Scheme 49

Oxazole formation is also observed when isoxazolones **248** (2000JCS(P1) 3212) and **250** (2008CHE113) undergo a pyrolytic loss of $CO_2$ to afford products **249** and **251**, respectively.

## 4.6 Isothiazoles, Thiazoles, and Ring-Fused Derivatives

FVP of the imines **252**, formed from aromatic aldehydes and 2-methylthioaniline, provides a convenient synthesis of 2-arylbenzothiazoles **253** (Scheme 50; 2008TL4145). Pyrolysis of the imine **254** from 2-allyloxybenzaldehyde similarly results in radical cyclization to afford the benzothiophene **255** in a good yield (2012OBC623). Loss of the *N*-amino group in the allylthiotriazoles **256** initiates a multi-step radical process leading to the thiazolotriazoles **257** in moderate yields (2001JCS(P1)424). Decarboxylation of the thiazoloisoindolones **258** occurs at 600 °C to give the products **259** without racemization (2004T3949).

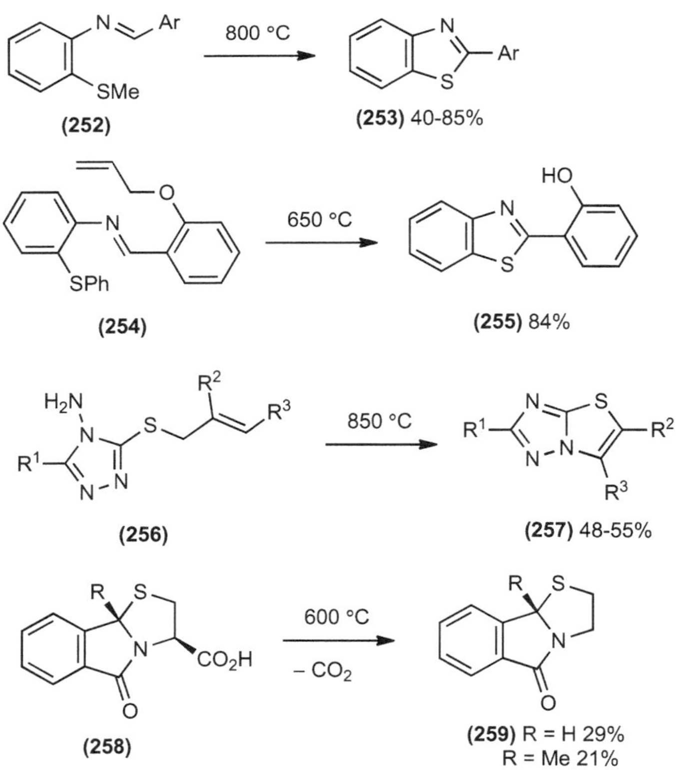


Scheme 50

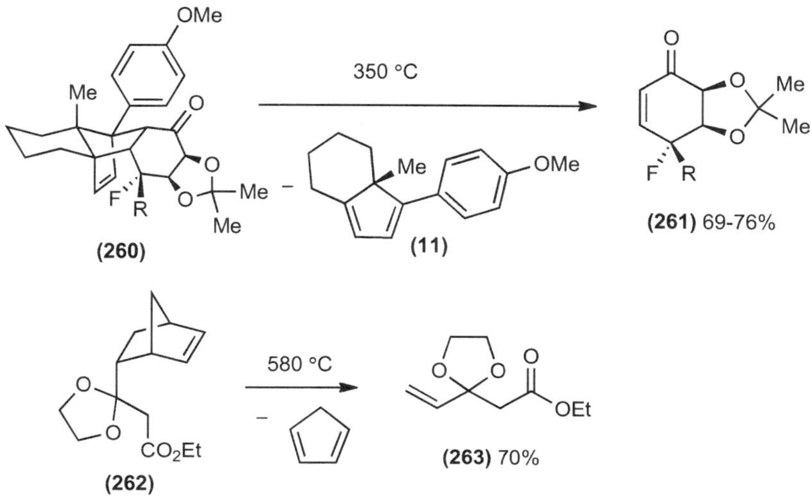


Scheme 51

## 4.7 Dioxolanes and Ring-Fused Derivatives

The retro-Diels–Alder reaction of adduct **260** by FVP at 350 °C is the final step in the construction of the dioxolane product **261** using diene **11** as a chiral template (Scheme 51; 2004ARK(ii)4). A retro-Diels–Alder reaction of adduct **262** was used to prepare the dioxolane **263** for use in total synthesis (2010T6331).

## 4.8 Dithioles and Ring-Fused Derivatives

Flow pyrolysis of dialkyl polysulfides may sometimes give high yields of 1,2-dithiol-3-thiones as illustrated by the formation of product **265** from

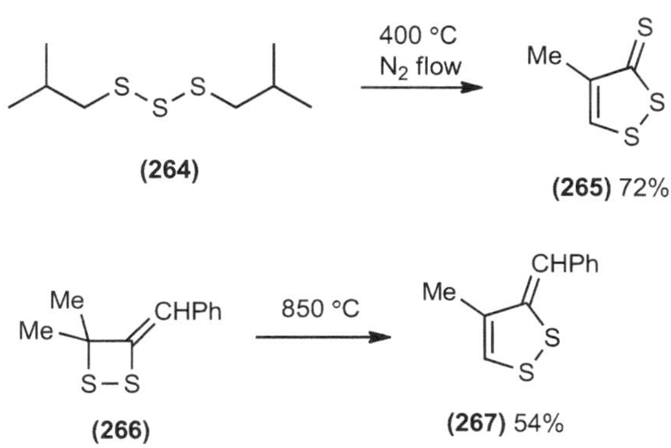


Scheme 52

diisobutyl trisulfide **264** (Scheme 52; 2003RJOC752). The similar benzylidenedithiole **267** is formed by FVP of the dithietane **266** (2006HCA991).

### 4.9 Triazoles and Ring-Fused Derivatives

An interesting interconversion is observed between the isomeric triazolobenzotriazines **268** and **269** under FVP conditions below 550 °C (Scheme 53; 2005T7489). At higher temperatures decomposition is observed. Pyrolysis of the 1,2,3-triazole *N*-oxide **270** unexpectedly affords the di(triazolyl) ethane **271** (2010ARK(x)34).

### 4.10 Oxadiazoles

FVP of the *N*-phenylbenzamidoxime **272** gives a low yield of the unexpected 1,3,4-oxadiazole product **273** in a process involving generation of a carbodiimide which then undergoes nitrile oxide cycloaddition (Scheme 54; 2009JAAP369). FVP provides a way of converting the precursor **274** into the parent 1,3,4-oxadiazole **275** (2012ARK(v)75).

## 5. SIX-MEMBERED RINGS

### 5.1 Pyridines, Piperidines, and Ring-Fused Derivatives

Mechanistic studies have been carried out on the formation of pyridines by ring expansion of pyrrolyl-1-methyl radicals formed by FVP of 1-aryloxymethylpyrroles at 700 °C (2015UP1). Isotopic labelling gives results consistent with a "walk" mechanism, and dimethylpyrrole precursor **276** gives 2,5- and 2,6-dimethylpyridines as shown in Scheme 55. The study

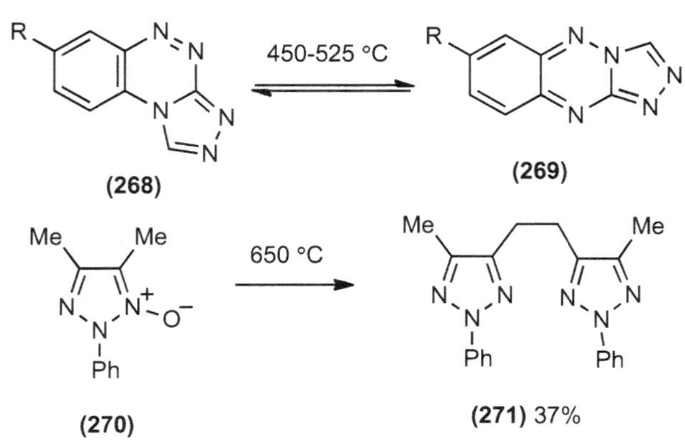


Scheme 53

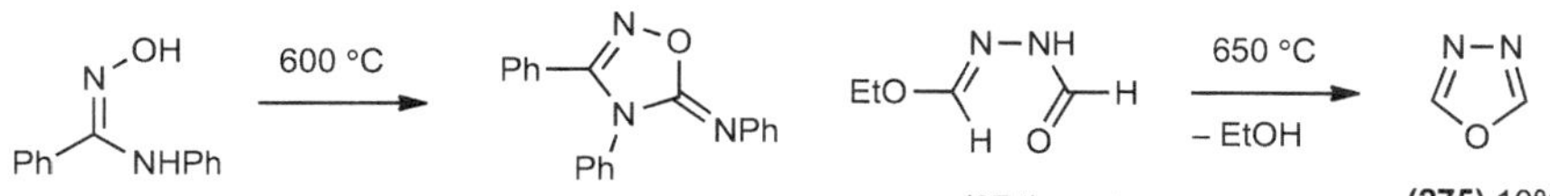


**(272)** **(273)** 28% **(274)** **(275)** 10%

Scheme 54

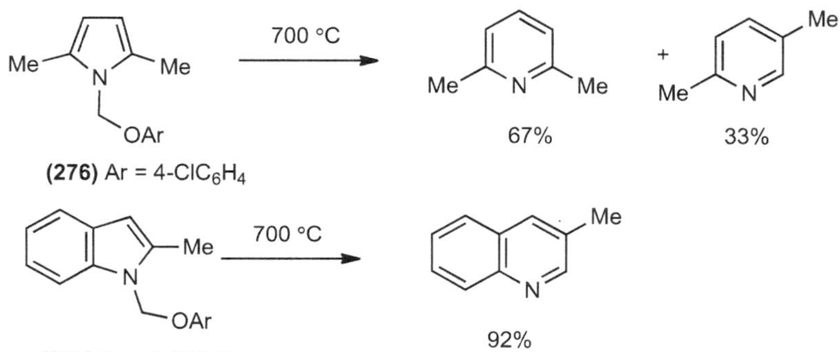


**(276)** Ar = 4-ClC$_6$H$_4$

**(277)** Ar = 4-ClC$_6$H$_4$

Scheme 55

was also extended to the indole derivative **277** which gave exclusively 3-methylquinoline (Scheme 55; 2015UP1).

FVP of the pyrrolyl radical precursor **278** leads to the pyrroloquinoline product **279** in a moderate yield (Scheme 56; 2015UP4). Ylides such as **280** derived from anthranilic acid, which were designed to cyclize to indoles following a loss of Ph$_3$PO, instead form quinolines **281** due to transposition of the reactive site from N to C in the intermediate *N*-methylaminyl radical (2008JOC9781). A range of spiro cyclopropane piperidines **282** are found to

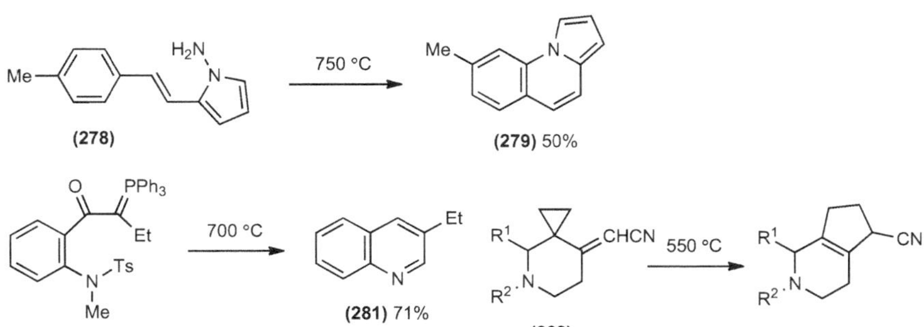


**(278)** **(279)** 50%

**(280)** **(281)** 71% **(282)** **(283)** 60-77%

Scheme 56

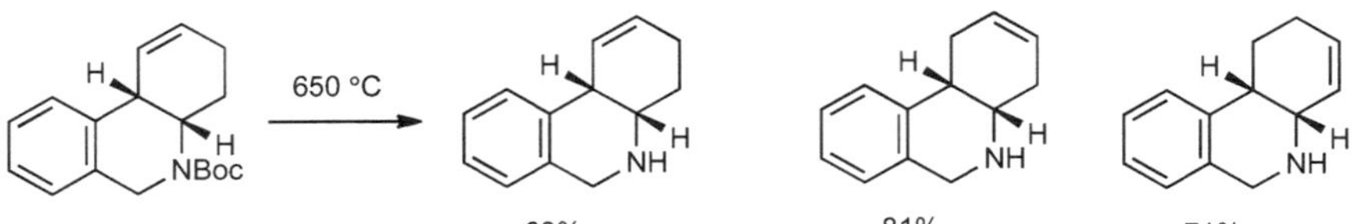


**(284)** 60% 81% 71%

**Scheme 57**

undergo ring expansion to cyanocyclopentapiperidine **283** upon FVP at 500 °C (2001SL433).

Convenient removal of the *N*-*tert*-butoxycarbonyl protecting group in hydrogenated phenanthridine derivatives such as **284** is achieved without double bond isomerization by FVP at 650 °C (Scheme 57). The isomeric compounds shown in Scheme 57 can also be formed clearly from their respective Boc derivatives in this way, in contrast to conventional deprotection using trifluoroacetic acid which results in double bond isomerization (2009SL1587).

FVP over $MoO_3$ at 550 °C results in efficient aromatization of tetrahydro derivatives to give quinolines and isoquinolines as illustrated by the examples **285**–**287** (Scheme 58; 2015UP9).

Subjecting the pyrrolopyrimidine **288** to high-temperature FVP results in isomerization to give the more thermodynamically stable pyrrolopyridines **289** and **290**. Theoretical modeling supports the involvement of a spiro intermediate (Scheme 59; 2015UP5). The isomerization of 2-methoxypyridine and analogs such as **291** into the corresponding *N*-methylpyridones **292** occurs under relatively mild reaction conditions, and the corresponding ethoxy compounds may also eliminate ethene as illustrated by conversion of **293** into **294** (2003AJC913). Pyrolysis of 2,3,5,6-tetrafluoro-4-mercaptopyridine **295**

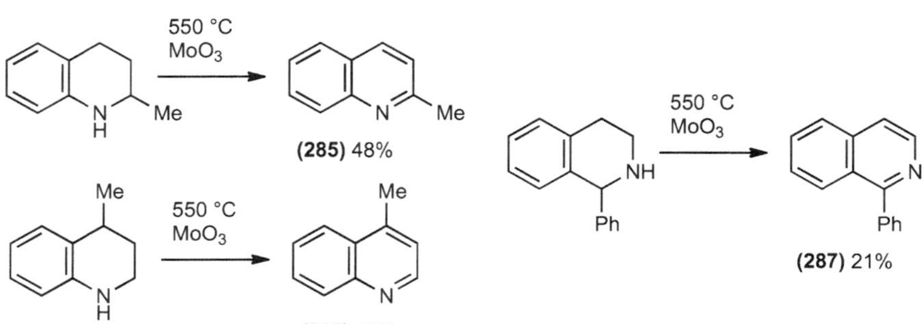


**(285)** 48% **(286)** 40% **(287)** 21%

**Scheme 58**

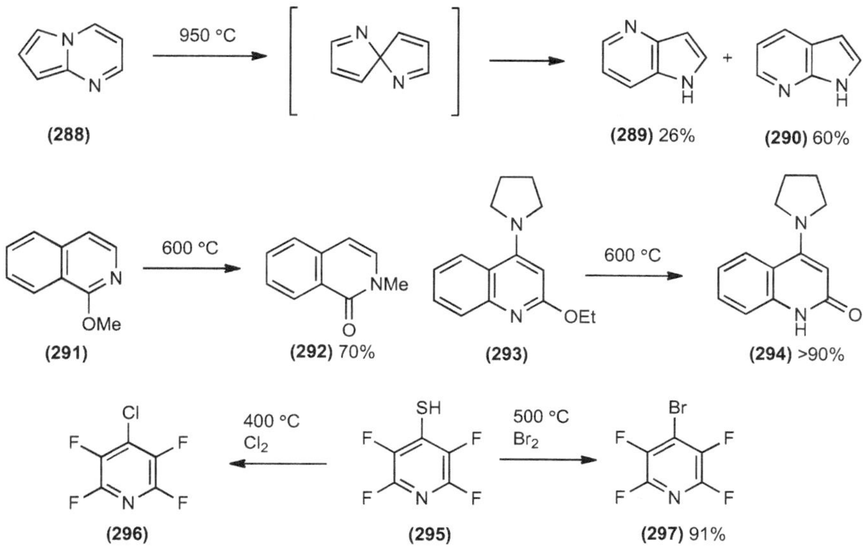


Scheme 59

in a stream of chlorine or bromine gives the halogenation products **296** or **297**, respectively (2005RJOC1647).

Extrusion of CO from the pyrrolinediones **298** is followed by ketene cyclization to give quinolones **299** (Scheme 60; 2002JCS(P1)1232). Unexpected formation of pyridine products is observed from pyrolytic decarboxylation of isoxazolone-containing cinnamic acid derivatives such as **300**: the ethyl ester gives **301** (X = OEt), in addition to a pyrrole product (see

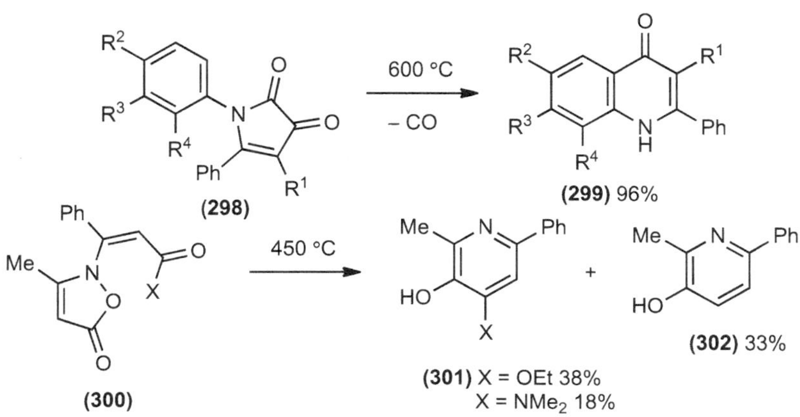


Scheme 60

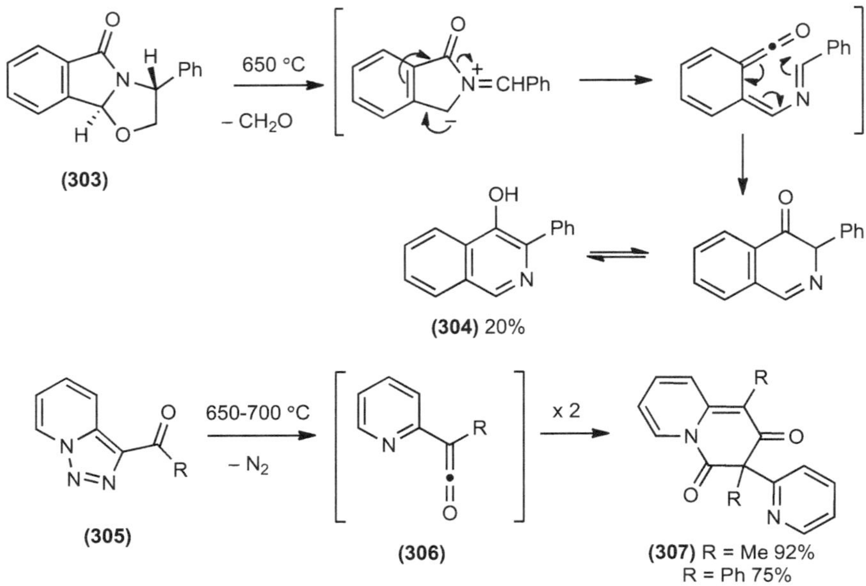


**Scheme 61**

Scheme 9), while the dimethyl amide gives both **301** (X = $NMe_2$) and the deaminated product **302** (2004AJC455).

In an unusual process, FVP of compound **303** results in a loss of formaldehyde and formation of 3-phenyl-4-hydroxyisoquinoline **304** by way of the azomethine ylide, the ring of which opens and recyclizes (Scheme 61; 2015UP11). The acyltriazolopyridines **305** undergo a loss of $N_2$ followed by a Wolff rearrangement to give pyridylketenes **306** which then undergo an unusual [4 + 2] dimerization to afford the pyridylquinolizinedione products **307** (2002JCS(P1)1366).

Substituted alkenyl and aryl acrylamides **308** and **310** undergo electrocyclic ring closure at 800 °C to produce the dihydropyridone derivatives **309** and **311**, respectively, in moderate yields (Scheme 62; 2002SC875). The unsaturated ester **312** also undergoes electrocyclization followed by in situ aromatization to give the tricyclic product **313** (2006OBC407).

Pyrolysis of the allylthio ester **314** proceeds by radical generation, cyclization with rearrangement, and loss of sulfur to afford the pyrroloquinoline **315** (Scheme 63; 2006OBC2446). Similar processes are observed but without rearrangement for both allyl ether **316** which gives **317** and *N*-benzylaniline **318** which gives **319** (2009OBC5173).

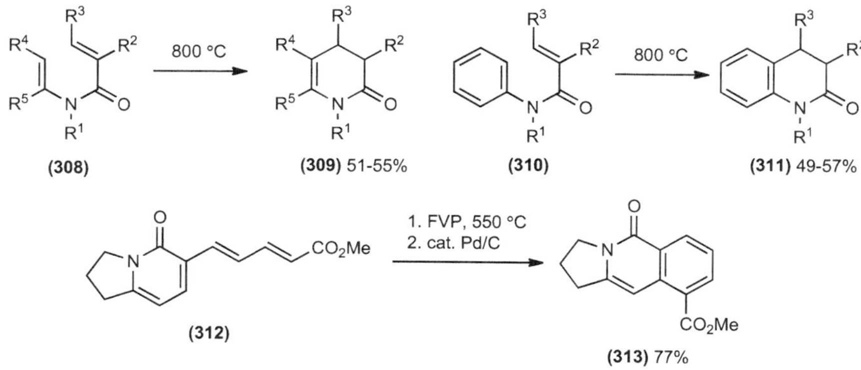


Scheme 62

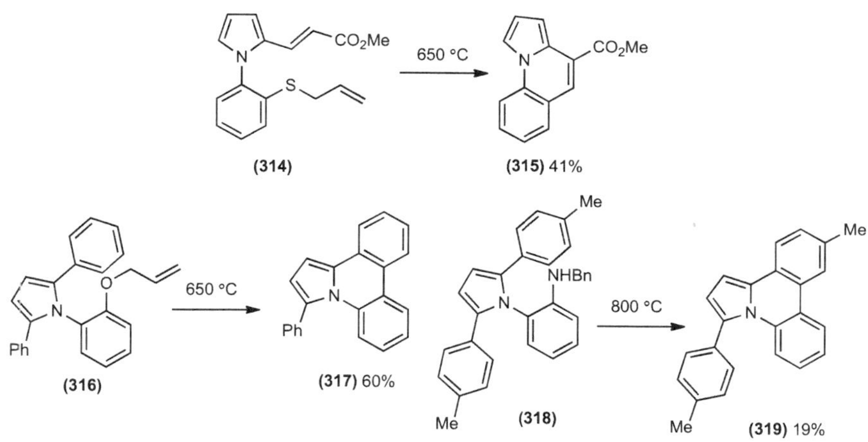


Scheme 63

The thermal cyclization of the indolyl oxime esters **320** at 650 °C to form indolopyridines **321** proceeds by a series of electrocyclic processes with loss of acetic acid (Scheme 64; 2008JOC8124). Access to the isomeric indolopyridones **323** and **325** is conveniently achieved, respectively, by high-temperature FVP of the indolylacrylate **322** and low-temperature FVP of the indolinylacrylate **324** through tungsten oxide to bring about aromatization (2009S2171).

FVP of the aminomethylene derivatives of Meldrum's acid has been widely exploited as a route to ring-fused pyridones. The simple aryl derivatives **326** give quinolones **327** in moderate-to-excellent yields (Scheme 65; 2009SL1847) although there can be some complications, such as

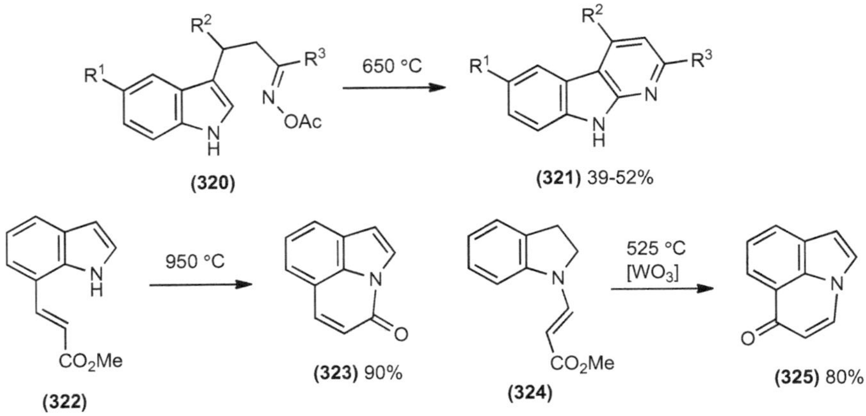


(320) (321) 39-52%

(322) (323) 90% (324) (325) 80%

**Scheme 64**

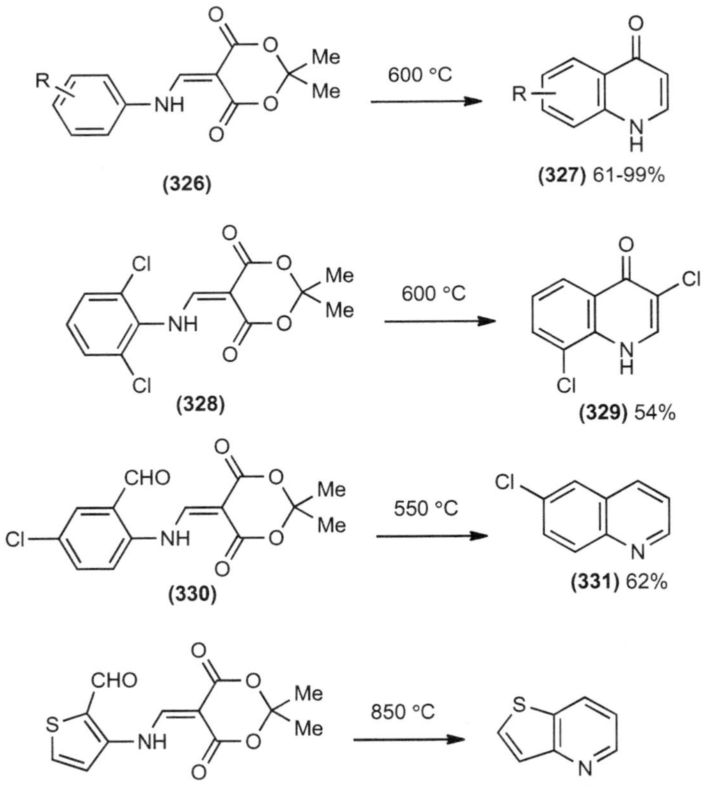


(326) (327) 61-99%

(328) (329) 54%

(330) (331) 62%

(332) (333) 57%

**Scheme 65**

(1) the unexpected chlorine migration in substrates such as **328** to give **329** and (2) the loss of $CO_2$ from formyl substrates such as **330** and **332** to give quinoline and thienopyridine products **331** and **333**, respectively (2010 NJC236).

FVP of the dimethylamino pyridylamino substrates **334** and **336** and subsequent trapping of the products with dimethylamine gives isomeric naphthyridones **335**, **337**, and **338** (Scheme 66; 2009AJC1631).

Aminomethylene Meldrum's acid derivatives **339** containing a pyrazole group give good yields of the pyrazolopyridones **340** at 500 °C, and higher temperature pyrolysis with an *N-tert*-butyl group present leads to concomitant deprotection to give **341** (Scheme 67; 2015S242). The pyridazinone compounds **342** similarly afford pyridopyridazinediones **343** but for the isomeric compounds **344** the expected products **345** are accompanied by pyrrolopyridazines **346** (2010NJC236).

## 5.2 Pyrans and Ring-Fused Derivatives

As noted in Section 2.2, when an epoxycyclopentenone **7** was formed thermally, it largely isomerized into the isomeric pyranone **8** (Scheme 3; 2000 TL9189). This is a general process and FVP of the benzo-, naphtho-, and benzofuro-cyclopentene epoxides **347**, **349**, and **351** leads to the isocoumarins **348**, and analogs **350** and **352**, the last of which is related to the natural product coriandrin (Scheme 68; 2000TL3677).

When the epoxytriones **353** were subjected to FVP, the retro-Diels–Alder loss of cyclopentadiene was accompanied by similar ring expansions to give the benzoquinone-fused pyranones **354** along with the indanone-fused decarbonylation products **355** (Scheme 69; 2009JOC1598).

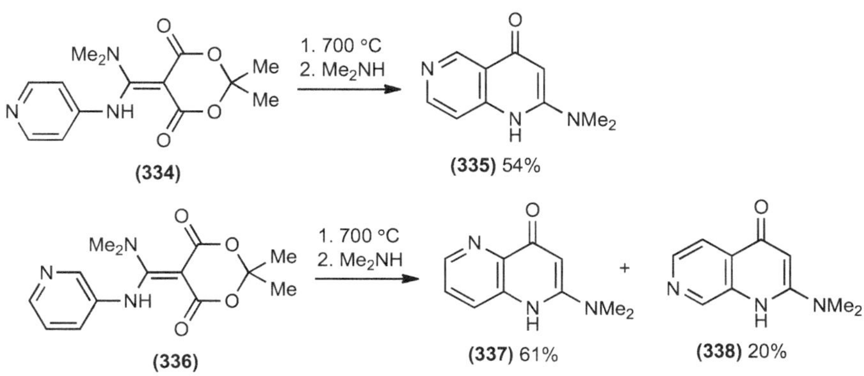


Scheme 66

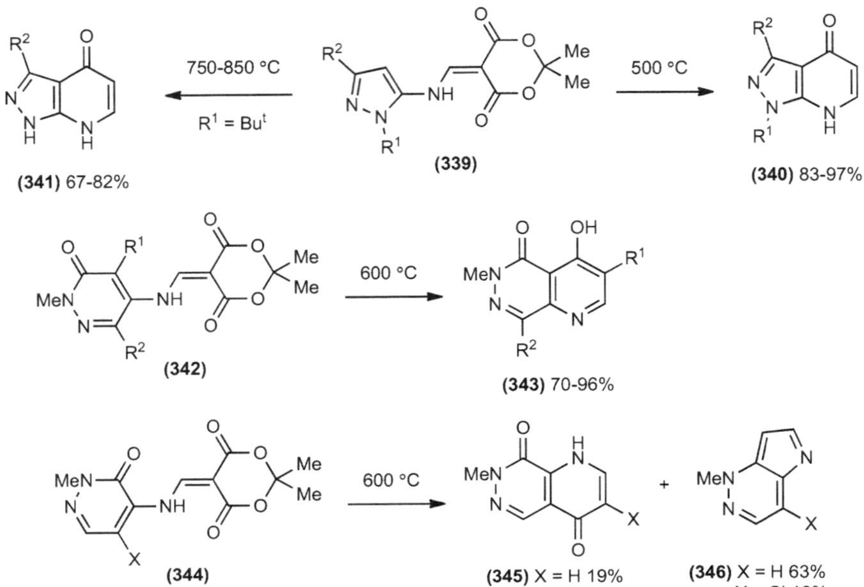


**(341)** 67-82% **(339)** **(340)** 83-97%

**(342)** **(343)** 70-96%

**(344)** **(345)** X = H 19%, X = Cl 34% **(346)** X = H 63%, X = Cl 16%

**Scheme 67**

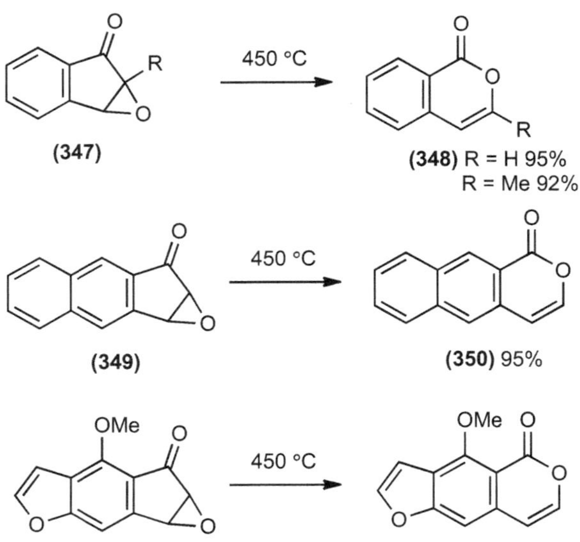


**(347)** **(348)** R = H 95%, R = Me 92%

**(349)** **(350)** 95%

**(351)** **(352)** 90%

**Scheme 68**

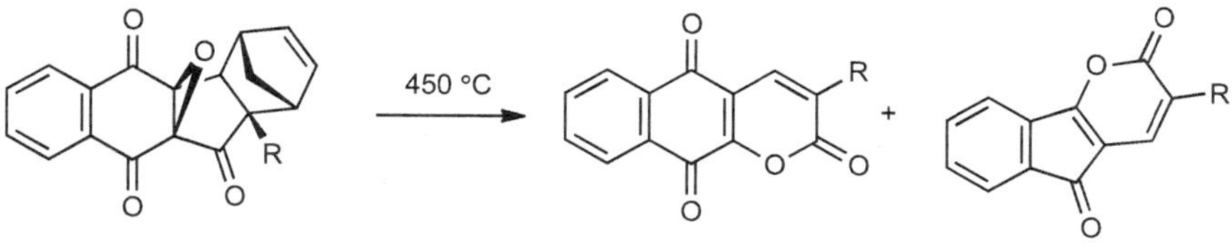

(353) (354) R = H 41% R = Me 42% (355) R = H 53% R = Me 52%

Scheme 69

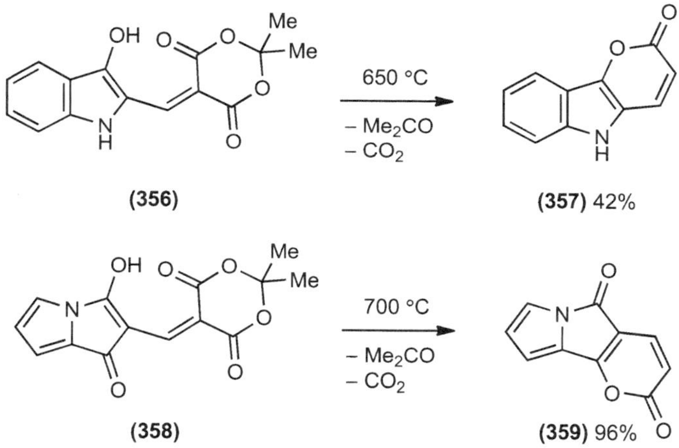

(356) (357) 42%

(358) (359) 96%

Scheme 70

Pyrolysis of Meldrum's acid derivatives with a suitably placed hydroxyl group can lead to the formation of ring-fused pyranones, as illustrated by the reactions of **356** (2010S1361) and **358** (2010OBC4383) to give the indole- and pyrrolizinone-fused products **357** and **359**, respectively (Scheme 70).

The furanones **360** undergo cyclization with loss of HCl to give alkynylcoumarins **361** in moderate yields (Scheme 71; 2014JOC65).

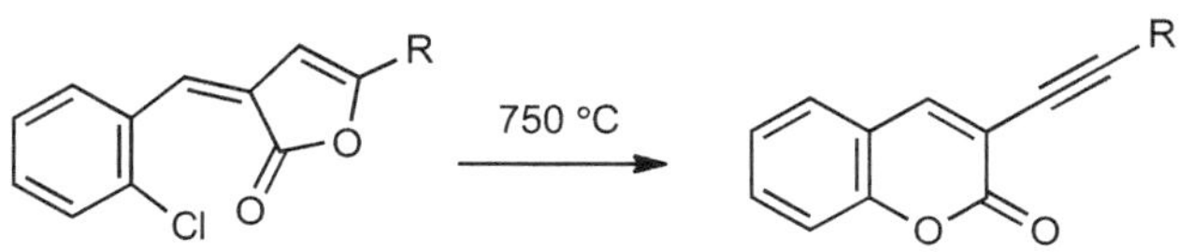

(360) (361) R = Me 52% R = Ph 40%

Scheme 71

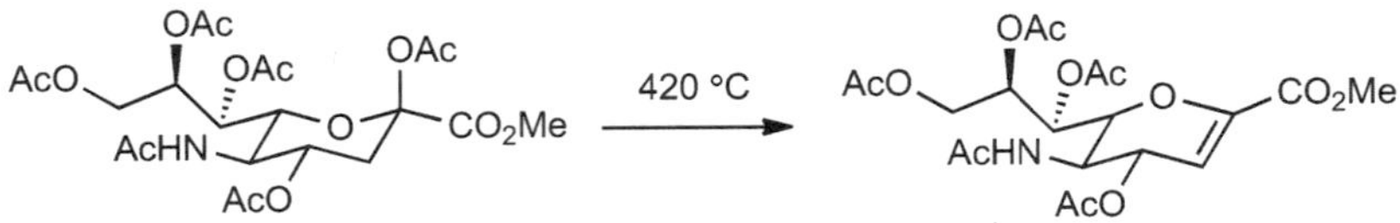

(362) (363) 95%

Scheme 72

A rare example of the use of FVP in carbohydrate chemistry is provided by the high-yielding elimination of acetic acid from the tetrahydropyran **362** to give the 3,4-dihydro-2*H*-pyran **363**, an intermediate in the synthesis of the influenza drug Relenza (zanamivir) (Scheme 72; 2009JOC4357).

## 5.3 Thiopyrans and Ring-Fused Derivatives

Compound **364** suffers a net dehydration upon FVP at 850 °C to give the benzothiapyran-4-thione **365** in a high yield (Scheme 73; 2006HCA991). Pyrolysis of the *tert*-butylthiocinnamate **366** gives a low yield of the benzothiapyran-2-one **367** (2007ARK(xi)64).

## 5.4 Diazines and Ring-Fused Derivatives

Mechanistic studies have been carried out on the formation of diazines by ring expansion of azolyl-1-methyl radicals formed upon FVP of 1-aryloxymethylpyrazoles and 1-aryloxymethylimidazoles at 700 °C (2015UP1). The pyrazole compound **368** gives exclusively pyrimidine, while the imidazole compound **369** gives both pyrazine and pyrimidine in the ratio shown in Scheme 74 (2015UP1). The effect of methyl substituents is demonstrated by the results for the imidazoles **370** and **371** (Scheme 74).

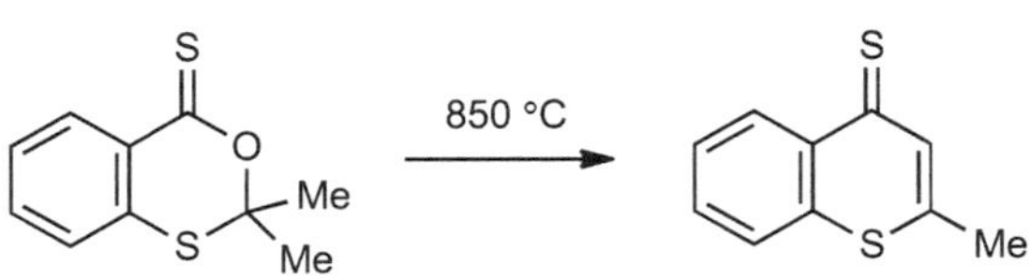

(364) (365) 89%

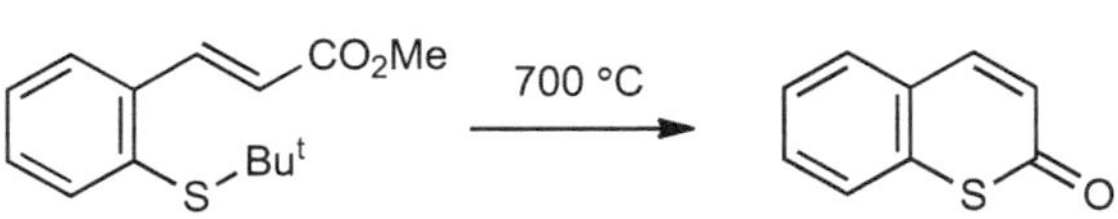

(366) (367) 28%

Scheme 73

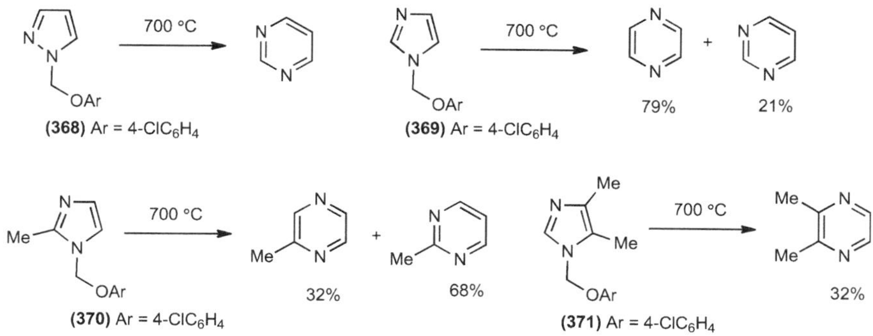


Scheme 74

FVP of both the iminyl radical precursor **372** and the pyrrolyl radical precursor **373** leads to a mixture of the same isomeric pyrroloquinoxalines **374** and **375** pointing to the involvement of a common spiro radical intermediate (Scheme 75; 2015UP4).

The hydroxypicolinic acid **376** loses water and CO upon pyrolysis to give a reactive ketene which dimerizes to the piperazinedione structure **377** (Scheme 76; 2002TL5285).

Subjecting the imidazopyrimidine **378** to high-temperature FVP results in isomerization to give the thermodynamically more stable pyrrolopyrazine **380**. The involvement of a spiro intermediate (Scheme 77) is supported both by theoretical modeling and by the fact that the imidazopyridazine **379** gives the same product under these conditions (2015UP5).

Pyrolysis of the azolyl malonamates **381** and **383** proceeds with rearrangement and loss of ethanol to give the thiazolo- and triazolo-pyrimidinedione

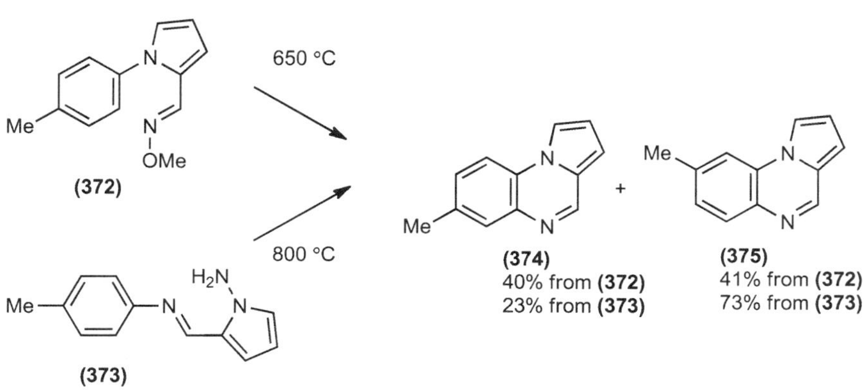


Scheme 75

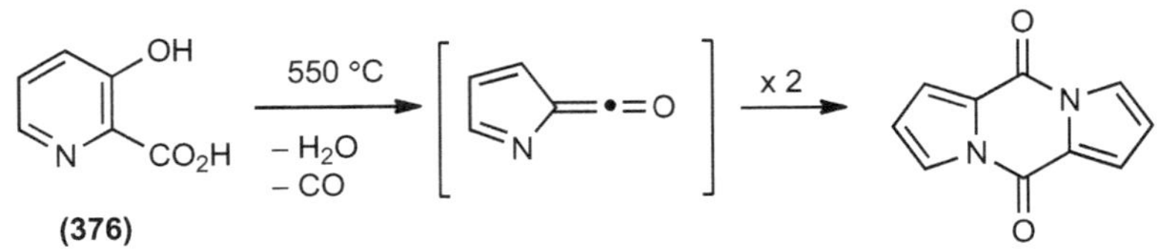


(376)

(377) 38%

Scheme 76

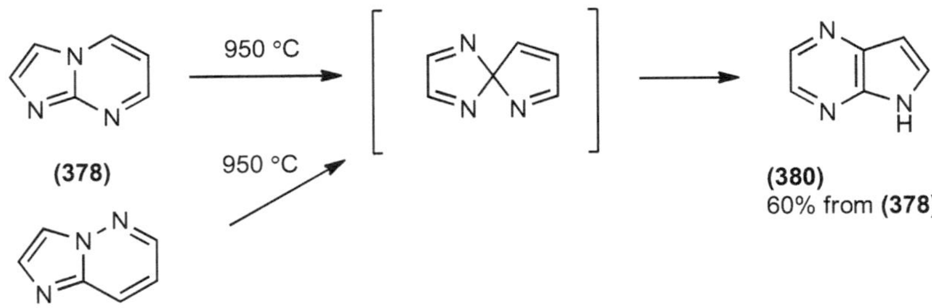


(378)

(379)

(380)
60% from (378)

Scheme 77

products **382** and **384**, respectively (Scheme 78; 2003ARK(x)262). The pyrimidinylamidine ester **385** behaves similarly to afford the pyrimidopyrimidinone product **386** (2004AJC577).

Pyrolysis of aminomethylene Meldrum's acid derivatives bearing a nitrogen heterocycle on the amino group results in ready cyclization as illustrated by conversion of **387** into **388** (2009SL1847) and **389** into **390** (2007SL2205; Scheme 79).

Convenient access to pyrazolo[2,3-*b*]pyridazinones is provided by cyclization of pyrazolylaminomethyleneketenes which are generated either from Meldrum's acid derivatives, such as **391** and **392**, or from a range of pyrazolylaminoacrylates, such as **393** and **394**, (Scheme 80; 2015UP8). It should be noted that these are the thermodynamic products formed at higher pyrolysis temperatures, and FVP of the same precursors at lower temperatures gives the isomeric mesomeric betaines with a pyrazolotriazine structure as described in Section 5.9.

## 5.5 Oxazines and Ring-Fused Derivatives

Various new thermal cyclization processes leading to 1,3-oxazines have been described including pyrolysis of (1) the benzisothiazolone **395** to give **396**

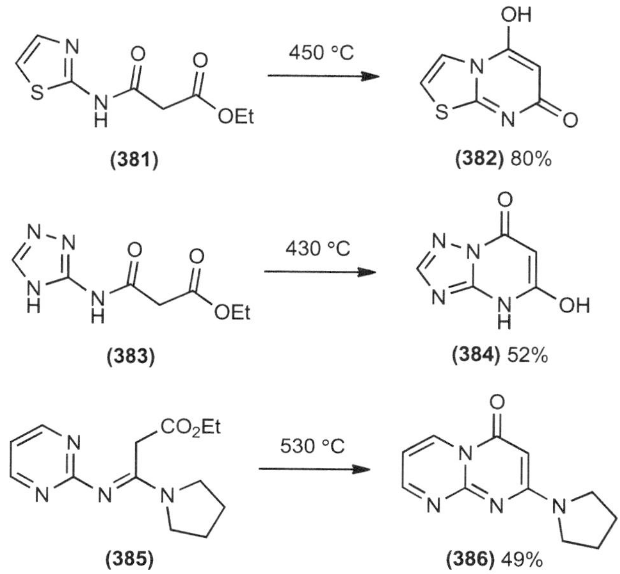


Scheme 78

(2000JCS(P1)3212) and (2) the acylnaphthotriazines **397** to give **398** (2005T10507; Scheme 81). The pyrrolinedione **399** loses CO on pyrolysis to give the iminoketene **400** which forms the dimer **401** with a 1,3-oxazine structure (2004OBC3518). The pyrrolooxazine **403** is a result of thermal $SO_2$ extrusion from the sulfone **402** and electrocyclization of the resulting azafulvenium methylide onto the benzoyl group (2001JCS(P1)1795).

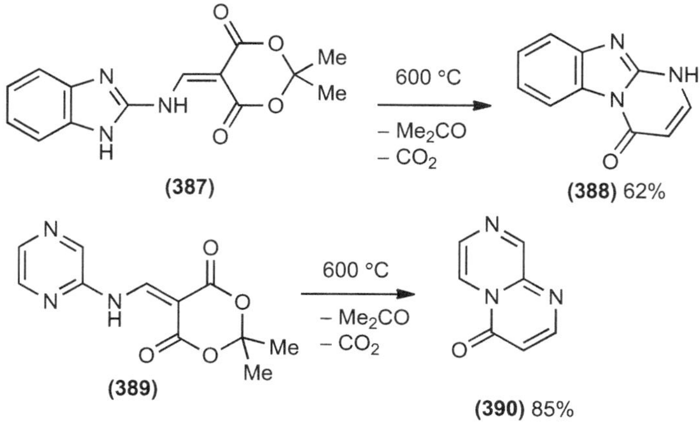


Scheme 79

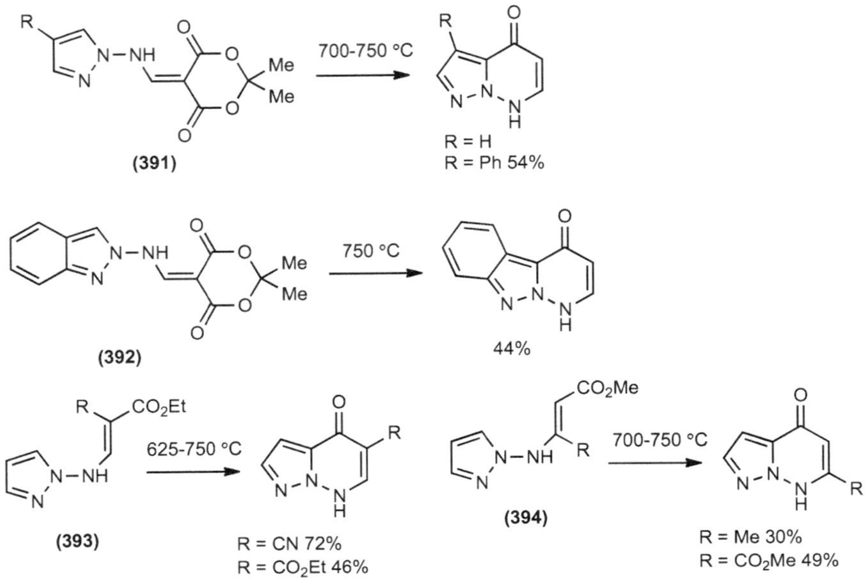


**Scheme 80**

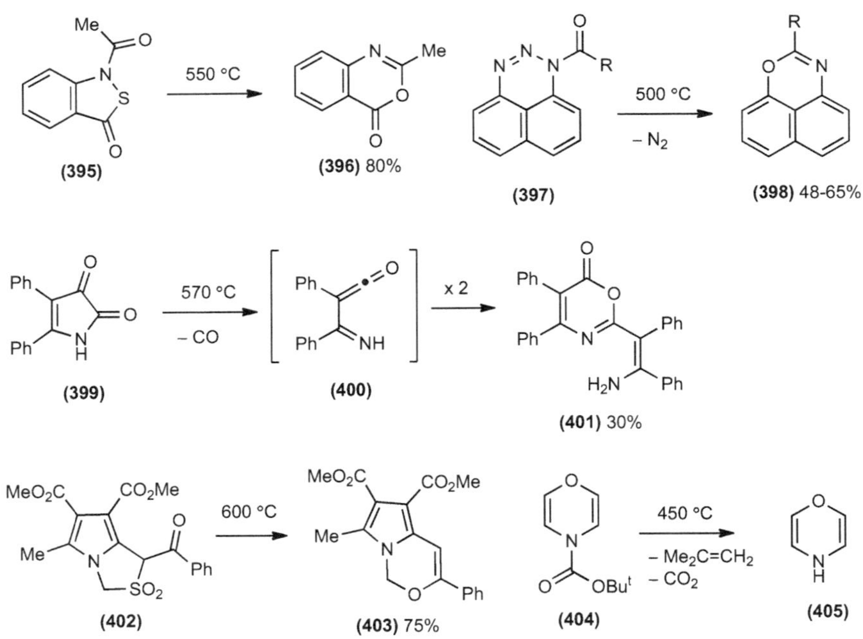


**Scheme 81**

Finally in this section, the unstable parent 1,4-oxazine **405** has been prepared for the first time by FVP deprotection of the *tert*-butoxycarbonyl derivative **404** and has been spectroscopically characterized (2013CC 11367).

## 5.6 Benzoxaphosphinines

FVP of both the 2-methoxyphenyl thiocarbonyl-stabilized ylide **406** and the 2-chlorophenyl carbonyl-stabilized analog **407** gives a variety of volatile products but leaves a residue of the 1,4-benzoxaphosphininium salt **408** in the inlet tube (2014H(88)1135; Scheme 82).

## 5.7 Dioxinones

Thermal extrusion of CO from the furandione **409** gives methoxycarbonyl(pivaloyl) ketene 410 which dimerizes to the dioxinone **411** in a high yield (Scheme 83; 2001T6757). Co-pyrolysis of the methyl ester **409** with the *tert*-butyl ester **412** gives the hetero-dimer **413** with a different dioxinone structure (2002JCS(P1)599).

## 5.8 Dithiins

A rare example of this type is provided by the conversion of the benzo[*d*][1,3]oxathiin-4-thione **35** into the dithiin-4-one **36** upon FVP at 570 °C (Section 3.1; Scheme 6; 2008HCA766).

## 5.9 Triazines and Ring-Fused Derivatives

When pyrazolylaminomethyleneketenes are generated either from Meldrum's acid derivatives such as **391** and **392** or from a range of pyrazolylaminoacrylates such as **393** and **394** at lower temperatures, the kinetic products with mesomeric betaine structures containing a 1,2,3-triazine are formed in moderate to good yields (Scheme 84; 2015UP8). The formation of the thermodynamic products at higher pyrolysis temperatures is discussed in Section 5.4.

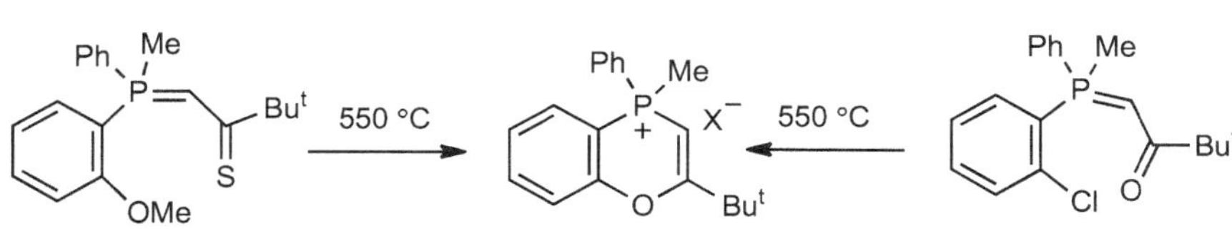


**(406)** **(408)** 20% **(407)**

Scheme 82

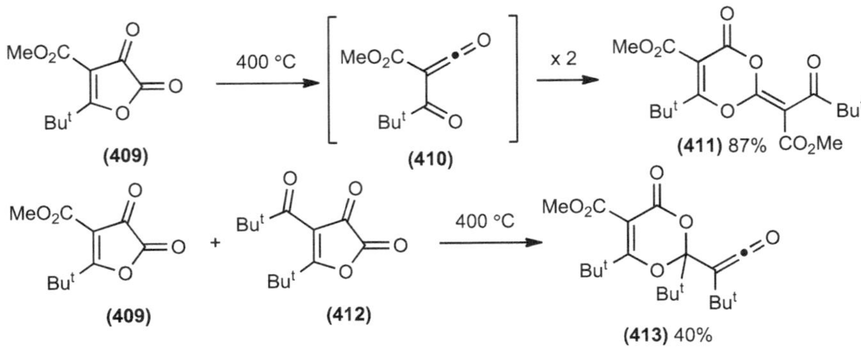


Scheme 83

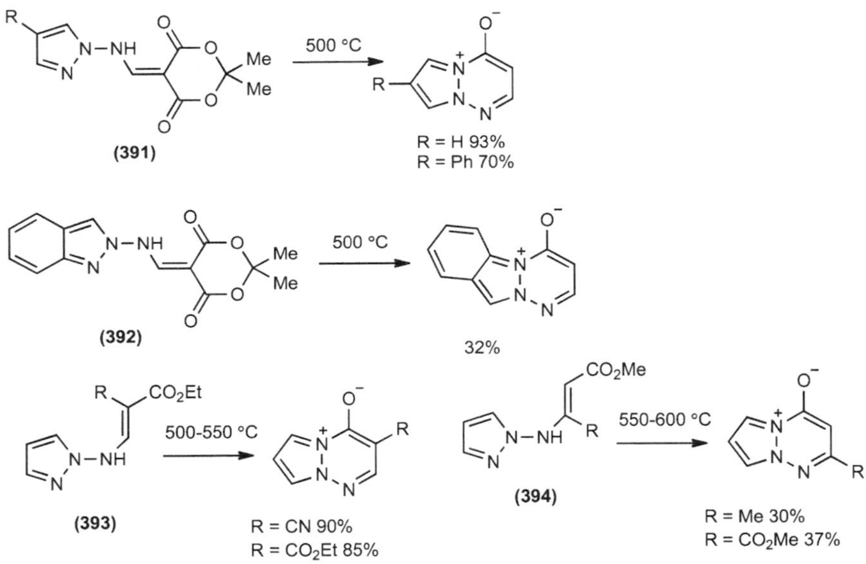


Scheme 84

# 6. SEVEN-MEMBERED RINGS

## 6.1 Azepines and Ring-Fused Derivatives

The pyrolysis of benzotriazolylacetophenone **414** has been examined carefully under conditions of catalytic FVP. Pyrolysis over the mesoporous molecular sieve Al-MCM-41 gives a higher yield and product selectivity with respect to the dibenzazepinone **415** (Scheme 85; 2007OL2179,

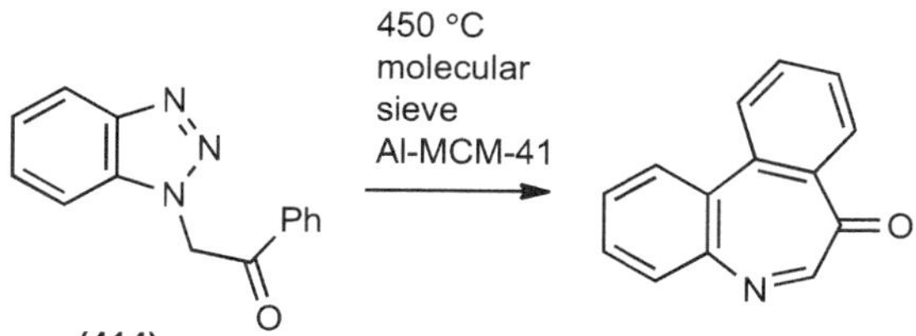


**(414)** **(415)** 87%

**Scheme 85**

2010MI98). In more recent work (2013ACSC1020) barium tungstate $BaWO_4$ has also proved to be an effective catalyst in this role.

Electrocyclic ring opening of the cyclobutene ring in the adduct **416** occurs upon FVP at 500 °C to give the bridged azepine structure **417** (Scheme 86; 2000CEJ3706).

The amino acid-derived stabilized phosphonium ylides **418** undergo double bond isomerization and condensation with loss of ethanol to afford the azepinedione ylides **419** in low yields (Scheme 87; 2001TL141).

The strained isoxazolines **420**, formed by nitrone cycloaddition to cyclopropylidenecyclobutane, undergo ready ring expansion to give the hexahydroazepinones **421** upon FVP at 600 °C (Scheme 88; 2001EJO3789).

## 6.2 Oxepines and Ring-Fused Derivatives

Electrocyclic ring opening of the cyclobutene rings in adducts **422** and **424** produces the bridged oxepine products **423** and **425** in high yields (Scheme 89; 2000CEJ3706).

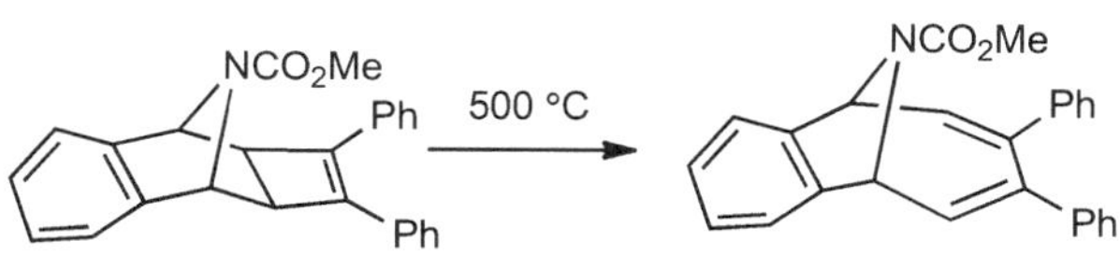


**(416)** **(417)** 96%

**Scheme 86**

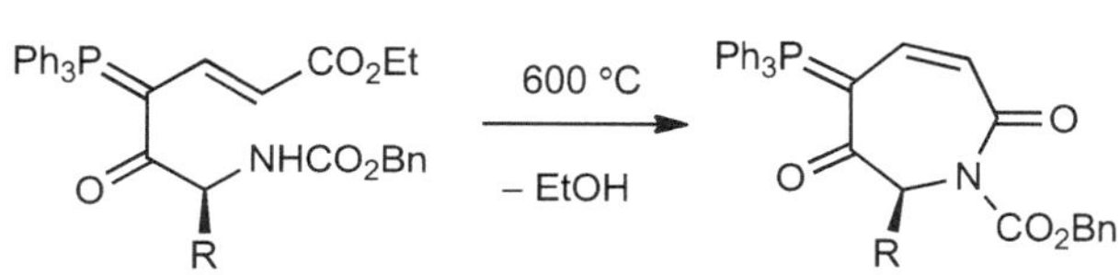


**(418)** R = H, Me, Pr$^{i}$ **(419)** 13-26%

**Scheme 87**

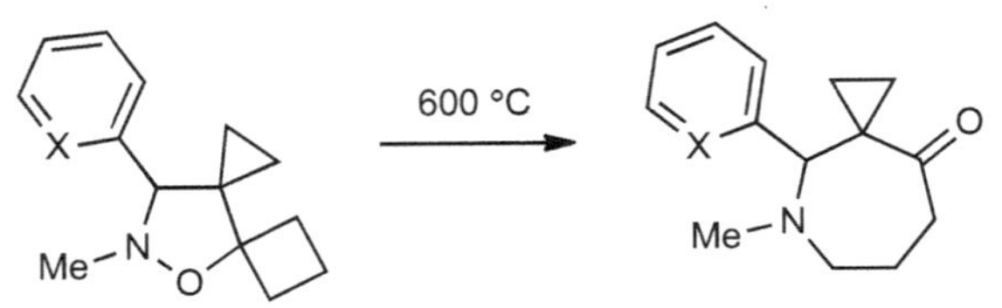


**(420)**

**(421)** X = CH 32%
X = N 30%

**Scheme 88**

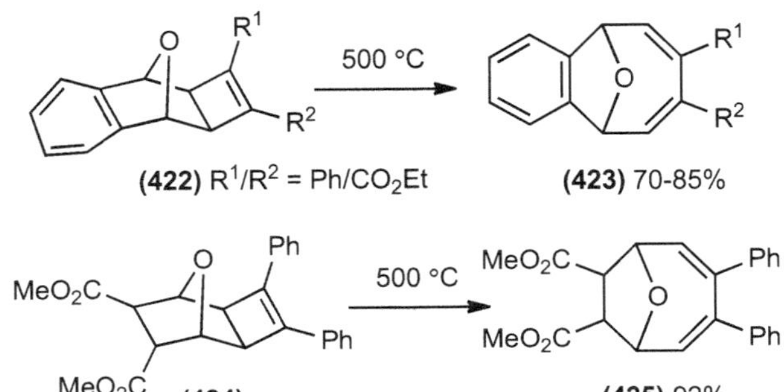


**(422)** $R^1/R^2$ = Ph/$CO_2Et$

**(423)** 70-85%

**(424)**

**(425)** 92%

**Scheme 89**

In a remarkable process, compound **426** loses nitrogen and the resulting carbene undergoes a Wolff rearrangement to give the ketene which then undergoes [2 + 2]-cycloaddition with the C=N double bond under the influence of the "chiral tether" to give the β-lactam **427** containing an oxepane ring in high diastereomeric excess (Scheme 90; 2003TL3115).

## 6.3 Dioxepines and Ring-fused Derivatives

Very high stereoselectivity caused by the chiral tether is observed when the ketene derived from FVP of diazo ester **428** followed by a Wolff rearrangement adds intramolecularly to give dioxepane-containing cyclobutanone **429** (Scheme 91; 2003TL3115).

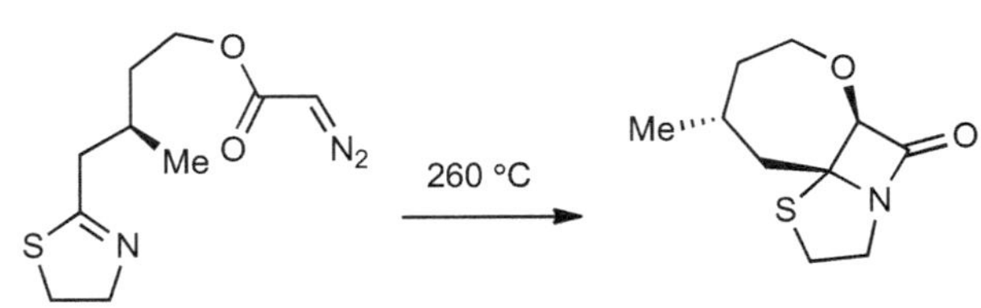


**(426)**

**(427)** 15-40%, 86% d.e.

**Scheme 90**

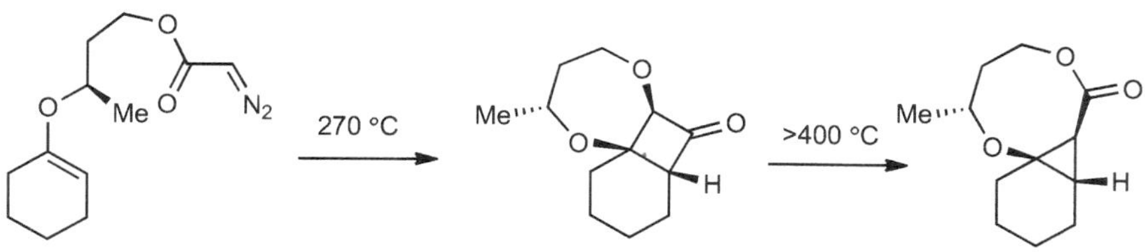


(428)

(429) 70%, > 99% d.e.

(430)

Scheme 91

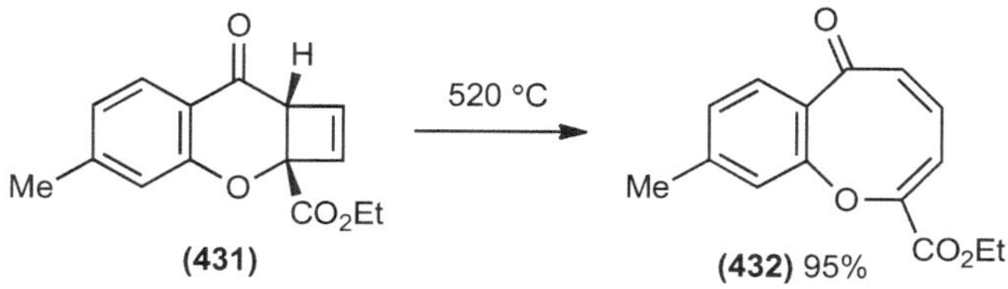


(431)

(432) 95%

Scheme 92

# 7. EIGHT-MEMBERED AND LARGER RINGS

## 7.1 Benzoxocanes

Convenient access to the benzoxocanone **432** is obtained by electrocyclic ring opening of the cyclobutene ring in compound **431** upon FVP at 520 °C (Scheme 92; 2004TL9653).

## 7.2 Dioxocanes

In contrast to formation of the product **429** at 270 °C(Section 6.3), when a chiral diazo ester such as **428** is pyrolyzed at a higher temperature the alternative isomeric product **430** containing a dioxocanone ring is formed by a rearrangement of the initial product **429** (Scheme 91; 2003TL3115).

## 7.3 Trithiacyclododecanes

A rare example of this type is provided by the formation of compound **34** in 30% yield by FVP of 2-mercaptonicotinic acid at 550 °C (Section 3.1; Scheme 6; 2002TL5285).

# 8. CONCLUSION AND OUTLOOK

There has been a remarkable progress in this area over the last 15 years. Specialist groups have continued to develop new thermal processes and extended the scope of previously known ones. At the same time, a significant number of routine applications of FVP in the course of target-directed synthesis have been described. For the method to be more widely adopted, it is perhaps useful to identify the main types of reaction where it consistently delivers predictable results. The following reactions are commonly employed:

- retro-Diels–Alder reactions, thus allowing the Diels–Alder reaction to be used as both a protection strategy and a chiral template strategy;
- isomerization involving ring expansion with relief of ring strain;
- extrusion of $N_2$ from diazo compounds, azides, triazoles, and tetrazoles to give products derived from the corresponding carbenes and nitrenes;
- extrusion of $SO_2$ from cyclic sulfones, although SO may occasionally be lost instead;
- extrusion of $CO_2$ from isoxazolones and lactones, although in the latter case CO may sometimes be lost instead;
- extrusion of $Ph_3PO$ from carbonyl-containing phosphonium ylides, usually to give alkynes;
- formation of aryl radicals from nitroarenes;
- formation of heteroatom-centred radicals by homolysis of the bond in groups such as O–Me, O–allyl, S–Me, S–allyl, N–benzyl, N–tosyl, N–amino, and N–OMe;
- formation and cyclization of alkylideneketenes (propadienones) from either Meldrum's acid derivatives (loss of $CO_2$ and acetone) or acrylates (loss of alcohol);
- cyclization to form five- and six-membered ring products with elimination of water, HCl, methanol, or ethanol.

By designing precursors capable of undergoing several of the aforementioned reaction types in sequence, elegant cascade processes leading to polycyclic products are possible. The potential of FVP over solid reagents and catalysts has only been recognized in a few cases, and this is surely an area set for future expansion.

## ACKNOWLEDGMENTS

We are grateful to Dr Lilian McNab for assistance in obtaining access to the unpublished work of Prof. Hamish McNab and to Cambridge Display Technology Ltd and the EPSRC (UK) for a studentship to Y.B.

## REFERENCES

1980MI1 R.F.C. Brown, *Pyrolytic methods in organic chemistry*, Academic Press: New York (1980).
1981AHC(28)231 C. Wentrup, *Adv. Heterocycl. Chem.*, **28**, 231 (1981).

| | |
|---|---|
| 1982RTC317 | U. Wiersum, *Rec. Trav. Chim. Pays-Bas*, **101**, 317 (1982). |
| 1982RTC365 | U. Wiersum, *Rec. Trav. Chim. Pays-Bas*, **101**, 365 (1982). |
| 1984MI31 | U. Wiersum, *Aldrichim. Acta*, **17**, 31 (1984). |
| 1992PHC1 | R.A. Aitken, I. Gosney, and J.I.G. Cadogan, *Prog. Heterocycl. Chem.*, **4**, 1 (1992). |
| 1993PHC1 | R.A. Aitken, I. Gosney, and J.I.G. Cadogan, *Prog. Heterocycl. Chem.*, **5**, 1 (1993). |
| 1996COS373 | H. McNab, *Contemp. Org. Synth.*, **3**, 373 (1996). |
| 1997MI1 | Y. Vallée, *Gas-Phase Reactions in Organic Synthesis*, Gordon and Breach: Amsterdam (1997). |
| 1997MI473 | R.A. Aitken, A.W. Thomas, *Chemistry of the functional groups*, **Suppl. A3**, S. Patai (ed.), Wiley, New York, 473 (1997). |
| 1999CR1163 | A.J.H. Klunder, J. Zhu, and B. Zwanenburg, *Chem. Rev.*, **99**, 1163 (1999). |
| 2000ARK(v)798 | R.A. Aitken, *Arkivoc*, **v**, 798 (2000). |
| 2000CEJ3706 | D.-J. Huang, D.K. Rayabarapu, L.-P. Li, T. Sambaiah, and C.-H. Cheng, *Chem. Eur. J.*, **6**, 3706 (2000). |
| 2000HCA1022 | M. Nagel and H.-J. Hansen, *Helv. Chim. Acta*, **83**, 1022 (2000). |
| 2000JCS(P1)3212 | W.K. Janowski, R.H. Prager, and J.A. Smith, *J. Chem. Soc., Perkin Trans. 1*, 3212 (2000). |
| 2000MOL1130 | J.C. Pommelet, F. Jourdain, and H. Dhimane, *Molecules*, **5**, 1130 (2000). |
| 2000TL3677 | D. Mal, M. Bandyopadhyay, S.K. Ghorai, and K. Datta, *Tetrahedron Lett.*, **41**, 3677 (2000). |
| 2000TL9189 | A.J. Derksen, A.C.L.M. van der Waals, M.H.A. van Ham, M. van Dongen, J.H. Borkent, R. de Gelder, A.J.H. Klunder, and B. Zwanenburg, *Tetrahedron Lett.*, **41**, 9189 (2000). |
| 2001AGE393 | G. Mlostoń, J. Romański, H.P. Reizenauer, and G. Maier, *Angew. Chem. Int. Ed.*, **40**, 393 (2001). |
| 2001ARK(vii)88 | M. Cox, R.H. Prager, and D.M. Riessen, *Arkivoc*, **vii**, 88 (2001). |
| 2001EJO3789 | A. de Meijere, M. von Seebach, S.I. Kozhushkov, R. Boese, D. Bläser, S. Cicchi, T. Dimoulas, and A. Brandi, *Eur. J. Org. Chem.*, 3789 (2001). |
| 2001JCS(P1)424 | D.D.J. Cartwright, B.A.J. Clark, and H. McNab, *J. Chem. Soc., Perkin Trans. 1*, 424 (2001). |
| 2001JCS(P1)1795 | O.B. Sutcliffe, R.C. Storr, T.L. Gilchrist, and P. Rafferty, *J. Chem. Soc., Perkin Trans. 1*, 1795 (2001). |
| 2001S2059 | A.E.M. Gaber and H. McNab, *Synthesis*, 2059 (2001). |
| 2001SL228 | R.A. Aitken and A.N. Garnett, *Synlett*, 228 (2001). |
| 2001SL433 | A. Brandi, S. Cicchi, M. Brandl, S.I. Kozhushkov, and A. De Meijere, *Synlett*, 433 (2001). |
| 2001T6757 | A. Stadler, K. Zangger, F. Belaj, and G. Kollenz, *Tetrahedron*, **57**, 6757 (2001). |
| 2001TA1913 | P. Kongsaeree, P. Meepowpan, and Y. Thebtaranonth, *Tetrahedron Asymm.*, **12**, 1913 (2001). |
| 2001TL141 | R.A. Aitken, G.M. Buchanan, N. Karodia, T. Massil, and R.J. Young, *Tetrahedron Lett.*, **42**, 141 (2001). |
| 2002ARK(viii)125 | H. McNab and M. Morrow, *Arkivoc*, **viii**, 125 (2002). |
| 2002JCS(P1)533 | R.A. Aitken, N. Karodia, T. Massil, and R.J. Young, *J. Chem. Soc., Perkin Trans. 1*, 533 (2002). |
| 2002JCS(P1)548 | A.E.M. Gaber, G.A. Hunter, and H. McNab, *J. Chem. Soc., Perkin Trans. 1*, 548 (2002). |
| 2002JCS(P1)599 | B.C. Wallfisch, F. Belaj, C. Wentrup, C.O. Kappe, and G. Kollenz, *J. Chem. Soc., Perkin Trans. 1*, 599 (2002). |

2002JCS(P1)1232 V.V. Ramana Rao and C. Wentrup, *J. Chem. Soc., Perkin Trans. 1*, 1232 (2002).

2002JCS(P1)1366 C. Plüg, A. Kuhn, and C. Wentrup, *J. Chem. Soc., Perkin Trans. 1*, 1366 (2002).

2002MI389 M.M. Baradarani and R.H. Prager, *Iran. J. Sci. Technol.*, **26**, 389 (2002).

2002SC875 S. Lesniak and B. Pasternak, *Synth. Commun.*, **32**, 875 (2002).

2002SL593 I. González-Temprano, N. Sotomayor, and E. Lete, *Synlett*, 593 (2002).

2002TL5285 C.-H. Chou, S.-J. Chiu, and W.-M. Liu, *Tetrahedron Lett.*, **43**, 5285 (2002).

2003AJC913 T. Lister, R.H. Prager, M. Tsaconas, and K.L. Wilkinson, *Aust. J. Chem.*, **56**, 913 (2003).

2003ARK(x)262 W.J. Peláez, I.T. Gafarova, and G.I. Yranzo, *Arkivoc*, **x**, 262 (2003).

2003RJOC752 N.A. Korchevina, E.N. Sukhomazova, and E.N. Deryagina, *Russ. J. Org. Chem.*, **39**, 752 (2003).

2003TL3115 T. Sugimura, T. Tei, and T. Okuyama, *Tetrahedron Lett.*, **44**, 3115 (2003).

2004AJC455 M. Cox, M. Dixon, T. Lister, and R.H. Prager, *Aust. J. Chem.*, **57**, 455 (2004).

2004AJC577 A. Clark and R.H. Prager, *Aust. J. Chem.*, **57**, 577 (2004).

2004ARK(ii)4 D. Meibom, H. Weinmann, and E. Winterfeldt, *Arkivoc*, **ii**, 4 (2004).

2004COC1071 G.I. Yranzo and E.L. Moyano, *Curr. Org. Chem.*, **8**, 1071 (2004).

2004JOC4477 G.F. Busscher, S. Groothuys, R. de Gelder, F.P.J.T. Rutjes, and F.L. van Delft, *J. Org. Chem.*, **69**, 4477 (2004).

2004MI19 H. McNab, *Aldrichim. Acta*, **37**, 19 (2004).

2004OBC3518 L. George, P.V. Bernhardt, K.-P. Netsch, and C. Wentrup, *Org. Biomol. Chem.*, **2**, 3518 (2004).

2004T3949 T.M.V.D. Pinho e Melo, C.I.A. Santos, A.M.d'A. Rocha Gonsalves, J.A. Paixão, and A.M. Beja, *Tetrahedron*, **60**, 3949 (2004).

2004TL3889 T.M.V.D. Pinho e Melo, M.I.L. Soares, A.M.d'A. Rocha Gonsalves, and H. McNab, *Tetrahedron Lett.*, **45**, 3889 (2004).

2004TL9653 S.K. Sabui and R.V. Venkateswaran, *Tetrahedron Lett.*, **45**, 9653 (2004).

2005BMC5299 K.L. Burgess, M.S. Corbett, P. Eugenio, N.J. Lajkiewicz, X. Liu, A. Sanyal, W. Yan, Q. Yuan, and J.K. Snyder, *Bioorg. Med. Chem.*, **13**, 5299 (2005).

2005JOC6629 T.M.V.D. Pinho e Melo, M.I.L. Soares, A.M.d'A. Rocha Gonsalves, J.A. Paixão, A.M. Beja, and M.R. Silva, *J. Org. Chem.*, **70**, 6629 (2005).

2005OL31 K.L. Burgess, N.J. Lajkiewicz, A. Sanyal, W. Yan, and J.K. Snyder, *Org. Lett.*, **7**, 31 (2005).

2005RJOC1647 V.E. Platonov, A.M. Maksimov, K.V. Dvornikova, and P.V. Nikul'shin, *Russ. J. Org. Chem.*, **41**, 1647 (2005).

2005T7489 W.J. Peláez, G.I. Yranzo, C. Gróf, Z. Riedl, and G. Hajós, *Tetrahedron*, **61**, 7489 (2005).

2005T10507 H. Al-Awadi, M.R. Ibrahim, H.H. Dib, N.A. Al-Awadi, and Y.A. Ibrahim, *Tetrahedron*, **61**, 10507 (2005).

2005TL2475 A. Sanyal, Q. Yuan, and J.K. Snyder, *Tetrahedron Lett.*, **46**, 2475 (2005).

2005TL3093 S. Leśniak and B. Pasternak, *Tetrahedron Lett.*, **46**, 3093 (2005).

2006EJO3721 G. Maier, H.P. Reizenauer, J. Romański, H. Petzold, and G. Mlostoń, *Eur. J. Org. Chem.*, 3721 (2006).

2006HCA991 T. Drewnowski, S. Leśniak, G. Mloston, R. Siedlecka, and J. Skarzewski, *Helv. Chim. Acta*, **89**, 991 (2006).

2006JOC659 Y. Uchiyama, N. Kano, and T. Kawashima, *J. Org. Chem.*, **71**, 659 (2006).

| | |
|---|---|
| 2006OBC407 | M. Babjak, A. Kanazawa, R.J. Anderson, and A.E. Greene, *Org. Biomol. Chem.*, **4**, 407 (2006). |
| 2006OBC2446 | J.I.G. Cadogan, C.M. Hewage, H. McNab, A.D. MacPherson, I.S. Nicolson, D. Reed, and I.H. Sadler, *Org. Biomol. Chem.*, **4**, 2446 (2006). |
| 2006OBC2912 | D.J. Aldous, A.S. Batsanov, D.S. Yufit, A.J. Dalençon, W.M. Dutton, and P.G. Steele, *Org. Biomol. Chem.*, **4**, 2912 (2006). |
| 2006TL791 | T.M.V.D. Pinho e Melo, M.I.L. Soares, and A.M.d'A. Rocha Gonsalves, *Tetrahedron Lett.*, **47**, 791 (2006). |
| 2006TL4377 | S. Jones and I. Wilson, *Tetrahedron Lett.*, **47**, 4377 (2006). |
| 2007ARK(xi)64 | A. Brown, J.I.G. Cadogan, A.D. MacPherson, and H. McNab, *Arkivoc*, **xi**, 64 (2007). |
| 2007ARK(xi)85 | H. McNab, D. Reed, I.D. Tipping, and R.G. Tyas, *Arkivoc*, **xi**, 85 (2007). |
| 2007JOC8760 | H. McNab and R.G. Tyas, *J. Org. Chem.*, **72**, 8760 (2007). |
| 2007OL2179 | E.L. Moyano, P.L. Lucero, G.A. Eimer, E.R. Herrero, and G.I. Yranzo, *Org. Lett.*, **9**, 2179 (2007). |
| 2007SL2205 | N.A. Al-Awadi, I.A. Abdelhamid, A.M. Al-Etaibi, and M.H. Elnagdi, *Synlett*, 2205 (2007). |
| 2007T1833 | T.M.V.D. Pinho e Melo, M.I.L. Soares, and C.M. Nunes, *Tetrahedron*, **63**, 1833 (2007). |
| 2007T12621 | R. Slamet and D. Wege, *Tetrahedron*, **63**, 12621 (2007). |
| 2008CHE113 | J. Khalafy, A. Poursattar Marjani, and A.R. Molla Ebrahimlo, *Chem. Heterocycl. Compd.*, **44**, 113 (2008). |
| 2008CJC119 | K.A. Kolmakov and A.J. Kresge, *Can. J. Chem.*, **86**, 119 (2008). |
| 2008EJO2998 | J. Romański, H.P. Reizenauer, H. Petzold, W. Weigand, P.R. Schreiner, and G. Mlostoń, *Eur. J. Org. Chem.*, 2998 (2008). |
| 2008HCA766 | T. Drewnowski, A. Chrostowska, S. Leśniak, A. Dargelos, and S. Khayar, *Helv. Chim. Acta*, **91**, 766 (2008). |
| 2008JOC6642 | L.A. Crawford, H. McNab, A.R. Mount, and S.I. Wharton, *J. Org. Chem.*, **73**, 6642 (2008). |
| 2008JOC8124 | F. Portela-Cubillo, B.A. Surgenor, R.A. Aitken, and J.C. Walton, *J. Org. Chem.*, **73**, 8124 (2008). |
| 2008JOC9781 | R.A. Aitken and L. Murray, *J. Org. Chem.*, **73**, 9781 (2008). |
| 2008OBC2334 | L.A. Crawford, N.C. Clemence, H. McNab, and R.G. Tyas, *Org. Biomol. Chem.*, **6**, 2334 (2008). |
| 2008S1676 | X.L.M. Despinoy, H. McNab, and R.G. Tyas, *Synthesis*, 1676 (2008). |
| 2008T9745 | M.I.L. Soares, S.M.M. Lopes, P.F. Cruz, R.M.M. Brito, and T.M.V.D. Pinho e Melo, *Tetrahedron*, **64**, 9745 (2008). |
| 2008TL4145 | C.-H. Chou, P.-C. Yu, and B.-C. Wang, *Tetrahedron Lett.*, **49**, 4145 (2008). |
| 2009AJC1631 | D. Lecoq, B.A. Chalmers, R.N. Veedu, D. Kvaskoff, P.V. Bernhardt, and C. Wentrup, *Aust. J. Chem.*, **62**, 1631 (2009). |
| 2009ARK(xiv)200 | W.R. Mitchell and R.M. Paton, *Arkivoc*, **xiv**, 200 (2009). |
| 2009CEJ5482 | S.I. Wharton, J.B. Henry, H. McNab, and A.R. Mount, *Chem. Eur. J.*, **15**, 5482 (2009). |
| 2009CRC378 | R. Umeda, H. Fukuda, K. Miki, S.M. Abdur Rahman, M. Sonoda, and Y. Tobe, *C. R. Chimie*, **12**, 378 (2009). |
| 2009JAAP369 | A.E.M. Gaber and H. McNab, *J. Anal. Appl. Pyrolysis*, **86**, 369 (2009). |
| 2009JOC1598 | S.R. De, S.K. Ghorai, and D. Mal, *J. Org. Chem.*, **74**, 1598 (2009). |
| 2009JOC4278 | A.P. Gaywood and H. McNab, *J. Org. Chem.*, **74**, 4278 (2009). |
| 2009JOC4357 | E.J. Horn and J. Gervay-Hague, *J. Org. Chem.*, **74**, 4357 (2009). |

2009NJC2402 R.A. Aitken and A.N. Garnett, *New J. Chem.*, **33**, 2402 (2009).
2009OBC2187 X.L.M. Despinoy and H. McNab, *Org. Biomol. Chem.*, **7**, 2187 (2009).
2009OBC4936 H. McNab, M. Morrow, S. Parsons, D.A. Shannon, and K. Withell, *Org. Biomol. Chem.*, **7**, 4936 (2009).
2009OBC5173 J.I.G. Cadogan, B.A.J. Clark, D. Ford, R.J. MacDonald, A.D. MacPherson, H. McNab, I.S. Nicolson, D. Reed, and C.C. Sommerville, *Org. Biomol. Chem.*, **7**, 5173 (2009).
2009PS(184)1269 T. Drewnowski, S. Leśniak, A. Chrostowska, A. Dargelos, and S. Khayar, *Phosphorus, Sulfur, and Silicon Relat. Elem.*, **184**, 1269 (2009).
2009S2171 H. McNab, D.J. Nelson, and E.J. Rozgowska, *Synthesis*, 2171 (2009).
2009S2531 L. Hill, G.A. Hunter, S.H. Imam, H. McNab, and W.J. O'Neill, *Synthesis*, 2531 (2009).
2009S2535 L. Hill, S.H. Imam, H. McNab, and W.J. O'Neill, *Synthesis*, 2535 (2009).
2009SL1587 L.R. Donaldson, D. Haigh, and A.N. Hulme, *Synlett*, 1587 (2009).
2009SL1847 L. Hill, S.H. Imam, H. McNab, and W.J. O'Neill, *Synlett*, 1847 (2009).
2009T6364 S. Leśniak, R.B. Nazarski, and B. Pasternak, *Tetrahedron*, **65**, 6364 (2009).
2009T6382 R. Jongkol, R. Choommongkol, B. Tarnchompoo, P. Nimmanpipug, and P. Meepowpan, *Tetrahedron*, **65**, 6382 (2009).
2010ARK(x)34 W.R. Mitchell and R.M. Paton, *Arkivoc*, **x**, 34 (2010).
2010MI98 E.L. Moyano, G.A. Eimer, P.L. Lucero, C.M. Chanquia, E.R. Herrero, and G.I. Yranzo, *Appl. Catal. A: General*, **373**, 98 (2010).
2010NJC236 A.P. Gaywood, L. Hill, S.H. Imam, H. McNab, G. Neumajer, W.J. O'Neill, and P. Mátyus, *New. J. Chem.*, **34**, 236 (2010).
2010OBC2961 M. Black, J.I.G. Cadogan, and H. McNab, *Org. Biomol. Chem.*, **8**, 2961 (2010).
2010OBC4383 H. McNab, J. Montgomery, S. Parsons, and D.G. Tredgett, *Org. Biomol. Chem.*, **8**, 4383 (2010).
2010OBC5166 A.P. Gaywood and H. McNab, *Org. Biomol. Chem.*, **8**, 5166 (2010).
2010S923 L.A. Crawford, H. McNab, A.R. Mount, J. Verhille, and S.I. Wharton, *Synthesis*, 923 (2010).
2010S1361 A.P. Gaywood and H. McNab, *Synthesis*, 1361 (2010).
2010T6331 H. Miyatake-Ondozabal and A.G.M. Barrett, *Tetrahedron*, **66**, 6331 (2010).
2010TA2719 S.A. Hasbullah and S. Jones, *Tetrahedron Asymm.*, **21**, 2719 (2010).
2011AGE1720 X. Zeng, H. Beckers, H. Willner, and J.F. Stanton, *Angew. Chem. Int, Ed.*, **50**, 1720 (2011).
2011CEJ10071 K. Banert, F. Köhler, A. Melzer, I. Scharf, G. Rheinwald, T. Rüffer, H. Lang, R. Herges, K. Hess, N. Ghavtadze, and E.-U. Würthwein, *Chem. Eur. J.*, **17**, 10071 (2011).
2011JCCS301 C.-H. Chou, Y.-T. Hsueh, and B.-C. Wang, *J. Chin. Chem. Soc.*, **58**, 301 (2011).
2011JPC(A)5435 J.B. Henry, S.I. Wharton, E.R. Wood, H. McNab, and A.R. Mount, *J. Phys. Chem. A*, **115**, 5435 (2011).
2011OBC4353 J.C. Ball, R. Gleave, and S. Jones, *Org. Biomol. Chem.*, **9**, 4353 (2011).
2012ARK(v)75 K.M. Aitken and R.A. Aitken, *Arkivoc*, **v**, 75 (2012).
2012ASJC2884 D. Setamdideh, S. Khanahmadzadeh, and N. Khorshidi, *Asian J. Chem.*, **24**, 2884 (2012).
2012OBC623 R. Leardini, H. McNab, D. Nanni, A.G. Tenan, and A. Thomson, *Org. Biomol. Chem.*, **10**, 623 (2012).
2012T1299 P.L. Lucero, W.J. Peláez, Z. Riedl, G. Hajós, E.L. Moyano, and G.I. Yranzo, *Tetrahedron*, **68**, 1299 (2012).

| | |
|---|---|
| 2013ACSC1020 | G. Lener, R.E. Carbonio, and E.L. Moyano, *ACS Catal.*, **3**, 1020 (2013). |
| 2013CC11367 | R.A. Aitken, K.M. Aitken, P.G. Carruthers, M.-A. Jean, and A.M.Z. Slawin, *Chem. Commun.*, **49**, 11367 (2013). |
| 2013CEJ14983 | T.Y. Vu, A. Chrostowska, T.K.X. Huynh, S. Khayar, A. Dargelos, K. Justyna, B. Pasternak, S. Leśniak, and C. Wentrup, *Chem. Eur. J.*, **19**, 14983 (2013). |
| 2013JOC628 | M.I.L. Soares, C.M. Nunes, C.S.B. Gomes, and T.M.V.D. Pinho e Melo, *J. Org. Chem.*, **78**, 628 (2013). |
| 2013OBC6650 | C. Masusai, D. Soorukram, C. Kuhakarn, P. Tuchinda, C. Pakawatchai, S. Saithong, V. Reutrakul, and M. Pohmakotr, *Org. Biomol. Chem.*, **11**, 6650 (2013). |
| 2013OJC1443 | C. Azimi, F. Sepehraddin, and H. Tahazadeh, *Oriental J. Chem.*, **29**, 1443 (2013). |
| 2013T722 | S. Leśniak, B. Pasternak, K. Justyna, T.Y. Vu, T.K.X. Huynh, S. Khayar, A. Dargelos, and A. Chrostowska, *Tetrahedron*, **69**, 722 (2013). |
| 2013T3646 | W.J. Peláez and T.M.V.D. Pinho e Melo, *Tetrahedron*, **69**, 3646 (2013). |
| 2014AJC1150 | C. Wentrup, *Aust. J. Chem.*, **67**, 1150 (2014). |
| 2014AJC1228 | S.-A.G. Abel, M.O. Eglinton, J.K. Howard, D.J. Hunt, R.H. Prager, and J.A. Smith, *Aust. J. Chem.*, **67**, 1228 (2014). |
| 2014AJC1288 | R.A. Aitken, C. Hauduc, M.S. Hossain, E. McHale, A.L. Schwan, A.M.Z. Slawin, and C.A. Stewart, *Aust. J. Chem.*, **67**, 1288 (2014). |
| 2014EJO2933 | W.J. Peláez, A.J. Pepino, G.A. Argüello, and T.M.V.D. Pinho e Melo, *Eur. J. Org. Chem.*, 2933 (2014). |
| 2014EJO3020 | K. Justyna, S. Leśniak, R.B. Nazarski, M. Rachwalski, T.Y. Vu, T.K.X. Huynh, S. Khayar, A. Dargelos, A. Chrostowska, and C. Wentrup, *Eur. J. Org. Chem.*, 3020 (2014). |
| 2014H(88)1135 | R.A. Aitken, G. Barker, L.P. Cleghorn, E.J. Reid, and S.S. Roberts, *Heterocycles*, **88**, 1135 (2014). |
| 2014JOC65 | R. Koch, H.M. Berstermann, and C. Wentrup, *J. Org. Chem.*, **79**, 65 (2014). |
| 2015RSCA19379 | R.A. Aitken and A.O. Oyewale, *RSC Adv.*, **5**, 19379 (2015). |
| 2015S242 | M. Mackay, A. Nortcliffe, H. McNab, and A.N. Hulme, *Synthesis*, **46**, 242 (2015). |
| 2015UP1 | K.A. Johnston, J.F. McLellan, H. McNab, and T.W. Muir, unpublished results (2015). |
| 2015UP2 | A.P. Gaywood, H. McNab, and L. McNab, unpublished results (2015). |
| 2015UP3 | A.R. Maguire, H. McNab, and C. Twomey, unpublished results (2015). |
| 2015UP4 | S. Borthwick, J. Foot, M. Ieva, H. McNab, L. McNab, E.J. Rozgowska, and A. Wright, unpublished results (2015). |
| 2015UP5 | K.A. Johnston, H. McNab, J. Montgomery, W.J. O'Neill, and S.I. Wharton, unpublished results (2015). |
| 2015UP6 | D.D.J. Cartwright and H. McNab, unpublished results (2015). See: D.D.J. Cartwright, PhD Thesis, University of Edinburgh, (2002). |
| 2015UP7 | N.C. Clemence and H. McNab, unpublished results (2015). See: N.C. Clemence, PhD Thesis, University of Edinburgh, (2005). |
| 2015UP8 | H. McNab and E.J. Rozgowska, unpublished results (2015). See: E.J. Rozgowska, PhD Thesis, University of Edinburgh, (2011). |
| 2015UP9 | G. McDougald and H. McNab, unpublished results (2015). See: G. McDougald, PhD Thesis, University of Edinburgh, (2000). |

2015UP10 P. Ianarelli and H. McNab, unpublished results (2015). See: P. Ianarelli, PhD Thesis, University of Edinburgh, (2008).
2015UP11 K. Boyle and H. McNab, unpublished results (2015). See: K. Boyle, PhD Thesis, University of Edinburgh, (2004).

CHAPTER THREE

# De Novo Synthesis of 3-Pyrrolin-2-Ones

**Erin T. Pelkey*, Sarah J. Pelkey, Jessica G. Greger**
Department of Chemistry, Hobart and William Smith Colleges, Geneva, NY, USA
*Corresponding author: E-mail: pelkey@hws.edu

## Contents



*Advances in Heterocyclic Chemistry*, Volume 115
ISSN 0065-2725
http://dx.doi.org/10.1016/bs.aihch.2015.04.001

## Abstract

This review presents a systematic survey of the literature (through the end of 2014) for de novo syntheses of 3-pyrrolin-2-ones from acyclic precursors or via transformation of other cyclic systems such as pyrroles, pyrrolidinone, and maleimides. One-component intramolecular cyclization approaches, two-component intermolecular cyclization approaches, and multicomponent intermolecular cyclization approaches are treated separately and organized based on the location(s) of the new bonds that are formed in the construction of the 3-pyrrolin-2-one ring system. 3-Pyrrolin-2-ones make important synthetic targets due to their occurrence in natural products, their diverse biological activity, and their utility as building blocks for the preparation of other materials.

**Keywords:** 3-Pyrrolin-2-ones; Aldol condensation; Claisen condensation; Metathesis; Multicomponent reactions; Pyrroles; Pyrrolidinones; Tetramic acids

## 1. INTRODUCTION

The purpose of this monograph is to provide a comprehensive review of the de novo syntheses of 3-pyrrolin-2-ones that have been reported through the end of 2014. 3-Pyrrolin-2-one (5*H*-pyrrol-2-one) (**1**) and 4-pyrrolin-2-one (3*H*-pyrrol-2-one) (**3**) are tautomers of 2-hydroxypyrrole (**2**) (Scheme 1). 3-Pyrrolin-2-ones have also been referred to as 1,5-dihydro-2*H*-pyrrol-2-ones, pyrrol-2-ones, γ-lactams, or butyrolactams. Some early synthetic reports represented these heterocycles as 2-hydroxypyrroles (1932MI1, 1937LAC265, 1942MI1, 1943LAC144). Later, NMR analysis showed that the lactam tautomers were actually present, δ 4.0–4.2 for the methylene of parent **1** and δ 2.9–3.2 for the methylene of parent **3**

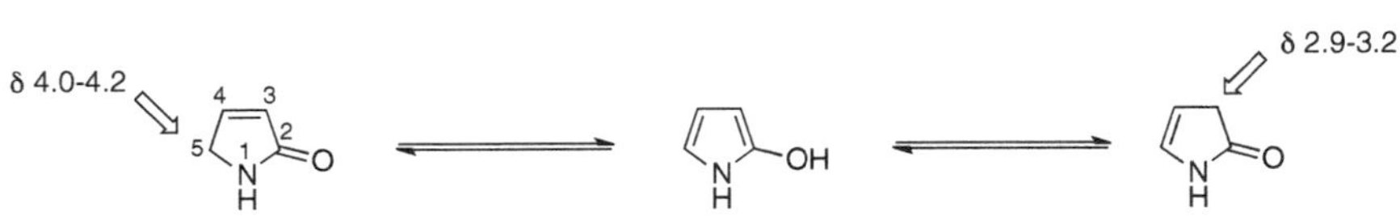


**1**
**3-pyrrolin-2-one**

**2**
**2-hydroxypyrrole**

**3**
**4-pyrrolin-2-one**

**Scheme 1** Tautomerism of 3-pyrrolin-2-ones.

(1962LAC165, 1965JOC3824, 1971MI1, 1978MI1), with the 3-pyrrolin-2-one tautomer **1** predominating regardless of the synthetic approach (1964JCS5999, 1970T4073, 1979JOC2798, 1980PT(1)227, 1989JA5346, 2002TL5057, 2007MI2).

A review article on 3-pyrrolin-2-ones (referred to therein as pyrrol-2-ones) appeared in 1972 by Rio and Masure (1972BSCF4598). Two review articles that described the reactivity (2004MI1) and synthesis (2003MI1) of 3-pyrrolin-2-ones by Egorova and Timofeeva have appeared more recently. In 2014, Ye and Gagosz summarized synthetic routes to γ-lactams (which included some coverage of 3-pyrrolin-2-ones) that involved transition metal catalysis (2014OBC1833). An important subclass of 3-pyrrolin-2-ones are 4-hydroxy-3-pyrrolin-2-ones **4** (Figure 1), and this structure is the major tautomer of a class of compounds known collectively as tetramic acids. Tetramic acids have received significant attention due in part to the large number of biologically active natural product congeners, and this class of heterocycles has also been reviewed (1995CR1981, 2007MI1). This review focuses on the synthesis and chemistry of derivatives of **1** with partial coverage of tetramic acids (enolic tautomers). Benzo[*c*]-fused 3-pyrrolin-2-ones (e.g., isoindolin-1-one **5** (Figure 1)) are also beyond the primary scope of this review and are only discussed sparingly.

Synthetic interest in 3-pyrrolin-2-ones arises from their occurrence in nature, diverse biological activity, and utility as building blocks for the preparation of nitrogen heterocycles and γ-amino acid derivatives. This review systematically (based on bond connections) describes de novo syntheses of 3-pyrrolin-2-ones. It is our hope that this review will serve as an informative reference for all synthetic chemists investigating this important class of compounds. In a separate monograph, we intend to follow up with a review on the reactivity of 3-pyrrolin-2-ones.

## 1.1 History and Parent

Early (pre-1940) interest in the synthesis of 3-pyrrolin-2-ones mainly derived from their presence as terminal groups in oligopyrrole bile pigments.

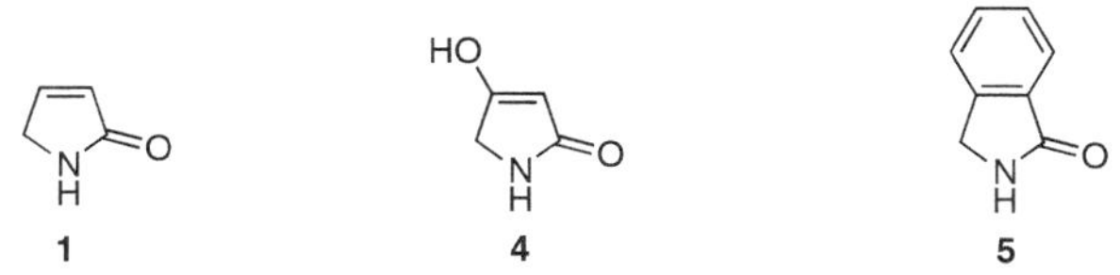

**1** **3-pyrrolin-2-one** | **4** **tetramic acid** | **5** **isoindol-1-ones**

**Figure 1**

An excellent synopsis of the history of the synthesis of 3-pyrrolin-2-ones by Atkinson and coworkers appeared in 1964 (1964JCS5999). At that time, there were four main synthetic routes to 3-pyrrolin-2-ones: (1) oxidation of pyrroles (1932MI1, 1937LAC265, 1942MI1); (2) oxidation of pyrrole-2-carboxaldehydes (1947MI1); (3) hydrolysis of 2-bromopyrroles (1943 LAC144); and (4) ring syntheses (1956LAC198; Scheme 2). All of these methods are discussed in the corresponding sections of the present monograph and include both historical examples and more recent examples (e.g., "synthesis from pyrroles" and "type 'ae' cyclizations"). These strategies remain among the most used methods for de novo 3-pyrrolin-2-one synthesis.

The first reported syntheses of the parent 3-pyrrolin-2-one **1** and isomeric 4-pyrrolin-2-one **3** can be traced back to 1951 and 1949, respectively. Several syntheses of **1** (and/or **3**) are summarized in Table 1. A quick glance of Table 1 reveals many discrepancies over the years (oil vs solid and different melting points), which makes it difficult to pinpoint the first successful synthesis of compound **1**. Grob and Ankli decarboxylated 4-pyrrolin-2-one **6** under vacuum and reportedly obtained **3**, which was characterized by elemental analysis (1949HCA2010; Table 1, Eqn (1)). Two years later, Langenbeck and Boser treated γ-phthalimido crotonate **7** with hydrazine and reported **1** as a solid product characterized by elemental analysis (1951CB526; Table 1, Eqn (2)). Bordner and Rapaport prepared 3-pyrrolin-2-one **9** by treatment of furan with ethyl azidoformate; subsequent deprotection of the carbamate by treatment with sodium hydroxide led to compound **1** which was characterized by $^{1}$H NMR (1965JOC3824; Table 1, Eqn (6)). The $^{1}$H NMR spectrum of **1** contained the distinctive methylene protons that typically appear in the δ 4.0–4.2

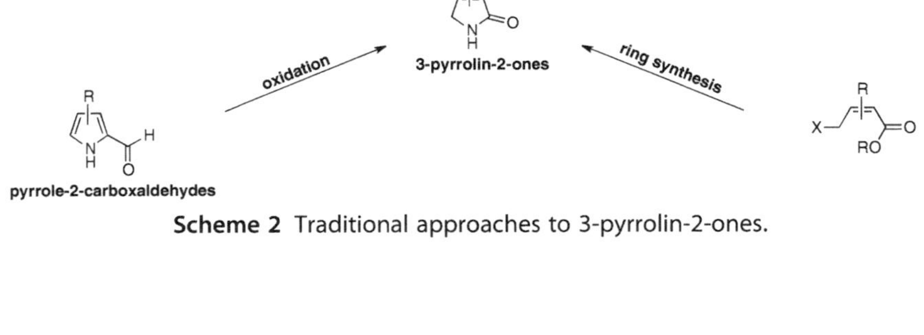


**Scheme 2** Traditional approaches to 3-pyrrolin-2-ones.

**Table 1** Summary of synthetic approaches to parent 3-pyrrolin-2-one (1)

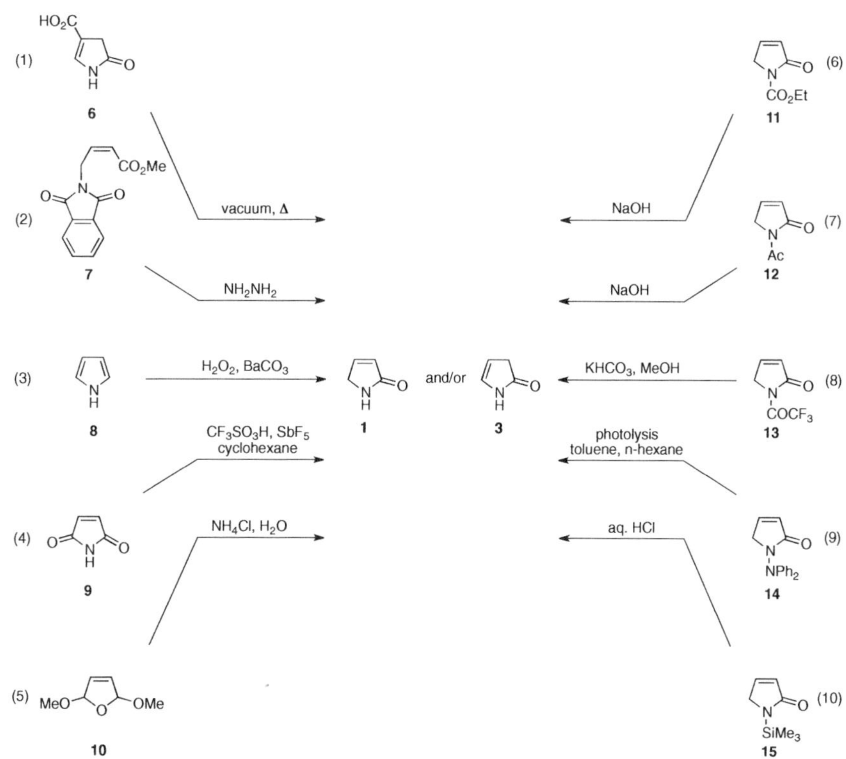


(*Continued*)

**Table 1** Summary of synthetic approaches to parent 3-pyrrolin-2-one (1)—cont'd

| Eqn | Product description | Product | C5 methylene signal of 1 | NMR solvent | References |
|---|---|---|---|---|---|
| (1) | Oil | **3**? | — | — | (1949HCA2010) |
| (2) | Solid (MP 140 °C) | **1**? | — | — | (1951CB526) |
| (3) | Solid | **1**:**3** (90:10) | δ 4.05 | $d_6$-acetone | (1970T4073) |
| (4) | —* | **1** | δ 4.24 | $d_6$-acetone | (2006EJOC4861) |
| (5) | Oil | **1**:**3** (88:12) | δ 4.08 | $CDCl_3$ | (2007MI2) |
| (6) | Solid (MP 165 °C) | **1**? | δ 4.2 | $CDCl_3$ | (1965JOC3824) |
| (7) | Solid | **1** | δ 4.08 | $CDCl_3$ | (1978MI1) |
| (8) | Oil[§] | **1**:**3** (90:10) | δ 4.1 | $CDCl_3$ | (1979JOC2798) |
| (9) | Not reported | **1** | — | — | (2009SL2621) |
| (10) | —* | **1**:**3** (90:10) | — | — | (1989JA5346) |

*Neat products not isolable under these reaction conditions.
[§]Pure **1** was obtained by recrystallization from $CHCl_3$ and reportedly had MP 23–27 °C.

range; this is a very useful spectroscopic handle for identifying 3-pyrrolin-2-one products. Additional related strategies leading to **1** involving nitrogen deprotection reactions (Table 1, Eqns (7)–(10)) have been published in subsequent years (1978MI1, 1979JOC2798, 2009SL2621, 1989JA5346), although none appear as of yet to be preparative.

The definitive approach to preparing the parent 3-pyrrolin-2-one **1** that has stood the test of time was reported by Bocchi and coworkers (1969OPPI271, 1970T4073). They oxidized pyrrole **8** with hydrogen peroxide in the presence of barium carbonate (Table 1, Eqn (3)). This reaction tends to produce a mixture of **1** and **3** in a ratio of 9:1 as ascertained by $^{1}$H NMR spectroscopy (1970T4073, 2002TL5057). This reaction has been adapted by many others over the years for the preparation of compound **1** (1981MI2, 1992JOC3760, 2012OBC7313, 2014BMCL5727).

Koltunov and Olah and coworkers reported an interesting reaction that generated **1** using superacid conditions (2006EJOC4861). Treatment of maleimide (**9**) with triflic acid/$SbF_5$ in the presence of cyclohexane gives **1** (Table 1, Eqn (4)).

Alves prepared **1** using a Cannizarro-type reaction of 2,5-dimethoxy-2,5-dihydroxyfuran (**10**) (2007MI2; Table 1, Eqn (5)). Treatment of **10** with ammonium chloride in the presence of HCl gives a crude product that contained both **1** and **3** in a ratio of 7:1. Column chromatography separated **1** and **3** and gave 0.68 g of **1**.

## 1.2 Natural Products

The 3-pyrrolin-2-one moiety is a structural component found in a large number of natural products including oligopyrrole plant pigments and heme metabolites (Figure 2), terpenoids (Figure 3), peptidics (Figure 4), and fused derivatives (Figure 7). Total syntheses of these natural products have helped contribute to the methodology that is available for the preparation of highly functionalized 3-pyrrolin-2-ones.

The most important classes of 3-pyrrolin-2-one natural products are oligopyrrole plant pigments and heme metabolites (1989MI1; Figure 2). There is a large body of literature that has been devoted to the total synthesis, synthetic approaches, structural analysis, and functional studies of 3-pyrrolin-2-one-containing oligopyrroles. Lightner and Boiadjiev published an excellent treatise on dipyrrinones, which are important building blocks that can be used to construct oligopyrroles (2006OPPI347). Some oligopyrrole natural products include biliburin (**13**) (1993T2185, 1998JOC6220, 1999JA9253, 2004T9017, 2008JOC2704), biliverdin (**14**) (2005CL800, 2006BCSJ1561,

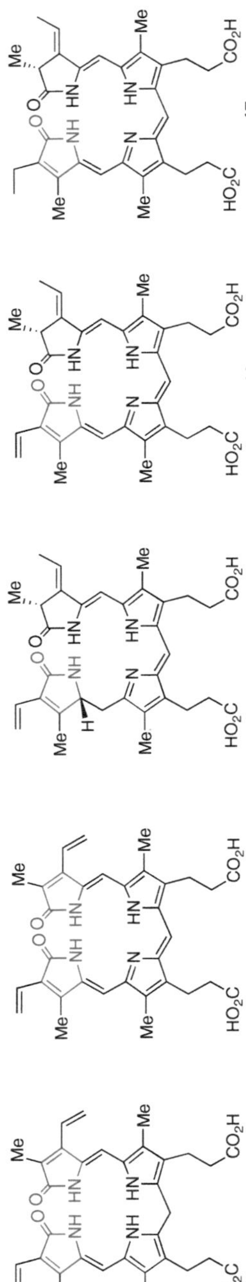


**13** **bilirubin**

**14** **biliverdin**

**15** **phycoerythrobilin**

**16** **phytochromobilin**

**17** **phytocyanobilin**

**Figure 2** Oligopyrrole natural products.

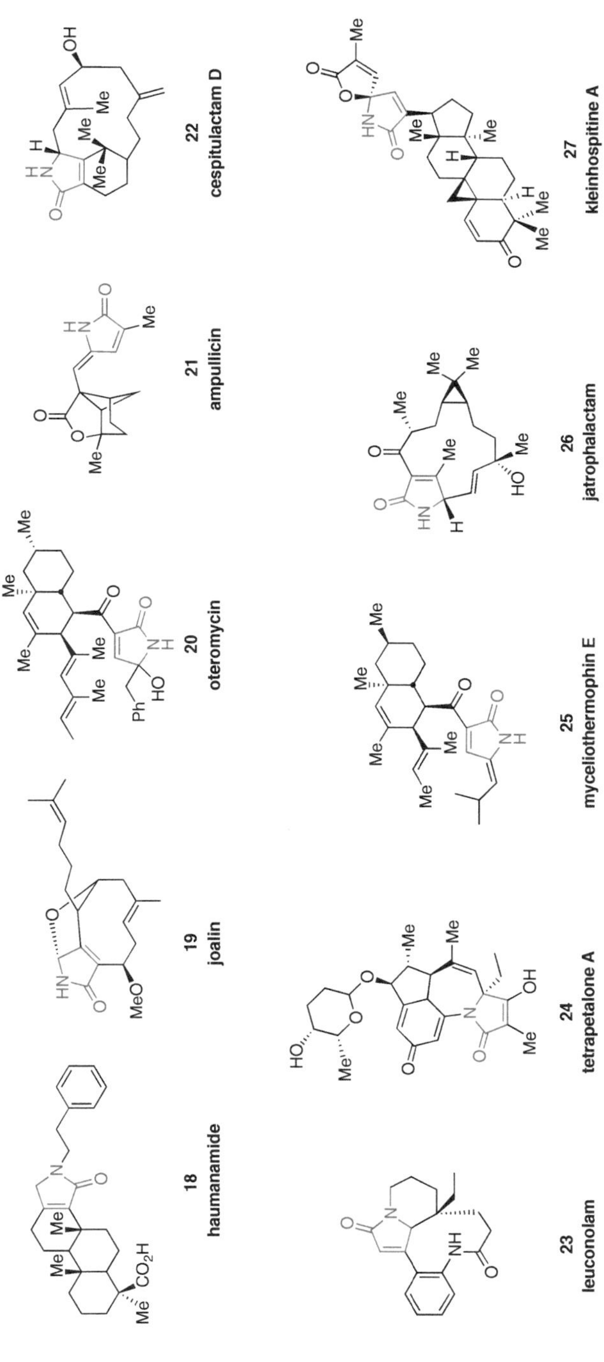


Figure 3 Terpenoid natural products.

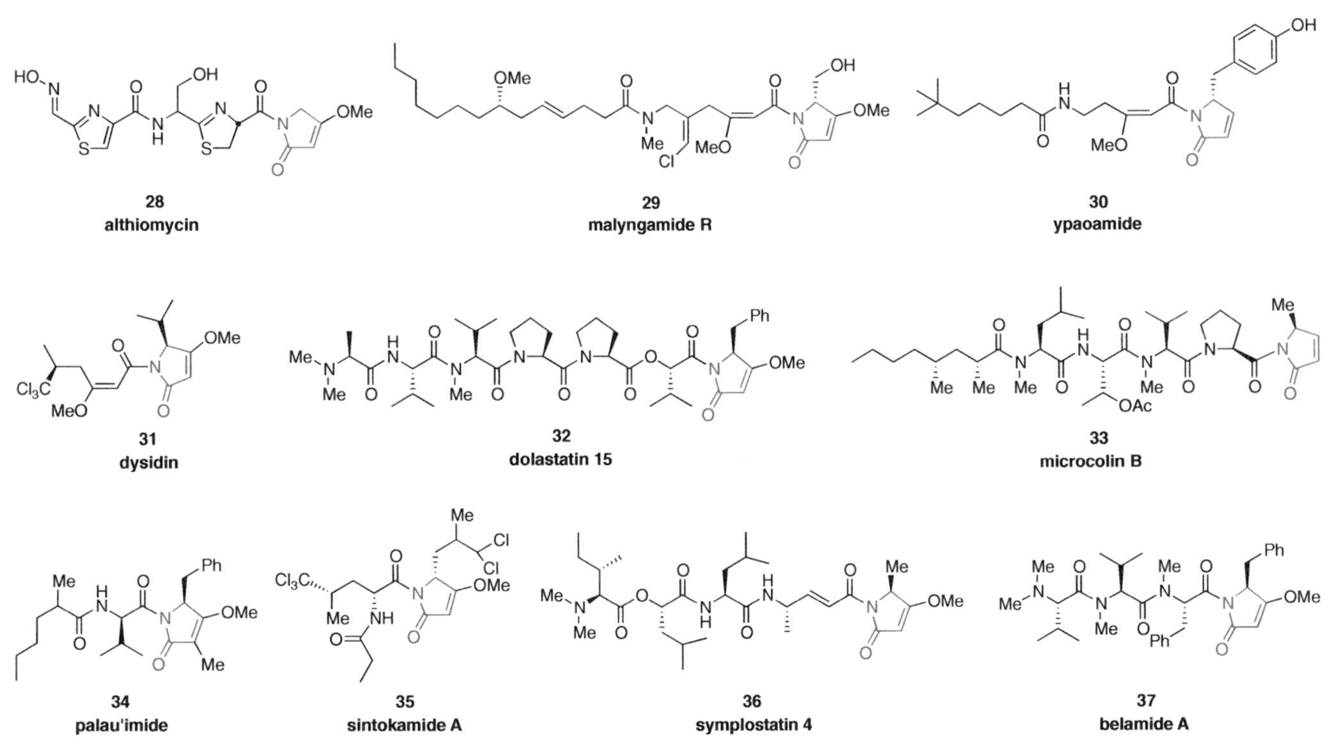


**Figure 4** Peptidic natural products.

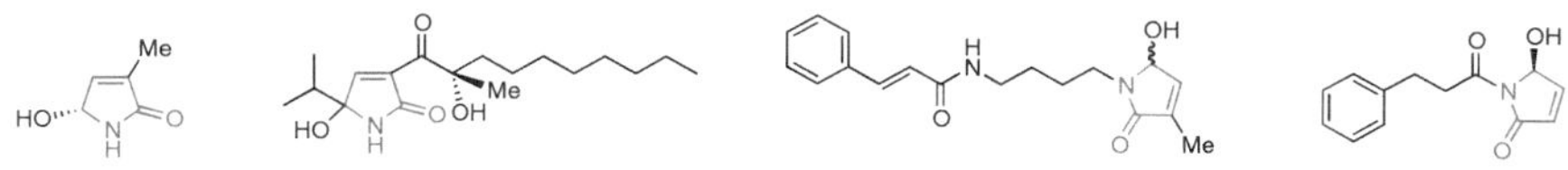

38 jatropham | 39 PI-091 | 40 gigantamide A | 41 chaplupyrrolidone A

**Figure 5** 5-Hydroxy-substituted 3-pyrrolin-2-one natural products.

2008CL198), phycoerythrobilin (**15**) (1978JA5928, 1978CB486), phytochromobilin (**16**) (1999SL901, 2000JOC205, 2000JOC8478, 2008JA11303), and phycocyanobilin (**17**) (1978JOC283, 2007EJOC1287) with selected synthetic efforts noted in the references.

A wide variety of 3-pyrrolin-2-one-containing terpenoid natural products have been isolated (Figure 3). Selected examples of terpenoid natural products include spongolactams (e.g., **18**) (1992TL1147, 2007JOC7190), joalin (**19**) (1993PT(1)1545), oteromycin (**20**) (1995JOC7040), ampullicin (**21**) (1990ABC813), echinophyllins (2000JNP1576, 2000TL2939), ZG-1494α (1996JAB967, 2000JNP768), talaroconvolutins (2000JNP768), ascosalipyrrolidinone A (2000JOC6412), pyrodysinoic acid (2001JNP1486, 2009JNP1691), plakinamine G (2002JNP1206), sarcotrines (2002 JNP1307), pyrrocidine A (2002TL1633), cespitulactams (e.g., **22**) (2005 TL7893, 2007JNP1961), leuconolam (**23**) (1984TL3483, 1986TL2501, 1989T7899, 2013JNP957, 2014JNP327), vitexlactam A (2002TL5131), turraparvin D (2003PH1173), tetrapetalone A (**24**) (2003TL7417), irregularasulfate (2007JNP1812), myceliothermophins (e.g., **25**) (2007CEJ6985), codinaeopsin (2008OL4149), fasciospongines (2007OL3037, 2009 JNP319), curdionolide C (2009HCA1665), jatrophalactam (**26**) (2009 OL5522), coscinolactams (2009T2905, 2014JNP1396), radianspenes (2012 PH190), kleinhospitine A (**27**) (2013OL2734), ircinianins and oxoircinianins (2013OBC4695, 2014TL4716), limonoids (2014JNP983), pyrrolocins (2014JNP2537), and phaeusmane H (2014EJOC5540). A good amount of synthetic effort has been directed toward the synthesis of 3-pyrrolin-2-one terpenoid natural products. The total synthesis of oteromycin (**20**) has been recently accomplished by Uchiro and coworkers (2013TL506); Snider and Bobbianna also reported an approach to the ring system of oteromycin (2004JOC8952). Nicolaou and coworkers synthesized the structurally-related myceliophermolins (e.g., **25**) (2014ACIE10970). Bermejo and coworkers published the total synthesis of ampullicin (**21**) and isoampullicin (1995TL7889, 1996TL5809, 2001JOC8287). The total synthesis of leuconolam (**23**) was reported by the research groups of Banwell

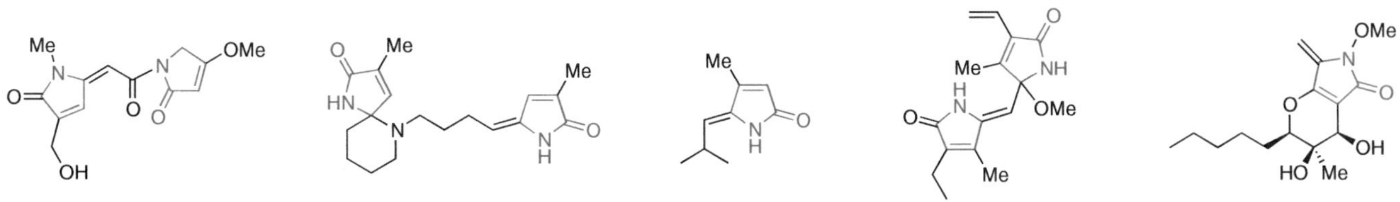


**42** **pukeleimide A** · **43** **pandamarine** · **44** **pulchellalactam** · **45** **rollipyrrole** · **46** **phaeosphaeride A**

**Figure 6** 5-Alkylidene-3-pyrrolin-2-one natural products.

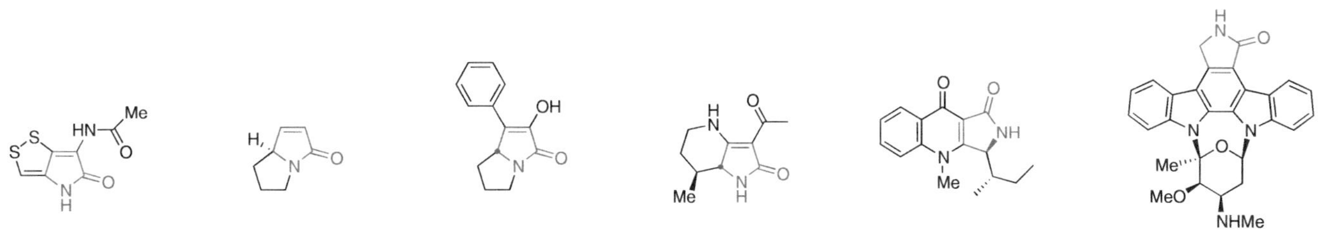

47 holomycin; 48 (*S*)-pyrrolam A; 49 phenopyrrozin 1; 50 laccarin; 51 quinolactacin A2; 52 staurosporine

**Figure 7** Fused 3-pyrrolin-2-one natural products.

(2006ARK(iii)163), Hoye (2013CS2262), and Zhu (2013JA19127). Frontier and coworkers reported a synthesis of tetrapetalone A methyl ether aglycone (2014ACIE9334); Sarpong and Marcus reported a synthesis of tetracyclic core of **24** (2010OL4560). Basabe and coworkers reported the synthesis of methyldecarboxy spongolactams (2010T2422).

A wide range of peptidic 3-pyrrolin-2-one natural products have been isolated and many of these have also been synthesized (Figure 4). Selected examples of peptidic 3-pyrrolin-2-one natural products include althiomycin (**28**) (1957JAB195, 1963JA1430, 1970JAB271, 1974JAB897, 1975CC121, 1975JAB286, 1982JAB635), malyngamides (e.g., **29**) (1979JA240, 2000 JNP965, 2007CAJ114, 2013PHL183), ypaoamide (**30**) (1996TL6263), pukeleimide C (1979TL2007), majusculamides (1988PH3101), dysidin (**31**) (1977HCA660, 1979JA240, 1985JOC4771, 1993JOC6336), dolastatin 15 (**32**) (1989JOC6005, 1993T9151, 1999JOC405), mirabimides (1991T2087, 1994JA8116), microcolins (e.g., **33**) (1992JNP613), palau'imide (**34**) (2002T7959), sintokamides (e.g., **35**) (2008OL4947), gallinamide (2009JNP14), symplostatin 4 (**36**) (2009CBC1634), belamide A (**37**) (2006TL3387), and leopolic acid A (2012TL6300).

Total syntheses of the antibiotic althiomycin (**28**) (1984TL2009, 1985BCSJ352, 1996BMCL1543, 2002BMCL561), malyngamides (e.g., **29**) (2007T3217, 2009JOC4149), ypoamide (**30**) (2009JOC7457), dysidin (**31**) (1984HCA1783, 2011CAJ2260), dolastatin 15 (**32**) (1991JA6692, 1992T4115, 1994BMCL1947, 1994T12097), immunosuppressive microcolins (e.g., **33**) (1996JOC3534, 1997JOC5542, 2010TA895), palau'imide (**34**) (2010CC5319, 2012OL3862), sintokamide A (**35**) (2010ACIE9702), sintokamide C (2010OL1100), gallinamide (2011CEJ13544, 2014JMC 10557), symplostatin 4 (**36**) (2010OL5576), and belamide A (**37**) (2011CEJ958) have been reported. The majority of the peptidic 3-pyrrolin-2-one natural products that have been isolated and synthesized contain 4-methoxy groups (methyl tetramates).

A few simple (nonsteroidal, nonpeptidic) 3-pyrrolin-2-one natural products that have been isolated contain a 5-hydroxy group (Figure 5). Hydroxy-substituted 3-pyrrolin-2-one natural products include jatropham (**38**) (1973MI1, 1980H(14)1073, 1992PH2767), PI-091 (**39**) (1990MI1), axinellamide (1995TL5851), gigantamide (**40**) (2007JNP1640), and chaplupyrrolidinone A (**41**) (2013PHL350). Glucoside-substituted (1987PH582, 1996PH1225) and pyrrolidinone-substituted (1992PH1084) analogs of **38** have also been characterized. Total syntheses of jatropham (**38**) (1981 H(16)1157, 1981H(16)1987, 1995TL4201, 1999TA4469, 2000TL9859),

PI-091 (**39**) (1995TL5551, 1996JOC2845, 1997JOC1918, 1999MI1, 2012BMCL4765), and gigantamide (**40**) (2013S2888) have been reported.

Another theme for simple (nonsteroidal, nonpeptidic) 3-pyrrolin-2-one natural products is provided by those that contain an alkylidene group at the 5-position (Figure 6). 5-Yliden-3-pyrrolin-2-one natural products include pukeleimides (e.g., **42**) (1979TL2003, 1979TL2007), pandamarine (**43**) (1992AJC1903, 1993PH1159), pulchellalactam (**44**) (1998JAB515), rollipyrrole (**45**) (2001TL7907), and phaeosphaeride A (**46**) (2006OL4067). Pulchellalactam (**44**) demonstrated potent activity as a CD45 protein tyrosine phosphatase inhibitor, and this prompted several groups to develop total syntheses (2002JOC4702, 2003T6221, 2004H(63)1013, 2004S1560, 2006 SC2253, 2007SC1503, 2007S1882). Total syntheses of pukeleimide A (**42**) (1985TL3617, 1993PT(1)2581, 2005SL453), phaeosphaeride A (**46**) (2011 OBC5825, 2012JOC9659), and phaeosphaeride B (2014JOC4043) have also been reported.

3-Pyrrolin-2-ones fused with a disulfide ring, specifically 4*H*-[1,2] dithiolo[4,3-*b*]pyrrol-5-ones, are known as pyrrothines; these compounds tend to display inhibitory activity against Gram-positive and Gram-negative bacteria (2001MI1, 2010JNP1164, 2014NPR905). Pyrrothine natural products include holomycin (**47**) (1959HCA563, 1977JAB334, 1979 JAB549), thiolutin (1952JA6304, 1955JA2861, 2002JAB696), and the thiomarinols (1993JAB1834, 1994JAB851, 1995JAB907, 1997JAB449). A number of total syntheses of holomycin (**47**) (1962ACIE265, 1963JA647, 1963LAC168, 1964JA5654, 1974BCSJ1484), thiolutin (1992LAC473), and analogs (1977JOC2891, 2007BMC4601, 2007OBC344) have been reported.

Other fused 3-pyrrolin-2-one natural products (Figure 7) that have been elucidated include (*S*)-pyrrolam A (**48**) (1990LAC525), phenopyrrozins (e.g., **49**) (1995JAB1413, 2006JNP580), laccarin (**50**) (1996H(43)685), salinosporamide C (2005JOC6196), oxocyclostylidol (2006JNP1212), quinolactacins (e.g., **51**) (2000JAB1247, 2000JAB1252, 2001JAB109, 2001 JAB831, 2006JAB418, 2006OBC1512), pyranonigrins (2004JNP1532, 2007JNP1180, 2013PHL162), ligulachyroine A (2012PHL541), and cordylactam (2013PHL162). Total syntheses of pyrrolam A (**48**) (1995TL291, 1996T869, 1996T3757, 1997SL1179, 1997TA515, 1999TA3827, 2000 CPB1593, 2004JOC6105, 2007S663, 2007S1499, 2008HCA1500, 2010 T1274, 2012S2673, 2014OL3780), phenopyrrozin 1 (**49**) (2012TL1891), laccarin (**50**) (2007CC728), and quinolactacin A2 (**51**) (2003JOC4523, 2004TL8793) have appeared. Several indolocarbazole natural products include a fused 3-pyrrolin-2-one moiety (e.g., staurosporine (**52**)); the rich biology and chemistry of these compounds have been reviewed (2006NPR1007). Staurosporine (**52**) was originally isolated by Omura and co-workers in 1977 (1977JAB275, 1995JAB535) and has been synthesized by the research groups of Danishsefsky (1995JA552, 1996JA2825) and Wood (1996JA10656, 1997JA9652).

Neolamellarin B and ningalin C are 3,4-diaryl-3-pyrrolin-2-one natural products that have been isolated from *Dendrilla nigra* (2007JNP1741) and *Didemnum* sp. (1997JOC3254), respectively. Ningalin C has been synthesized by the research groups of Steglich (2000TL9477) and Ruchirawat (2002OL2633). A simple *N*-acyl-3-pyrrolin-2-one natural product was isolated from *Piper demeraranum* (1989JNP891). Spirocyclic 3,4-dichloro-3-pyrrolin-2-ones named armeniaspiroles were found to have antibacterial activity (2012BMCL6292).

## 1.3 Biologically-Active Materials

3-Pyrrolin-2-ones have proven to be useful scaffolds in the discovery of novel drug entities (Figure 8). A small number of examples (of many) of biologically-active 3-pyrrolin-2-ones arranged by type of activity include COX-II inhibitors (e.g., **53**) (2000BMCL1745, 2001MI2, 2002MI1, 2004MI2, 2009BMCL2270), VEGF-R inhibitors (e.g., **54**) (2008JMC3814), HIV-1 integrase inhibitors (e.g., **55**) (2005JMC8009, 2007BMCL6155, 2007BMC5487), HDAC inhibitors (e.g., **56**) (2012JMC10766, 2013CMC272), antitumor agents (e.g., **57**) (1997JMC391, 2000MI1, 2002JBC31243, 2007BMCL3426, 2012JMC9630), aspartyl protease inhibitors (2013JOC2661), β-secretase inhibitor (2013OPRD985), herbicides (2000MI1, 2005MI1), antibacterial agents (e.g., **58**) (2014BMCL1052, 2014EJMC127), antidepressants (e.g., **59**) (2006BMCL3906), antimalarials (e.g., **60**) (2013BMCL6167, 2013JMC73, 2014JMC10557), mineralicorticoid receptors (2013BMC5983), gland cell-permeable peptide mimics (e.g., **61**) (2013JOC4823), and neurological agents (2014JMC2334). This list represents just a small sample of what has been reported regarding drug discovery with 3-pyrrolin-2-ones.

## 1.4 Building Blocks in Synthesis

3-Pyrrolin-2-ones have served as important building blocks for a wide variety of functionalized pyrroles, saturated nitrogen heterocycles, and acyclic nitrogen-containing small molecules (Scheme 3). Examples of the uses for 3-pyrrolin-2-ones are described in this section.

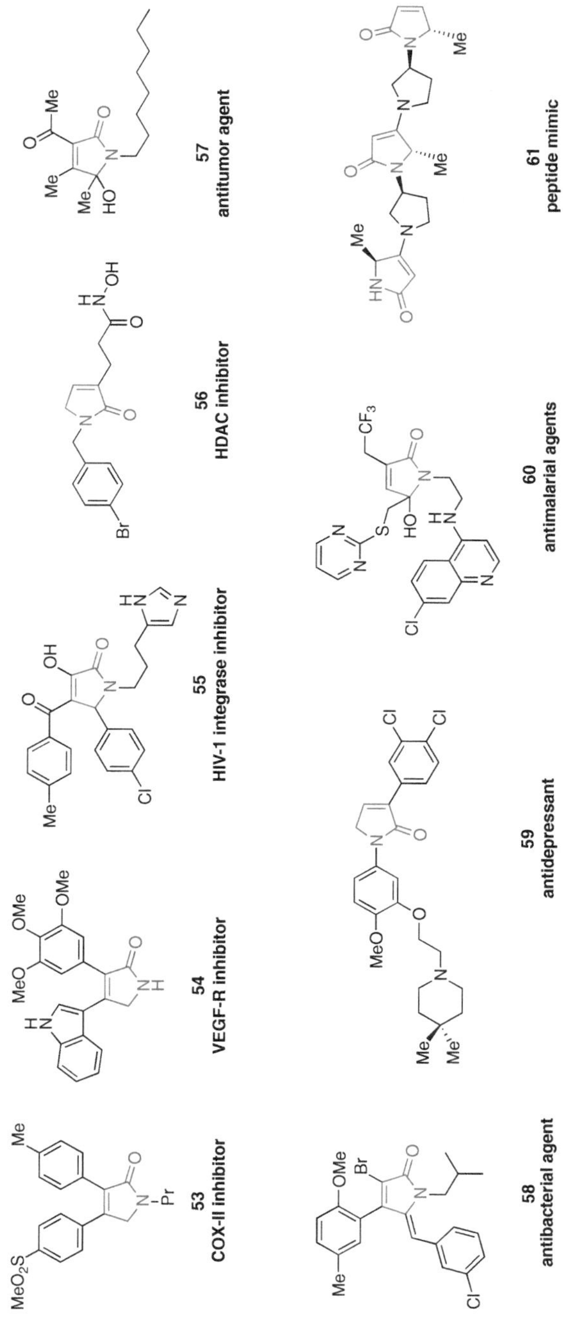


**Figure 8** Biologically-active 3-pyrrolin-2-one materials.

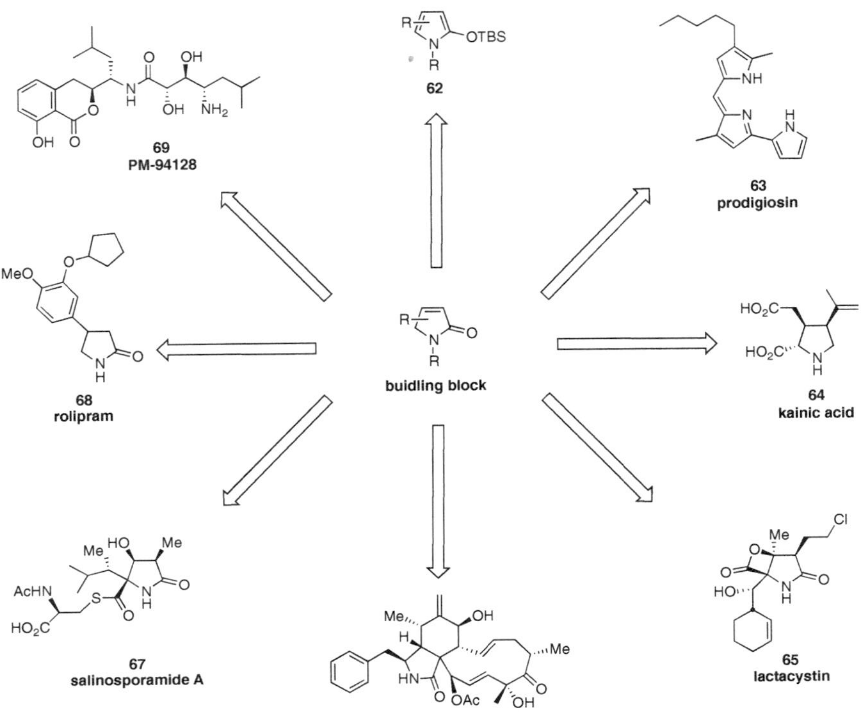


**Scheme 3** 3-pyrrolin-2-one applications in synthesis.

3-Pyrrolin-2-ones are precursors to 2-silyloxypyrroles **62**, which are compounds that have been elaborated into saturated nitrogen heterocycles and azacarbohydrates (1992JOC3760, 1995S607, 1997JMC168, 1999 SL1333, 2000CR1929, 2000CSR109, 2009SL1525, 2011CR3076, 2014T2595). 5-Alkylidene-3-pyrrolin-2-ones are precursors to 2-trifloxy-pyrroles that have been elaborated into the prodigiosins (e.g., **63**) (1996 SL513, 1999JOC8275, 2001CBC60, 2003ACIE3582, 2005ACIE2777, 2007JMC1528, 2007OL1879, 2009JA14579, 2013OBC3834, 2014 OBC4132). 3-Pyrrolin-2-ones have been utilized as key intermediates in the total syntheses of densely functionalized pyrrolidine and pyrrolidin-2-one natural products and medicinal agents including: the neuronal disease molecular probe kainic acid (**64**) (1998SL507, 2007SL1521, 2012 CEJ13274); the neurotrophic compounds lactacystin (**65**) (1994JA2139, 2002TL6609, 2006JOC1220, 2008JOC2041) and domoic acid (1982 JA3511, 2005T5713); the antitumor agent lucilactaene (2005ACIE3110,

2005JA16038); the antiinfluenza compound A-315675 (2002JOC5445, 2002JA4716); the actin polymerization inhibitor cytochalasin D (**66**) (1983JA5510, 1989T2417); the 20S proteasome inhibitors salinosporamide A (**67**) (2005JA8298, 2007CAJ20, 2007TL381) and omuralide (2007 JOC10108, 2008JOC2041); and the anti-depressant rolipram (**68**) (1993JOC36, 1997S559, 2006JA1464, 2007SL1557). A 3-pyrrolin-2-one intermediate proved useful in the synthesis of the cytotoxic chiral amine PM-94128 (**69**) (2009JOC7566). A chiral 5-alkoxymethyl-3-pyrrolin-2-one was a useful building block for the preparation of a series of stereospecifically deuterated glutamic acid derivatives (2011OL161).

## 2. INTRAMOLECULAR APPROACHES TO 3-PYRROLIN-2-ONES

There are exactly five different one-bond disconnections possible for the construction of 3-pyrrolin-2-ones. All five possible intramolecular approaches have appeared in the literature and are depicted in Scheme 4. For example, a type "a" cyclization involves a lactamization ring forming step between the N1 and C2 (carbonyl carbon). The most commonly utilized intramolecular approaches to 3-pyrrolin-2-ones are type "c" cyclizations, followed by type "a" and type "e" cyclizations. Only a small number of type "d" cyclizations are known and, to our knowledge, there are only three type "b" cyclization approaches in the literature.

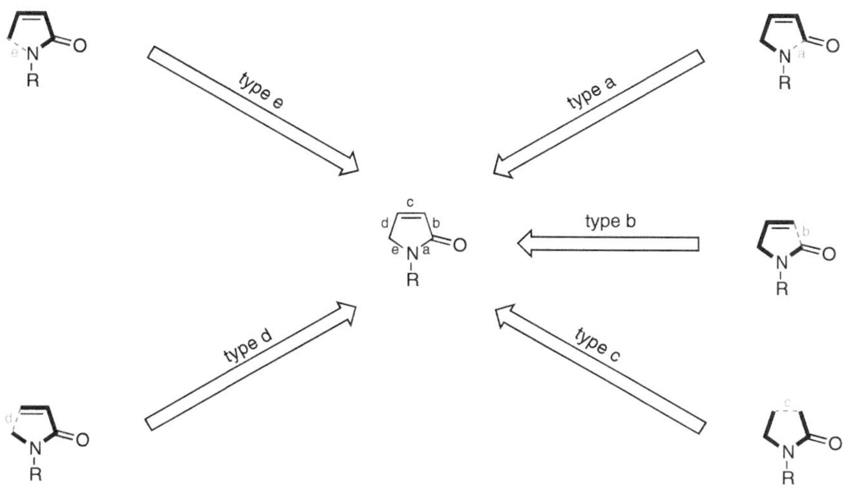


**Scheme 4** Intramolecular approaches to 3-pyrrolin-2-ones.

## 2.1 Type "a" Cyclizations

A direct approach to 3-pyrrolin-2-ones involves the intramolecular condensation reactions of *cis* α,β-unsaturated γ-aminocarboxylic acid derivatives **70** (Scheme 5). Different methods for the preparation of aminocarboxylates **70** have been developed, and the cyclization reactions leading to 3-pyrrolin-2-ones usually occur under mild conditions. A recent report by König and Heilmann and coworkers has emphasized the facile transformation of intermediates **70** into the corresponding 3-pyrrolin-2-ones as they tried and failed to avoid this tranformation using a tosyl protecting group on the nitrogen (2014EJOC3170); Kruse and coworkers also reported a facile transformation of derivative **70** (X = succinimide ester) into 3-pyrrolin-2-ones (1985T5241). The first reported synthesis of the parent 3-pyrrolin-2-one (**1**) followed this strategy; a free amine was generated by treatment of the corresponding phthalimido derivative with hydrazine (Table 1, Eqn (2); 1951CB526). More recently, γ-aminoesters have been prepared by reduction of the corresponding γ-azido 2-butenoate esters in the syntheses of pukeleimide A (**42**) (1993PT(1)2581) and azasteroids (1993SC1111). A quinolone-fused γ-aminoester was utilized in a total synthesis of quinolactacin A2 (**51**) (2004TL8793). The reduction of γ-nitroesters provided an entry to 5,5-disubstituted-3-pyrrolin-2-ones, which are building blocks in the synthesis of oligopyrrole macrocycles (1987PT(1)2027).

Grison and coworkers used a Horner–Wadsworth–Emmons reaction to prepare the requisite *cis* γ-amino α,β-unsaturated esters **73** (Scheme 6; 2001TL3831). Olefination of aminoaldehydes **71** with phosphonates **72** provides vinylogous aminoesters **73**. Boc deprotection followed by cyclization then gives 3-pyrrolin-2-ones **74**. The Koskinen research group have used similar sequences to prepare various 3-pyrrolin-2-one building blocks en route to the cytotoxic polyketide calyculin C (1994TL7417, 1998JOC92, 2010S2816). In unrelated studies, removal of Boc groups from other aminoesters (similar to **73**) results in the spontaneous formation of 3-pyrrolin-2-ones (2005TL3123, 2007JOC2547).

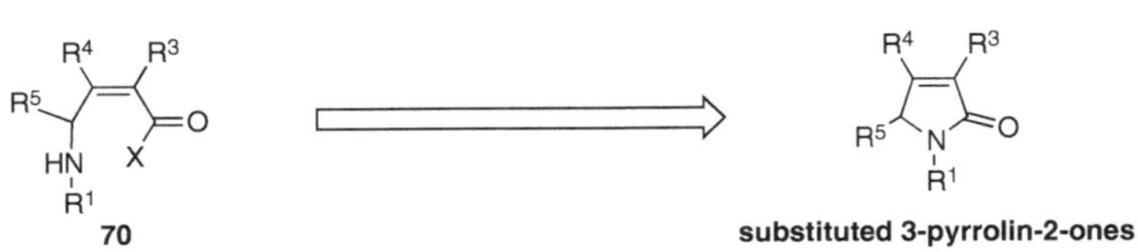


Scheme 5

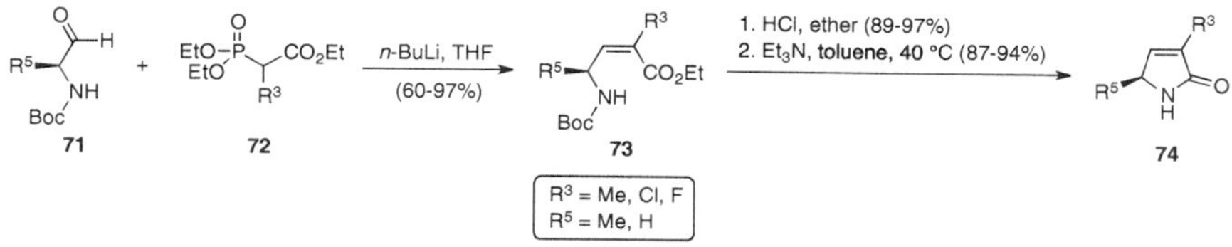

Scheme 6

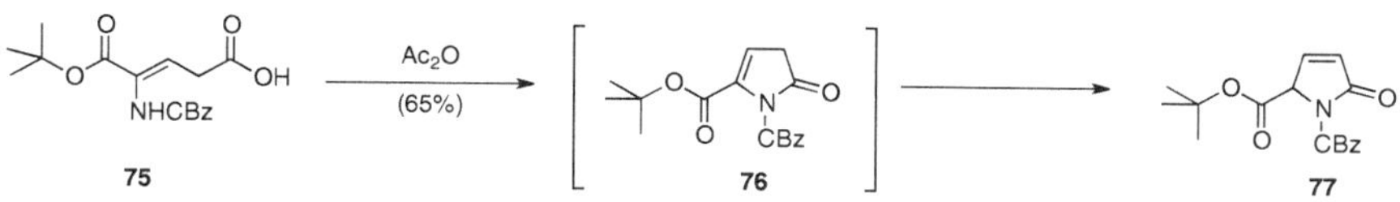

Scheme 7

Shin and coworkers investigated the chemistry of dehydroglutaric acid derivatives and discovered a route to 5-acyl-3-pyrrolin-2-ones (Scheme 7; 1985TL85). Treatment of the mono-*tert*-butyl glutarate **75** with acetic anhydride led to formation of the 3-pyrrolin-2-one **77** after activation of the carboxylic acid, cyclization, and double-bond isomerization of the 4-pyrrolin-2-one intermediate **76**. Ring closures of structurally-related alkylideneamine-substituted succinate derivatives have been investigated by Vasvári-Debreczy and coworkers (1978PT(1)795, 1980PT(1)227) and also by Liao and coworkers (2012EJOC5324).

Dieter and coworkers reported an enantioselective total synthesis of (*R*)-pyrrolam (**48**), which involved an intramolecular cyclization of a γ-aminoester as a key step (2002JOC847, 2004JOC6105; Scheme 8). Enantioselective metallation of a Boc-pyrrolidine (**78**) mediated by (−)-sparteine, followed by transmetallation to the cuprate, and cross-coupling with (*Z*)-3-iodopropenoate (**79**) gives the corresponding γ-aminoester (**80**). The best conditions for the subsequent deprotection/cyclization to (*R*)-pyrrolam (**48**) involve trimethylsilyl chloride and phenol.

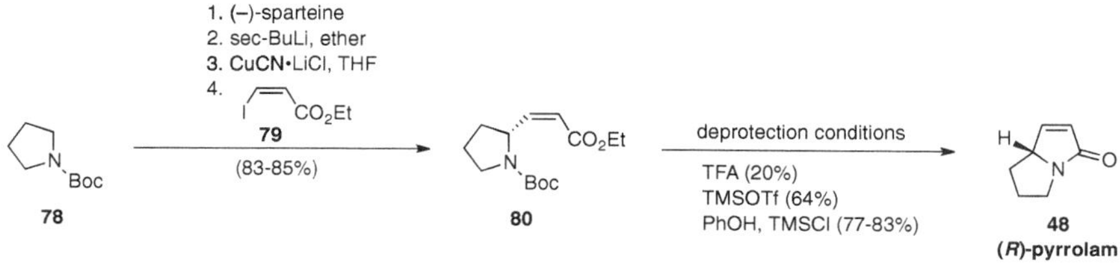

Scheme 8

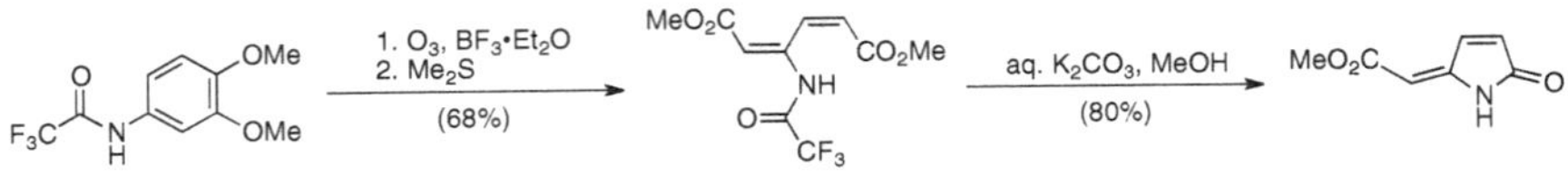


Scheme 9

Yoshida and coworkers used an ozonolysis of aniline derivative **81** to produce γ-aminoester derivative **82** (2005H(65)1481; Scheme 9). Subsequent hydrolysis of trifluoroacetamide **82** led to the alkylidene-substituted 3-pyrrolin-2-one **83**.

Palacios and coworkers used a ring-opening reaction of a 2*H*-azirine to generate *cis* γ-amidoesters (Scheme 10; 2011JOC9472). Treatment of 2*H*-azirine **84** with the diethyl malonate derivative **85** and sodium hydride gives the amidoester **86**. Subsequent heating of **86** in the presence of sodium hydride then led to the formation of 5-phosphoryl 3-pyrrolin-2-one **87**. This overall transformation can also be accomplished in one step without isolation of intermediate **86**.

Corriu and coworkers used a Michael addition to γ-aminoynoate esters to provide access to *cis* γ-aminoenoate esters (Scheme 11; 1984JOMCC61, 1987TL2963, 1993T4603). Treatment of γ-aminoynoate ester **88** with butylcopper followed by quenching with different electrophiles gives γ-aminoesters **89**. Desilylation of **89** followed by cyclization of the intermediate free amine produced 4-butyl-3-pyrrolin-2-ones **90**. An analogous sequence involving stannylcuprates was deployed by Reginato and coworkers in the synthesis of 4-tributylstannyl-3-pyrrolin-2-ones (1995T2129, 1998T10227).

Vallée and coworkers found that Zn-mediated reduction of γ-hydroxylaminoynoate esters gives intermediate γ-aminoesters that subsequently cyclize into 3-pyrrolin-2-ones (Scheme 12; 1997TL5503, 1999SL602). This transformation involves the reduction of the hydroxylamine functionality, stereoselective reduction of the ynoate ester, and subsequent cyclization to the lactam. Hydroxylamine **93** was prepared by the addition of the anion derived from ethyl propiolate (**91**) to nitrone **92**. Treatment of **93** with zinc in acetic acid led to reduction of both the hydroxylamine

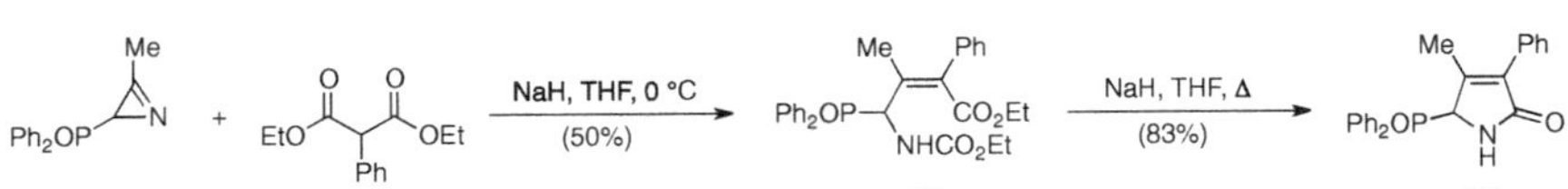


Scheme 10

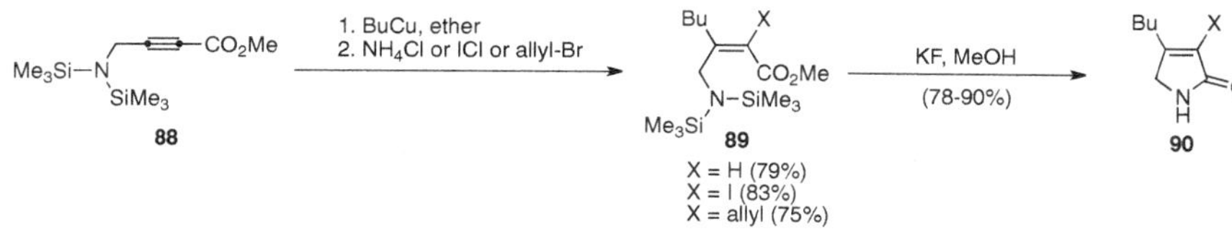


Scheme 11

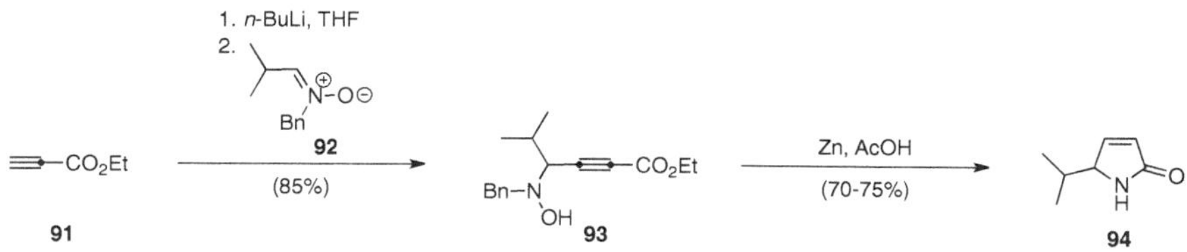


Scheme 12

moiety and the alkyne moiety and subsequent cyclization to the 3-pyrrolin-2-one **94**. This methodology was also used by Hanessian and coworkers to make a 1,5-disubstitued 3-pyrrolin-2-one utilized in the synthesis of the antiinfluenza compound A-315675 (2002JA4716). They also utilized an alternative endgame sequence involving a Lindlar reduction followed by a Mo-mediated reduction of the hydroxylamine.

Trost and Roth have used ruthenium chemistry to generate complex 3-pyrrolin-2-ones (1999OL67; Scheme 13). An Alder-ene reaction between alkene **95** and alkyne **96** catalyzed by Ru catalyst **97** provided highly functionalized γ-aminoester **98**. Cyclization of the latter mediated by $(Bu_3Sn)_2O$ then gives chiral 3,5-dialkyl-3-pyrrolin-2-one **99**. Koskinen and Passiniemi investigated a similar cyclization and found it necessary to use $(Bu_2SnCl)_2O$ in place of $(Bu_3Sn)_2O$ (2010S2816). The Trost research group later prepared 4-alkynyl-3-pyrrolin-2-ones from amine-substituted ynenoates using TMSOTf-mediated cyclization (2011JA740).

A well-explored method for the preparation of 3,4-disubstituted 3-pyrrolin-2-ones, first reported by Plieninger and coworkers, involves the

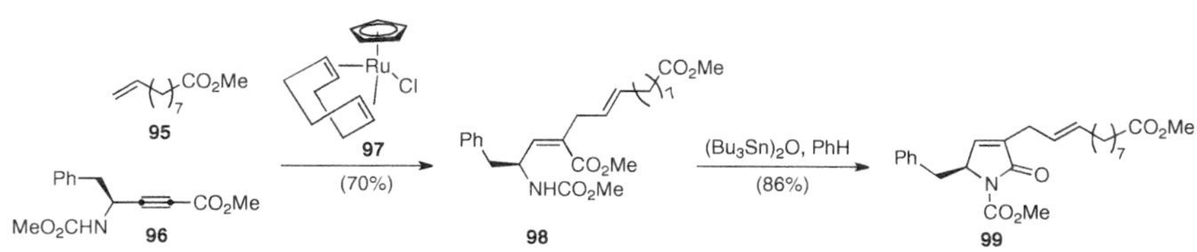


Scheme 13

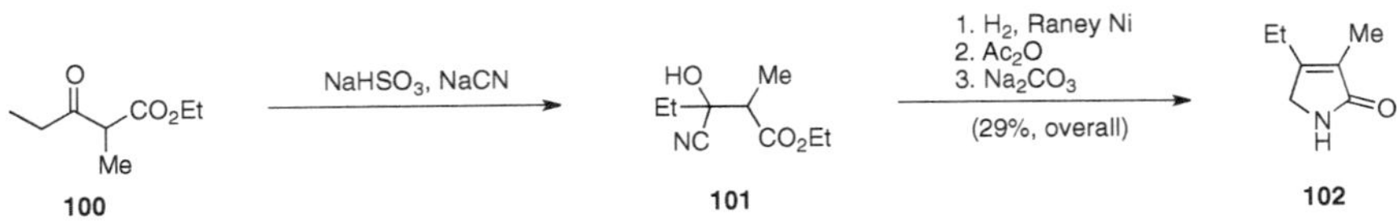

Scheme 14

Raney Ni-mediated reduction of β-cyanohydrin esters (1956LAC198, 1964LAC60, 1978CB486, 1989HCA518). Illustrated by recent example by the Rapoport group, cyanohydrin **101** was prepared by the treatment of β-ketoester **100** with sodium bisulfite followed by sodium cyanide (Scheme 14; 1984JA2645, 1991JA8024). Subsequent reduction of the nitrile with Raney Ni in the presence of acetic anhydride then gives the 4-ethyl-3-methyl-3-pyrrolin-2-one (**102**), which is an important building block in the synthesis of plant bile pigments. In a separate report, **102** was prepared in this fashion by Gurjar and coworkers and used in the synthesis of the blood glucose lowering agent glimepiride (2003TL4853). Improved yields for the reduction step have been reported by Bhandari and Sharma (2004IJC(B)2467). Reduction of an *o*-cyanobenzoate derivative mediated by sodium borohydride and cobaltous chloride, as reported by Beccalli and coworkers, provided staurosporinone (the aglycone of **52**) (1998 T6909). Finally, Raney Ni-mediated reductions of β-cyanobenzoates have been utilized by the Hudkins (2003JHC135, 2007JMC433) and Prudhomme (2008BMC4419) research teams to prepare heterocycle-modified staurosporionone derivatives.

The Ohe group have investigated novel Cu-catalyzed cyclizations of cyano enones en route to highly substituted 3-pyrrolin-2-ones (2007JOMC579, 2008JOC9174). Treatment of cyano enone **103** with copper triflate in the presence of furan gives 5,5-diaryl 3-pyrrolin-2-one **105** (Scheme 15). The reaction presumably proceeds via a dehydrative cyclization to carbocationic intermediate **104**, which is trapped by an aromatic nucleophile to give the observed product.

In an attempt to prepare fused pyrrolidinones such as **108**, Balci and coworkers inadvertently prepared fused 3-pyrrolin-2-ones (Scheme 16;

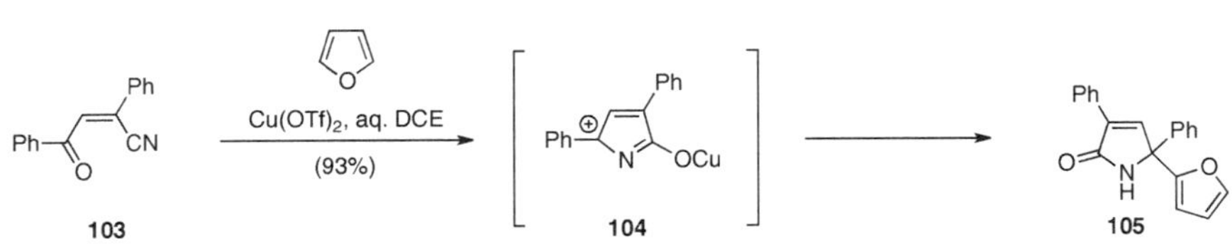

Scheme 15

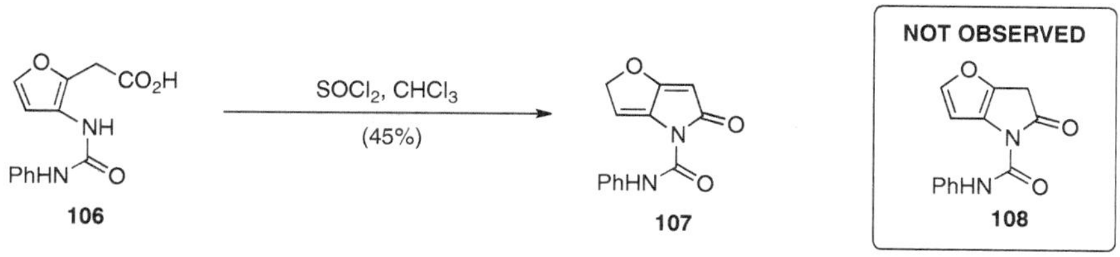


Scheme 16

2014T5993). Treatment of carboxylic acid **106** with thionyl chloride led to the formation of **107** via a cyclization and subsequent tautomerization of the furan ring; the desired product **108** was not observed. The structure of the tautomer **107** was confirmed unambigously by X-ray crystallography.

Meldrum's acid (**110**) has been used as a building block for the preparation of tetramic acids (2008CSR789). One of the earlier reports came from Jouin and coworkers, where coupling of Boc amino acids (**109**) with Meldrum's acid (**110**), mediated by isopropenyl chloroformate and DMAP, gives 1,3-dione intermediates **111** (Scheme 17; 1987PT(1)1177). Heating **111** in ethyl acetate (or $CH_3CN$) then gives the tetramic acids **112**. Others have used isopropenyl chloroformate as a coupling agent in the synthesis of tetramic acids (1996TA2365, 2010ACIE9702). Other coupling reagents explored for this sequence include DCC (1994JHC1145, 2001PT(1)1421), 1,1′-carbonyldiimidazole (1996JHC825), isopropyl chloroformate (2002JOC4702, 2007TL827), EDC (2006OL2103, 2007OBC3486, 2009JOC5032), and BOP-Cl (1993TL6705). This sequence has been utilized in the total syntheses of dolastatin 15 (**32**) (1992T4115), detoxinine (1988H(27)2843), and pulchellalactam (**44**) (2002JOC4702). This methodology has also been adapted to the solid phase (2003BMCL2505, 2005SL2763).

## 2.2 Type "b" Cyclizations

Kambe and coworkers investigated an interesting intramolecular (type "b") selenoamidation of allenes as a novel route to 3-pyrrolin-2-ones (Scheme 18; 2009EJOC3141). Treatment of allene **113** with catalytic $Pd(PPh_3)_4$ in

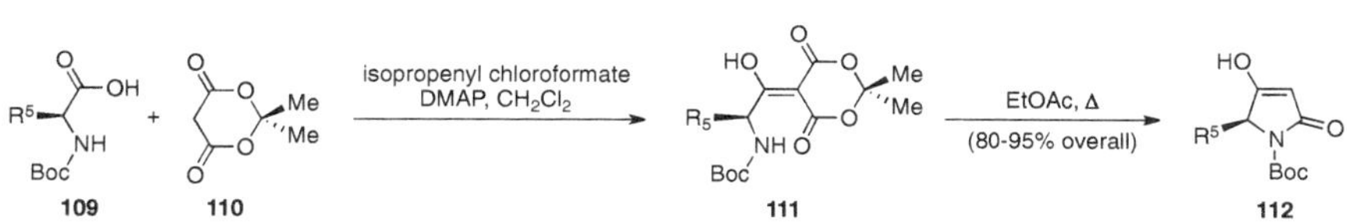


Scheme 17

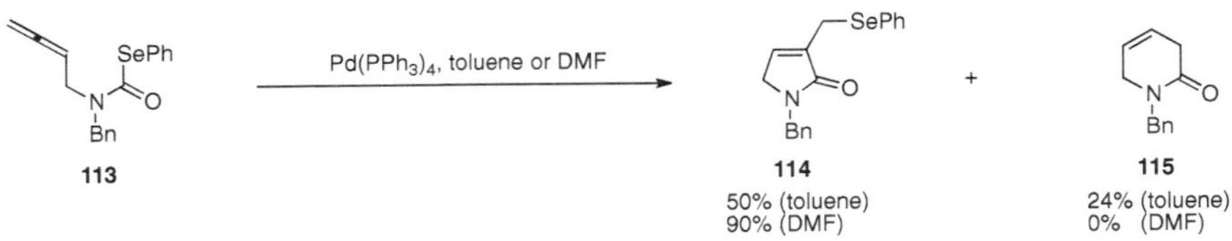


Scheme 18

toluene gives a mixture of 3-pyrrolin-2-one **114** and piperidine **115**; changing the solvent to DMF led to the formation of **114** as the only product.

A related type "b" cyanoamidation leading to 3-pyrrolin-2-ones was reported by the Takemoto group (Scheme 19; 2006OL2711). Treatment of the alkyne **116** with catalytic $Pd(PPh_3)_4$ in xylene gives an equal mixture of 3-pyrrolin-2-one **117** and alkylidene lactam **118**. The mechanism presumably involves a 5-*exo*-trig oxidative addition of cyanide and the amide carbonyl across the alkyne; isomerization of **118** under the reaction conditions then leads to the formation **117**.

Henin and coworkers investigated an unusual cross-coupling reaction involving chloroformamides, which gives 3-pyrrolin-2-ones as the minor products (Scheme 20; 1986TL6339). Heating homoallylic chloroformamide **119** in the presence of $Pd(OAc)_2$ led to a mixture of the 3-pyrrolin-2-one **120** and the α-methylene pyrrolidinone **121** in a 1:3 ratio, respectively. Alternative reaction conditions and/or substrates give a lower percentage yield of the 3-pyrrolin-2-one.

## 2.3 Type "c" Cyclizations

Highly substituted 3-pyrrolin-2-ones, especially congeners-containing substitution at C-3 and/or C-4, are often prepared utilizing intramolecular aldol-like condensations of α-amidoketones **122** (Scheme 21). Jacobi and coworkers prepared a 3,4-dialkyl-3-pyrrolin-2-one (Scheme 21: $R^3 = CH_2CH_2SeAr$; $R^4 = Me$) by treating an α-amidoketone **122** with *t*-BuOK; this was a key step in a synthesis of the pyrromethenone moiety

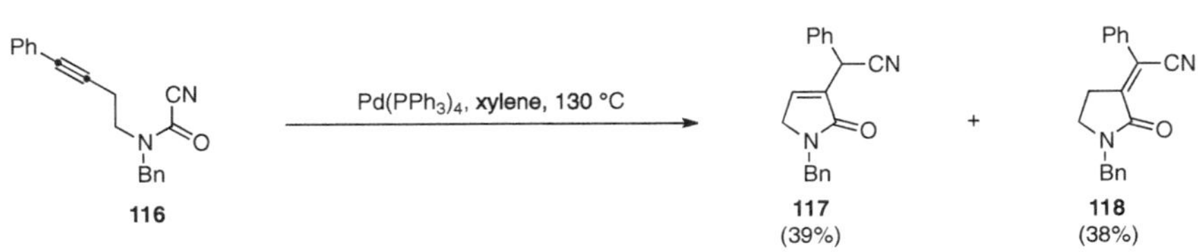


Scheme 19

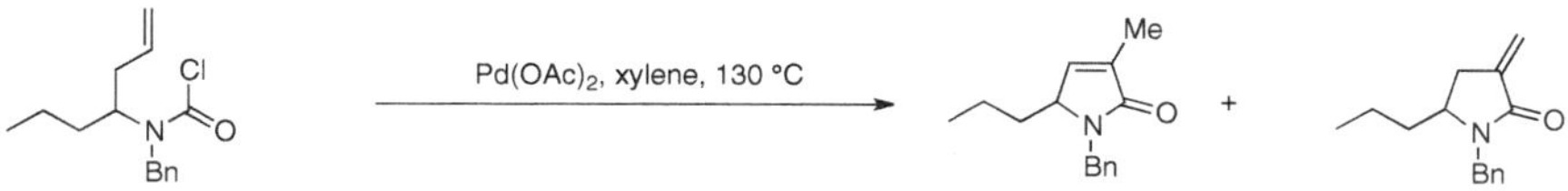


119 120 (19%) 121 (57%)

Scheme 20

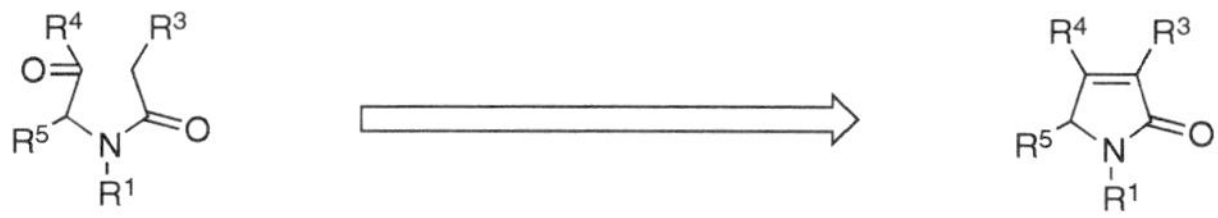


122 **substituted 3-pyrrolin-2-ones**

Scheme 21

found in the linear tetrapyrroles phytochrome and phycocyanin (2000JOC8478). The Sarpong group used this reaction to prepare a spirocyclic 3-pyrrolin-2-one en route to the Kopsia indole alkaloids (2012OL648). Verniest and de Kimpe used a similar reaction to prepare 4-aryl-5-methyl-3-pyrrolin-2-ones (2003SL2013). Pal and coworkers prepared 3,4-diaryl-3-pyrrolin-2-ones (Scheme 21: $R^3$ and $R^4$ = Ar) by treating α-amidoketones **122** with potassium carbonate (2002SL947, 2004T3987). The same reaction utilizing DBU as a base produced 3,4-diarylmaleimides via a cyclocondensation and facile oxidation. Interestingly, Bosch and coworkers utilized DBU in a similar cyclocondensation and obtained the corresponding 4-hydroxypyrrolidine; subsequent treatment of the latter with *p*-TsOH gives 3,4-diaryl-3-pyrrolin-2-ones, which were evaluated as selective COX-II inhibitors (2000BMCL1745). Other bases that have been utilized in aldol cyclocondensations to 3,4-diaryl-3-pyrrolin-2-ones and other aryl-substituted 3-pyrrolin-2-ones include *t*-BuOK (1994S25, 2001MI2, 2001MI3, 2008JMC3814, 2009BMCL2270, 2010JMC5108, 2013BMC5983, 2013MI1), NaOMe (2009CPB979), $Et_3N$ (1987IJC(B)63), KOH (to 3-alkyl-4-aryl-3-pyrrolin-2-ones) (2007BMCL1908), and NaOEt (to 5-alkylidene 3-aryl-3-pyrrolin-2-ones (1983JHC687).

Aldol condensations of α-amidoketones have been used to prepare 3,4-bis(indol-3-yl)-3-pyrrolin-2-ones; these compounds are important precursors in the synthesis of staurosporinone (**52**) (1983H(20)469), differentially *N*-substituted staurosporinones (1994S25, 1999S275, 2002OL2005), and a benzofuran analog of staurosporinone (2001MI3). Fukuyama and coworkers prepared 3,4-bis(indol-3-yl)-3-pyrrolin-2-one **124** by treatment of the bisindole ketone **123** with DBU and molecular sieves; **124** is an

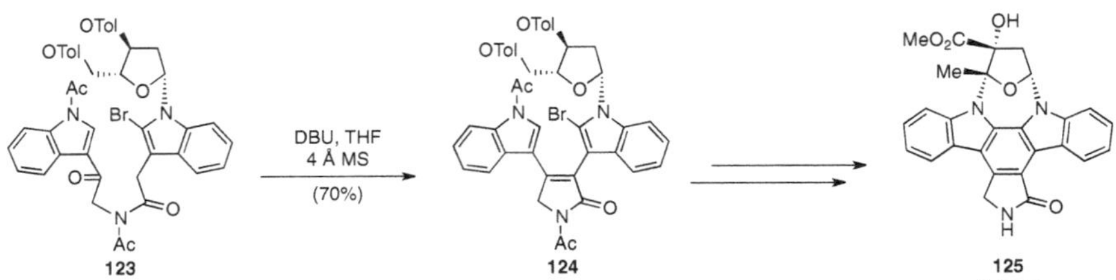


**Scheme 22**

important intermediate in a total synthesis of the PKC inhibitor K-252a (**125**) (Scheme 22; 1999JA6501).

A number of natural products contain a 3-acyl-3-pyrrolin-2-one moiety such as oteromycin (**20**) (1995JOC7040). In addition, several natural product have been prepared using 3-acyl-3-pyrrolin-2-one building blocks; examples include fusarin C (1984CC122) and epolactaene (1995 JAB733). Intramolecular aldol-type condensation reactions provide a good route to 3-acyl-3-pyrrolin-2-ones (Scheme 23; 2004JOC8952). Snider and Neubert treated the 1,3-dicarbonyl derivative **126** with sodium hydroxide to give a mixture of the cyclic tautomers **127** and **128**. Subsequent oxidation with dimethyldioxirane (DMDO) gives the 5-hydroxy-3-pyrrolin-2-one **129**. Schmidlin and Tamm prepared the aldehyde moiety of **126** by hydrolysis of an acetal (1980HCA121), while Snider and Neubert utilized a modified Moffat oxidation of the corresponding primary alcohol (2004JOC8952).

The only solid-phase approach to 3-pyrrolin-2-ones (disregarding tetramic acids) was reported by Miller and coworkers (1999JCC223; Scheme 24). Cross-coupling amino alcohols **130** with resin-bound 1,3-dicarbonyls **131** followed by a Cr-mediated oxidation gives the amideketones **132**. Intramolecular aldol condensations of **132** give resin-bound 3-pyrrolin-2-ones **133** and subsequent treatment with TFA gives the 3-carboxy-3-pyrrolin-2-ones **134**. Several solid-phase approaches to tetramic acids have appeared where an ester moiety is attached to a resin; intramolecular Dieckmann-like condensations then give tetramic acids (1998JOC4808,

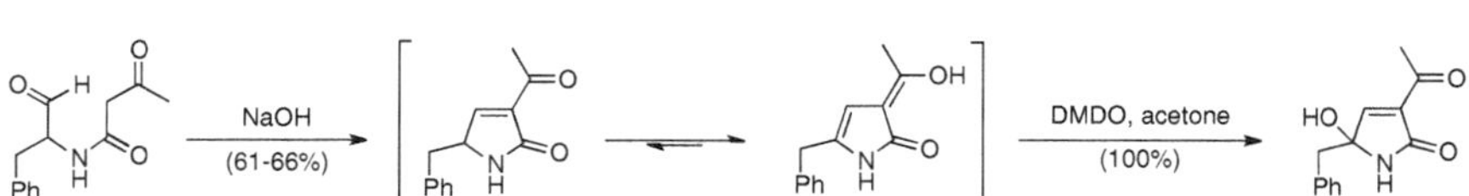


**Scheme 23**

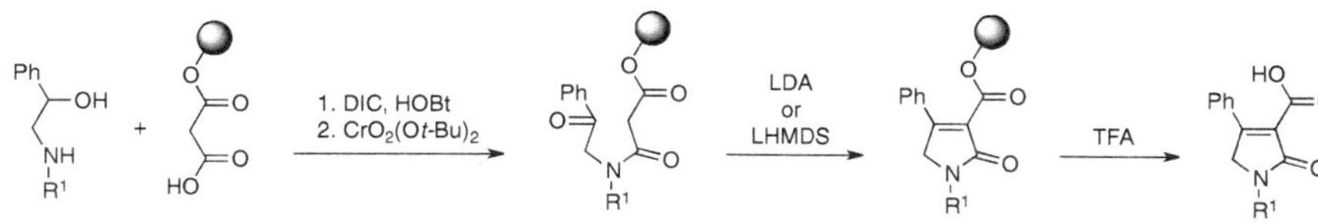

130 131 132 133 134

Scheme 24

1998TL4369, 1999OPRD177, 2003BMCL2505, 2005OL5521, 2005 SL2763, 2006BMCL2205, 2006JCC480).

A number of solution-phase approaches to 3-substituted tetramic acids have employed intramolecular Dieckmann-like condensations. Pelkey and coworkers prepared 3-phenyltetramic acids **137** via a cyclocondensation involving amidoesters **136** mediated by potassium *tert*-butoxide (Scheme 25; 2009JHC447, 2014JOC8049); this work was based on the pioneering work of King and MacMillan (1950JA1236). Amidoesters **136** were prepared by DCC-mediated coupling of commercially available arylacetic acids (**135**) with ethyl glycinate followed by protection with $Boc_2O$. Tetramic acids **137** were later converted into the corresponding triflates **138** by treatment with triflic anhydride; the triflates proved to be useful building blocks for the preparation of 3,4-diaryl-3-pyrrolin-2-ones via Suzuki–Miyaura cross-coupling reactions. Many others have prepared 3-substituted tetramic acids: selected examples include 3-acyltetramic acids (1964JA5654, 1985 JOC1344, 1985JA1777, 1988TL5829, 1994H(38)1839, 2005JMC5932, 2008BMCL1840, 2009SL2487, 2014JHCE25), 3-cyanotetramic acids (1985JHC1599, 1997ACIE2454), 3-methylthiotetramic acids (1985 LAC1095), 3-chlorotetramic acids (1979HCA872, 2003SL1007), 3-benzothiadiazine-substituted tetramic acids (2005OL5521, 2006BMCL2205), 3-aryltetramic acids (1950JA1236, 1997S305, 2007BMCL1908, 2007 S2317, 2007TL7259, 2009CPB979, 2009TL1705, 2010TA1758, 2013 MAC3895, 2014TL2142), 3-aryloxytetramic acids (2011BMC4903), and *N*-hydroxy-3-aryltetramic acids (2002H(57)881). Zali-Boeini and

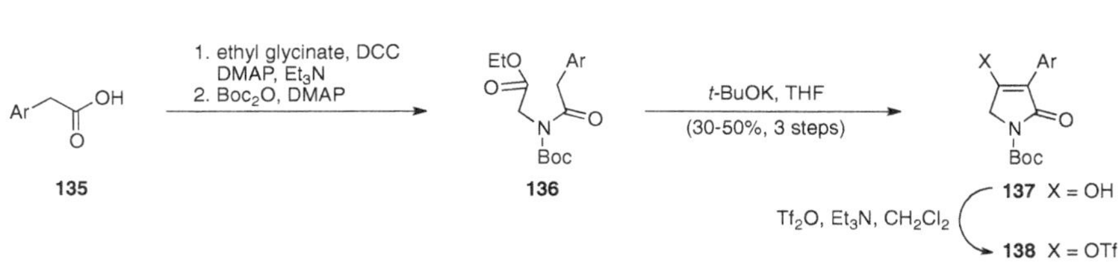

135 136 137 X = OH 138 X = OTf

Scheme 25

coworkers used a cyano electrophile in place of an ester electrophile; this provided access to 4-amino-3-pyrrolin-2-ones (2012JOC5808).

Ring-closing alkene metathesis (RCM) is a powerful tool that has been employed in the synthesis of functionalized 3-pyrrolin-2-ones from the corresponding *N*-allyl acrylamides. This is perhaps the most common route for the de novo synthesis of highly substituted 3-pyrrolin-2-ones. Not surprisingly, an RCM approach to 3-pyrrolin-2-ones was first reported by the Grubbs research group with the preparation of *N*-benzyl-3-pyrrolin-2-one (1993JA9856; see also 2001JA11312). An early application to a complex 3-pyrrolin-2-one was reported by Nakagawa and coworkers in the total synthesis of the natural product (*S*)-pyrrolam (**48**) (Scheme 26; 1997SL1179, 2000CPB1593). They transformed the chiral pyrrolidine **139** into diene **140** in three steps using the reagents shown in Scheme 26. Treatment of diene **140** with Grubbs' first-generation Ru catalyst **141** then provided (*S*)-pyrrolam **48**. Others have also prepared (*R*)-pyrrolam (**48**) using an RCM strategy (2010T1274). An RCM approach to the natural product pulchellalactam (**44**) has recently been communicated by Chavan and coworkers (2007SC1503). RCM reactions have been investigated in the preparation of a wide variety of 3-pyrrolin-2-ones including chiral 5-substituted 3-pyrrolin-2-ones (1996SL67, 2004TL719, 2005SL1734, 2006JOMC5336, 2007JOC7443, 2007OBC2357, 2008JOC6239, 2011 OBC6761, 2013HCA1564, 2013CEJ16746), chiral 5,5-disubstituted 3-pyrrolin-2-ones (2006JOMC5336, 2006OL939), chiral *N*-substituted 3-pyrrolin-2-ones (2005TA2235, 2010BMCL1924), 3-acylated 3-pyrrolin-2-ones (2006SL2807), 3-alkylated 3-pyrrolin-2-ones (2011BMCL1218, 2012BMCL4189, 2012JMC10766, 2014CMC649), 4-vinyl 3-pyrrolin-2-ones (2004JOC6305, 2008TL7334, 2011MI2), fluorinated 3-pyrrolin-2-ones (2004TL57, 2004TL959, 2006EJOC1166, 2014ACSC2374, 2014 EJOC5777), 3-carbamoyl 3-pyrrolin-2-ones (2004JCC684), *N*-alkyl 3-pyrrolin-2-ones (2011SL37, 2000JOC2204, 2008OL597), *N*-arylated

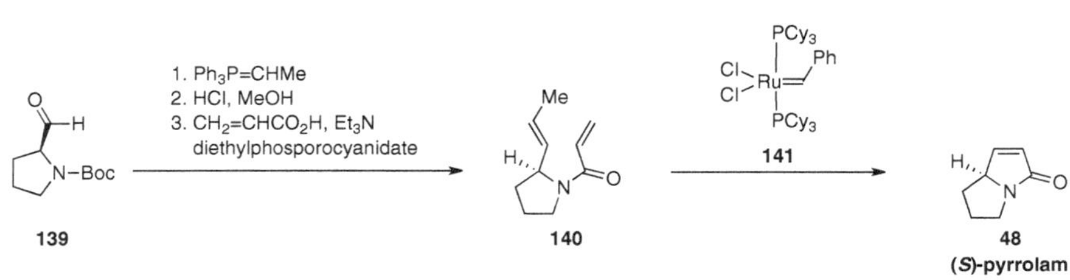


Scheme 26

3-pyrrolin-2-ones (2001MI6, 2012S3523), *N*-carbamoylated 3-pyrrolin-2-ones (2003SC223, 2007OBC2357, 2007SC1503, 2012MI1), *N*-sulfonyl 3-pyrrolin-2-ones (2000JA8168, 2001ACIE4251, 2002ACIE2602, 2002 JA4954, 2005JA4510, 2007ASC1066, 2008CC4312, 2010CC806), *N*-amino 3-pyrrolin-2-ones (2006S3490), *N*-hydroxy 3-pyrrolin-2-ones (2012TL4114), and fused 3-pyrrolin-2-ones (2009OL2892). RCM reactions leading to 3-pyrrolin-2-one intermediates have been coupled with other reactions in cascade sequences to give complex heterocycles (2011ACIE5188, 2012OL5022, 2013OL1986).

Hoveyda and coworkers have investigated RCM reactions with recyclable Ru catalysts (2002JA4954), dendritic Ru catalysts (2000JA8168), and polymer-supported Ru catalysts (2001ACIE4251, 2005JA4510). Polymer-supported Ru catalysts have also been investigated by others in the context of RCM-type syntheses of 3-pyrrolin-2-ones (2002ACIE2602, 2007ASC1066, 2008CC4312, 2010CC806).

Intramolecular Wittig-like reactions have been utilized to form 3-pyrrolin-2-ones from activated α-amidocarbonyl compounds. In an early example, Stork and Matthews prepared 3,4-dialkyl-3-pyrrolin-2-ones via a modified Horner–Wadsworth–Emmons reaction (1970CC445). More recently, Tilve and coworkers used an intermolecular Wittig reaction to prepare (*S*)-pyrrolam (**48**) (Scheme 27; 2008HCA1500). Treatment of the bromoacetamide **142** with triphenylphosphine followed by sodium hydride gives the product **48**, presumably via the ylide intermediate **143**. Intramolecular Wittig and Horner–Wadsworth–Emmons reactions have been used to generate a variety of 3-pyrrolin-2-ones including steroidal-substituted 3-pyrrolin-2-ones (1978HCA977), quinoline-fused 3-pyrrolin-2-ones (2008T4387), cyclohexane-fused 3-pyrrolin-2-ones (2003 TL7485), and 4-cyclopropyl-substituted 3-pyrrolin-2-ones (2004H(63) 699, 2009BMC6879).

A novel Wittig-like route to 5-substituted 4-alkoxy-3-pyrrolin-2-ones (alkyl tetramates) utilizing a resin-bound cumulated ylide has been reported by the Schobert group (Scheme 28; 2004OBC3524, 2005T2301).

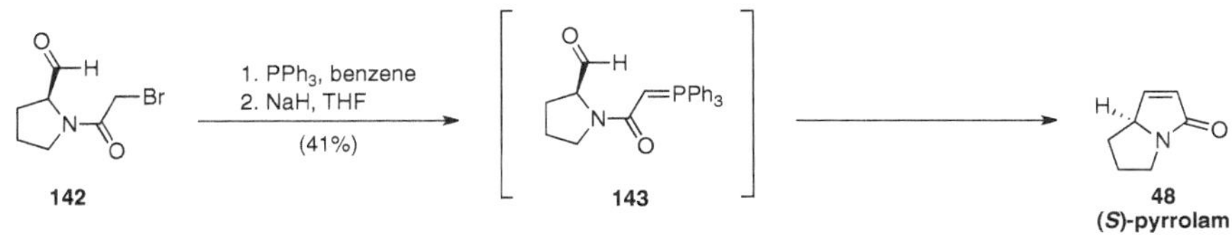


Scheme 27

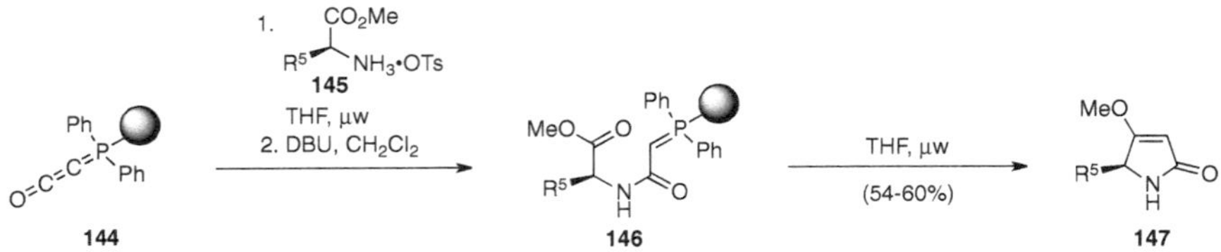


Scheme 28

Treatment of the cumulated ylide **144** with aminoesters **145** under microwave conditions followed by DBU gives the resin-bound ylides **146**. Heating the latter in THF produced methyl tetramates **147** via a Wittig cyclative cleavage. This strategy was also reported by Schobert to complete a total synthesis of pyrrolam (**48**) (2007S1499). In a separate study, Pudelová and Krchnák reported a solid-phase synthesis of 1,3,4-trisubstituted 3-pyrrolin-2-ones (2009JCC851).

Marinelli and coworkers prepared highly functionalized 3-pyrrolin-2-ones using a base-promoted cyclization of *N*-propargylamides (Scheme 29; 2006S2019). For example, propargylamine **148** was acylated with ethyl malonyl chloride (**149**) to give the propargylamide **150** which upon treatment with cesium carbonate gives the 3-pyrrolin-2-one **151**. This transformation seems to require full substitution of the methylene position of the 3-pyrrolin-2-one to work. In a separate paper by the same research group, a similar transformation leading to 3-pyrrolin-2-ones was also investigated utilizing an electrochemical method (2007EJOC2430). Park and coworkers use a related propargylamide cyclization to prepare pyrrole-fused 3-pyrrolin-2-ones (2014ACIE1346).

An intramolecular Pd-mediated Heck-type approach to 3-pyrrolin-2-ones has been disclosed by Yang and Shea (1995MI1; Scheme 30). Treatment of *N*-allyl-2-bromoacetamide **152** with palladium acetate and triphenylphosphine in acetonitrile gives 4-methyl-3-pyrrolin-2-one **153** by a 5-*exo-trig* cyclization onto a terminal alkene, reductive elimination, and isomerization

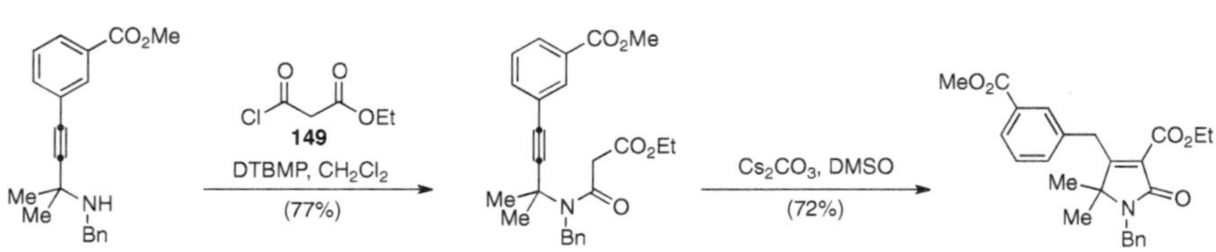


Scheme 29

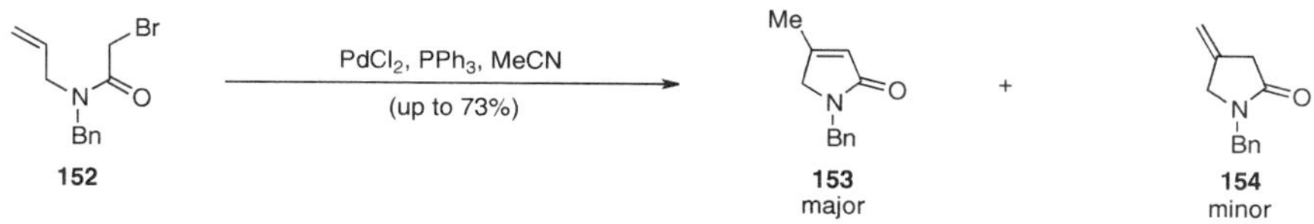


**Scheme 30**

of the exocyclic alkene. While optimizing the reaction conditions, the unisomerized alkylidene pyrrolidinone **154** was also obtained as a minor product. Mori and coworkers also investigated this same Heck reaction with the corresponding iodides (1985T5465). Clark and coworkers prepared fused 3-pyrrolin-2-ones using a Cu-mediated 5-*endo* cyclization of enamines (2001MI5). The de Meijere research group investigated a tandem Heck cyclization and Diels–Alder approach that allowed for the synthesis of fused 3-pyrrolin-2-ones from *N*-allyl propenamides (2001EJOC1673).

Wang prepared a variety of 3-pyrrolin-2-ones using Pd-catalyzed cycloisomerizations of enynes (Scheme 31; 2010JOC582). For example, treatment of enyne **155** with palladium acetate in the presence of the aryl bromide **156** gives 3-pyrrolin-2-one **157**. Others have investigated cycloisomerizations of enynes and obtained 3-pyrrolin-2-ones under certain conditions (2007OL1717, 2010ASC893). Tang and coworkers investigated the Rh-mediated cycloisomerizations of 1,5-diynes and enynes, which produced 3-pyrrolin-2-ones and cyclopropane-fused pyrrolidinones, respectively (2013JA8201). Parsons and coworkers investigated Bergman-type cycloisomerizations of enediynes for the preparation of fused 3-pyrrolin-2-ones (2010T6526).

Mitchell and coworkers prepared the 3-pyrrolin-2-one building block (**161**) as part of a synthetic strategy aimed at the natural product muironolide A (Scheme 32; 2013SL1861). They treated allylamine **158** with methyl malonyl chloride (**159**) followed by diazotization with *p*-acetamidobenzenesulfonyl azide (*p*-ABSA) gives diazoester **160**. Heating **160** in benzene

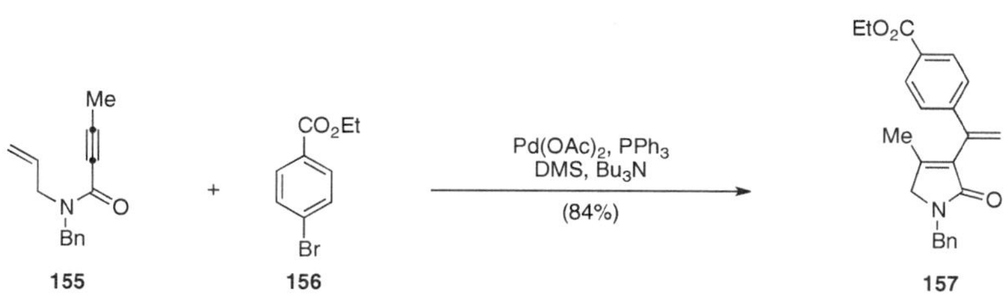


**Scheme 31**

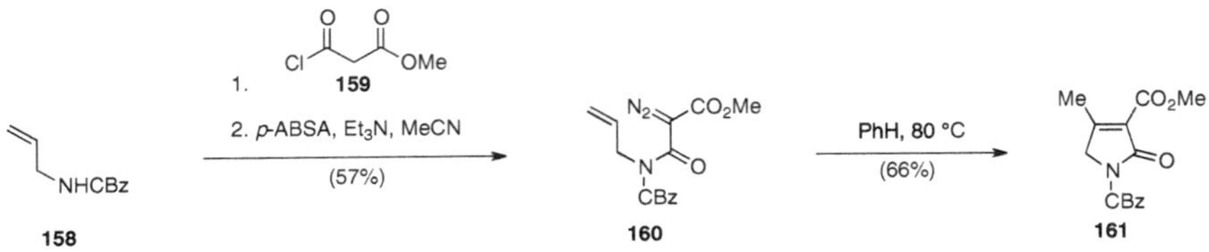


Scheme 32

produced 3-pyrrolin-2-one **161**. Others have examined Rh-mediated cyclization reactions of diazo compounds leading to 3-pyrrolin-2-ones and other heterocycles (1996JOC2179, 1998BMC1273).

1,3-Dipolar cycloaddition–cycloreversion reactions involving isomünchnones sometimes lead to the formation of substituted 3-pyrrolin-2-ones as minor by-products (1994JOC2447, 1994TL7159, 1996T3247, 1998BMC1273).

Tamura and coworkers studied Pummerer-type cationic cyclizations including a variant that led to a 3-pyrrolin-2-one (Scheme 33; 1981 TL4301). Treatment of the α-sulfonylamide **162** with trifluoroacetic anhydride led to the formation of 3-pyrrolin-2-one **164** and pyrrolidinone **165** following a Pummerer-type reaction manifold that includes the cation intermediate **163**.

A few radical type "c" cyclization approaches to 3-pyrrolin-2-ones have been reported. Clark and coworkers treated enamide **166** with cerium ammonium nitrate (CAN); this led to the formation of 3-pyrrolin-2-one **167** via an oxidative 5-*endo-trig* cyclization (Scheme 34; 2003OL2063). The latter comprises the heterocyclic core of the platelet inhibitor PI-091 (**39**). A domino radical cyclization to 3-pyrrolin-2-ones that could serve as rigid peptide mimetics has been reported (2003JOC1552, 2005SL83); the mechanism includes a transfer–elimination of a dibenzylamine group. Radical cyclization of a benzofuran-substituted amide led to the formation of a 3-pyrrolin-2-one via ring opening of a spirocyclic intermediate (2004TL8931).

Pattenden and coworkers investigated radical cyclization reactions of *N*-propargyl bromoacetamides mediated by tributyltin hydride; in some

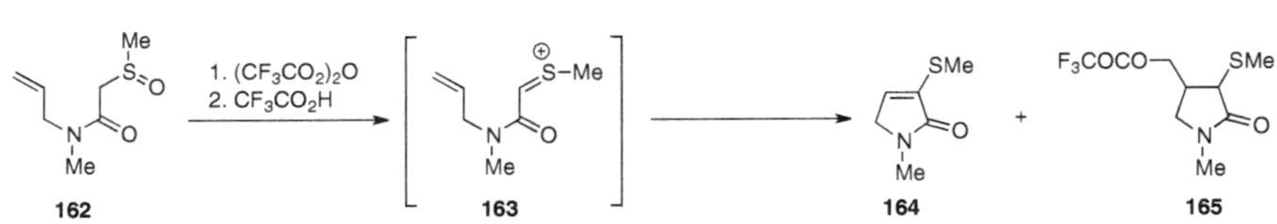


Scheme 33

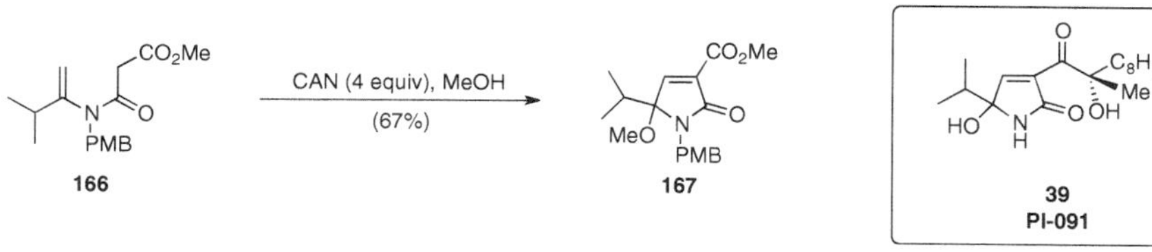


Scheme 34

cases, 3-pyrrolin-2-ones were obtained as products (1989TL7469). Separately, the groups of Ghelfi, Clark, and Ishibashi have investigated radical cyclization approaches to 3-pyrrolin-2-ones and halopyrrolidinones involving monohalo-, dihalo-, and trihalo-substituted acetamide substrates. In one of the early studies, treatment of the dichloroacetamide **168** with copper chloride led to the formation of 3-pyrrolin-2-one **169** via a radical atom transfer cyclization (Scheme 35; 2001TL1999). The reaction required an electron-withdrawing group on the amide (Boc or Ts) as the reaction failed with a benzyl protecting group. Additional cyclization substrates that have been studied include *N*-vinyl dichloroacetamides (2003T6221, 2008ARK(xiv)7), *N*-allyl dichloroacetamides (2003T1147, 2004T11493, 2006T746, 2008S3131, 2009T1481, 2010SC1040, 2011S1267, 2011 T408), *N*-allyl trichloroacetamides (2006TL6263, 2010T1357), and *N*-propargyl bromoacetamides (2012JOC6778). In each of these studies, 3-pyrrolin-2-ones were obtained after based–mediated elimination reactions of the halopyrrolidinone intermediates.

Stephenson and Tucker discovered an approach to fused 3-pyrrolin-2-ones via a tandem Ir-mediated radical cyclization and Cope rearrangement (Scheme 36; 2011OL5468). Treatment of cyclopropane-substituted propargylamide **170** with an iridium reagent and triethylamine led to the formation of fused 3-pyrrolin-2-one **172**. The reaction manifold that explains the outcome of the reaction includes a radical cyclization to the spirocyclic intermediate **171**, Cope rearrangement and re-aromatization.

Recently, the Pettus group has developed a novel synthesis of tetramic acid derivatives via Sm-mediated cyclizations of bromoacetamido esters

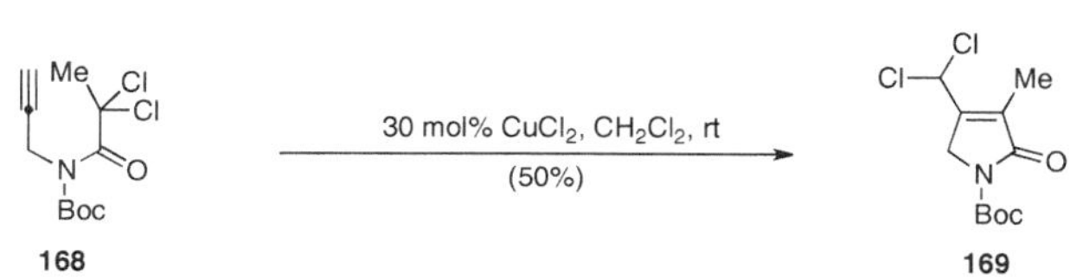


Scheme 35

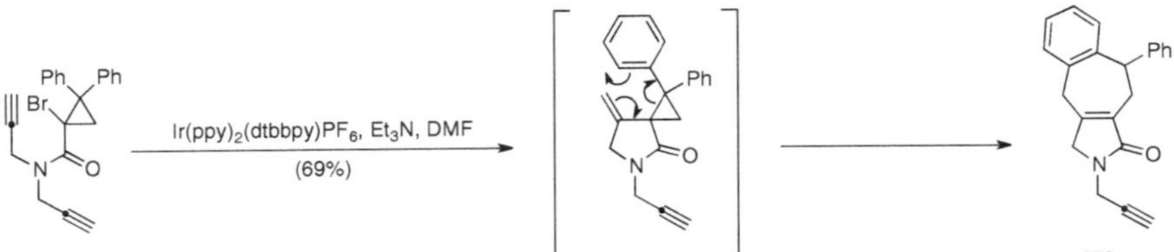


170 171 172

Scheme 36

(Scheme 37; 2012OL3862). Treatment of the bromoacetamide **173** with samarium iodide and HMPA followed by diazomethane gives the methyl tetramate **174**. The latter intermediate allowed completion of a formal synthesis of palau'imide (**34**). Pettus and coworkers later used this methodolgy in an approach to tetrapetalone (**24**) (2014OL1294).

## 2.4 Type "d" Cyclizations

The first type "d" cyclization that is discussed involved an intramolecular Michael addition reported by Asaoka and coworkers (Scheme 38; 2004H(63)1009). Treatment of thio-substituted propenamide **175** with LDA produced the 3-pyrrolin-2-one **177** presumably via elimination of the sulfanyl group from the intermediate Michael adduct **176**. Interestingly, this transformation was completely stereoselective (the opposite enantiomer of **175** gives the opposite enantiomer of **177**). Hess and Burton investigated a Zn-mediated intramolecular Michael addition involving a propynamide leading to 3-pyrrolin-2-ones (2011ASC2966); Pawar and Pigge investigated a Au-catalyzed variation with 5-pyridine-substituted 3-pyrrolin-2-ones (2013TL6067).

Clayden and coworkers reported intramolecular Michael-addition–fragmentation leading to 3-vinyl- and 3-aryl-substituted 3-pyrrolin-2-ones (2004OL609). For example, treatment of the carboxamide **178** (DEB = 1,1-diethylbutylcarbonyl) with LDA gives 3-pyrrolin-2-one **180** through a fragmentation of the initial Michael adduct **179** (Scheme 39).

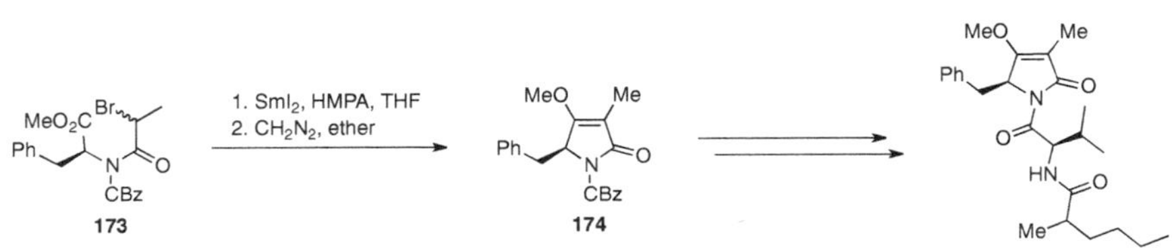


173 174 34 palau'imide

Scheme 37

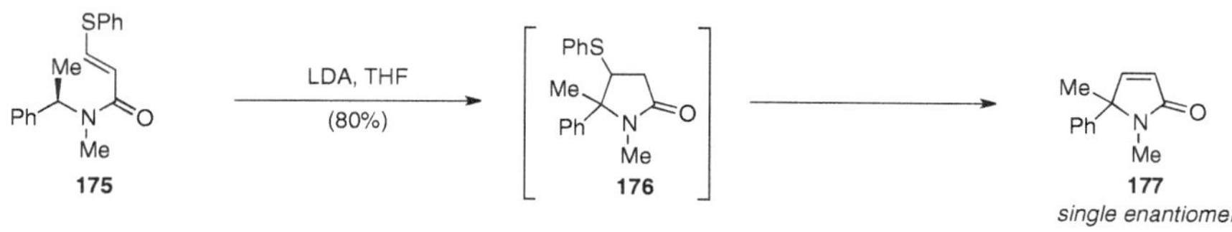


Scheme 38

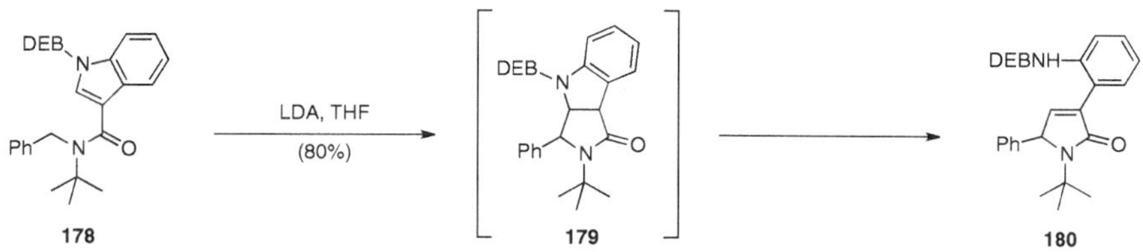


Scheme 39

Parameswarappa and Pigge developed a type "d" intramolecular aldol-like approach to highly substituted 3-pyrrolin-2-ones (Scheme 40; 2012JOC8038, 2014OBC1090). Treatment of pyridine-substituted 1,3-dicarbonyl derivative **181** with ethyl chloroformate and sub-stoichiometric Ti(O*i*Pr)$_4$ followed by TFA gives the 3-pyrrolin-2-one **183**. The mechanism probably involves acylation of the pyridine nitrogen followed by deprotonation of the benzylic position to give intermediate **182**, which then undergoes an aldol condensation. The same group also examined Au-catalyzed intramolecular Michael additions using structurally-related substrates (2013TL6067).

Gilbert and Blackburn showed that the generation of an alkylidenecarbene intermediate is a viable route to 3-pyrrolin-2-ones (Scheme 41; 1986JOC3656). Treatment of the 2-oxopropanamide **184** with dimethyl diazomethyl(phosphonate) (**185**) and *t*-BuOK gives the 3-pyrrolin-2-one **187**. The 3-pyrrolin-2-one ring is formed by an intramolecular C—H insertion reaction involving alkylidenecarbene intermediate **186**.

Zang and Larock reported an interesting iodo-mediated *ipso*-cyclization of alkynyl carboxamides leading to give spirocyclic 3-pyrrolin-2-ones (Scheme 42; 2005JA12230); this work built upon the earlier work of

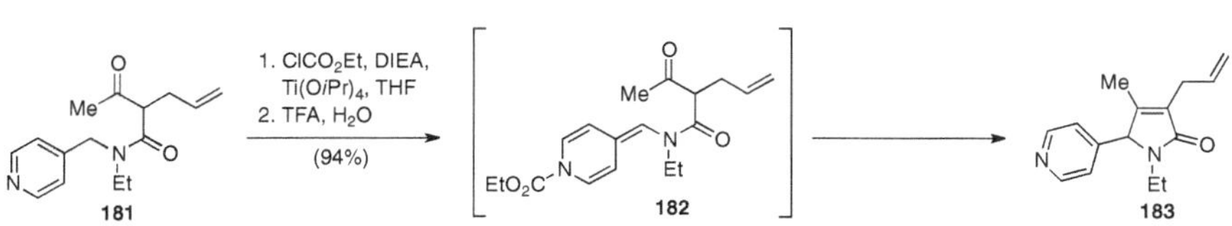


Scheme 40

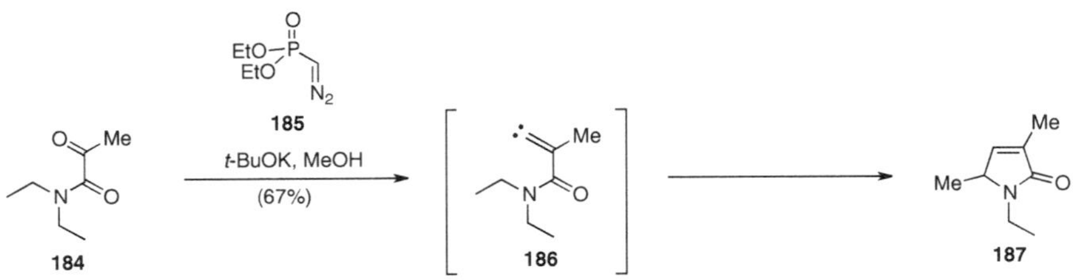


Scheme 41

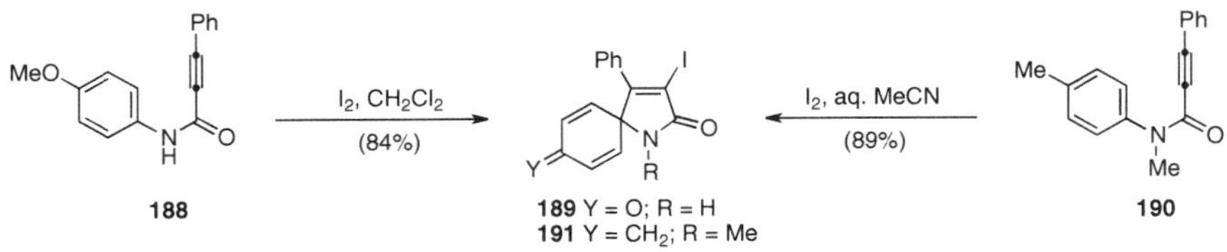


Scheme 42

Fanghänel who investigated a thioether substrate (2003EJOC47). Treatment of alkyne **188** with iodine led to the spirocyclic 3-pyrrolin-2-one **189** via an electrophilic iodocyclization of the aromatic ring onto the alkyne. Li and coworkers extended this reaction manifold to tolyl substrates (e.g., **190**) and methylene-containing spirocyclic 3-pyrrolin-2-ones (e.g., **191**) (2008JOC3658). Unlike the prior work of Larock, Li reported that secondary amides failed with their system and the reaction required a tertiary amide. Li and others have explored a number of variants to this reaction including *para*-unactivated arenes (2008OL1063, 2009S891, 2009S2407, 2012 JOC2837), quinolones (2011SL2657), and quinolines (2013TL1344). Enantioselective additions to spirocyclic cyclohexadienones that are structurally related to **191** have been explored by the groups of Gaunt (2011CS1487) and You (2012CC6363).

Continuing with this structural theme, Li and coworkers extended the electrophilic trigger in the reaction manifold to spirocyclic 3-pyrrolin-2-ones: new variations included bromocyclization (2012JOC2837), thiocyano-cyclization (and iodo and bromo) mediated by electrophilic fluoride (2008JOC9008), acyl-cyclization mediated by a peroxide (2014JOC4582), and phosphoryl-cyclization mediated by silver (2014CC13998; Scheme 43). Treatment of alkynylamide **192** with aldehyde **193** and *tert*-butyl hydrogen peroxide (TBHP) led to the acylated 3-pyrrolin-2-one **194**; a similar reaction with phosphonate **195** and silver nitrate led to the phosphonate-substituted 3-pyrrolin-2-one **196**.

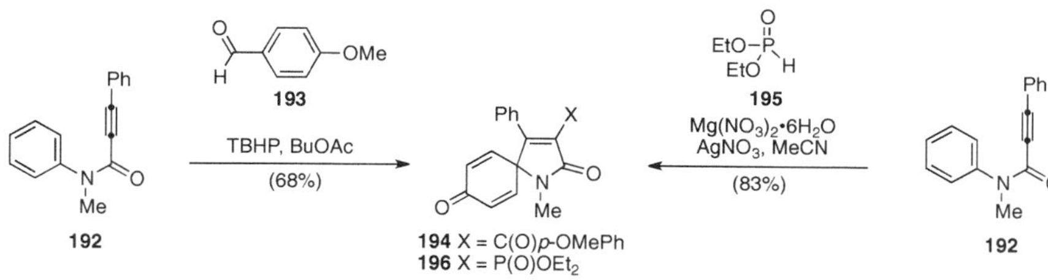


Scheme 43

The Vanderwal group has developed a novel tandem cycloaddition/metathesis sequence leading to tricyclic 3-pyrrolin-2-ones (Scheme 44; 2012OL5566). This reaction extended the work of Himbert and Henn who first reported a cycloaddition process involving unactivated benzene rings leading to novel fused 3-pyrrolin-2-ones (1982ACIE620). Heating alkyne **197** in a microwave in the presence of $K_2CO_3$ led to the formation of **199** via a cycloaddition of the allene intermediate **198**. Treatment of **199** with Grubbs–Hoveyda catalyst **200** led to the formation of the tricyclic 3-pyrrolin-2-one **201** via a metathesis-rearrangement process.

## 2.5 Type "e" Cyclizations

The Dittami group developed a cyclocondensation approach to 5-hydroxy-3-pyrrolin-2-ones from γ-oxoenamides (Scheme 45; 1995TL4197, 1995 TL4201). For example, irradiation of (*E*)-enamide **202** produced 5-hydroxy-3-pyrrolin-2-one **204** via the isomerized intermediate (*Z*)-enamide **203**. An alternate approach to (*Z*)-enamides, and subsequently to 5-hydroxy-3-pyrrolin-2-ones, was discovered by Nagoa and coworkers (2000TL2419). They utilized the hydrolysis of α-hydroxy-α-(β-propiolamide)malonates to produce 5-carboethoxy-5-hydroxy-3-pyrrolin-2-ones, inhibitors of cathepsin B.

Geffken and coworkers reported an intramolecular DCC-mediated cyclocondensation of *N*-alkoxy-α-aryl-α-hydroxybutenamides, which led

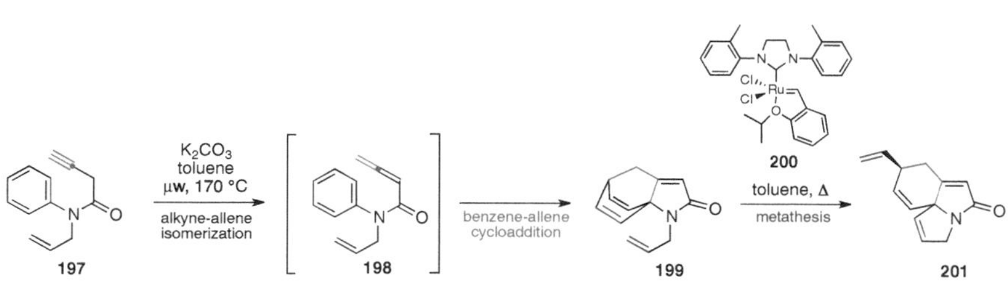


Scheme 44

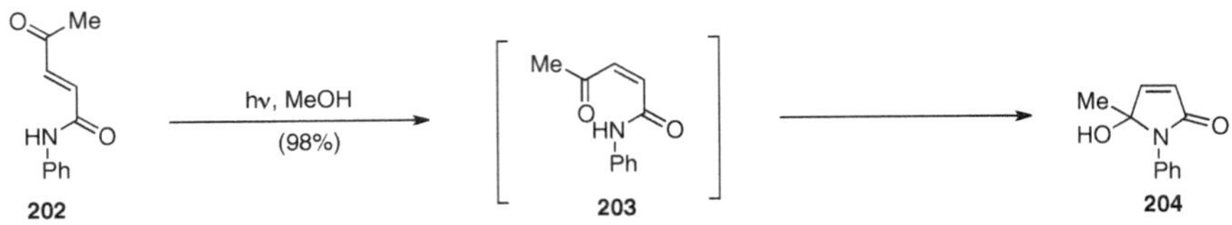

Scheme 45

to the formation of *N*-alkoxy-3-aryl-3-pyrrolin-2-ones (1996MI1). They later found a different route to 3-pyrrolin-2-ones via oxathiazolidinone intermediates also generated from *N*-alkoxybutenamides (1997MI2).

One of the oldest reports of a synthesis of a highly substituted 3-pyrrolin-2-one involved a type "e" cyclization. Bashour and Lindwall treated γ-oxo-β-hydroxycarboxamide **205** with HCl and obtained a product presumed to be 5-hydroxy-3-pyrrolin-2-one **206** (Scheme 46; 1935JA178). Given the similarity of reactions that have been reported since then (e.g., Scheme 42), this probably was the correct structural assignment. Moon later reported a synthesis of 4-acyl-3-pyrrolin-2-ones via thermal cyclization of γ-ketocarboxamides (1977JOC2219).

Intramolecular cyclocondensations of γ-oxocarboxamides lead to 3-pyrrolin-2-ones via isomerization of the initially formed 4-pyrrolin-2-ones. Micheli and coworkers utilized this strategy to prepare 1,3-diaryl-3-pyrrolin-2-ones that are 5-$HT_{2C}$ inhibitors (2006BMCL3906); while Shrestha-Dawadi and Lugtenburg prepared $^{13}$C-labeled 3-cyano-4-methyl-3-pyrrolin-2-ones from the corresponding γ,γ-dimethoxybutanamides (2007EJOC1294). Ojima has explored a Rh-catalyzed hydrocarbonylation of 3-butenamides which leads to a mixture of 4-methyl-3-pyrrolin-2-one and 3,4-dihydro-2-pyridone depending on the regioselctivity of the hydrocarbonylation (1991JOC2024).

The De Kimpe group prepared 4-aryl-3-pyrrolin-2-ones starting with cyclobutanones (Scheme 47; 2004SL1059). Treatment of cyclobutanone **207** with isopropylamine followed by sodium methoxide led to the amide **208** via a ring-opening reaction and substitution–elimination.

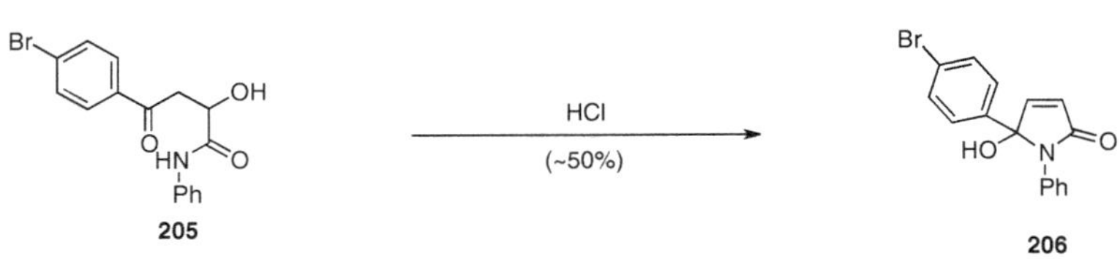

Scheme 46

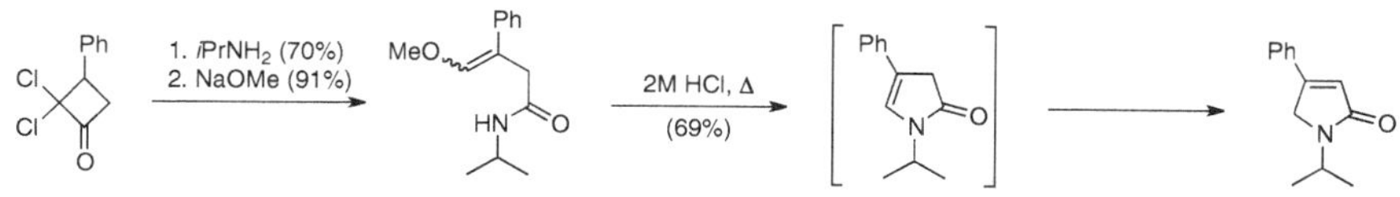


207 208 209 210

Scheme 47

Acid-mediated cyclocondensation of **208** then leads to the 3-pyrrolin-2-one **210** via the corresponding 4-aryl-4-pyrrolin-2-one intermediate **209**.

Prandi and coworkers investigated cyclization reactions involving dienamides; they found differential reactivity based on the substitution pattern of the dienamide (Scheme 48; 2011OBC2535). For example, treatment of dienamide **211** with TFA led to the formation of 3-pyrrolin-2-one **212**. The same reaction with the methyl-containing substrate **213** led to the formation of imino ether **214**. The structures of the 3-pyrrolin-2-one products were confirmed by ROESY NMR experiments; the structure of one of the imino ether products (*N*-naphthyl) was confirmed by single crystal X-ray analysis.

Intramolecular nucleophilic substitutions of γ-halocarboxamides provide a different category of the type "e" route to 3-pyrrolin-2-ones. Lautens and coworkers generated γ-iodocarboxamides via a Lewis acid-mediated ring opening of methylenecyclopropylamides (e.g., **215**; Scheme 49; 2003JA4028). Treatment of **215** with magnesium iodide led to the 4-methyl-3-pyrrolin-2-one **217** via the iodo intermediate **216**. In a related nucleophilic cyclization, Tabei prepared 1-aryl-3-pyrrolin-2-ones by treating 4-bromo-3-acetoxybutananilides with DBU (1981H(16)795).

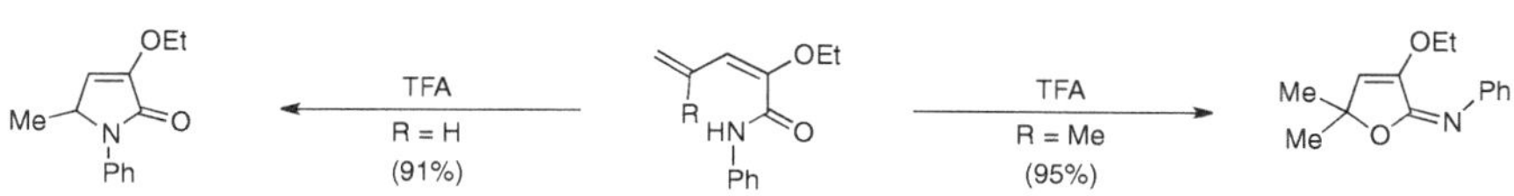


212 211 R = H 213 R = Me 214

Scheme 48

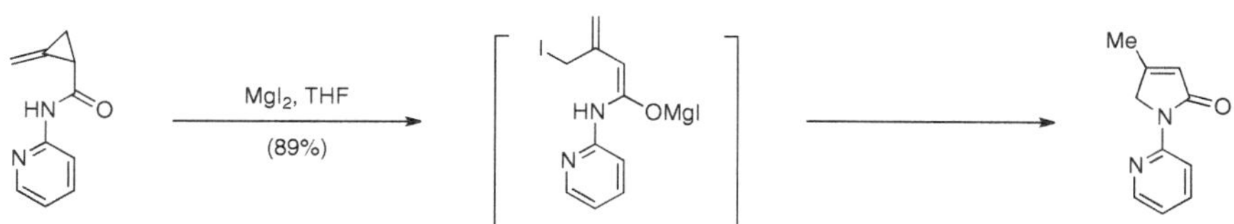


215 216 217

Scheme 49

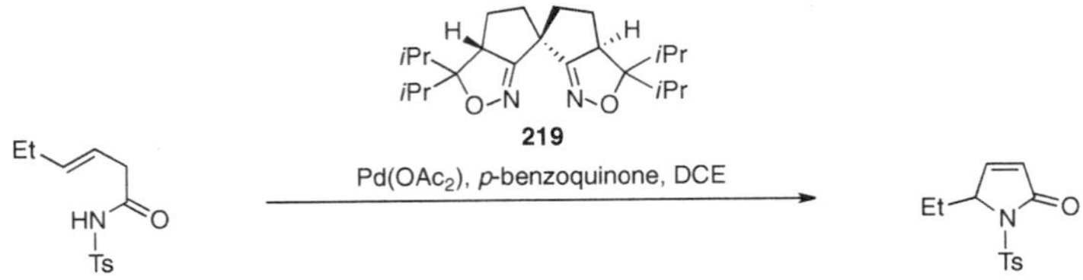


218 220

**Scheme 50**

The Sasai research group found an intramolecular Heck-type approach to simple 3-pyrrolin-2-ones (Scheme 50; 2010CC9064). For example, treatment of β,γ-unsaturated amide **218** with palladium acetate in the presence of the isoxazoline ligand **219** and *p*-benzoquinone gives 3-pyrrolin-2-one **220** via a 5-*endo-trig* cyclization.

Ma and Xie discovered a route to 4-bromo-5-hydroxy-3-pyrrolin-2-ones via electrophilic 5-*endo-dig* cyclizations of 2,3-alkadienamides (Scheme 51; 2000OL3801, 2005T251). Treatment of dienamide **221** with $CuBr_2$ led to the formation of 5-hydroxy-3-pyrrolin-2-one **223**. Presumably, the latter arose from a facile hydroxylation of intermediate 3-pyrrolin-2-one **222** under the reaction conditions. Interestingly, a more recent study involving a similar 2,3-alkadieneamide substrate has led to the 3-pyrrolin-2-one without the 5-hydroxy group (2014CEJ2154). Brandsma and coworkers studied a related Ag-mediated cyclization of dienamides that led to a mixture of dihydrofurans and 3-pyrrolin-2-ones (2002TL1569).

Abarbri and coworkers reported an approach to 5-alkylidene-3-pyrrolin-2-ones via iodine-mediated electrophilic cyclizations of (*Z*)-alk-2-en-4-ynamides (Scheme 52; 2004TL2063). Treatment of enyne **224** with iodine monochloride led to a stereoisomeric mixture of 5-(iodoalkylidene)-3-pyrrolin-2-ones **225** and **226**. The stereoselectivity of the reaction decreased with substituted enynes (R ≠ H) compared to unsubstituted enynes (R = H). In a different study by the Yoshimatsu group, an anionic cyclization of alkynylamides led to the formation 5-(phenylselenomethylene)-3-pyrrolin-2-ones (1996PT(1)1839). The group of Ohe and Uemera

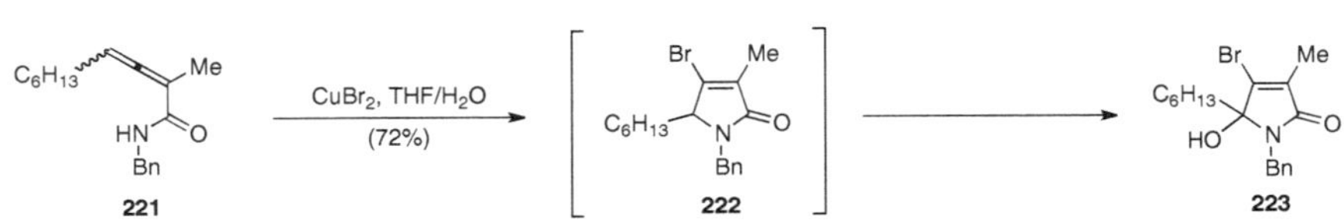


221 222 223

**Scheme 51**

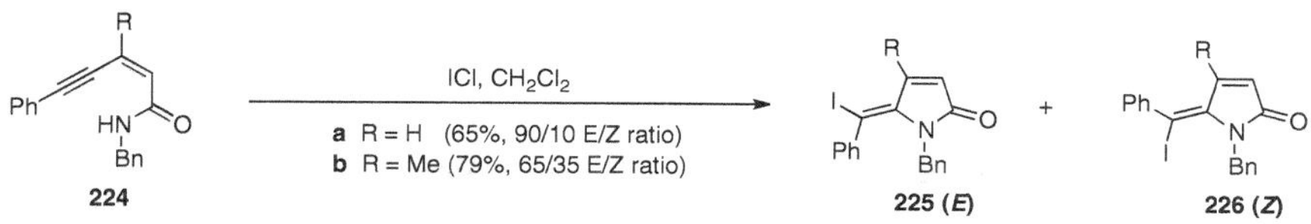


Scheme 52

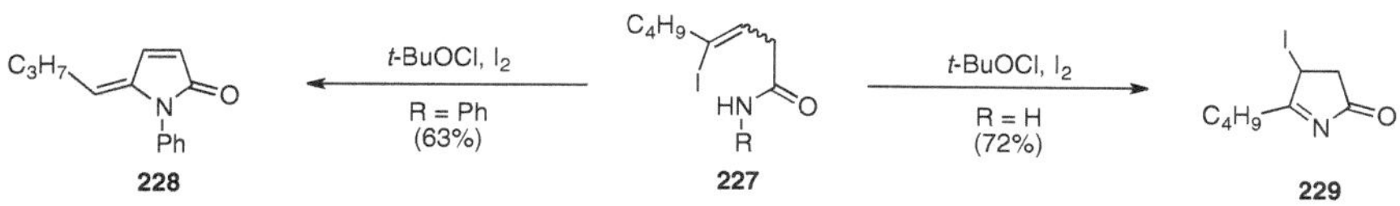


Scheme 53

reported a Rh-mediated tandem cyclization–cyclopropanation of enynes, which gives 5-cyclopropyl-substituted 3-pyrrolin-2-ones (2003OL2615).

Finally, a 5-*endo* radical cyclization approach to 5-alkylidene-3-pyrrolin-2-ones has been reported by Tang and Li (Scheme 53; 2004OL3229). Treatment of 4-iodo-3-octenoamide **227** (R = Ph) with *t*-butyl hypochlorite gives 5-(butylidene)-3-pyrrolin-2-one **228**; the same reaction with the corresponding primary amide **227** (R = H) gives the iminoketone **229**.

## 3. INTERMOLECULAR APPROACHES TO 3-PYRROLIN-2-ONES

There are exactly 10 different two-bond disconnections possible for the construction of 3-pyrrolin-2-ones. Eight of these intermolecular approaches have appeared in the literature and are depicted below (Scheme 54). For example, a type "ac" cyclization could combine a Dieckmann-type condensation (to form bond "c") and a lactamization (to form bond "a"). The most commonly utilized intermolecular approaches to 3-pyrrolin-2-ones involve type "ae" cyclizations followed closely by type "ab" (carbonylations) and type "ac" cyclizations. Only a small handful of reports of the other types of intermolecular approaches have been reported.

### 3.1 Type "ab" Cyclizations

Carbonylative coupling reactions (e.g., from **230**) and [4+1] cycloadditions (e.g., from **231**) are two metal-mediated methods for synthesizing 3-pyrrolin-2-ones from four atom components and CO (Scheme 55). A review of Fe-mediated reactions leading to dihydropyrrolones by Rück-Braun and

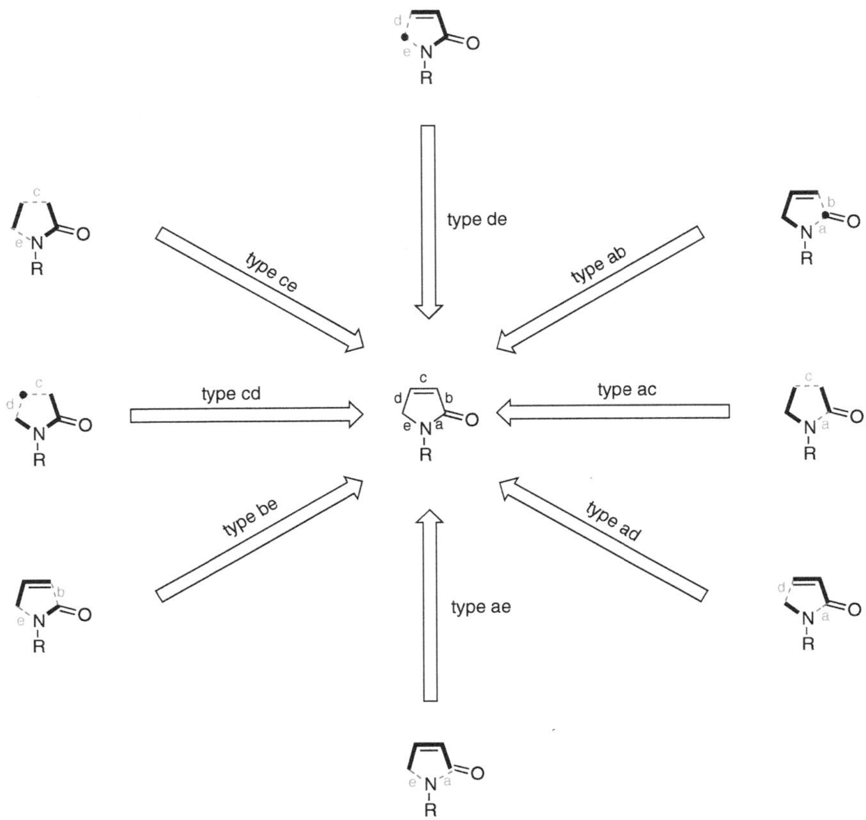


**Scheme 54** Intermolecular approaches to 3-pyrrolin-2-ones.

Amrhein has appeared (2000EJOC3961). Unsaturated metalloaminocarbenes substituted with chromium and iron have been investigated as precursors to 3-pyrrolin-2-ones by the groups of Rudler (1995JOMC127) and Rück-Braun (2000OM3527), respectively.

Crisp and Meyer reported one of the first carbonylative approaches to 3-pyrrolin-2-ones (Scheme 56; 1995T5585). Allylic alcohol **232** was transformed into allylic amine **233** in two steps: treatment of **233** with $Pd(PPh_3)_4$

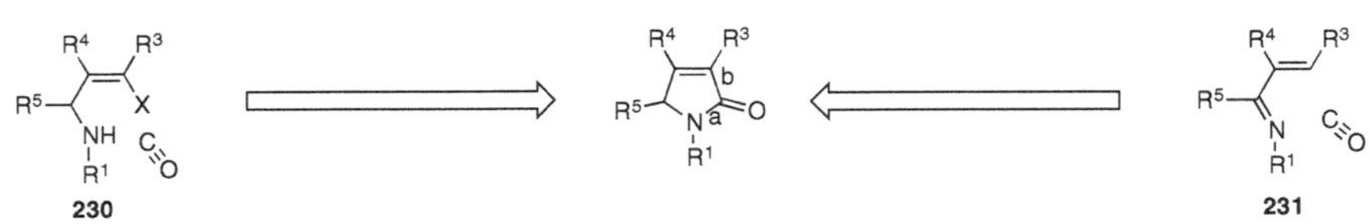

**Scheme 55**

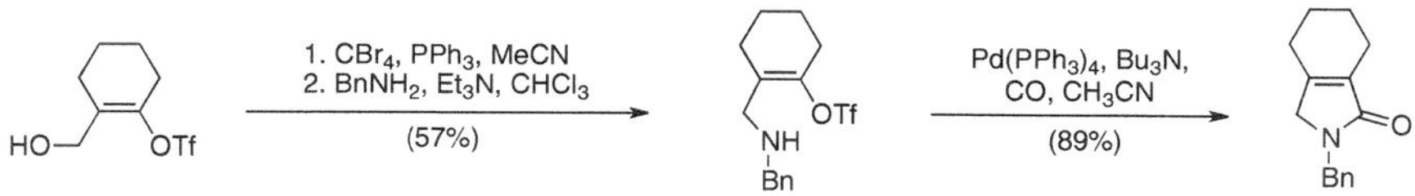


232 233 234

Scheme 56

and CO gives the fused 3-pyrrolin-2-one **234**. The Negishi group reported a similar carbonylation cyclization leading to 3-butyl-3-pyrrolin-2-ones utilizing structurally-related vinyl iodide starting materials (1995TL1771, 1995JA3422, 1996T11529). Perhaps the very first Pd-mediated carbonylative approach that gives a 3-pyrrolin-2-one was reported by Mori and coworkers (1983JOC4058). They treated a 3-iodo-3-butenamine with $Pd(OAc)_2$, $PPh_3$ and CO and obtained a 10:1 mixture of an alkylidene-substituted pyrrolidinone and a 3-pyrrolin-2-one.

Yoshida and coworkers used carbonylative cross-coupling reactions of vinyl boronates to prepare aryl-substituted 3-pyrrolin-2-ones (Scheme 57; 2006JA1464). Vinyl boronate **237** was prepared using a three-component coupling reaction of iodoarene **235**, allenylboronate **236**, and benzylamine. Treatment of **237** with palladium acetate and triphenylphosphine in the presence of CO gives the 4-aryl-3-pyrrolin-2-one **238**. The latter is a useful building block for the preparation of the antidepressant rolipram (**68**), via hydrogenation.

Metal-mediated cycloadditions involving carbon monoxide is a powerful strategy for preparing rings containing carbonyls. The [4+1] cycloaddition of α,β-unsaturated imines with CO leads to the formation of 3-pyrrolin-2-ones; Murai and coworkers investigation this transformation using $Ru_3(CO)_{12}$ (Scheme 58; 1999JA1758). Treatment of the unsaturated imine **239** with a Ru catalyst in the presence of CO led to 3-phenyl-3-pyrrolin-2-one **240**. A related sequence involving α-imino alkynes has been reported by Alper (2001JA10214).

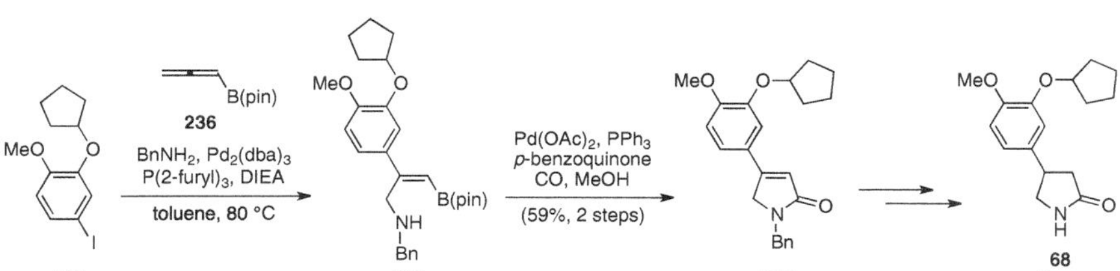


235 237 238 68 rolipram

Scheme 57

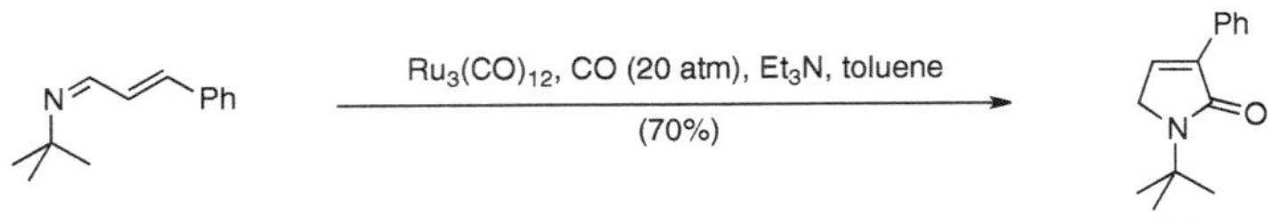


Scheme 58

Kang found that α-allenic sulfonamides can also serve as precursors to 3-pyrrolin-2-ones (Scheme 59; 2001OL2851). Treatment of allene **241** with a Ru catalyst and CO gives 3-methyl-3-pyrrolin-2-one **242** via a carbonylative cyclization. Grigg and coworkers extended this reaction to the preparation of fused 3-pyrrolin-2-ones (2011T4318). However, Dieter and coworkers found that without a high pressure of CO, dihydropyrroles are the main products of reactions involving allenylamides (2005JOC2109).

Gabrielle and coworkers developed a new synthesis of 4-amino-3-pyrrolin-2-ones utilizing a Pd-catalyzed oxidative aminocarbonylation of propargylamines (2005SL935). The Orito group synthesized unsymmetrical staurosporinone derivatives utilizing a Pd/Cu-catalyzed carbonylative cyclization of 5-aminomethylindolo[2,3-*a*]carbazoles (2007JOC2008).

Quai and coworkers utilized a [4+1] cycloaddition with isocyanide to prepare highly substituted 3-pyrrolin-2-ones (Scheme 60; 2004TL1413). Treatment of benzylidene-1,3-diketone **243** with tert-butyl isocyanide (**244**) gives 5-hydroxy-3-pyrrolin-2-ones **246** (Scheme 24). The authors propose an oxidative rearrangement mechanism for the conversion of initial adduct **245** to the observed product **246**. Similar reactions have been reported by others in the synthesis of 3-aryl-4-carboethoxy-5-hydroxy-3-pyrrolin-2-ones (2007T8987) and 3,5-diaryl-5-hydroxy-3-pyrrolin-2-ones (2007TL8056).

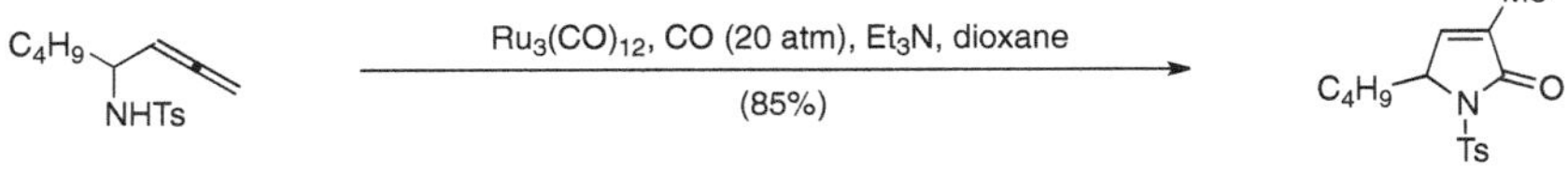


Scheme 59

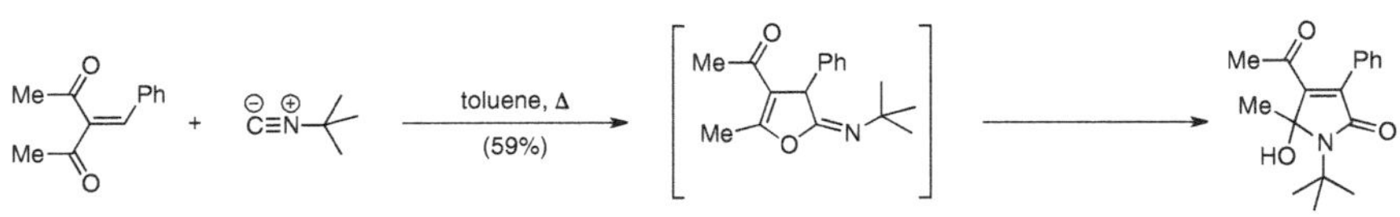


Scheme 60

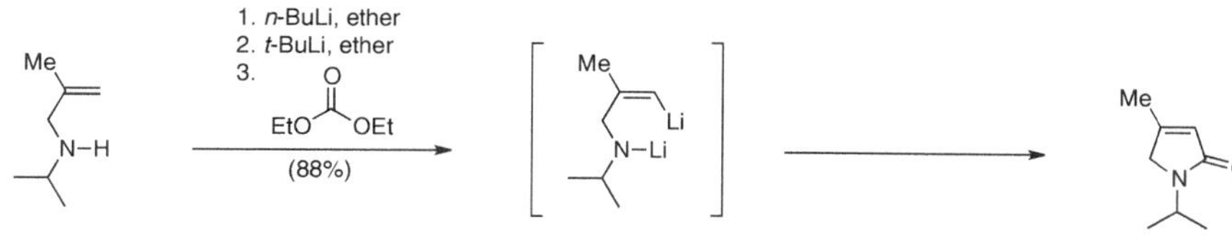


247 248 249

Scheme 61

Barluenga and coworkers utilized a dilithiation reaction involving allylic amines to prepare 3- and 4-substituted 3-pyrrolin-2-ones (1988TL4859, 1993TL7777). Sequential treatment of allylic amines (e.g., **247**) with *n*-butyllithium, *tert*-butyllithium and then diethyl carbonate gives 3-pyrrolin-2-ones (e.g., **249**), presumably via dilithiated intermediates such as **248** (Scheme 61). Yus and coworkers used $CO_2$ as an electrophile in a similar sequence used to prepare 1,5-dimethyl-4-phenyl-3-pyrrolin-one (1995 TA2081).

The Ogoshi group has explored the generation of aza-nickelacycles as precursors to a variety of nitrogen heterocycles including 3-pyrrolin-2-ones (Scheme 62; 2007ACIE4930, 2014CEJ4105). Mixing diphenylacetylene (**250**) and the benzaldimine (**251**) with Ni(cod)$_2$ and tricyclohexylphosphine led to the nickelapyrroline **252** in quantitative yield. Treatment of **252** with CO gives the 3,4,5-triphenyl-3-pyrrolin-2-one **253**.

## 3.2 Type "ac" Cyclizations

Aldol-type condensations are useful for making the "c" connection in 3-pyrrolin-2-ones, hence it makes sense to combine an aldol reaction with an amidation to form a straightforward intermolecular-type "ac" approach to 3-pyrrolin-2-ones. It turns out that not many aldol-lactamization approaches have been reported for making (non-tetramic acid) 3-pyrrolin-2-ones. One example, reported by Yao and coworkers, involves the synthesis of lactam-terminated polyketides (Scheme 63; 2007BMCL3426). 1,3,5-Trisubstituted 3-pyrrolin-2-ones **257** were prepared by treating esters

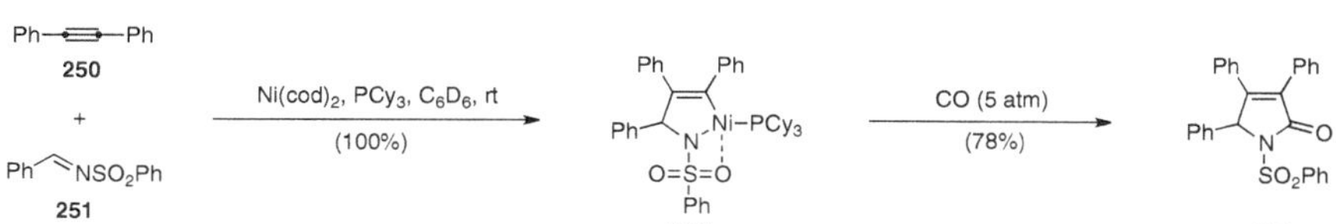


250 251 252 253

Scheme 62

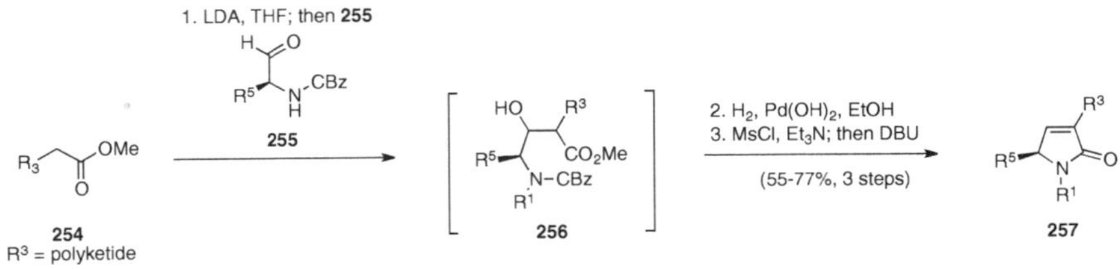


Scheme 63

**254** with LDA followed by α-amino aldehyde derivatives **255**. This aldol reaction produces γ-aminoester intermediates **256**, which are converted into **257** by a sequence that includes a hydrogenation–cyclization, mesylation, and elimination.

Bossio and coworkers utilized a four-component reaction that included a type "ac" cyclization for the construction of highly substituted 3-pyrrolin-2-ones (Scheme 64; 1999H(50)463). Treatment of α-aminoketone **258**, α-cyanoacetic acid (**259**), cyclohexyl isocyanide (**260**), and aldehydes **261** with potassium carbonate gives 3-cyano-3-pyrrolin-2-ones **263** via the amide intermediates **262**.

A venerable approach to 3-acyltetramic acids has become known as the Lacey–Dieckmann cyclization as originally reported by Lacey in 1954 (1954JCS850). In this reaction, Dieckmann cyclization substrates (e.g., **266**) are generated from the reaction of amines with diketene (**265**). This strategy was employed by the Rhinehart group to make building blocks for tirandamycin (Scheme 65; 1978JA4225). Treatment of the α-aminosuccinate (aspartate) **264** with diketene (**265**) followed by sodium ethoxide produced the 3-acetyltetramic acid **267** via Dieckmann cyclization of the polycarbonyl intermediate **266**. This strategy has been employed subsequently by many others: simple examples include 3-acetyltetramic acids (1984CPB4197), 3-unsubstituted tetramic acids via decarboxylation (1990PT(1)611), and N-unsubstituted 3-acyltetramic acids (2005 JMC5932).

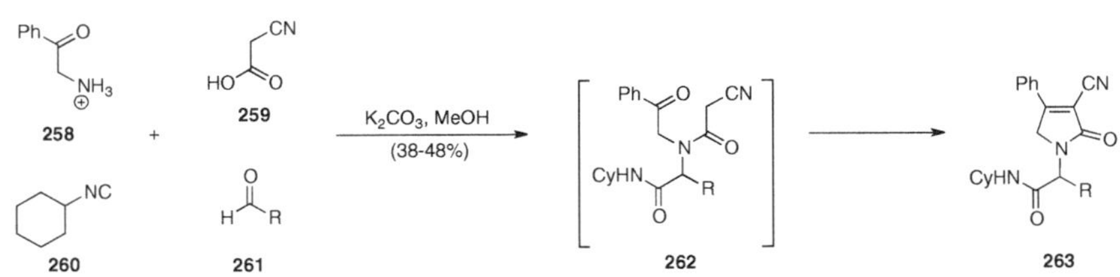


Scheme 64

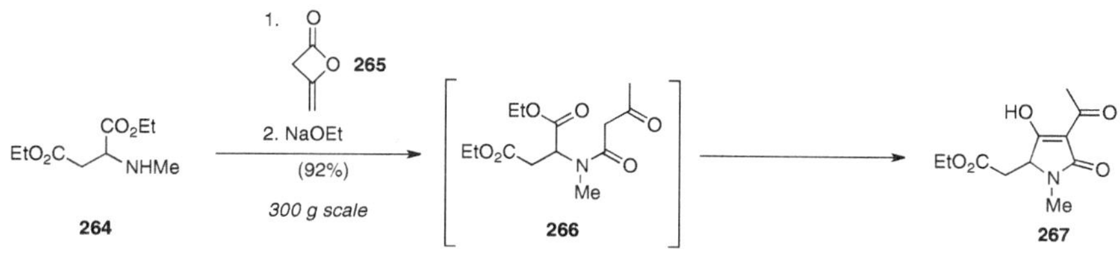


Scheme 65

A variation of the Lacey–Dieckmann cyclization which gives 3-aryltetramic acids was developed by Le Gall and coworkers (Scheme 66; 2010EJOC1142). For example, they treated glycinate **268** with the arylacetate **269** in the presence of *t*-BuOK and obtained the 3-aryltetramic acid **271**. The reaction proceeds through amidation followed by Dieckmann cyclization of amide intermediate **270**. They were also able to obtain *N*-unsubstituted 3-aryltetramic acids by removal of the 2,4-dimethoxybenzyl group on the nitrogen. Lee and coworkers developed a similarly modified Lacey–Dieckmann cyclization to 3-cyanotetramic acids (2008 JMC1487); they used a cross-coupling reaction between cyanoacetic acid and an aminoester to form the Dieckmann cyclization substrate.

In another variation of the Lacey–Dieckmann cyclization that reverses the order of bond formation ("c" bond first then the "a" bond), the Igglessi-Markopoulou group has used *N*-hydroxysuccinimide esters as activating groups toward Claisen condensations (Scheme 67; 1996CC1323, 1998PT(1)2443). Treating methyl acetoacetate (**273**) with *N*-hydroxysuccinimide glycinate **272** and *t*-BuOK gives 3-acetyltetramic acid **275** via lactamization of Claisen condensation intermediate **274**. This transformation has been exploited in the synthesis of 3-dienoyltetramic acids (1997PT(1) 3543), *N*-acetyltetramic acids (2001JHC1203), chiral tetramic acids (1999 TA1873, 2010EJOC5989), *N*-methyltetramic acids (2003EJOC4593), and 3-benzyloxycarbonyltetramic acids (2012T4506). Other types of activated esters that have been used in this transformation include *p*-nitrophenyl esters (1985JHC1599) and benzotriazolyl ester (2001SL1653).

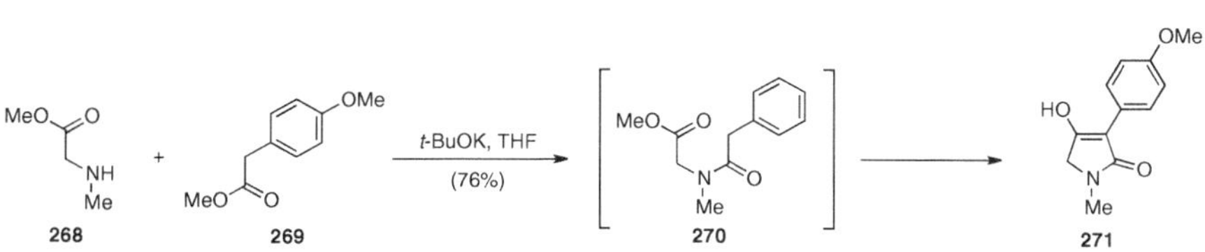


Scheme 66

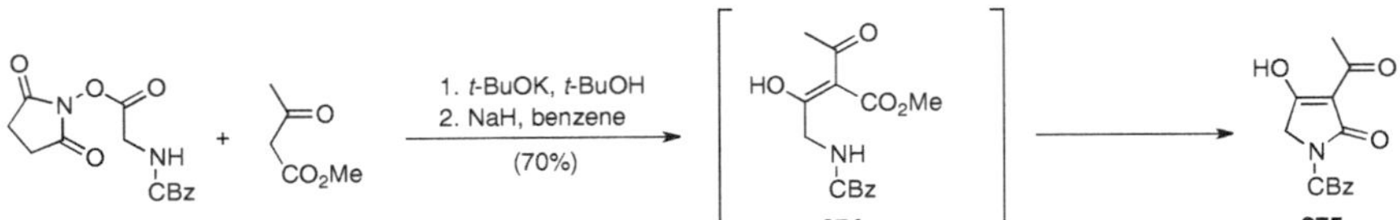

272 273 274 275

Scheme 67

The cyclocondensation of 1,2-dicarbonyl compounds with β-aminocarboxylate derivatives provides access to 4-acyl-3-hydroxy-3-pyrrolin-2-ones (regioisomers of 3-acyltetramic acids) (Scheme 68). This strategy was originally reported by Southwick and coworkers (1949JA2532, 1953JA3413, 1956JOC1087) and used by others to prepare 4-ethoxycarbonyl-3-hydroxy-3-pyrrolin-2-ones (1991JMC1011, 1991JMC1721, 2008BMCL1623). In an illustrative recent example, Pace and coworkers have treated α-aminosuccinate **276** with diethyl oxalate **277** in the presence of sodium ethoxide and obtained 3-hydroxy-3-pyrrolin-2-one **278** (2008BMCL3865). Li and Lyle prepared fused 5-alkylidene-3-hydroxy-3-pyrrolin-2-ones, anticancer analogs of the pyrrothine class of natural products (e.g., holomycin (**47**)), using related synthetic methodology (2007BMC4601). They used a cyclocondensation between imines and oxalyl chloride to give a 3-hydroxy-3-pyrrolin-2-one.

Barluenga unexpectedly obtained 3-pyrrolin-2-ones (instead of the corresponding pyridinones) in a reaction involving diethyl fumarate (**280**) (Scheme 69; 1981S200). Treatment of **280** with imine **279** in the presence of $AlCl_3$ gives the 3-pyrrolin-2-one **281** (Scheme 69; 1981S200). The structure of **281** was confirmed by NMR analysis of its $LiAlH_4$ reduction product. Mechanistically, the transformation proceeds via a Michael addition by the enamine tautomer of **279** to the unsaturated ester **280** followed by an intramolecular amidation.

Schobert and coworkers developed a novel approach to 4-methoxy-3-pyrrolin-2-ones (methyl tetramates) (e.g., **284**) involving Wittig

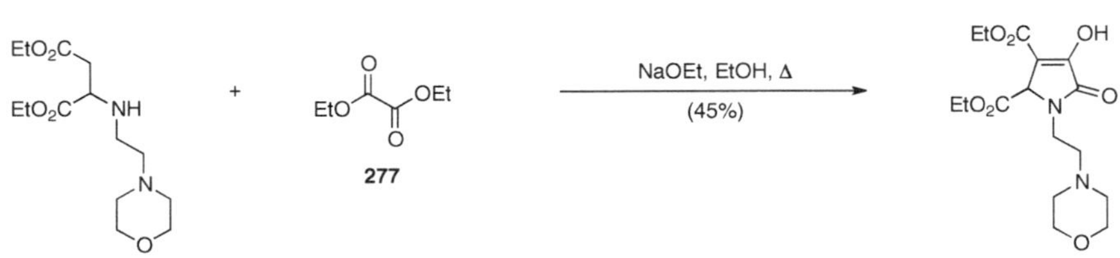

276 277 278

Scheme 68

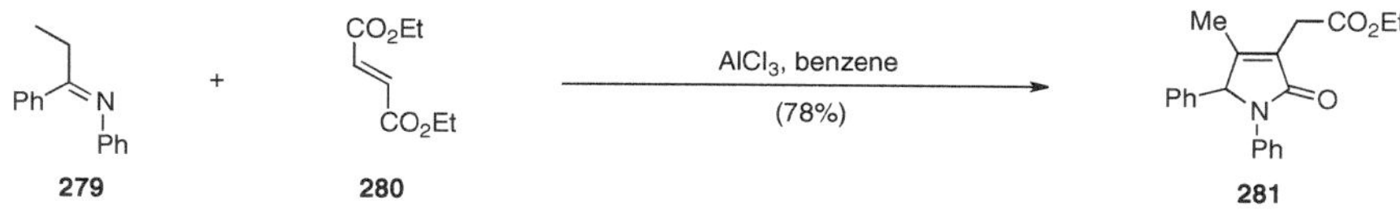


Scheme 69

cyclocondensation reactions between α-amino ester derivatives (e.g., **282**) and keteneylidene(triphenyl)phosphorane (**283**) (Scheme 70; 1996PT(1) 2799). Cumulated Wittig reagent **283** (Bestmann's reagent) has been used to prepare a variety of tetramates including ethyltetramates (2002 MI2), fused tetramates (2007OL4939), chiral tetramates (2012T697), benzyl tetramates (2006S3902, 2013JOC2455), and pyrrolinone-pyrrolidines (2011JA12350, 2012OBC921, 2013JOC4823).

Andrus and coworkers synthesized a simple chiral 3-pyrrolin-2-one (non-tetramate) using Bestman's reagent **283** and an aldehyde in place of an ester (Scheme 71; 1997JOC5542). Treatment of **283** with chiral aminoaldehyde **285** gives 5-methyl-3-pyrrolin-2-one **286**. Stork and Nakamura used an intermolecular Wittig-lactamization sequence in the synthesis of a 3-hydroxo-3-pyrrolin-2-one building block en route to cytochalasin D (**66**) (1983JA5510). Berkes and coworkers prepared an O-fused tetramate from the Wittig-lactamization of an α-aminolactone (2011SL1631).

Shindo and coworkers have used the addition of ynolates to aldehydes to prepare functionalized 3-pyrrolin-2-ones; the reaction proceeds via strained oxetenoxide intermediates (Scheme 72; 2006CEJ524). Treatment of ynolate **288** with the α-aminoketone **287** followed by thionyl chloride produced **291**. A suggested mechanism for this transformation involves formation of the oxetenoxide intermediate **289**, ring opening to the

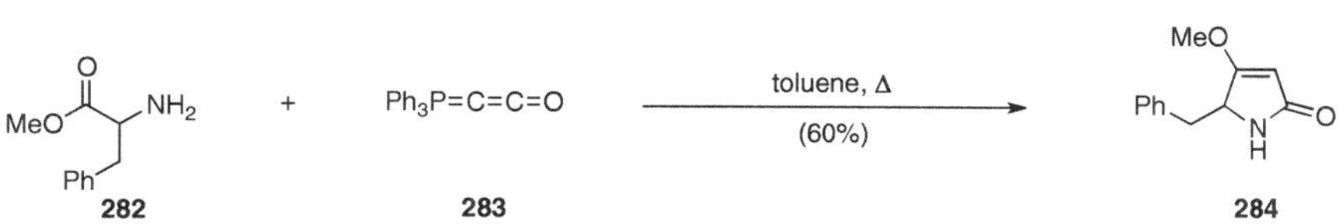


Scheme 70

Scheme 71

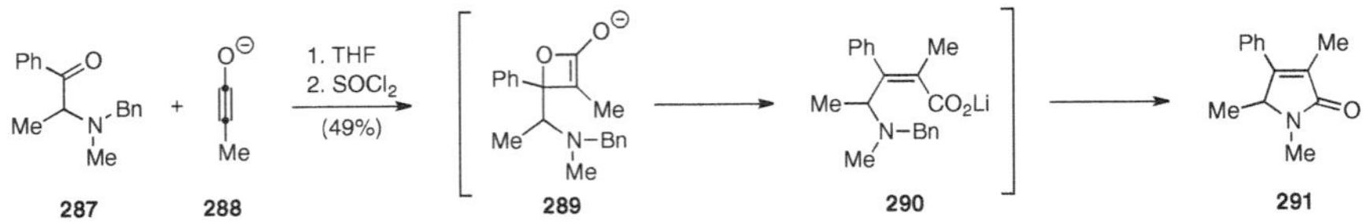


Scheme 72

carboxylate **290**, and a debenzylative-lactamization mediated by thionyl chloride, which ultimately leads to **291**.

The Langer group has developed a number of domino reactions, involving 1,3-dicarbonyl anions, for the synthesis of heterocycles including 3-pyrrolin-2-ones (2001SL1437, 2004EJOC1897, 2004SL2779, 2005SL453). With regard to the latter, treatment of ethyl acetoacetate (**292**) with the oxalic acid derivative **293** leads to the formation of the 5-alkylidene-3-pyrrolin-2-one **295** (Scheme 73; 2001SL1437). The reaction proceeds through the formation of a furanoid intermediate **294** followed by a Dimroth rearrangement to give **295**.

In separate but closely timed reports, Li and coworkers (2014OL3704) and Tang and coworkers (2014TL6455) reported a new approach to 3-pyrrolin-2-ones using Rh-catalyzed [3+2] cycloadditions of ketene silyl acetals and 1,2,3-triazoles (Scheme 74). From the work of Li, treatment of the 1,2,3-triazole **296** with ketene silyl acetal **297** and $Rh_2(OAc)_4$ gives the 3-pyrrolin-2-one **298**.

## 3.3 Type "ad" Cyclizations

Combining a Michael addition and a lactamization is a recipe for developing type "ad" intermolecular approaches to 3-pyrrolin-2-ones. Nájera and

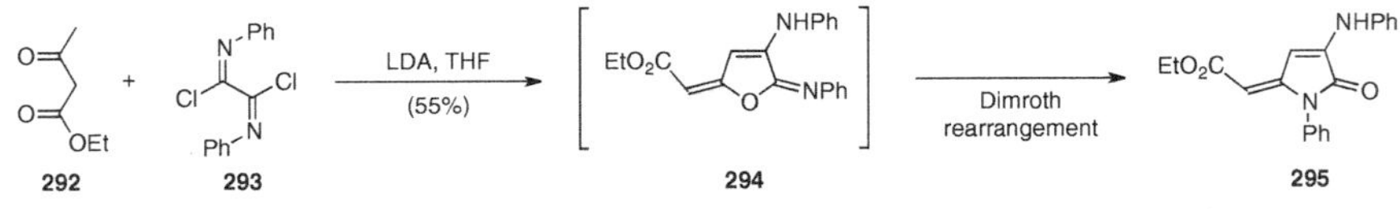


Scheme 73

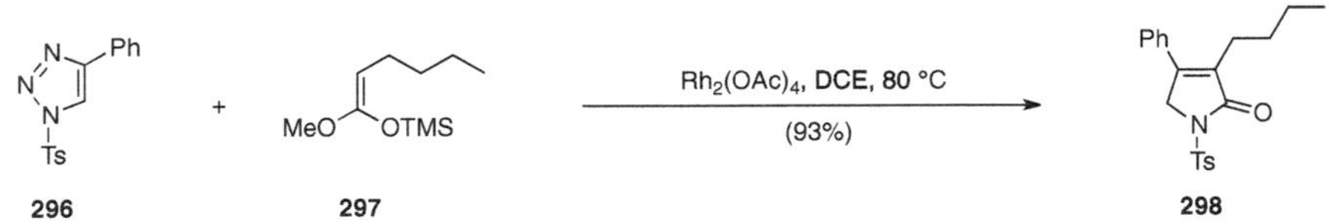


Scheme 74

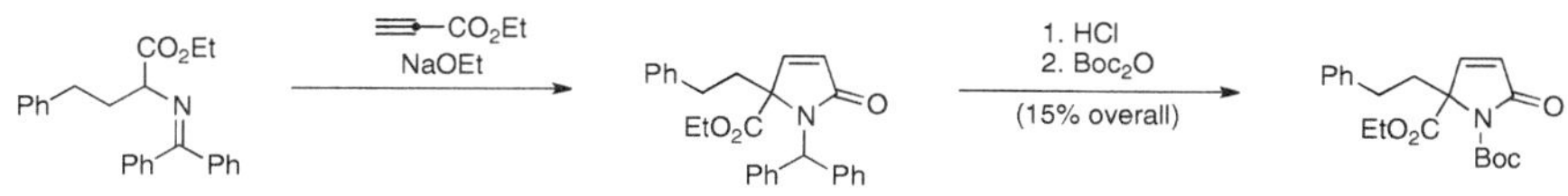

299 300 301

Scheme 75

coworkers used this strategy to prepare 5,5-disubstituted-3-pyrrolin-2-ones (Scheme 75; 1998T9447). They treated α-imino ester **299** with sodium ethoxide followed by ethyl propiolate to obtain 3-pyrrolin-2-one **300** via a Michael addition/amidation sequence; the latter was then converted into the Boc-protected 3-pyrrolin-2-one **301**. Schäfer and coworkers discovered a domino sequence to either 3-pyrrolin-2-ones or 4-pyrrolin-2-ones by the combination of trimethyl aconitate and imines (2001ACIE4212).

Xu and Lu have reported a phosphine-catalyzed cyclocondensation between imines **302** and dimethyl acetylenedicarboxylate (DMAD) (**303**) which leads to the formation of 4-carbomethoxy-3-methoxy-3-pyrrolin-2-one **305** (Scheme 76; 1998JOC5031). They propose a mechanism for this transformation that includes a Michael addition by the phosphine, addition of the resulting zwitterion to imine **302**, intramolecular amidation, and Michael addition–elimination involving intermediate **304** and methoxide. Similar multicomponent-type "ad" sequences involving DMAD (**303**) leading to 3-pyrrolin-2-ones have also been reported by Alizadeh and coworkers (2013HCA1991) and by Asghari and coworkers (2008TL1824).

Recently, Jia, Li and coworkers have investigated multicomponent reactions that lead to the formation of tricyclic 3-pyrrolin-2-ones (Scheme 77; 2014OL5604). Heating tropone (**306**), the allene **307**, and isopropyl

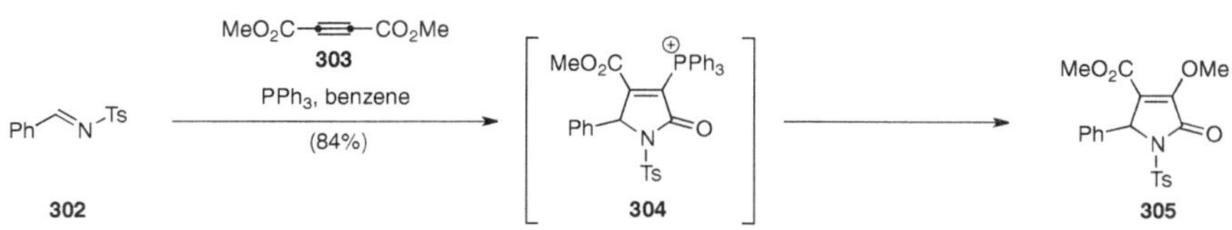

302 304 305

Scheme 76

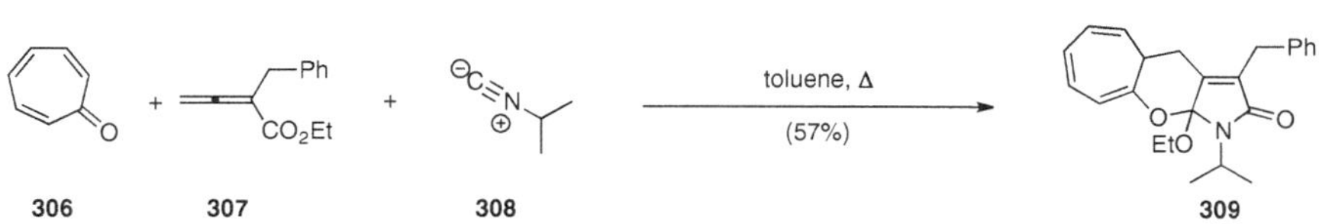

306 307 308 309

Scheme 77

isocyanide (**308**) gives the 3-pyrrolin-2-one **309**. The mechanism of this reaction includes an isocyanide addition to the allene ("d" bond) and amidation ("a" bond) among other steps leading to the categorization as a type "ad" approach to 3-pyrrolin-2-ones.

## 3.4 Type "ae" Cyclizations

A straightforward approach to 3-pyrrolin-2-ones involves the cyclocondensation of γ-halobutenoates **310** with primary amines (Scheme 78). This general strategy has been utilized by several research groups to prepare a variety of 4-aryl-3-pyrrolin-2-ones (1981MI1, 1990JOC3880, 1992MI1). These compounds have served as precursors to 3-substituted-γ-amino acid derivatives (GABA analogs) (1981AJC2641, 1987JMC743, 1991EJMC395, 2001SC1481) and 4-arylpyrrolidinones (inhibitors of cardiac cAMP phosphodiesterase) (1993JMC1041). Treatment of sulfoxonium analogs of **310** ($X = (SOMe_2)^+$) with primary amines gives 4-aryl-1-methyl-3-pyrrolin-2-ones (1976BCSJ3341) and 1-aryl-4-methyl-3-pyrrolin-2-ones (1982MI1).

Vaccher and coworkers prepared γ-bromo-β-arylbutenoates (e.g., **312**) by bromination of the corresponding β-arylbutenoates (e.g., **311**) (Scheme 79; 1987JMC743, 1991EJMC395, 2001SC1481). Treatment of **312** with ammonia gives the 4-(thien-2-yl)-3-pyrrolin-2-one (**313**). The Inagaki group have used this approach to prepare a 4-cyclopropyl-3-pyrrolin-2-one, which is a synthetic precursor to the antibacterial agent DQ-113 (2003JMC1005, 2003ARK112). Carroll and coworkers adapted this strategy to prepare dihydropyridine-fused analogs of 3-pyrrolin-2-ones, which are inhibitors of bladder contractions (2004JMC3180). Amri and coworkers prepared 3-phosphonate 3-pyrrolin-2-ones (2010SC766).

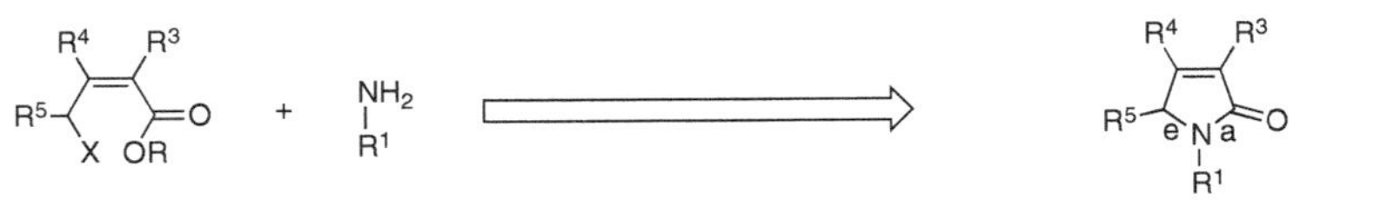


**310** **substituted 3-pyrrolin-2-ones**

Scheme 78

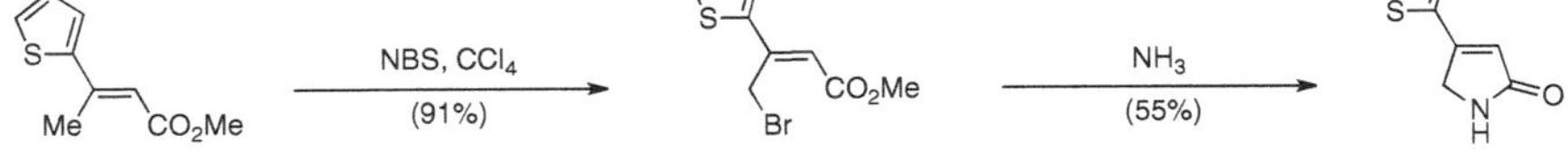


**311** **312** **313**

Scheme 79

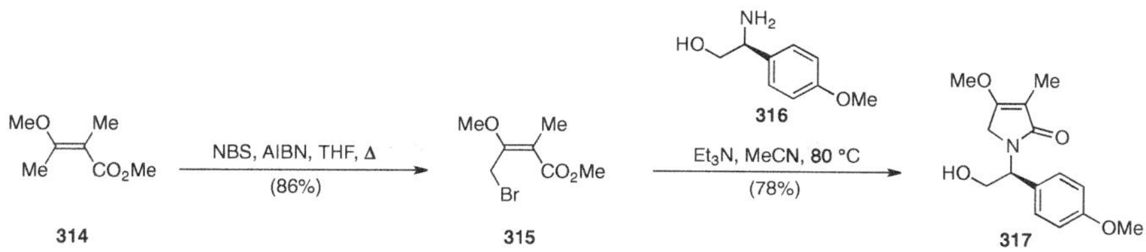


Scheme 80

4-Methoxy-3-pyrrolin-2-ones (methyl tetramates) have been prepared by the cyclocondensation of γ-halo-β-alkoxybutenoates **310** (e.g., $R^4$ = OMe) with primary amines (Scheme 78). This strategy was pioneered by the groups of Pinnick (1984TL1871), Pattenden (1985TL3617, 1993PT(1)2581), Jones (1986TL5285), and McGarrity (1992S391, 1993JOC4010). Recent applications include the preparation of *N*-benzyl tetramates (2003OL4341, 2007TL2819), *N*-aryl tetramates (2014OL4384), and O-fused tetramates (2014TL4753). Huang and coworkers utilized this strategy to prepare *N*-chiral tetramates (Scheme 80; 2004SL247, 2010CC5319). Regioselective bromination of unsaturated ester **314** with NBS and AIBN gives allylic bromide **315** which upon treatment with chiral amine **316** gives the chiral methyl tetramate **317**.

The Painter group developed another approach to tetramates (e.g., **320**) that involved the reductive amination of γ-oxobutenoates (e.g., **319**) in the presence of primary amines (Scheme 81; 2002S869). Unsaturated dicarbonyl **319** was prepared in three steps by the phosphine-catalyzed Michael addition of benzyl alcohol to butynoate **318**, desilylation, and oxidation of the resulting primary alcohol with 2-iodoxybenzoic acid (IBX). Treatment of **319** with *tert*-butylamine followed by sodium borohydride gives the *O*-benzyl tetramate **320** via a reductive amination—lactamization.

Two different groups independently developed imination—Michael addition—lactamization approaches to 3-amino-3-pyrrolin-2-ones. Palacios reported that the treatment of α-ketoester **321** with two equivalents of *p*-methylaniline (**322**) in the presence $Ti(OEt)_4$ and sulfuric acid gives the

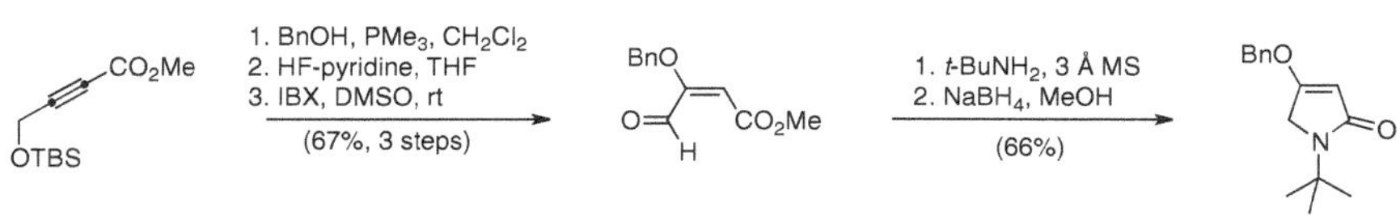


Scheme 81

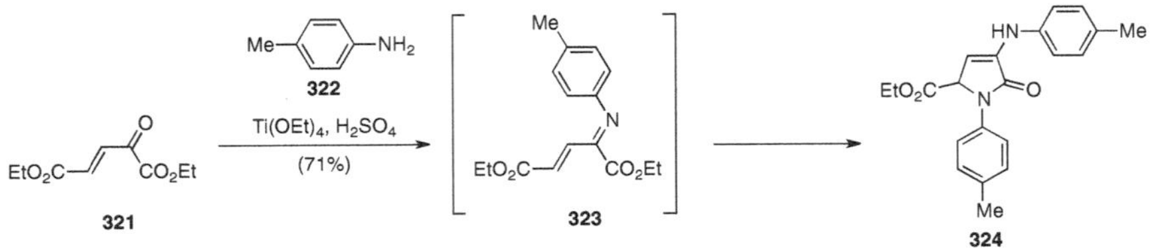


Scheme 82

3-amino-3-pyrrolin-2-one **324** via the imine intermediate **323** (Scheme 82; 2006EJOC2843). Chen and coworkers prepared 1,5-diaryl-3-amino-3-pyrrolin-2-ones using the same reaction although a Lewis acid catalyst was not required (2006JHC949).

Cossío and coworkers found a route to 3-pyrrolin-2-ones by condensation of amines with γ-ketoacids (1,4-dicarbonyls) (Scheme 83; 2010JOC7435). For example, microwave irradiation of ketoacid **325** and benzylamine (**326**) in the presence of acetic acid gives 3-pyrrolin-2-one **327**; this reaction did not proceed in the absence of microwave irradiation. Montforts and Schwartz used a condensation of an aldehyde-ester 1,4-dicarbonyl to prepare a 3-pyrrolin-2-one building block en route to bonellin dimethyl ester (oligopyrrole bile pigment) (1991LAC709). Peterson and coworkers obtained simple 3-pyrrolin-2-ones by condensation of unsaturated 1,4-dialdehydes with primary amines (1997MI3).

Thebtaranonth and coworkers reported a relatively simple route to 4-carboalkoxy-3-pyrrolin-2-ones from diene diesters (Scheme 84; 1987 TL6675). Treatment of diene **328** with benzylamine (**326**) gives a 3:1

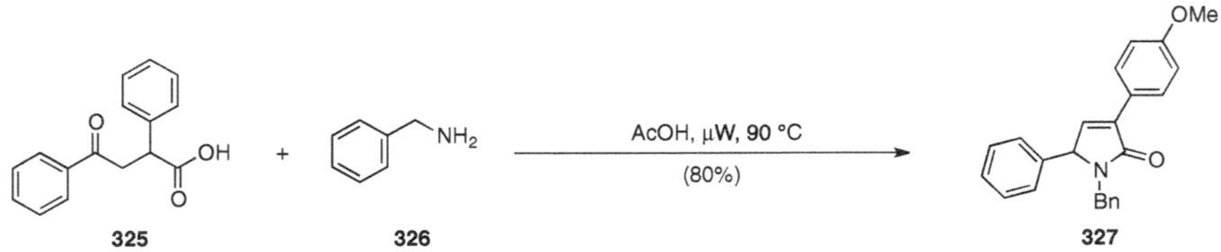


Scheme 83

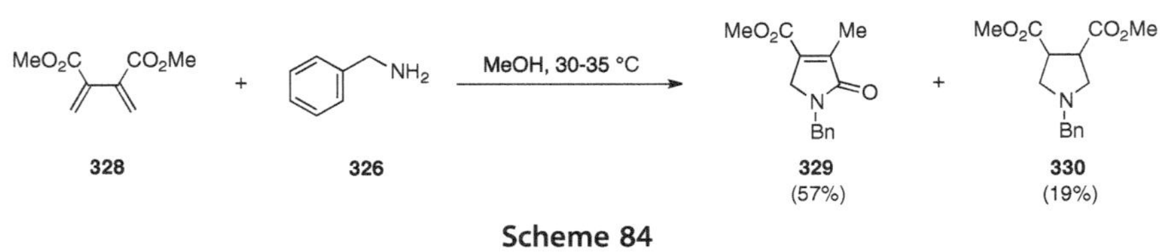


Scheme 84

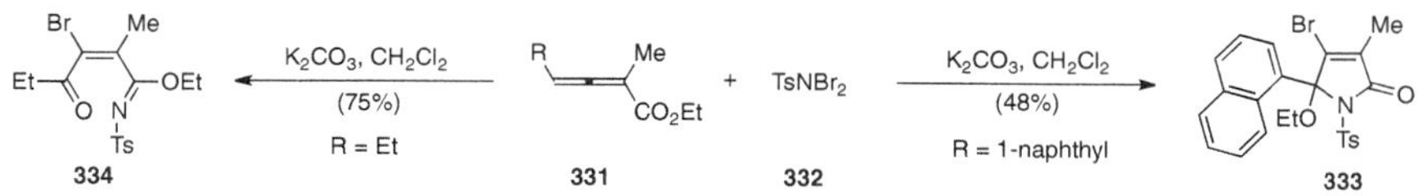


Scheme 85

mixture of 3-pyrrolin-2-one **329** and pyrrolidine **330**. Product **329** arises from a Michael addition, lactamization, and double-bond isomerization; product **330** arises from a double Michael addition.

While investigating the reactivity allenoates (e.g., **331**), Huang and Shen unexpectedly obtained the highly functionalized 3-pyrrolin-2-one **333** (Scheme 85; 2007JOC3961). Treatment of **331** (R = 1-naphthyl) with $TsNBr_2$ (**332**) in the presence of $K_2CO_3$ gives **333**; treatment of the related **331** (R = ethyl) under the same conditions led to imidic ethyl ester **334**. The authors did not suggest a mechanism or explain the diffential reactivity.

## 3.5 Type "be" Cyclizations

A few type "be" intermolecular approaches to 3-pyrrolin-2-ones have appeared in the literature. Zhou, Li, and coworkers applied a Rh-catalyzed hydroamidation reaction in a synthesis of 5-alkylidene-3-pyrrolin-2-ones (e.g., **338**) (Scheme 86; 2013OL1814). Treatment of oxime **335** and the arylisocyanate **336** with a Rh catalyst gives the 3-pyrrolin-2-one **338**. The mechanism of the reaction probably includes the formation of the hydroamidation intermediate **337** and subsequent condensation and elimination of *O*-methylhydroxylamine.

Rosas and coworkers developed two related approaches to 3-pyrrolin-2-ones that employ metal-mediated cyanation reactions of ynones (1997OM2726, 2005OM4893). Regarding the first approach, treatment of ynone **339** with potassium cyanide in the presence of a Ni catalyst gives 5-hydroxy-3-pyrrolin-2-one **341** via cyanide-derived Michael adduct **340** (Scheme 87; 1997OM2726).

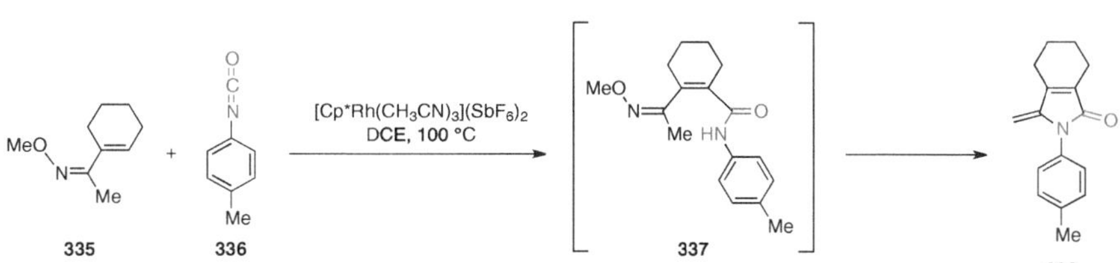


Scheme 86

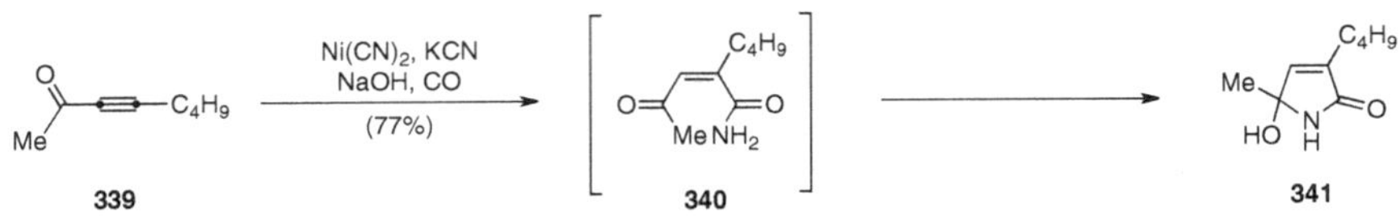


Scheme 87

Alizadeh and coworkers have utilized isocyanates in a novel three-component type "be" cyclization approach to highly substituted 3-pyrrolin-2-ones (Scheme 88; 2006S3431). Treatment of dibenzoylacetylene (**342**) with tosyl isocyanate (**343**) and diethylamine gives the 1-tosyl-3-pyrrolin-2-ones **345**. The authors propose a mechanism that includes a Michael addition by the amine, nucleophilic attack onto the isocyanate by the enamine (producing intermediate **344**), addition to the ketone, and tautomerization.

## 3.6 Type "cd" Cyclizations

As far as we can tell, there is only one intermolecular type "cd" approach to 3-pyrrolin-2-ones that has appeared in the literature. Aoyama and Shiori investigated the use of lithium trimethylsilyldiazomethane **347** to generate alkylidenecarbenes from carbonyl compounds; alkylidene carbenes are useful intermediates that can be used to form new rings including 3-pyrrolin-2-ones (1996H(42)75). For example, treatment of the α-ketoamide **346** with **347** in the presence of magnesium bromide gives the fused 3-pyrrolin-2-one **349** via alkylidenecarbene intermediate **348** (Scheme 89).

## 3.7 Type "ce" Cyclizations

Highly complex 3-pyrrolin-2-ones can be obtained from intermolecular cyclizations that combine aldol condensations (to form the "c" bond) and nucleophilic additions by amide nucleophiles (to form the "e" bond). This strategy was pioneered by Howard and coworkers for the synthesis of 3-cyano-3-pyrrolin-2-ones by the cyclocondensation of β-cyanoamides and 1,2-diketones (1959JA4355, 2005AJC882). Exploiting this methodology,

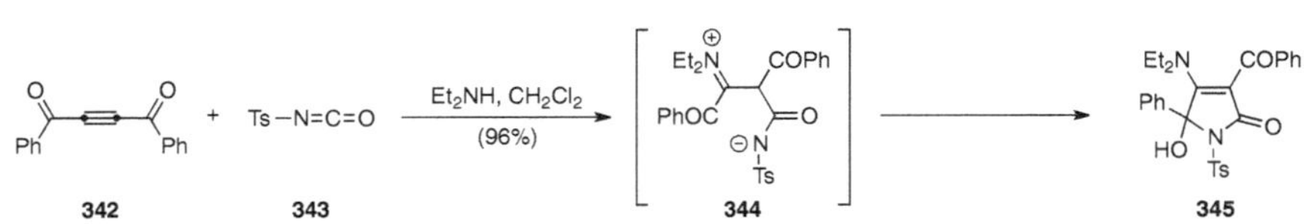


Scheme 88

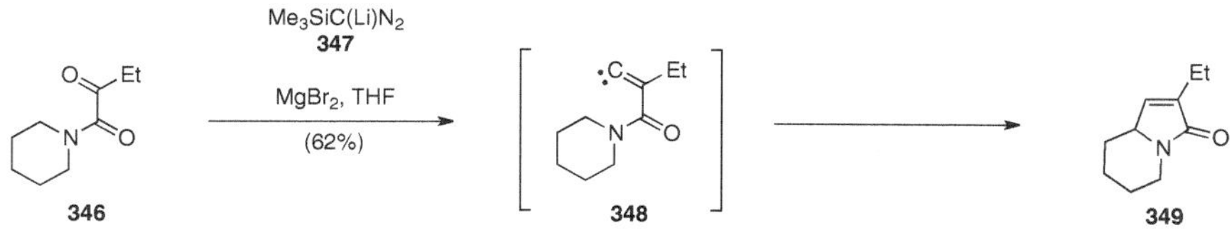


Scheme 89

Osada and coworkers synthesized a large number of 3-acyl-3-pyrrolin-2-ones, which are analogs of the natural product epolactaene (Scheme 90; 1997JMC391). Treatment of diketone **350** with β-ketoamide **351** under buffered conditions produces the 3-acyl-3-pyrrolin-2-one **352**. This methodology was also utilized to produce MT-21 (**57**); these compounds were evaluated as pharmaceutical agents for the treatment of cancer and neurodegenerative disease (1997JMC391, 2000MI1, 2002JBC31243).

Reversing the position of the aldol electrophile, Enders and coworkers developed a type "ce" approach to complex 3-pyrrolin-2-ones that involved a quadruple cascade process (Scheme 91; 2014S1539). Treatment of cinnamaldehyde (**353**) and the α-ketoamide **354** with pyrrolidine catalyst **355** and NaOAc gives the fused 3-pyrrolin-2-one **357**. The mechanism of the reaction includes a Michael addition by the amide followed by aldol condensation to generate the 3-pyrrolin-2-one intermediate **356**; subsequent Michael and aldol transformations with a second equivalent of **353** then gives the fused cyclopentene ring.

Yavari and Bayat used similar functionality and replaced the aldol condensation with a Wittig reaction to prepare 3-ethoxy-3-pyrrolin-2-ones in a type "ce" fashion (Scheme 92; 2002SC2527). Treatment of

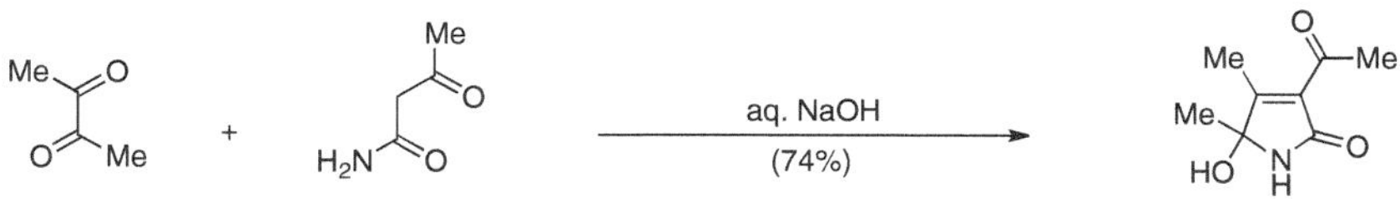


Scheme 90

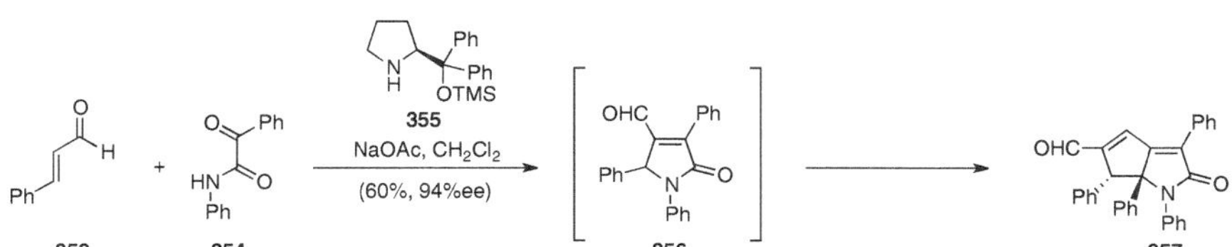


Scheme 91

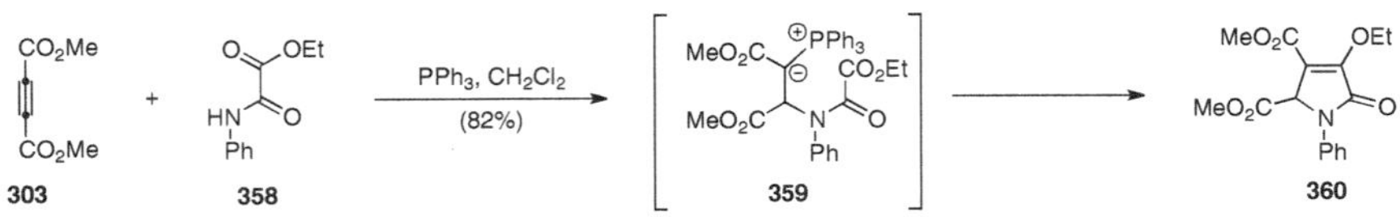


Scheme 92

DMAD (**303**) with oxamate **358** in the presence of triphenylphosphine gives the 3-ethoxy-3-pyrrolin-2-one **360** (2002SC2527); this reaction presumably proceeds via the Michael adduct **359**, which then undergoes an intramolecular Wittig reaction.

Li and Yu developed a novel and surprising route to 3-pyrrolin-2-ones using nitrostyrenes as the nitrogen source (Scheme 93; 2011S1831). Mixing 2,4-pentanedione (**361**) with nitrostyrene (**362**) in aqueous DMSO led to the formation of 5-hydroxy-3-pyrrolin-2-one **364**. They followed the reaction by NMR in order to ascertain the mechanism and proposed carboxamide **363** as an intermediate in the reaction pathway.

## 3.8 Type "de" Cyclizations

As far as we know, there are only three reported type "de" approaches to 3-pyrrolin-2-ones. Kollár and coworkers obtained 3-methyl-3-pyrrolin-2-one (**366**) utilizing a Rh-catalyzed hydroformylation reaction involving methacrylamide (**365**) (Scheme 94; 1990JOMC375).

The Katsumara research group has investigated a tandem Stille–Heck cyclization sequence in the formation of 3-pyrrolin-2-ones (Scheme 95; 2012TL837). Treatment of stannane **367** with iodoamide **368** gives the 3-pyrrolin-2-one **370**. The reaction presumably proceeds through the diene

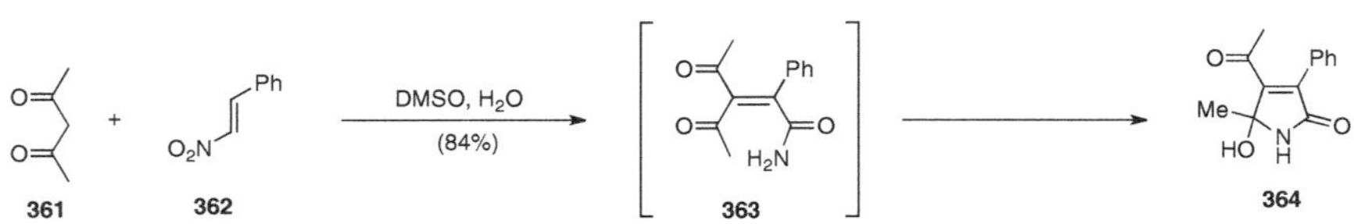


Scheme 93

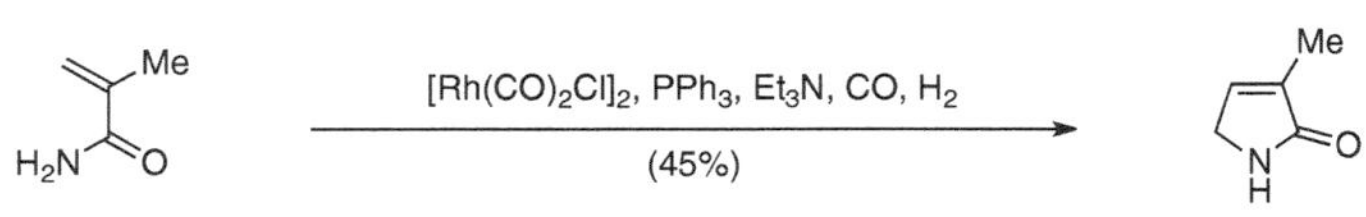


Scheme 94

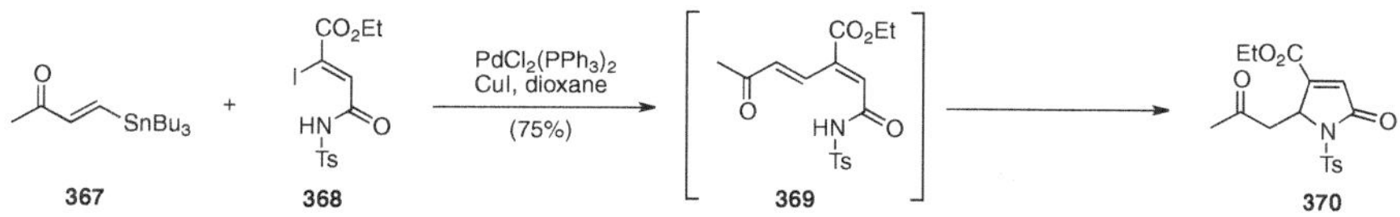


Scheme 95

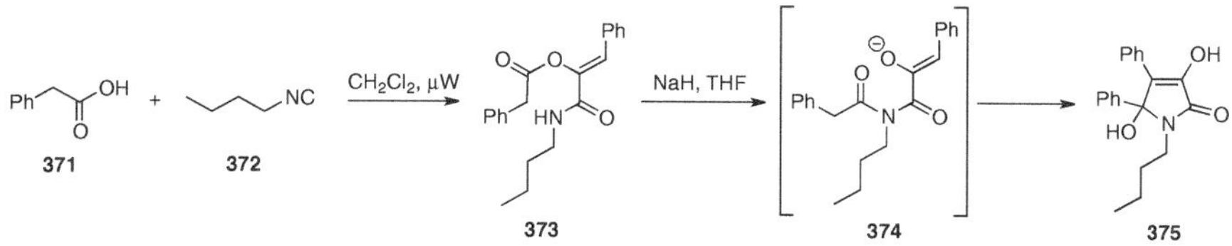


Scheme 96

Stille coupling product **369** followed by a Pd-mediated 5-*exo-trig* cyclization.

Finally, in an investigation of Passerini-like reaction products, Basso and coworkers reported a type "de" synthetic approach to 3-hydroxy-3-pyrrolin-2-ones (Scheme 96; 2009OL4068). The Passerini-like reaction between two equivalents of phenylacetic acid (**371**) and butyl isocyanide (**372**) under microwave irradiation gives the amide **373**. Treatment of **373** with base then gives the 3-hydroxy-3-pyrrolin-2-one **375** via intermediate **374**.

# 4. MULTICOMPONENT APPROACHES TO 3-PYRROLIN-2-ONES

Only a few multicomponent approaches to 3-pyrrolin-2-ones, where the atoms found in the 3-pyrrolin-2-one ring arise from three (or four) building blocks, have appeared in the literature (Scheme 97). There are exactly 10 different three-bond disconnections (three-component reactions) possible for the construction of 3-pyrrolin-2-ones; five of these have been reported. In addition, there is one four-component approach to 3-pyrrolin-2-ones (type "acde").

## 4.1 Type "abd" Cyclizations

The aza-Pauson–Khand reaction ([2+2+1] cycloaddition), where one of the π-bond-containing components contains a nitrogen atom in place of a carbon atom (e.g., imine in place of alkene), is a powerful strategy for preparing 3-pyrrolin-2-ones and fused 3-pyrrolin-2-ones. One of the earliest

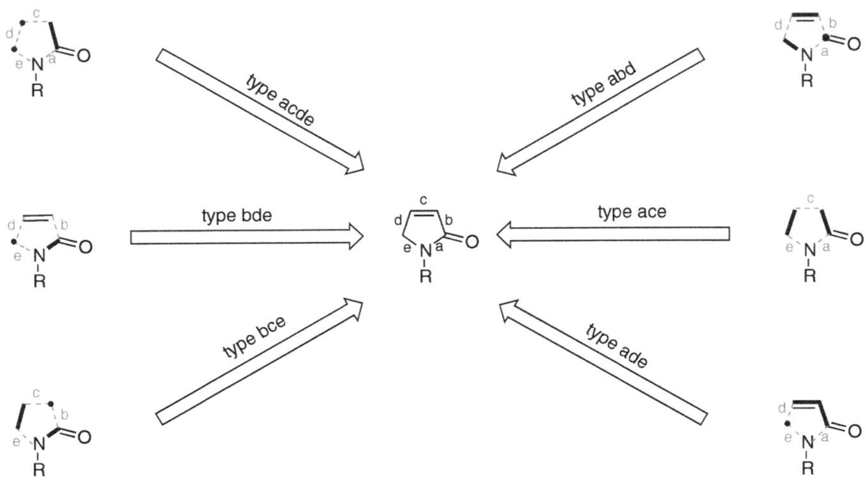


**Scheme 97** Multicomponent approaches to 3-pyrrolin-2-ones.

examples, reported by Murai and coworkers, involves an intramolecular aza-Pauson–Khand reaction (Scheme 98; 1999JOMC(579)177). Treatment of yne-imine **376** with catalytic $Ru_3(CO)_{12}$ and CO led to cyclopentane-fused 3-pyrrolin-2-one **377**. Mukai and coworkers used an intramolecular aza-Pauson–Khand reaction to prepare a fused 3-pyrrolin-2-one intermediate en route to physostigmine; the nitrogen component in this transformation was a carbodiimide (2006OL83).

Murai and coworkers also reported one of the earlier intermolecular aza-Pauson–Khand reactions to give substituted 3-pyrrolin-2-ones (e.g., **380**) (Scheme 99; 2000S925). The Ru-catalyzed [2+2+1] cycloaddition between imine **378**, 1,2-diphenylacetylene (**379**) and CO gives the 3-pyrrolin-2-one **380**. Relatively good regioselectivity was also observed with unsymmetrical alkynes. Rudler and coworkers investigated a Cr-mediated cycloaddition involving **379** and an aminocarbene in a unique approach to 3-pyrrolin-2-ones (1999EJOC3315).

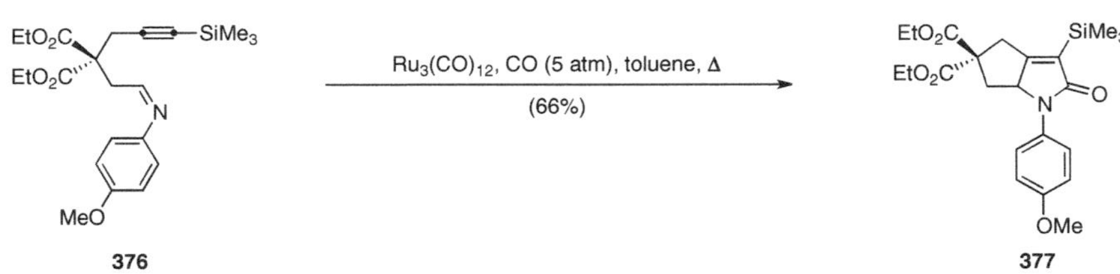


**Scheme 98**

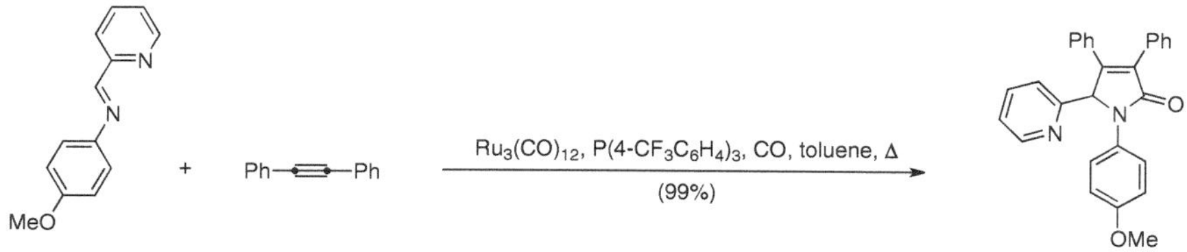


378 379 380

Scheme 99

The Ogoshi group has studied the use of Ni catalysts and nickel acycle intermediates in the synthesis of 3-pyrrolin-2-ones (2007ACIE4930, 2014CEJ4105, 2014JA15877). In an illustrative example of a Ni-catalyzed [2+2+1] cycloaddition, they treated imine **381**, 1,2-diphenylacetylene (**379**), and phenyl formate (**382**) with Ni(cod)$_2$ and PCy$_3$ and obtained the 3-pyrrolin-2-one **383** (Scheme 100; 2014JA15877). Interestingly, phenyl formate (**382**) served as a synthetic equivalent to CO.

The Micalizio group has developed a highly regioselective [2+2+1] annulation methodology to 3-pyrrolin-2-ones that uses an alcohol functionality as a directing group (Scheme 101; 2007ACIE3912). Treatment of imine **384** and 3-pentyn-1-ol (**385**) with Ti(O*i*Pr)$_4$ and cyclopentylmagnesium bromide led to the formation of the azatitanacyclic intermediate **386**. Subsequent addition of $CO_2$ gives the 3-pyrrolin-2-one **387**. This work was an extension of the earlier work Sato and coworkers, who developed Ti-mediated [2+2+1] annulation methodology (1997TL6849).

The Ryu group investigated a radical [2+2+1] cycloaddition approach to 3-pyrrolin-2-ones using amidines as the nitrogen source in place of imines (Scheme 102; 2013JA1006). Treatment of DBU (**388**) and

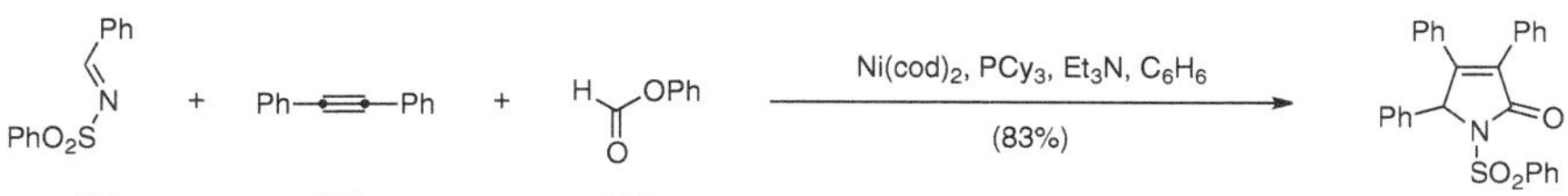


381 379 382 383

Scheme 100

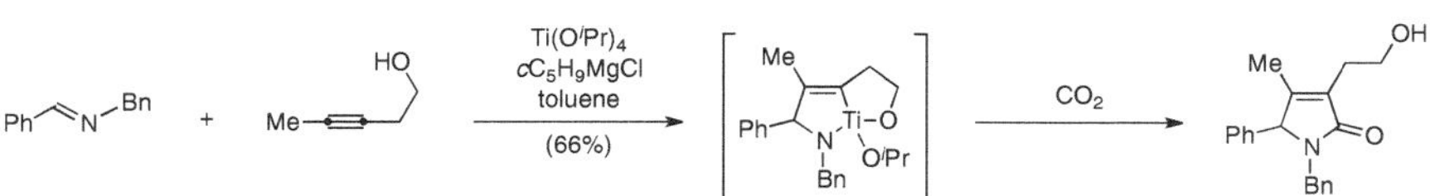


384 385 386 387

Scheme 101

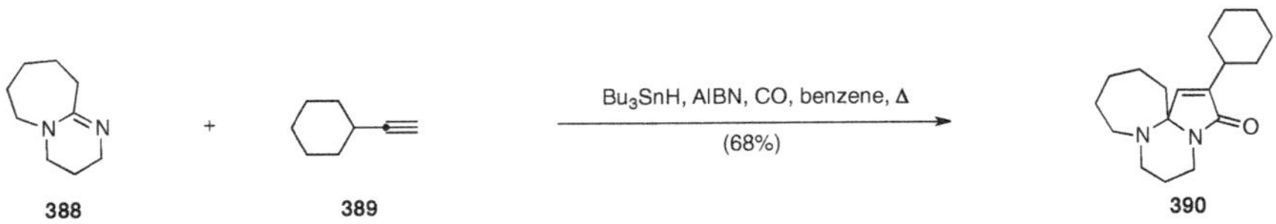


Scheme 102

cyclohexylacetylene (**389**) with $Bu_3SnH$ and CO led to the formation of the tricyclic 3-pyrrolin-2-one **390**.

## 4.2 Type "ace" Cyclizations

Anary-Abbasinejad and coworkers reported an interesting three-component type "ace" cyclization leading to substituted 3-pyrrolin-2-ones that combined an amidation ("a" bond), Michael addition ("e" bond), and intramolecular Wittig reaction ("e" bond) (Scheme 103; 2010SC1350). Treatment of 2-amino-4-methylpyridine **391**, DMAD (**303**), and ethyl chlorooxoacetate (**392**) with triphenylphosphine gives the 3-pyrrolin-2-one **394** via the ylide intermediate **393**.

Dömling and coworkers adapted a four-component Ugi-type reaction sequence to the synthesis of highly substituted 3-pyrrolin-2-ones (Scheme 104; 2004OL39). The combination of allyl isocyanide (**395**), primary amine **396**, α-phosphonoacetic acid **397**, and α-ketoaldehyde **398** gives the 5-carboxamido-3-pyrrolin-2-one **399**. The two-step process involves an Ugi multicomponent reaction followed by an intramolecular Horner/Wadsworth/Emmons cyclization.

## 4.3 Type "ade" Cyclizations

A large number of three-component reactions to 3-hydroxy-3-pyrrolin-2-ones have been reported due to their interesting biological activity as antitumor agents (2012JMC9630), HIV-1 integrase inhibitors (2007BMC5487, 2010JCC84), West Nile virus protease inhibitors (2013CMC1554), antifungal agents (2012BMC1240), antiinflammatory agents (2008MI1), and

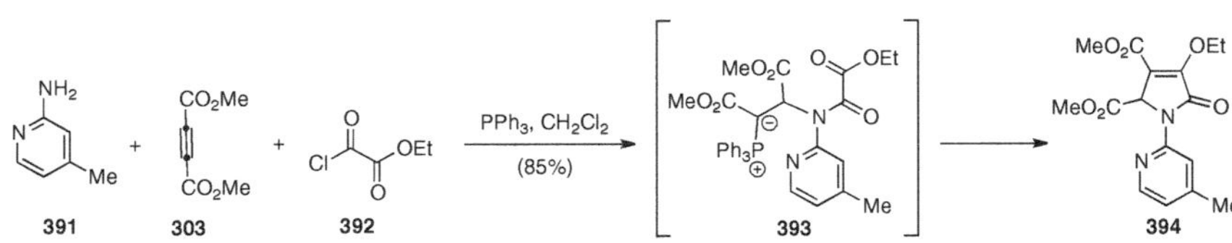


Scheme 103

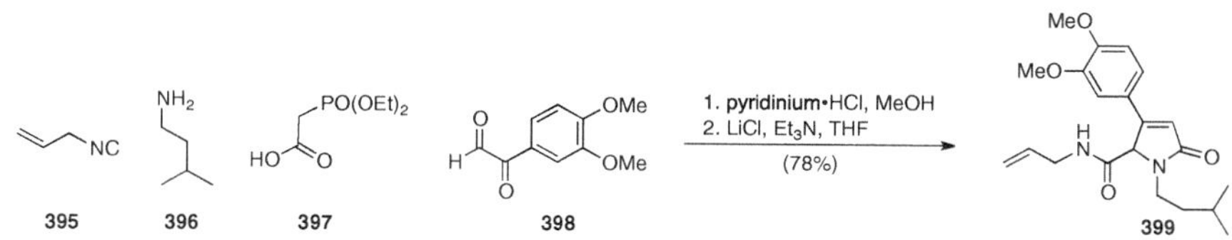


395 396 397 398 399

Scheme 104

neurological agents (2014JMC2334). A convergent approach to 3-hydroxy-3-pyrrolin-2-ones involves type "ade" cyclizations. For many of these multicomponent synthetic approaches, the three components include an aldehyde, a 1,2-dicarbonyl reagent, and a primary amine (Scheme 105; 2006T6018, 2007BMC5487, 2008MI1, 2008MI2, 2008MI3, 2009MI1, 2010MI1, 2012BMC1240, 2012JMC9630, 2013CMC1554, 2014JMC2334). The mechanism of the transformation involves amidation (amine + carboxylic acid derivative), imination (amide + aldehyde), and condensation onto the imine (enolate/enol + imine).

Liotta and Traynelis and coworkers prepared a wide range of indole-substituted 3-hydroxy-3-pyrrolin-2-ones (Scheme 106; 2014JMC2334). For example, treatment of pyruvate **400**, benzaldehyde (**401**), and tryptamine (**402**) with pyridinium *p*-toluenesulfonate (PPTS) gives the 3-hydroxy-3-pyrrolin-2-one **403**.

This chemistry can also be used to prepare 3-amine-substituted 3-pyrrolin-2-ones by using two equivalents of amine (2009JCC685, 2008JMC5833). An

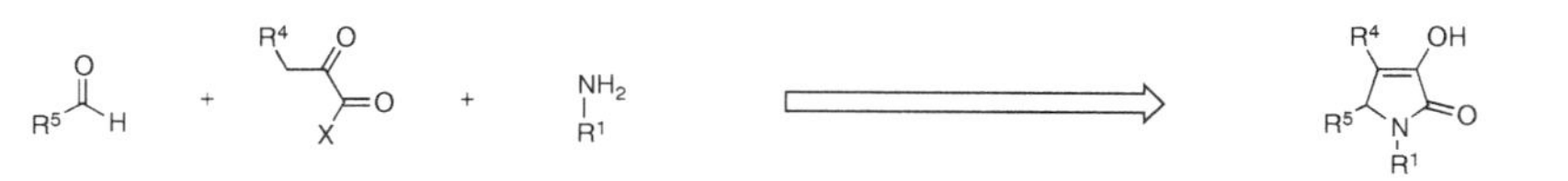

aldehydes 1,2-dicarbonyls primary amines 3-hydroxy-3-pyrrolin-2-ones

Scheme 105

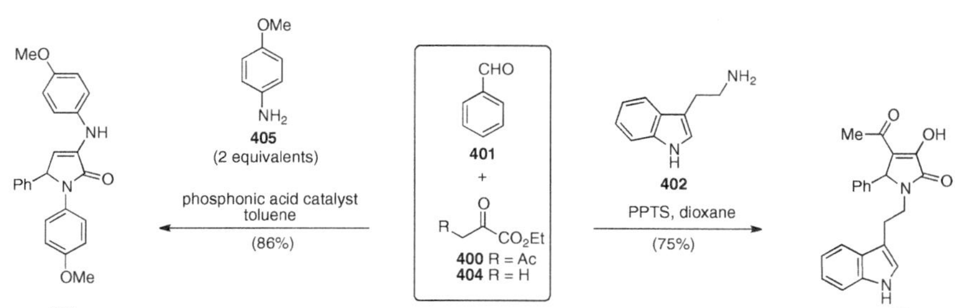


406 403

Scheme 106

example described by Luo, Cheng, and coworkers illustrates the amine case (Scheme 106; 2008EJOC4350). Treatment of **401**, ethyl pyruvate (**404**), and two equivalents of *p*-anisidine (**405**) gives the 3-arylamino-3-pyrrolin-2-one **406**. This type of strategy to 3-amine-substituted 3-pyrrolin-2-ones has also been adapted to the solid phase by Pendri and coworkers (2010JCC84).

A different three-component approach to 3-hydroxy-3-pyrrolin-2-ones was disclosed by Yan and coworkers, and this involves using DMAD (**303**) in place of pyruvate (Scheme 107; 2011EJOC2981). For example, they treated DMAD (**303**), benzaldehyde (**401**), and *p*-anisidine (**405**) with *p*-toluenesulfonic acid (*p*-TsOH) and obtained the 3-hydroxy-3-pyrrolin-2-one **407**. The same group explored oxindoles in place of aldehydes and obtained the corresponding spirocyclic oxindole 3-pyrrolin-2-ones (2012T8539).

By modifying the above reaction manifold by using two equivalents of the amine component, 3-amine-substituted 3-hydoxy-3-pyrrolin-2-ones can be obtained (2009JCC685, 2012TL2622, 2013T589, 2013TL1371). A recent example by Zhu and coworkers has provided an illustrative example (Scheme 107; 2012EJMC232). By mixing DMAD (**303**), benzaldehyde (**401**), and two equivalents of cyclohexylamine (**408**) with acetic acid the 3-pyrrolin-2-one **409** was obtained. It is possible to use two different amine components.

## 4.4 Type "bce" Cyclizations

A novel three-component reaction sequence leading to 5-hydroxy-3-pyrrolin-2-ones has been reported by Liang and coworkers (Scheme 108; 2007T8987). Mixing arylaldehyde **410**, 2,4-pentanedione (**361**), and cyclohexyl isocyanide (**260**) with piperidine gives the 3-pyrrolin-2-one **411**. The reaction proceeds via the formation of 2-aminofuran intermediates followed by a Dimroth-type rearrangement to the corresponding 3-pyrrolin-2-ones.

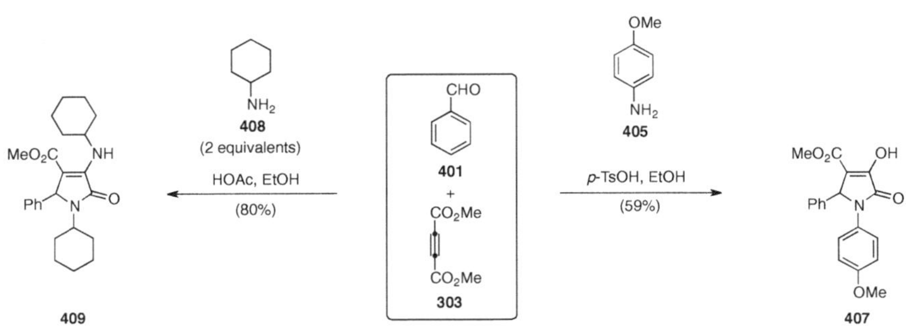


Scheme 107

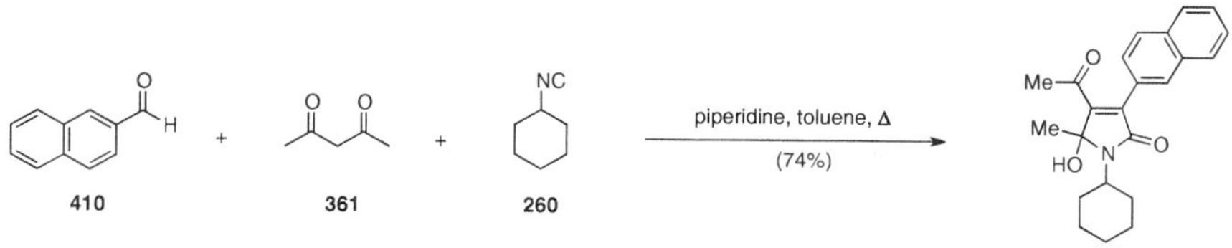


Scheme 108

Ghandi and coworkers used *o*-salicaldehyde as the aldehyde component in this reaction and obtained chromane-fused 3-pyrrolin-2-ones (2013T3054).

## 4.5 Type "bde" Cyclizations

The Iwasawa research group reported a novel multicomponent approach to highly substituted 5-methoxy-3-pyrrolin-2-ones including PI-091 (**39**), a platelet aggregation inhibitor (1997JOC1918, 1998JOC3164). The combination of Fischer carbene **412**, lithium acetylide **413**, and tosyl isocyanate (**343**) gives 5-methoxy-3-pyrrolin-2-one **414** (Scheme 109). More recently, Takahashi and coworkers have reported a similar multicomponent reaction that involved isocyanates, alkynes, acid chlorides, and diethylzirconocene (2007TL9199).

Kurahashi and Matsubara have used an isocyanate as both a nitrogen source and CO equivalent in a [2+2+1] cycloaddition to give 3-pyrrolin-2-ones (Scheme 110; 2010CC8055). Mixing methyl acrylate (**415**), 2-octyne (**416**), and phenyl isocyanate (**417**) with a Ni catalyst and carbene ligand gives a 5:1 mixture of the 3-pyrrolin-2-ones **418** and **419**, respectively.

Nair and coworkers have used a type "bde" multicomponent reaction sequence to prepare *N*-amino-3-pyrrolin-2-ones (2001CL738, 2002T8113).

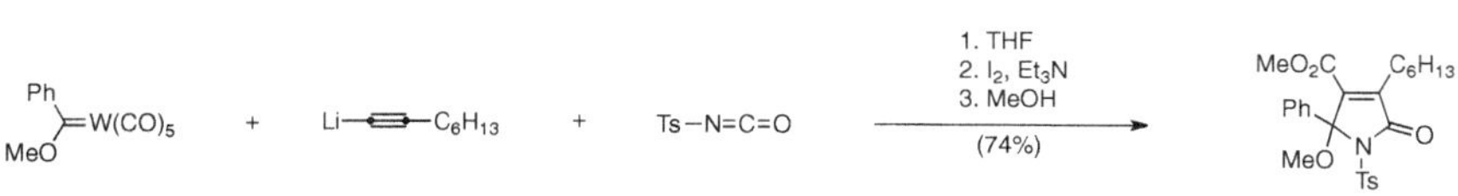


Scheme 109

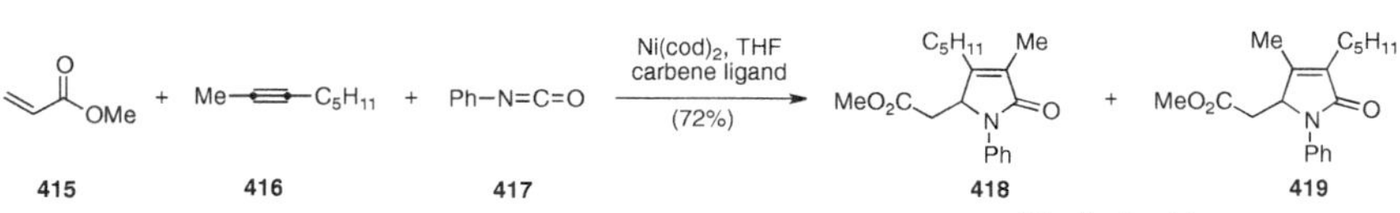


Scheme 110

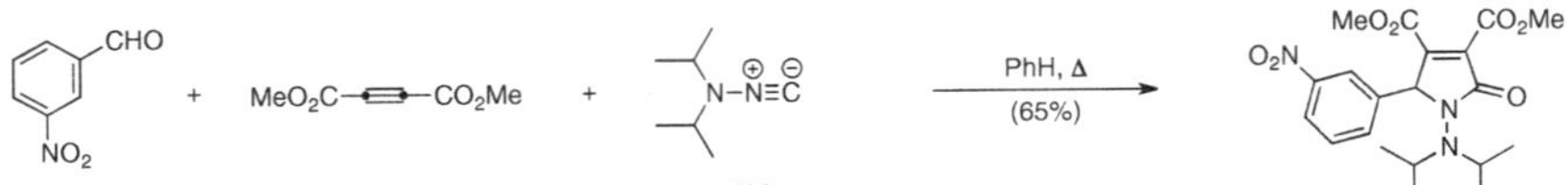


420 303 421 422

Scheme 111

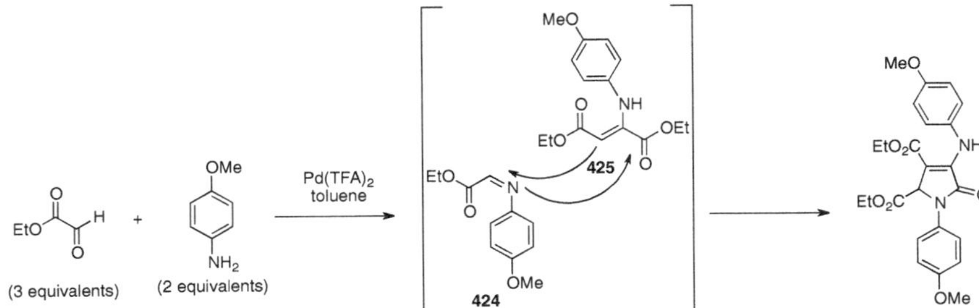


423 405 426

Scheme 112

Heating an aromatic aldehyde (e.g., **420**), DMAD (**303**), and the *N*-amino-isocyanide **421** gives *N*-amino-3-pyrrolin-2-ones (e.g., **422**) (Scheme 111). Presumably, this is another example of a multicomponent reaction leading to a 2-aminofuran intermediate, which is converted to a 3-pyrrolin-2-one via a Dimroth rearrangement.

## 4.6 Type "acde" Cyclizations

There is at least one multicomponent reaction that uses four different components that lead to formation of four of the five bonds within a 3-pyrrolin-2-one ring. Zeng, Ye, and coworkers developed a Pd-catalyzed type "acde" multicomponent approach to amine-substituted 3-pyrrolin-2-ones (Scheme 112; 2012OL5640). Mixing three equivalents of ethyl glyoxalate (**423**) and two equivalents of *p*-anisidine (**405**) with a Pd catalyst led to formation of the 3-arylamino-3-pyrrolin-2-one **426**. A suggested mechanism includes the Pd-mediated formation of imine **424** and enamine **425** and subsequent cyclocondensation leading to **426**.

# 5. SYNTHESIS OF 3-PYRROLIN-2-ONES FROM OTHER HETEROCYCLES

Heterocyclic ring systems are useful precursors to 3-pyrrolin-2-ones (Scheme 113). The two most common heterocyclic precursors to

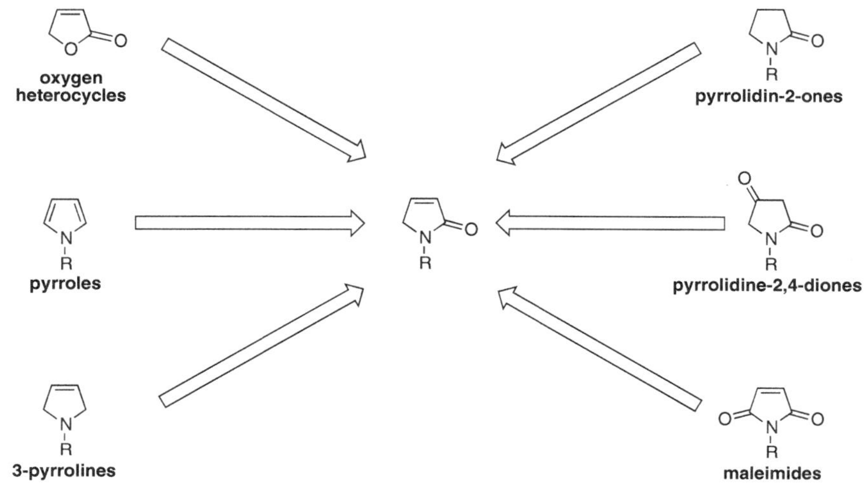


**Scheme 113** Synthesis of 3-pyrrolin-2-ones from other heterocycles.

3-pyrrolin-2-ones are pyrrolidin-2-ones and pyrroles. Elimination of leaving groups located at either C3 or C4 or even C5 of pyrrolidin-2-ones provides access to 3-pyrrolin-2-ones. There are a large number of methods to transform pyrroles into 3-pyrrolin-2-ones, including direct oxidation of simple pyrroles, oxidation of pyrrole-2-carboxaldehydes, and hydrolysis of 2-halopyrroles. This section describes the different methods that have been reported for the conversion of various heterocycles into 3-pyrrolin-2-ones organized by ring type.

## 5.1 Synthesis from Pyrrolidin-2-Ones

The majority of strategies for converting pyrrolidin-2-ones into 3-pyrrolin-2-ones fall into one of two categories (Scheme 114): (1) α-substitution of pyrrolidin-2-ones followed by elimination and (2) conversion of hydroxy groups located at C4 or C5 into leaving groups followed by elimination. Another well-developed strategy involves introducing the alkene functionality via a retro-cycloaddition; this method is often used for the preparation of chiral 3-pyrrolin-2-ones.

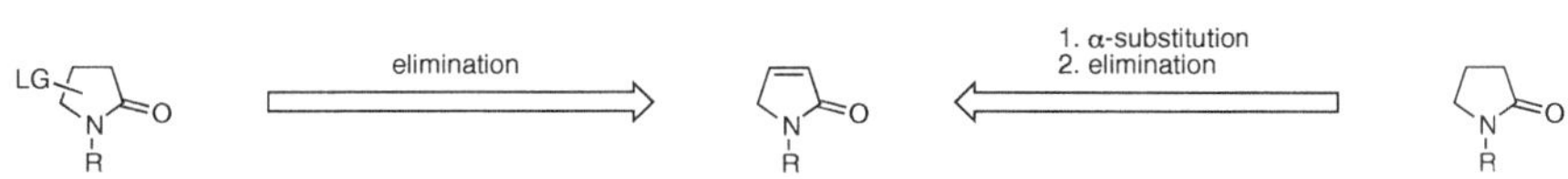


**Scheme 114**

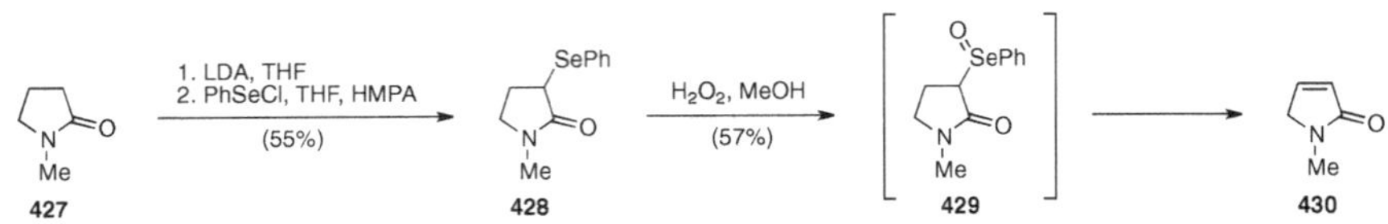


Scheme 115

Pyrrolidin-2-ones are converted into 3-pyrrolin-2-ones in three steps: (1) α-selenylation of an enolate; (2) oxidation of the alkyl aryl selenide to the selenoxide; and (3) *syn*-elimination of the selenoxide intermediate. Perhaps the first report of this methodology (although rarely cited as such) was by Zoretic and Soja in 1977 (Scheme 115; 1977JHC681). Selenylation of *N*-methylpyrrolidinone (**427**) was accomplished by deprotonation by LDA followed by treatment with benzeneselenyl chloride giving **428** (1976JOC3587). Oxidation of **428** with hydrogen peroxide gives the 3-pyrrolin-2-one **430** via oxidation of the selenium followed by the facile elimination of the selenoxide intermediate **429**. One of the earliest uses of this three-step methodology (and often cited as precedent by others) was by Ohfune and Tomita in their total synthesis of domoic acid (1982JA3511). Mun and Smith revisited this sequence in the preparation of 1-benzyl-3-pyrrolin-2-one (2007SC813). Unexpectedly, when the oxidation reaction was allowed to warm to room temperature they obtained 1-benzyl-5-hydroxy-3-pyrrolin-2-one as the major product. Evidently, the combination of hydrogen peroxide and PhSeOH is enough to introduce a hydroxy group to the 5-position of 3-pyrrolin-2-ones.

The selenylation/oxidation/elimination methodology has played an important role in the preparation of chiral pyroglutamate derivatives (5-carboxypyrrolidin-2-ones or 5-oxoprolines). Herdeis and Kelm utilized a trioxabicyclo[2.2.2]octane ester to protect the carboxylic acid functionality of a pyroglutamate (Scheme 116; 2003T217). This eliminated the need to change oxidation states and lowered the acidity of the α-proton of the carboxylic acid functionality. Treatment of pyrrolidin-2-one **431** with LiHMDS followed by benzeneselenyl chloride and then by *m*-CPBA gives the 3-pyrrolin-2-one **432**. Treatment of **432** with diethylcuprate followed

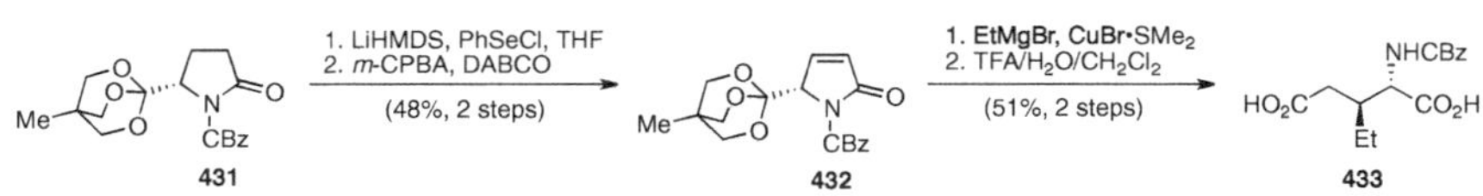


Scheme 116

by acid hydrolysis gives the CBz-protected glutamic acid derivative **433**. Nishiyama and coworkers investigated similar routes to glutamic acids and cyclopropane amino acids (2003CC776, 2005T8456).

The α-selenylation/oxidation/elimination methodology has been employed in the synthesis of several 3-pyrrolin-2-one natural products including microcolin B (1997JOC5542), pyrrolam A (**48**) (1997TA515, 2007S663), ampullicin (2001JOC8287), aglairubine (2002TL4609), and the myceliothermophins (e.g., **25**) (2014ACIE10970). In addition, this methodology has been utilized in the synthesis of the 3-pyrrolin-2-one moieties in analogs of microcolins A and B (2005BMCL4043) and the spirocyclic moiety of pandaramine (1995TL7705). Additional examples include the preparation of 3-pyrrolin-2-one intermediates en route to the total synthesis of nitrogen-containing natural products including: lactacystin (**65**) (1994JA2139, 2006JOC1220), omuralide (2007JOC10108), kainic acid (**64**) (1998SL507), kainic acid analogs (1991TL3057), pramanicin (1999JOC6005), swainsonine (1987CPB2140), ceramide sphingolipid (2010TL4317), cytochalasans (1982JOC1534, 1986TL2691, 1990HCA122), and cytochalasin-type alkaloids (1989T2417, 2013CAJ1243). Other applications include the preparation of the 3-pyrrolin-2-one intermediates en route to a variety of cyclic and acylic nitrogen-containing molecules including 3-alkylpyrrolidinones (1991TL1379, 1991MI1), 4-alkylpyrrolidinones (1990SL63), 3,4-dihydroxy-3-pyrrolidinones (1991TL6949, 1992CPB1925, 2005JOC3826, 2010SL2156, 2010TL4317), 3,4-fused pyrrolidinones (1995SC1265, 2001H(55)2099), 1,2-fused pyrrolidinones (1998T12361), chiral pyrrolidines (1994TL119, 2001JMC3074), 2-iminopyrrolidines (2002BMCL3337), glutamic acid derivatives (1991TL283, 1991TL5983), glutamine derivatives (2001T6353), aminocarbohydrates (2010SL2156), and chiral secondary amines (1989JA1524, 2010TL4317).

Zoretic and Soja also reported that sulfoxides could be eliminated to give 3-pyrrolin-2-ones (1977JHC681). The major difference between sulfoxides and selenoxides is that sulfoxides can be isolated and the elimination step leading to 3-pyrrolin-2-ones requires heating in a high boiling solvent. In an attempt to generate 3,4-dehydropyroglutamate **437** for use as a dienophile, Ezquerra and coworkers found the sulfoxide chemistry to be superior to selenoxide chemistry given the stability of the sulfoxides (Scheme 117; 1995T10107). Treatment of pyroglutamate **434** with LiHMDS followed by diphenyl disulfide and then by *m*-CPBA gives the sulfoxide **435**. Heating **435** in the presence of cyclopentadiene (**436**) gives the cycloadduct **438** (in 50% ee), which was formed by a cycloaddition involving in situ generated

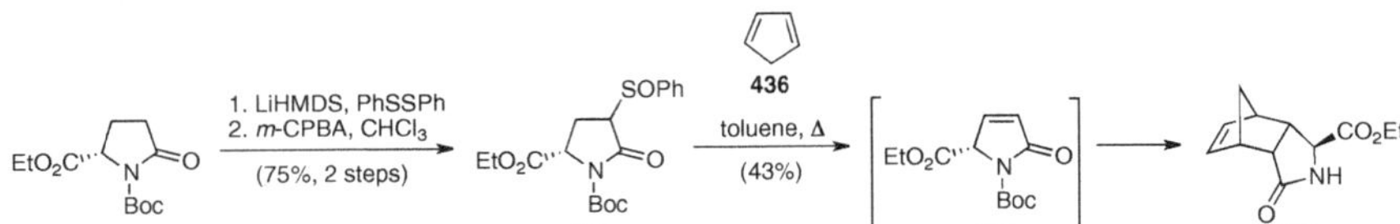


434 435 437 438

Scheme 117

intermediate **437**. Ezquerra and coworkers utilized sulfoxides chemistry to synthesize 5-substituted 3,4-dehydropyroglutamates (1998T9447).

Recently, Gallagher and coworkers investigated the use of sulfoxides to prepare chiral 5-substituted 3-pyrrolin-2-ones **440** and 5-substituted 3-phenylsulfanyl-3-pyrrolin-2-ones **441** from the corresponding sulfides **439** (Scheme 118; 2006OBC1868). The latter were formed via a Pummerer-type rearrangement mediated by trifluoroacetic anhydride. The Motherwell group investigated a fluoro-Pummerer reaction leading to 3-phenylsulfanyl-3-pyrrolin-2-ones utilizing difluoroiodotoluene (DFIT) (2001TL8523). Tanikaga and coworkers investigated the synthesis, oxidation, and elimination of 4-phenylthiopyrrolidin-2-ones (1987CC386).

Matsuo and Aizawa developed a novel one-pot dehydrogenation of carboxylic acid derivatives and included an example leading to a 3-pyrrolin-2-one (2005TL407). They treated 1-(phenylsulfonyl)pyrrolidin-2-one with LDA followed by *N*-*tert*-butylbenzenesulfinimidoyl chloride and obtained 1-(phenylsulfonyl)-3-pyrrolin-2-one in 53% yield.

Bonete and Nàjera prepared a 5-hydroxy-3-pyrrolin-2-one via a DBU-mediated elimination of a tosyl group from a 3-tosylpyrrolidin-2-one (1995T2763).

The Reiser group have exploited a radical ring-opening reaction to prepare the chiral 3-pyrrolin-2-one building block **443** (Scheme 119; 2006JOC2173). Treatment of the 3-bromopyrrolidin-2-one **442** with tributyltin hydride led to **443**, presumably via formation of the α-radical from the bromide followed by fragmentation of the cyclopropane ring. A number of stereoselective reactions have been investigated using building block **443** including the Pd-mediated cyclopropanation with diazonomethane giving the fused pyrrolidin-2-one **444**.

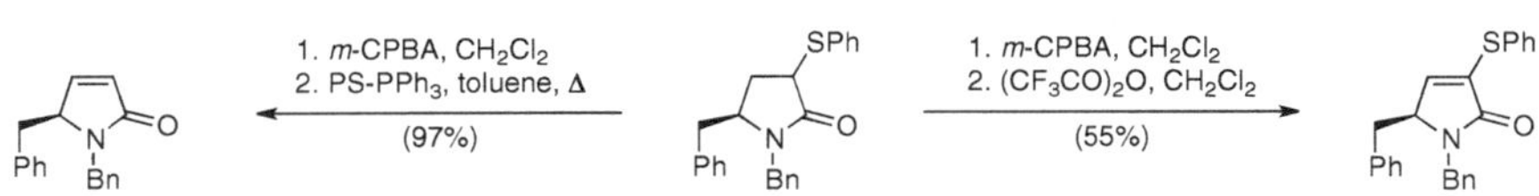


440 439 441

Scheme 118

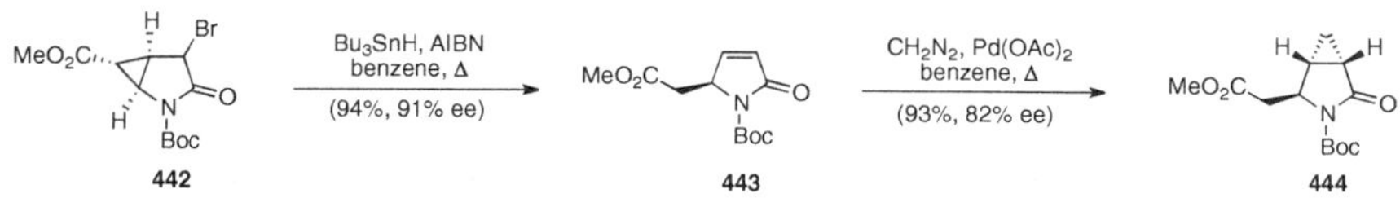


**Scheme 119**

4-Hydroxypyrrolidin-2-ones (parent = **445**) can serve as precursors to 3-pyrrolin-2-ones via transformation of the hydroxy group into a good leaving and subsequent elimination. Baker and Sifniades prepared 1-trifluoroacetyl-3-pyrrolin-2-one **13** starting from **445** (Scheme 120; 1979JOC2798). Treatment of **445** with trifluoroacetic anhydride gives the trifluoroacetate **446**; subsequent treatment of crude **446** with triethylamine then gives **13**. Hydrolysis of the trifluoroacetamide protecting group then gives a 9:1 mixture of the parent 3-pyrrolin-2-one (**1**) and 4-pyrrolin-2-one (**3**). Other bases have been used to eliminate 4-acetoxy groups including sodium bicarbonate (1978MI1), LDA (1983TL4755), triethylamine (1992T3313), sodium hydride (1992T3313), and DBU (lactone-leaving group) (2012T3336). Hiemstra and coworkers utilized a trichloroacetate-leaving group in their synthesis of a chiral 5-isopropoxy-3-pyrrolin-2-ones from a 4-hydroxy-5-isopropoxypyrrolidin-2-one (1992JOC1059, 1999JOC2567).

Tian and coworkers developed a large-scale synthesis of *N*-Boc-3-pyrrolin-2-one (**449**), which is an important precursor utilized in the generation of the important building block *N*-Boc-2-*tert*-butyldimethylsiloxypyrrole (**450**) (2002OPRD416) (Scheme 121). The key step is synthesis and elimination of a 4-mesyloxy group; this strategy for converting 4-hydroxypyrrolidinones into 3-pyrrolin-2-ones was first reported by Pinza and Pifferi (1978MI1, 1978S614). They converted 4-amino-3-hydroxybutyric acid (**447**) into

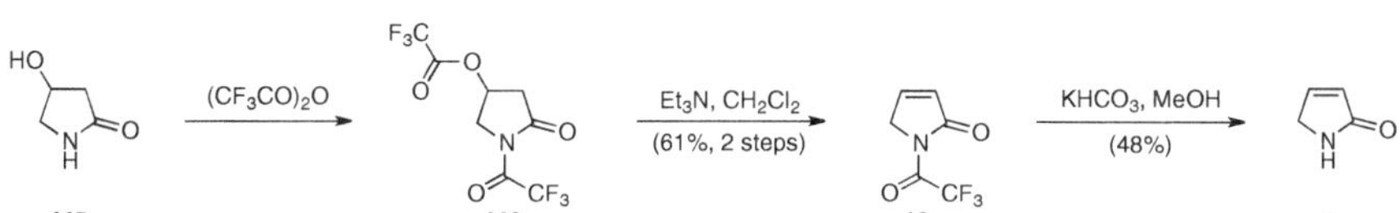


**Scheme 120**

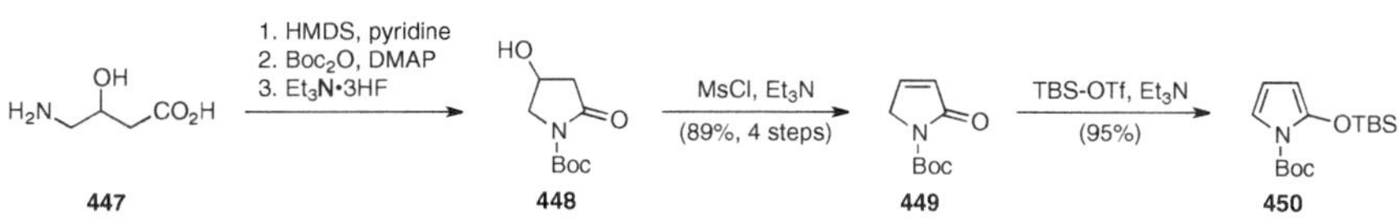


**Scheme 121**

4-hydroxypyrroldin-2-one **448** in three steps using a cyclodehydration. Treatment of **448** with mesyl chloride then gives **449** via mesylation and subsequent elimination. Similar eliminations involving mesylates (1995TL291, 1996T3757) and tosylates (1999TA3827) have been employed in the preparation of the natural product pyrrolam (**11**). Some additional recent applications of mesylate eliminations to give 3-pyrrolin-2-ones have been reported for the preparation of chiral 5-substituted 3-pyrrolin-2-ones (1996JOC3534, 1999JOC2567), chiral *N*-substituted 3-pyrrolin-2-ones (2011TL1253), 3-pyrrolin-2-one building blocks en route to rigidiusculamides (2012CAJ504), and spirocyclic 3-pyrrolin-2-ones (2013JOC2661, 2013OPRD985).

Oba and coworkers observed racemization when eliminating a mesylate derived from a 4,5-cis-substituted pyrrolidin-2-one en route to a chiral 3-pyrrolin-2-one (2009TL5053). In order to combat this problem, they inverted the stereochemistry of the hydroxy group using a Mitsunobu-like reaction (Scheme 122). Treatment of alcohol **451** with iodine, polymer-supported triphenylphosphine (PS-$PPh_3$), and imidazole gives the 3-pyrrolin-2-one **453** via intermediate iodide **452**.

Schmidt and coworkers found that a Boc carbonate can serve as a precursor to 3-pyrrolin-2-ones via a facile elimination (Scheme 123; 1993S216). Treatment of 4-hydroxypyrrolidin-2-one **454** with two equivalents of $Boc_2O$ in the presence of DMAP gives Boc carbonate **455**. Elimination of the carbonate group by treatment with $CsCO_3$ gives 3-pyrrolin-2-one **456**. This strategy has been employed in the total synthesis of ypoamide (**30**) (2009JOC7457), microcolin B (2010TA895), and rhazinilam (2014ARK(iii)256, 2014EJOC7865). Mattern found that 3-pyrrolin-2-ones can be obtained directly in one step by increasing the reaction time for the first step involving DMAP (1996TL291).

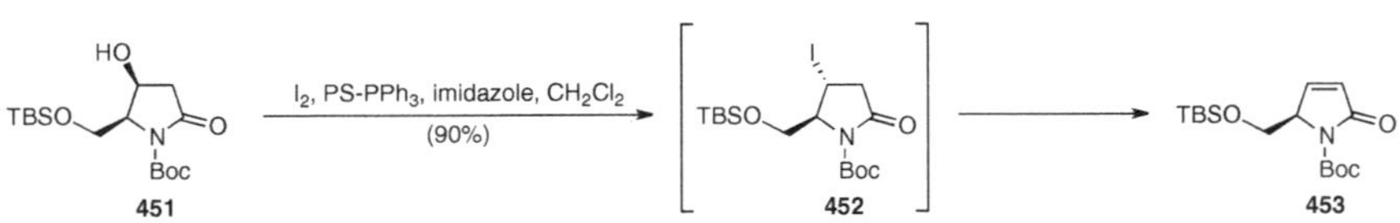


**Scheme 122**

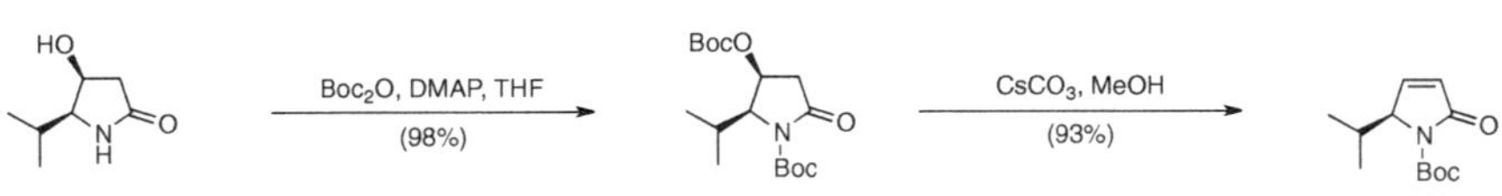


**Scheme 123**

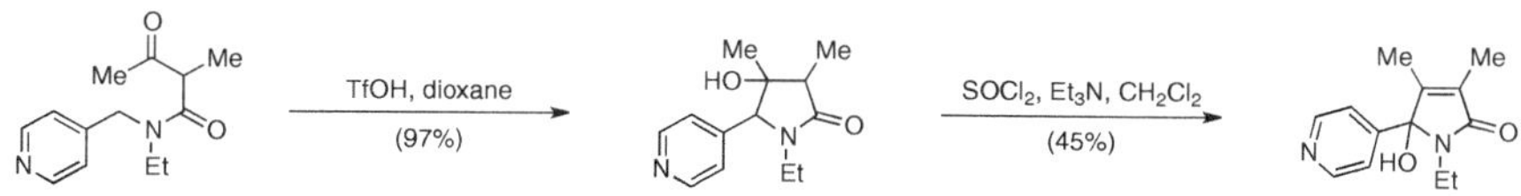


457 458 459

**Scheme 124**

Pigge and coworkers investigated the use of $SOCl_2$ as a dehydrating agent (Scheme 124; 2014OBC1090). The 4-hydroxypyrrolidin-2-one **458** was formed via an acid-mediated benzylic cyclization of ketoamide **457**. Treatment of **458** with $SOCl_2$ then gives the 5-hydroxy-3-pyrrolin-2-one **459** via a dehydration and subsequent benzylic oxidation. The use of thionyl chloride as a dehydrating agent was also mentioned by Lee and coworkers (2013JOC2661).

The acid-mediated dehydration of 5-hydroxypyrrolidin-2-ones **202** provides an entry to 3-pyrrolin-2-ones **205**, and several examples have been reported (1989JHC1713, 1996JOC4542, 2006BMCL3906, 2007TL1571). For example, treatment of 5-hydroxypyrrolidinone **460** with a variety of acid catalysts (Amberlyst resin gave the best results) gives 4-methyl-3-pyrrolin-2-one (**463**); the mechanism for this transformation probably involves the generation of the iminium intermediate **461**, deprotonation to 4-pyrrolin-2-one **462**, and tautomerization to **463** (Scheme 125; 2004S1560). Spieckamp and coworkers were among the first to study the generation of 5-hydroxypyrrolidin-2-ones by reduction of succinimides with sodium borohydride (1975T1437, 1978T179).

Pyroglutamics acids (5-oxoprolines) can serve as precursors to 3-pyrrolin-2-ones via a similar formation of iminium intermediates and isomerization. Haldar and Ray discovered a direct conversion of pyroglutamic acid **464** to the 5-ethoxy-3-pyrrolin-2-one **466** by treatment with CAN and ethanol (Scheme 126; 2008TL3659). The mechanism of the reaction likely involves the formation of iminium intermediate **465**. Rigo and coworkers observed that the treatment of pyroglutamic acids with trifluoroacetic

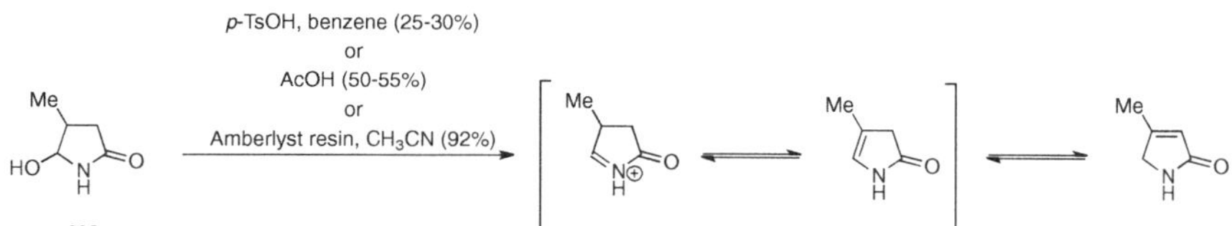


460 461 462 463

**Scheme 125**

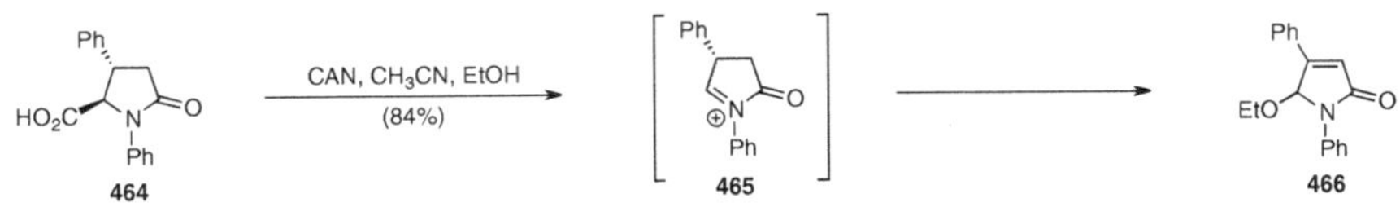


Scheme 126

anhydride followed by a Lewis acid led to the formation of 3-pyrrolin-2-ones via an acyliminium intermediate that was generated by a decarboxylation reaction (2000JHC143, 2002T9239, 2004TL9197).

Interesting building blocks that have been utilized to prepare a diverse array of 3-pyrrolin-2-ones are 4-chloromethyl-3-chloropyrrolidin-2-ones (e.g., **467** and **469**) (Scheme 127). These compounds are available from either the Ru-catalyzed cyclization (1993TL6821, 2011S1267) or radical cyclization (1999T5839) of *N*-allyl-α-perchloroamides. Slough found that treatment of **467** with a Ru catalyst led to the formation of the 3-pyrrolin-2-one **468** via elimination of the α-chloro group (1993TL6825). Ghelfi and coworkers investigated a wide range of transformations of **469**. For example, treatment of **469** with lithium methoxide led to the formation of the 5-methoxy-3-pyrrolin-2-one **470** (Scheme 127; 2001TL4573, 2003T1147, 2008S3131, 2009T1481, 2010SC1040, 2010T1357). Ghelfi and coworkers also explored the reactivity of **469** and related structures with amines which led to different nitrogen-substituted 3-pyrrolin-2-ones and pyrrolidin-2-ones (1999TL8595, 2004T11493).

Oxidation of 3-hydroxy-3-pyrrolidinones provides 3-hydroxy-3-pyrrolin-2-ones via transformation to the ketone and tautomerization to the enol. Puthukanoori and coworkers used this strategy to prepare phenopyrrozin **49** (Scheme 128; 2012TL1891). Alpha-hydroxylation of pyrrolidinon-2-one **471** gives phenopyrrozin 1 (**49**) in low yield. The reaction proceeds through the formation of hydroxyamide **472**, in situ oxidation to ketoamide **473**, and tautomerization to phenopyrrozin **49**. Feng and coworkers reported a transformation of 3-hydroxypyrrolidin-2-one to 3-hydroxy-3-pyrrolin-2-one via an oxidation with $Na_2Cr_2O_7$ (2009BMCL2112).

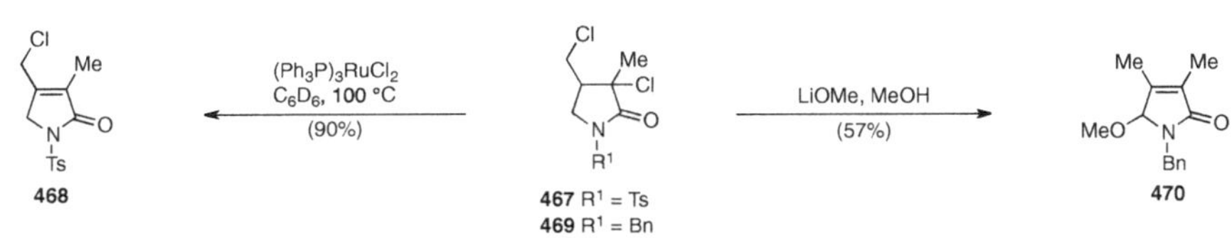


Scheme 127

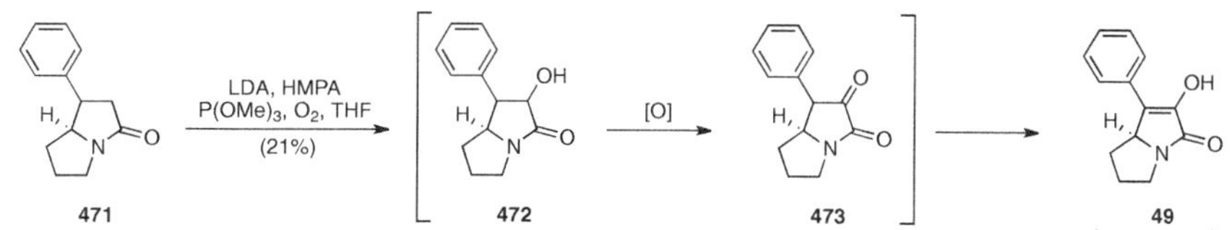


Scheme 128

Fabio and coworkers reported an interesting route to quinoline-substituted 3-pyrrolin-2-one **477** (Scheme 129; 2002JOC7319). Wittig reaction of pyrrolidinone ylide **474** with aldehyde **475** gives the alkylidene-pyrrolidinone **476**. A Pd-mediated Heck reaction with the aryl side chain then gives 3-pyrrolin-2-one **477**. Merreddy and coworkers reported a Rh-mediated isomerization of an alkylidene-substituted pyroglutamic acid derivative to the corresponding 3-pyrrolin-2-one (2011CC3219).

Koizumi and coworkers utilized cycloreversion reactions to generate 3-pyrrolin-2-one intermediates from tricyclic and tetracyclic intermediates (1991CL2135, 1992TA535, 1992CPB1670). The tricyclic scaffolds allow the stereoselective introduction of substituents into the 5-position, and the subsequent intermediates have been utilized in the synthesis of nitrogen-containing natural products. Koizumi and coworkers used a cycloreversion reaction to prepare 3-pyrrolin-2-one building blocks useful for the synthesis of indolizidine alkaloids (e.g., **480**) (Scheme 130; 1994PT(1)15). Flash-vaccum pyrolysis (FVP) of tetracycle **478** gives the fused 3-pyrrolin-2-one **479**, which can be converted to the indolizidine natural product laburnine (**480**). Cycloreversions involving oxabicyclo[2.2.1]heptenes occur under milder conditions (1992CPB1670). Snyder and coworkers studied cycloreversion reactions using chiral aminoanthracenes in the preparation of chiral 5-substituted 3-pyrrolin-2-ones (2005TL2475, 2005OL31); Jones and coworkers used this strategy to prepare (*R*)-pyrrolam A (**48**) (2011OBC4353, 2014OL3780).

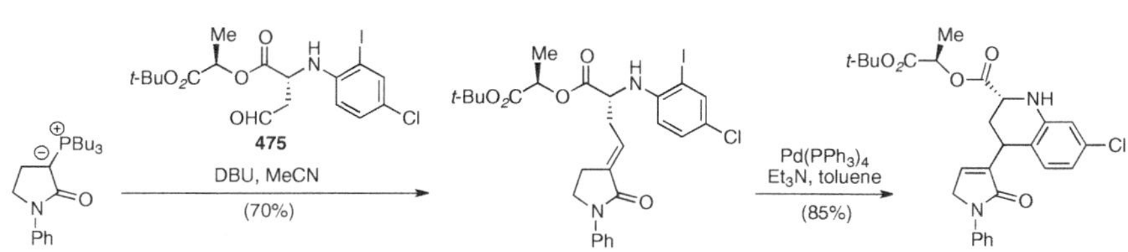


Scheme 129

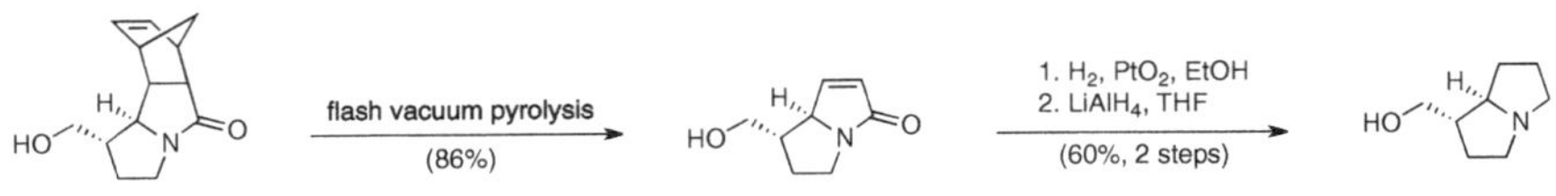


478 479 480

Scheme 130

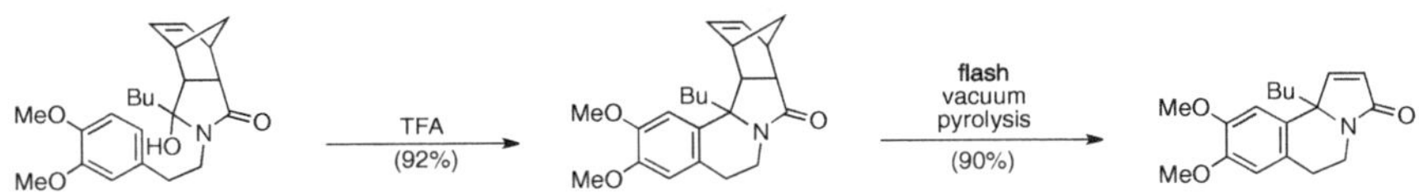


481 482 483

Scheme 131

Lete and coworkers have utilized FVP to prepare the isoquinoline-fused 3-pyrrolin-2-ones **483** from lactams **482** (Scheme 131; 1996TL7841, 1998T12361, 2002SL593, 2007ARK206). The latter were prepared by cyclization of the *N*-acyliminium ions generated by treatment of hydroxy-lactams **481** with TFA. Simpkins investigated a related sequence in an asymmetric synthesis of the erythrinan alkaloids (2003TL7803).

## 5.2 Synthesis from Pyrrolidin-2,4-Diones

Pyrrolidine-2,4-dione (**484**) and its tautomer, 4-hydroxy-3-pyrrolin-2-one or tetramic acid (**4**), are collectively known as tetramic acids (Scheme 132; 1995CR1981, 2007MI1). Reactivity of the hydroxy tautomer leads to 4-alkoxy-3-pyrrolin-2-ones; nucleophilic substitution at the ketone of the dione tautomer leads to a variety of 4-substituted 3-pyrrolin-2-ones.

Due to the prevalence of the 4-methoxy-3-pyrrolin-2-one moiety (methyl tetramates) in natural products (see Figure 4), a variety of methods have been developed to convert tetramic acids into methyl tetramates. In the context of the total synthesis of althiomycin (**28**), Inami and Shiba used diazomethane to convert **4** into methyl tetramate (**485**), which was accompanied by the formation of 2-methoxy isomer **486** (Scheme 133; 1984TL2009, 1985BCSJ352). Others have used diazomethane to prepare

484
pyrrolidin-2,4-dione

4
tetramic acid

Scheme 132

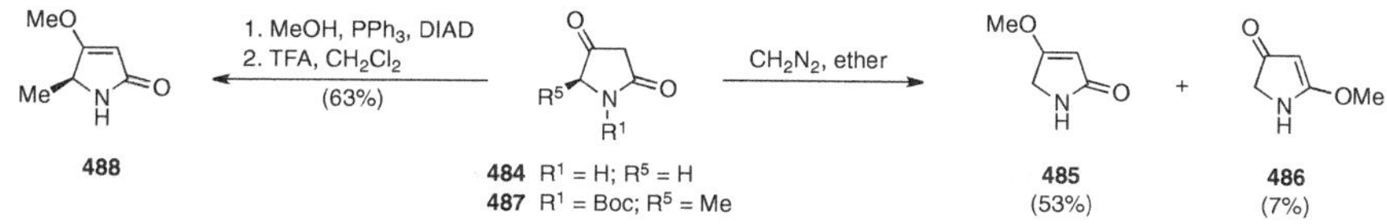

Scheme 133

methyl tetramates for the total synthesis of sintokamide C (2010OL1100) and palau'imide (**34**) (2012OL3862). Frontier and coworkers used (trimethylsilyl)diazomethane to make a methyl tetramate intermediate en route to tetrapetalone aglycon (2014ACIE9334).

A second method for preparing methyl tetramates involves using dimethylsulfate as the electrophile as seen en route to dolastatin 15 (**32**) (1991JA6692), mirabimide (1994JA8116), dysidin (**31**) (1984HCA1783), and plasminogen activator inhibitors (2009CPB979). A third method for preparing methyl tetramates (e.g., **488**) involves using a Mitsunobu reaction between tetramic acids (e.g., **487**) and methanol as reported by Payne and coworkers in the total synthesis of simplostatin (**36**) (Scheme 133; 2010OL5576). This strategy was also used en route to dolastatin 15 (**32**) (1992T4115, 1994T12097, 1994BMCL1947, 1999JOC405), malyngamide (2007T3217, 2009JOC4149), sintokamide A (**35**) (2010ACIE9702), dysidin (**31**) (2011CAJ2260), and gallinamide A (2011CEJ3544). Tønder and coworkers prepared ethyl tetramates by treatment of **484** with KHMDS, 18-crown-6, and ethyl tosylate (2006OL2103).

Acylation of tetramics acids occurs initially at O4 followed by migration to C3 which allows the preparation of 3-acyltetramic acids, a structural subunit of a number of natural products (2007MI1, 2011JOC1342). One of the first reports of this transformation was by Yoshii and coworkers (1987CPB4368) and has been systematically studied by Jeong and Moloney (Scheme 134; 2011JOC1342). Treatment of **484** and benzoic acid (**489**) with DCC/DMAP gives the 4-acyloxy-3-pyrrolin-2-one **490**; for *N*-substituted tetramic acids, the use of mixed anhydrides gives better yields

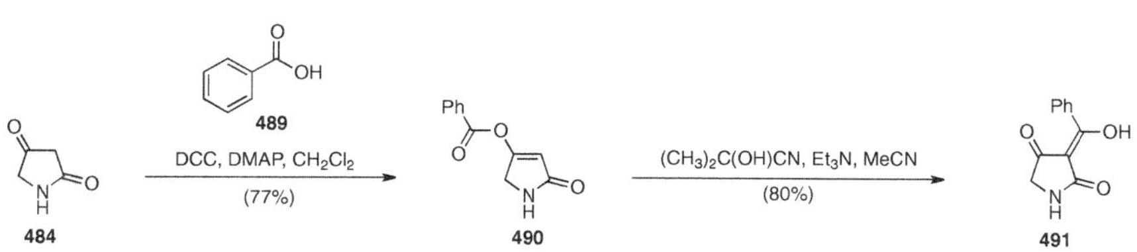

Scheme 134

Scheme 135

for alkyl carboxylic acids. Treatment of **490** with acetone cyanohydrin and trimethylamine gives the 3-acyltetramic acid **491**. This overall strategy (either DCC/DMAP or mixed anhydride) has been used in the total synthesis of 3-acyltetramic acid natural products by the research groups of Schobert (2005T2301, 2013JOC2455) and Yoda (2010SL2944, 2012JOC4391, 2012OL5142).

Tønder and coworkers condensed amino acid derivatives (e.g., **493**) with tetramic acids (e.g., **492**), mediated by acetic acid, to obtain 4-amino-3-pyrrolin-2-ones (e.g., **494**) (Scheme 135; 2007OBC2207). Huang and coworkers reported a similar transformation leading to 4-amino-3-pyrrolin-2-ones using chiral pyrrolidines (2003OL4341). Burgess used the same reaction conditions to make pyrrolinone–pyrrolidinones in their search for simple peptidomimetics (2011JA12350, 2013JOC4823). Tao and coworkers studied a multicomponent reaction involving *N*-benzyl tetramic acid with oxindole, aniline, and dicyanomethane, which led to the formation of complex heterocycle that contained a 4-amino-3-pyrrolin-2-one moiety (2014T8768).

Pinheiro and coworkers investigated the synthesis and reactions of oxime derivatives of tetramic acids (2009TL2402). Reduction of a tetramic acid oxime with sodium borohydride gives a 4-amino-3-pyrrolin-2-one.

Tetramic acid derivatives have been converted into 4-unsubstituted 3-pyrrolin-2-ones utilizing a two-step procedure developed by Decicco and Glover (Scheme 136; 1996JOC3534). Reduction of tetramic acid **495** with sodium borohydride gives the 4-hydroxypyrrolidine **496**; elimination of the mesylate derived from **496** then gives 5-methyl-3-pyrrolin-2-one **497**. Compound **497** is a building block used en route for a total

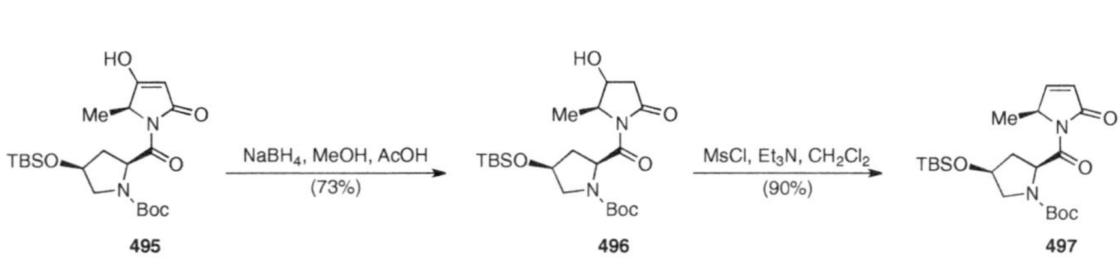


Scheme 136

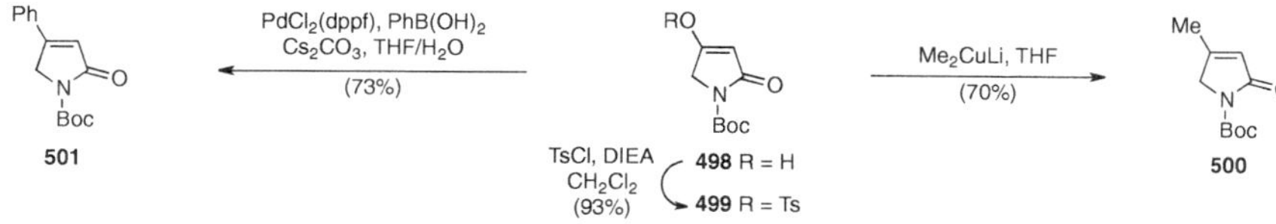


Scheme 137

synthesis of microcolin A. Feringa and coworkers have utilized this sequence to prepare simple 3-pyrrolin-2-ones (1999JOC2567).

Tetramic acid sulfonates are useful intermediates for the synthesis of 3-pyrrolin-2-ones. Li and coworkers prepared the 4-methyl-3-pyrrolin-2-one **500**, a precursor to pulchellalactam (**44**), by treatment of tetramic acid tosylate **499** with dimethylcuprate (Scheme 137; 2002JOC4702). Tosylate **499** was prepared by tosylation of tetramic acid **498**. The Pelkey group has investigated cross-coupling reactions of **499** with arylboronic acids, which give 4-aryl-3-pyrrolin-2-ones (e.g., **501**) (Scheme 137; 2007TL827).

The Pelkey research group have also studied the synthesis and cross-coupling reactions of 3-aryltetramic acid sulfonates (e.g., **503**) en route to 3,4-diaryl-3-pyrrolin-2-ones (e.g., **505**) (Scheme 138; 2007S2317, 2009JHC447, 2014JOC8049). Treatment of tetramic acid **502** with triflic anhydride followed by TFA gives tetramic acid triflate **503**. Pd-catalyzed cross-coupling of **503** with 4-methoxyphenylboronic acid (**504**) gives the 3,4-diaryl-3-pyrrolin-2-one **505**. Others have investigated similar cross-coupling reactions with 3-aryltetramic acid triflates (2011T10026) and bicyclic tetramic acid triflates (2002OL119, 2007BMCL1908, 2008SL2107).

## 5.3 Synthesis from Pyrrolidin-2,3-Diones

The tautomers pyrrolidine-2,3-dione (**506**) and 3-hydroxy-3-pyrrolin-2-one (**507**) are useful precursors to substituted 3-pyrrolin-2-ones (Scheme 139). The enol tautomer can be trapped as a triflate, and nucleophilic additions to the keto tautomer lead to 3-substituted 3-pyrrolin-2-ones.

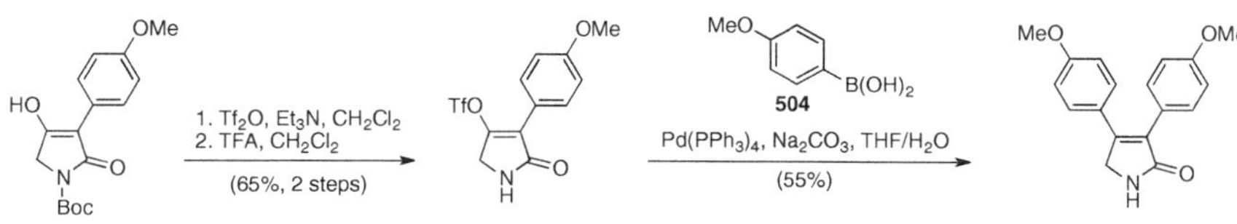


Scheme 138

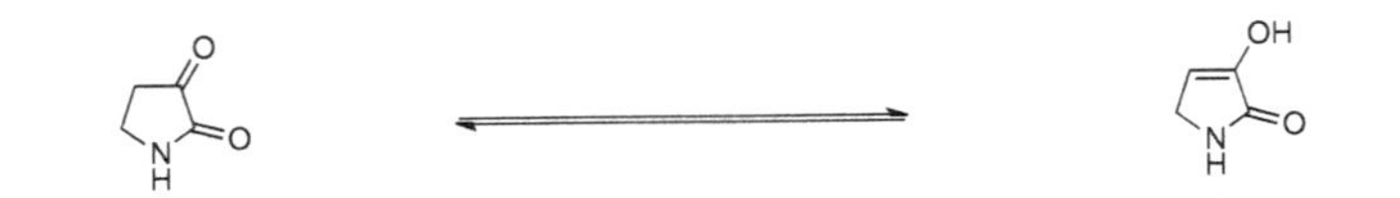

506
pyrrolidin-2,3-dione

507
3-hydroxy-3-pyrrolin-2-one

Scheme 139

Southwick and coworkers developed strategies for preparing *N*-alkylpyrrolidin-2,3-diones (1949JA2532, 1956JOC1087); they studied the chemistry of this class of molecules including bromination reactions leading to 4-bromo-3-hydroxy-3-pyrrolin-2-ones (1962JOC3075) and diazomethane reactions leading to 3-methoxy-3-pyrrolin-2-ones (1953JA3413).

The fused 2,3-pyrrolidinedione **509** has been transformed into the natural product pyrrolam (**48**) by Ohta and coworkers (Scheme 140; 1996T869). Ozonolysis of alkylidene-substituted pyrrolidin-2-one **508** gives **509**, which was converted into triflate **510** using triflic anhydride and DIEA. Pd-mediated radical reduction of **510** then gives pyrrolam (**48**).

Gallagher and coworkers have investigated cross-coupling reactions of 3-triflyloxy-3-pyrrolin-2-ones (e.g., **512**) (Scheme 141; 2007OBC2636). Treatment of 2,3-pyrrolidinedione **511** with *N*-phenyltriflimide gives triflate **512**. Suzuki cross-coupling reactions of **512** with phenylboronic acids under microwave irradiation give **513**; the yield and stereochemical fidelity of the transformation under microwave conditions were comparable to the standard thermal conditions.

Sundberg and coworkers prepared a 3-(3-pyrrolyl)-3-pyrrolin-2-one, a precursor to phosphodiesterase inhibitors, by treating 2,3-pyrrolidinedione with a 3-pyrrolyllithium reagent followed by an acid-mediated dehydration (1985JOC425, 1988JOC976).

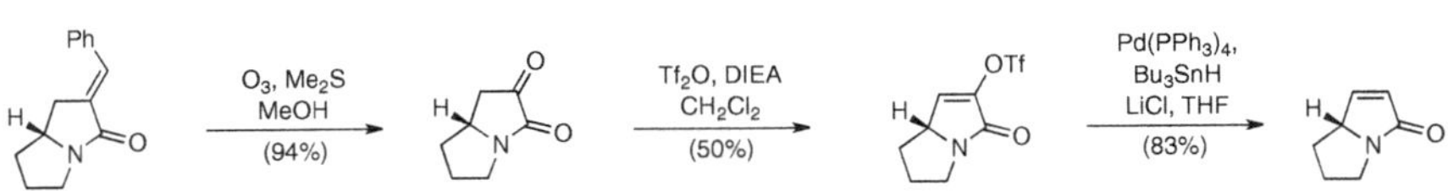


508 509 510

48
pyrrolam A

Scheme 140

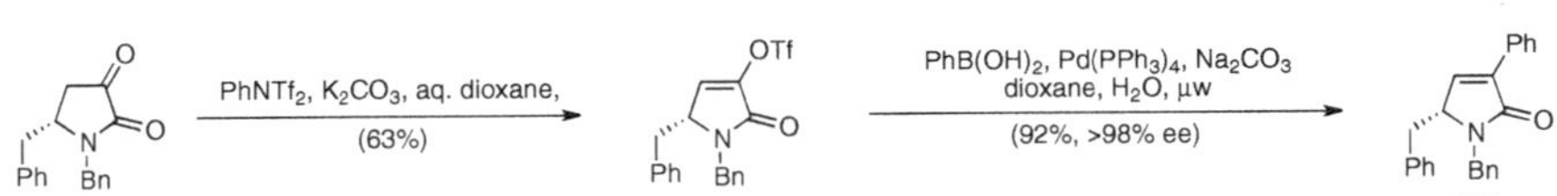


511 512 513

Scheme 141

Matsuo condensed 2,3-pyrrolidinediones with anilines, which gives 3-amino-3-pyrrolin-2-one derivatives (2005H(65)2451).

## 5.4 Synthesis from Maleimides

Maleimides (pyrrole-2,5-diones) can be converted into 3-pyrrolin-2-ones and 5-hydroxy-3-pyrrolin-2-ones by complete or partial reduction of one of the imide carbonyls, respectively. In the case of unsymmetrical substrates, the reaction requires a regioselective reduction of one of the two carbonyls; there has been success in this area although some substrates prove to be problematic (1994TL9135, 1999TA4469). Nucleophilic additions to maleimides by carbon-based nucleophilics also provide access to 3-pyrrolin-2-ones.

The reduction of maleimide **9** has been investigated using a variety of different conditions (Scheme 142). Gaviña and coworkers reported perhaps the first reduction of maleimide (**9**) (1988JA4017). Treatment of **9** with $LiAlH_4$ gives 5-hydroxy-3-pyrrolin-2-one **514** (yield not disclosed). Takabe and coworkers performed this reaction using Luche conditions ($NaBH_4$, $CeCl_3$) (2002PT(1)707). Koltunov studied the reduction of imides with superacid (2006EJOC4861); their treatment of **9** with superacid ($CF_3SO_3H$, $SbF_5$) gives fully-reduced 3-pyrrolin-2-one (**1**) in excellent yield.

The Nagasaka group researched the transformation of a substituted maleimide into a 5-hydroxy-3-pyrrolin-2-one while completing a total synthesis of jatropham (**38**) (1981H(16)1987). Subsequently, Takabe and coworkers investigated the regioselectivity of the reduction of unsymmetrical maleimides in a series of reports (Scheme 143; 1992PH1084, 2002PT(1) 707, 2004H(63)1013, 2007TL8956). Treatment of citraconimide (**515**) with sodium borohydride gives 5-hydroxy-4-methyl-3-pyrrolin-2-one (**516**) with high regioselectivity (1999TA4469). A lower yield was observed

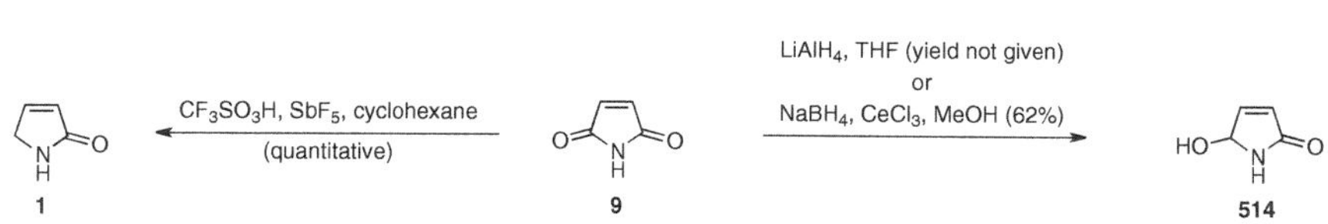


Scheme 142

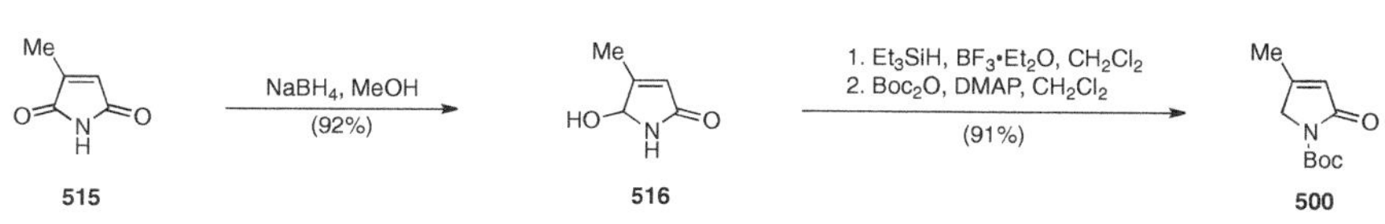


Scheme 143

using Luche conditions ($NaBH_4$, $CeCl_3$), while the bulkier reducing agent DIBAL reversed the regioselectivity. Further reduction of **516** with triethylsilane followed by Boc protection gives **500**, a precursor to pulchellalactam (**44**) (2004H(63)1013). Several additional examples of the transformation of maleimides to 5-hydroxy-3-pyrrolin-2-ones (2004S1560, 2004MI3, 2004OL577, 2013S2888) and chiral 5-hydroxy-3-pyrrolin-2-ones (obtained via an asymmetric resolution) (2008TA2115) have been described. Coster and coworkers reported that the reduction of 4-methoxymaleimides with sodium borohydride proceeded with high regioselectivity to give a 5-hydroxytetramate (2006JOC4703).

The transformation of maleimides into 3-pyrrolin-2-ones has been well studied in the context of the synthesis of the indolocarbazole alkaloids. Many indolocarbazole alkaloids are comprised of indolocarbazole rings that are fused with a 3-pyrrolin-2-one ring. One prominent example is staurosporine **52**, which is a potent inhibitor of protein kinase C (1995JAB535). The Clemmensen reduction has been utilized by several groups for the preparation of 3-pyrrolin-2-one-fused indolocarbazole natural products (1987JOC1177, 1994TL5555, 1995TL8383, 2003S497, 2007TL7399, 2012TL6861) and analogs (2001JHC591, 2007JMC433) from the corresponding maleimides. One of the earliest examples was reported by Raphael and Hughes (1983TL1441, 1990PT(1)2475). Treatment of maleimide **517** with Zn/Hg amalgam in the presence of HCl gives 3-pyrrolin-2-one **518** (Scheme 144).

Alternative two-step methods for the reduction maleimide-fused indolocarbazoles to 3-pyrrolin-2-one-fused indolocarbazoles have been reported (1993TL8361, 1995S1511, 1998JOC6053, 2000H(53)2119). Danishefsky and coworkers investigated the regioselectivity of this reduction with unsymmetrical substrates and found this to be a difficult problem. Their reaction conditions included treatment with l-selectride followed by benzeneselenol and toluenesulfonic acid (1994TL9135, 1995JA552, 1996JA2825).

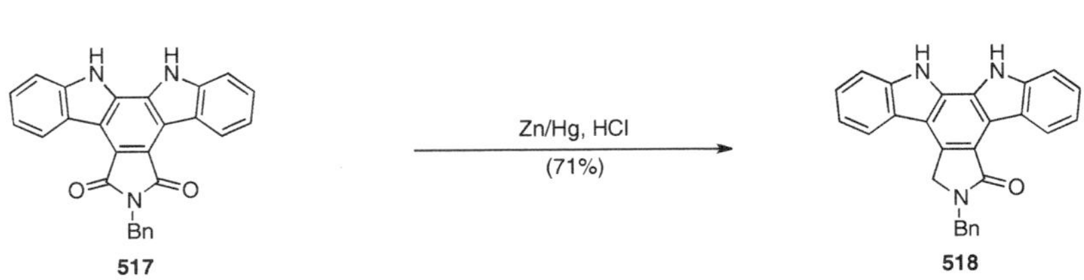


Scheme 144

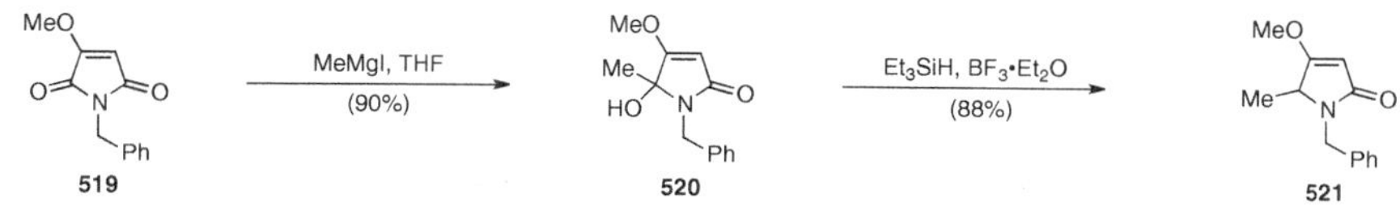


Scheme 145

Grignard additions to substituted maleimides have been studied by Pattenden and coworkers who obtained a mixture of regioisomers (1993PT(1)2567). Huang and coworkers have further studied the regioselectivity of Grignard addition reactions to unsymmetrical maleimides (Scheme 145; 2009SL297, 2010T172). For example, treatment of the 3-methoxymaleimide **519** with methylmagnesium iodide gives 5-hydroxy-3-pyrrolin-2-one **520** with excellent regiocontrol (directed by the methoxy group); subsequent reduction gives 3-pyrrolin-2-one **521** (2009SL297). Tamura and coworkers prepared the alkylidene 3-pyrrolin-2-one natural product phaeosphaeride A, via a Grignard addition by methylmagnesium bromide to a fused maleimide followed by dehydration (2011OBC5825). Daïch and coworkers investigated the addition of alkynyllithium to maleimides which gives 5-alkynyl-5-hydroxy-3-pyrrolin-2-ones (2001TL573). The latter are converted into 5-alkylidene-3-pyrrolin-2-ones upon treatment with *p*-toluenesulfonic acid.

Pattenden and coworkers employed a Horner–Wadsorth–Emmons reaction to transform maleimides into 5-alkylidene-3-pyrrolin-2-ones (Scheme 146; 1993PT(1)2567). For example, treatment of 3,4-dimethylmaleimide (**522**) with phosphorus ylide **523** gives the alkylidene ester **524** with high Z-stereoselectivity.

Izgu and Hoye reported an intramolecular cyclization between silyl ketene acetals and maleimides en route to the fused 3-pyrrolin-2-one natural product leuconolam (**23**) (Scheme 147; 2013CS2262, 2006OL5191). They found $MeAlCl_2$ to be a viable Lewis acid to promote the desired cyclization. In the event, treatment of model system maleimide **525** with $MeAlCl_2$ gives fused 5-hydroxy-3-pyrrolin-2-one **526**. Sterner and coworkers applied this cyclization to the synthesis of lactam analogs of galiellalactone (2012T3336).

Scheme 146

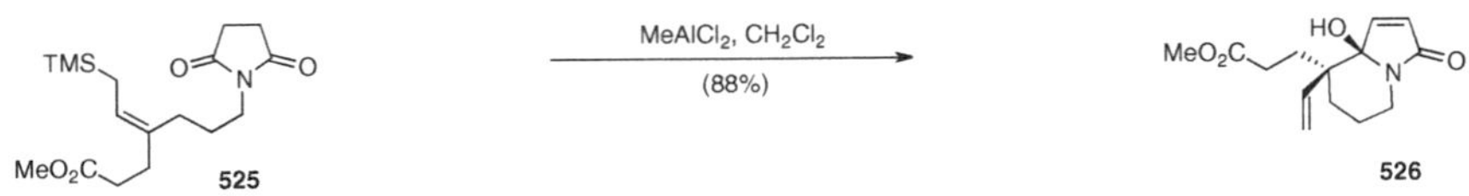


Scheme 147

## 5.5 Synthesis from 3-Pyrrolines

Oxidation of 3-pyrrolines provides another entry to the 3-pyrrolin-2-one ring system. Hayes and coworkers developed a three-step sequence for this transformation (Scheme 148; 2002TL2649). Treatment of 3-pyrroline **527** with TPAP and NMO gives cyclic imine **528**. Further oxidation with sodium hypochlorite followed by treatment with acetyl chloride gives the 3-pyrrolin-2-one **529**. They utilized this strategy to prepare lactacystin (**19**) (2002TL6609, 2008JOC2041) and analogs (2006OBC193).

Brasholz and Reissig demonstrated the direct oxidative conversion of a 3-pyrroline to a 3-pyrrolin-2-one using $Mn(OAc)_3$ and *t*-BuOOH (2007SL1294); Sarpong and coworkers used this reagent system in a complex alkaloid synthesis (Scheme 149; 2010OL4560). Treatment of fused 3-pyrroline **530** with $Mn(OAc)_3$ and *t*-BuOOH gives fused 3-pyrrolin-2-one **531**. Other reagent systems have been employed for the direct conversion of 3-pyrrolines into 3-pyrrolin-2-ones include $CrO_3/H_2SO_4$ (1998PT(1)667, 2004CC1422, 2005CEJ4227) and CuBr/*t*-BuOOH (2003JOC4523).

## 5.6 Synthesis from Pyrroles

One of the oldest approaches to 3-pyrrolin-2-ones involves the direct oxidation of pyrroles that contain at least one free α-position. The most commonly

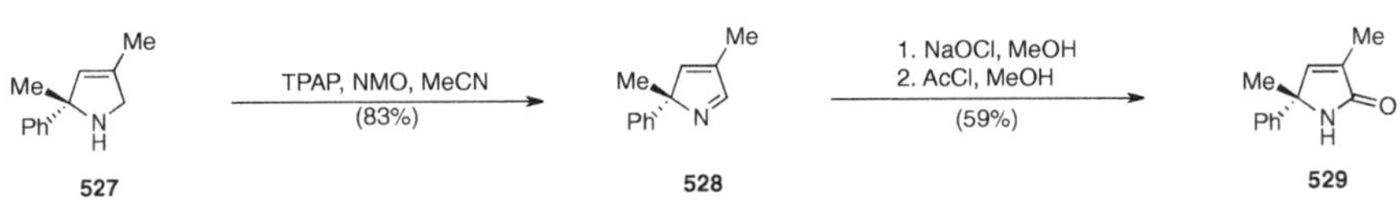


Scheme 148

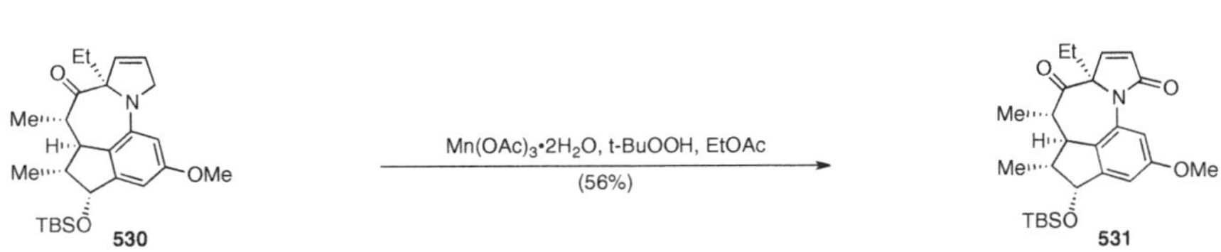


Scheme 149

used reagent for this transformation is hydrogen peroxide; selected early examples were reported by the groups of Fischer (1932MI1, 1937LAC265, 1942MI1), Plieninger (1962LAC165), Atkinson (1964JCS5999), Chierici (1967T737, 1969OPPI271, 1970T4073), Lightner (1976S335), Lugtenberg (1981MI2), and Bonnett (1981PT(1)322). Regarding the parent ring system, Bocchi and coworkers treated pyrrole (**8**) with hydrogen peroxide and barium carbonate (Scheme 150; 1970T4073). They obtained a 9:1 mixture of 3-pyrrolin-2-one (**1**) and 4-pyrrolin-2-one (**3**). Macdonald and coworkers used the same reaction conditions and also reported obtaining an ~9:1 ratio of products **1** and **3** (2002TL5057), while others report only the formation **1** with this oxidation (1981MI2, 1992JOC3760, 2012OBC7313, 2014BMCL5727).

The hydrogen peroxide oxidation of pyrroles has been explored extensively by the research groups of Lightner and Chen for the synthesis of symmetrical 3,4-dialkyl-3-pyrrolin-2-ones (1976S335, 1979T463, 1991JOC6027, 1993T2185, 2002SC1031, 2006T7043, 2007T11030), 3,4,5-trialkyl-3-pyrrolin-2-ones (1990S1062, 2009MI2), and 3,4-dialkoxy-3-pyrrolin-2-ones (2007MI3, 2009MI3). The Fukuyama group used a hydrogen peroxide oxidation of a 2,3,4-trisubstituted pyrrole to make the corresponding 3,4,5-trisubstituted 3-pyrrolin-2-one en route to a total synthesis of the dictyodendrins (2011CAJ560).

The hydrogen peroxide oxidation of unsymmetrical 3,4-dialkyl-3-pyrrolin-2-ones has been studied infrequently, probably due to the fact that the regioselectivity of this reaction is difficult to control. For example, as reported by Gärtner and coworkers, treatment of the unsymmetrical pyrrole **532** with hydrogen peroxide in the presence of pyridine gives nearly an equal (albeit separable) mixture of 3-pyrrolin-2-ones **533** and **534** (2008JA11303; Scheme 151). Other method for regioselectively

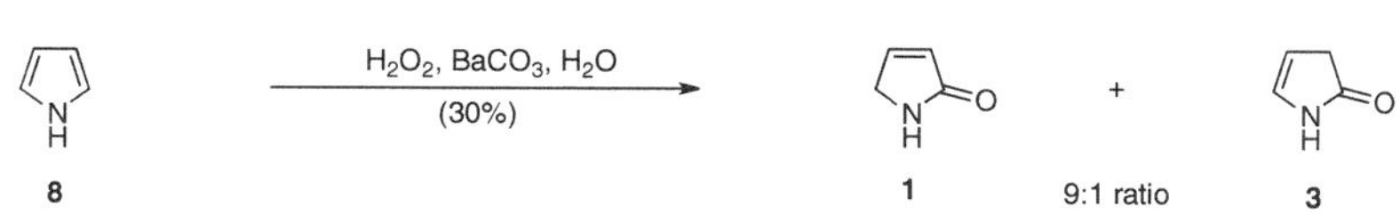


**Scheme 150**

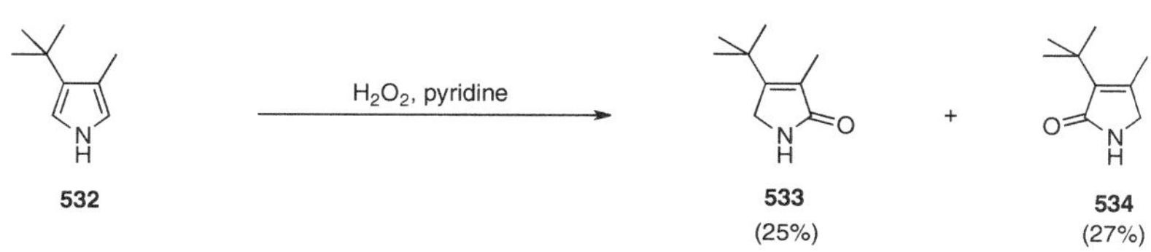


**Scheme 151**

accomplishing this type of transformation include the use of blocking groups (e.g., tosyl) or groups that promote *ipso*-type substitutions via hydrolysis (e.g., halogens) or oxidation (e.g., formyl), and these are discussed in detail later in this section.

The oxidation of simple pyrroles with two free α-positions to 3-pyrrolin-2-ones has been studied recently using hypervalent iodine oxidizing agents. Smith and coworkers investigated the oxidation of pyrrole and N-substituted pyrroles using Dess–Martin periodinane (**536**) (Scheme 152; 2013OL1714). For example, treatment of *N*-methylpyrrole (**535**) with 2.5 equivalents of **536** gives the 5-acyloxy-3-pyrrolin-2-one **537**. Reduction of the acyl group was accomplished by treatment with $Et_3SiH$ in the presence of $BF_3 \cdot Et_2O$, which gives *N*-methyl-3-pyrrolin-2-one (**538**). They also studied this oxidation reaction with *N*-(2-hydroxyethyl)pyrrole and found that the pyrrole ring was oxidized preferentially in the presence of a primary alcohol. Gültekin and Sentürk and coworkers studied the oxidation of *N*-tosylpyrrole with phenyliodine bis(trifluoroacetate) (PIFA) (2010BMC4468); treatment with one equivalent of PIFA gives *N*-tosyl-3-pyrrolin-2-one, whereas two equivalents of PIFA gives *N*-tosyl-5-hydroxy-3-pyrrolin-2-one.

Pyrroles undergo photooxidation with singlet oxygen to give a range of oxidized pyrrole products. Reid and de Mayo irradiated a dilute solution of pyrrole (**8**), using a 100 W tunsten lamp in the presence of the photosensitizer eosin, and obtained 5-hydroxy-3-pyrrolin-2-one as the major product (1962CI1576). The Lightner group has done extensive research into photooxidation of simple pyrroles including 3,4-diethylpyrrole (1971CC1099), 3,4-diethyl-2-methylpyrrole (1971TL4417), 2,4-dimethylpyrrole (1972JHC167), *N*-methylpyrrole (1976JA802), *N*-phenylpyrrole (1974JHC1097, and *N*-*tert*-butylpyrrole (1975JOC2724). Additional studies by others have looked at the photooxidation of *N*-alkylpyrroles (1967JOC3330), 2,3,4-trimethylpyrrole (1967TL3041), *N*-phenylpyrrole (1971JOC31), *N*-methylpyrrole (1999JOC 2567), *N*-(2-hydroxyethyl)pyrrole (2009JOC7274), and N-chiral 2-methyl-pyrroles (2002TA601). In most cases, the photooxidation of pyrroles leads to

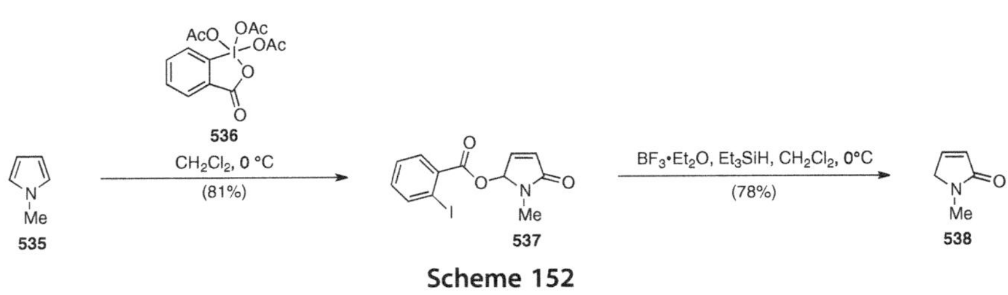


Scheme 152

5-hydroxy-3-pyrrolin-2-ones and/or 5-alkoxy-3-pyrrolin-2-ones. For example, Lightner and coworkers irradiated *N*-methylpyrrole (**535**) in the presence of Rose Bengal and obtained a mixture of 5-hydroxy-1-methyl-3-pyrrolin-2-one (**540**), 5-methoxy-1-methyl-3-pyrrolin-2-one (**541**), and *N*-methylmaleimide (**542**) (Scheme 153; 1976JA802). They studied the reaction using low-temperature NMR and obtained evidence for the formation of the endoperoxide intermediate **539**.

Introduction of different groups (e.g., formyl, halogen, carboalkoxy, and tosyl) at the α-position of the pyrrole ring leads to substrates that can undergo regioselective oxidations to 3-pyrrolin-2-one derivatives. Pyrrole-2-carboxaldehydes are converted into 3-pyrrolin-2-ones with regiospecific loss of the formyl group either by photooxidation (1973JHC273), treatment with $H_2O_2$ (1947MI1, 2000TL2825), or treatment of *m*-CPBA (1988PT(1) 1569, 2014MI1). This transformation has been exploited by Pelkey and coworkers for the preparation of 3,4-dialkylpyrroles (2006JOC6678) and 3,4-diarylpyrroles (2011JOC8203). Given the regioselective nature of the transformation, this strategy provides access to unsymmetrical 3,4-disubstituted pyrroles. For example, treatment of pyrrole-2-carboxaldehyde **543** with $H_2O_2$ in the presence of $NaHCO_3$ gives the 3-pyrrolin-2-one **544** (Scheme 154); the latter compound is the lactam analog of the selective COX-II inhibitor rofecoxib (2011JOC8203). The reaction mechanism involves a Dakin–West-type oxidative rearrangement to the formate ester and subsequent hydrolysis to the 2-hydroxypyrrole, which tautomerizes to the 3-pyrrolin-2-one. Helquist and coworkers (2009JMC6494) and Tsolomitis

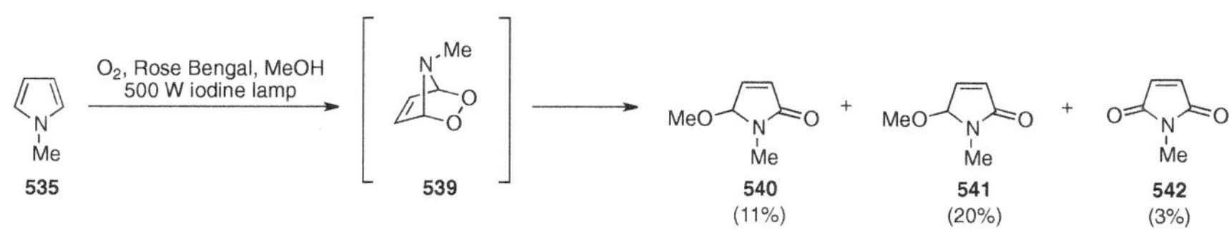


Scheme 153

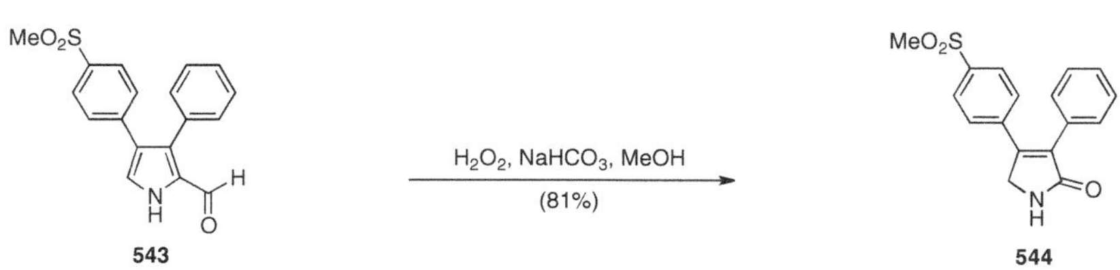


Scheme 154

(2004TL9353) have studied the hydrolysis of 2-acetoxypyrroles leading to 3-pyrrolin-2-ones.

Oxidation of *tert*-butyl pyrrole-2-carboxylates leads to 5-carboalkoxy-3-pyrrolin-2-ones with retention of the carboxylate group. Inomata and coworkers studied this transformation using chloranil as oxidant (2011H(82) 1157). The introduction of iodine provides another class of substrate (2-carboxalkoxy-5-iodopyrroles) that can be regioselectively oxidized to 3-pyrrolin-2-ones (2000BCSJ497, 2005CL800, 2010BCSJ1309). For example, Inomata and coworkers prepared the α-iodopyrrole **546** by treatment of pyrrole-2-carboxylate **545** with *N*-iodosuccinimide (NIS) (Scheme 155; 2010BCSJ1309). Regioselective oxidation of **546** was then accomplished by treatment with $Pb(OAc)_4$, which gives the 5-acetoxy-3-pyrrolin-2-one **547**.

Acid-mediated hydrolysis of 5-halopyrrole-2-carboxylates provides access to 5-carboxy-3-pyrrolin-2-ones, but this transformation often lacks the regioselectivity of other methods. This reaction was first reported by Siedel (1943LAC144). In a more recent example, Battersby and coworkers have treated 5-bromopyrrole-2-carboxylate **548** with sulfuric acid in methanol, and a mixture of 3-pyrrolin-2-ones **549** and **550** was obtained in low yield (Scheme 156; 1988PT(1)1557). These reaction conditions led to both hydrolysis of the bromine and deprotection of the *t*-butyl ester followed by decarboxylation. In a related reaction involving a symmetrical substrate, Lugtenberg and de Groot converted 5-bromo-2-trichloroacetyl-3,4-dimethylpyrrole into 3,4-dimethyl-3-pyrrolin-2-one via treatment with NaOH (1982MI2).

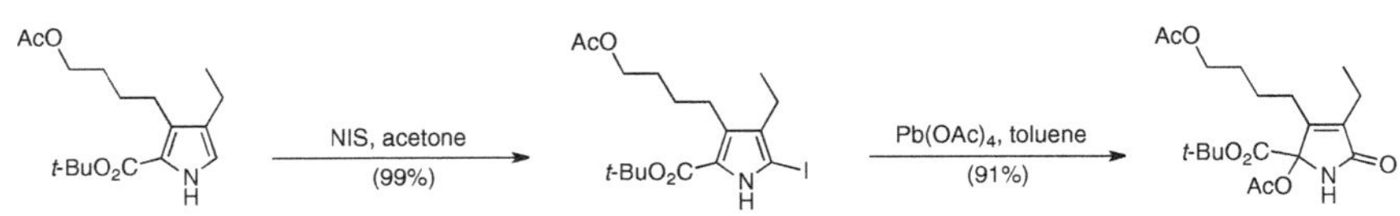


545 546 547

Scheme 155

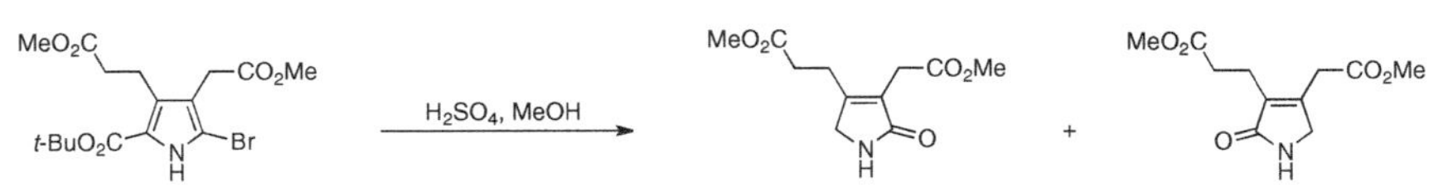


548 549 (13%) 550 (18%)

Scheme 156

In contrast to the previous example (Scheme 156), oxidative hydrolysis of 5-halo-2-tosylpyrroles proceeds with excellent regioselectivity. This transformation has been used extensively by the research groups of Inomata (1993CL1437, 1994BCSJ3088, 1994BCSJ3320, 1998CL455, 1999MI2, 1999SL901, 2001CL590, 2006BCSJ1561) and Lightner (1999JA9253, 2000T7869, 2001JHC527, 2006T7043, 2001MI7, 2001T7813, 2005JHC731, 2005JOC8417, 2005JOC8417, 2014MI1). Inomata and coworkers originally found that the acid-mediated hydrolysis of 2-tosylpyrroles lacked regioselectivity, and they obtained a mixture of regioisomeric 3-pyrrolin-2-ones (1993CL1437). They subsequently developed a three-step procedure for regioselectively converting 2-tosylpyrroles into 3-pyrrolin-2-ones (Scheme 157). Bromination of the 2-tosylpyrrole **551** with $PhN^{+}(CH_3)_3Br_3^{-}$ gives the 5-bromopyrrole **552**; subsequent hydrolysis of **552** with aqueous TFA produced 5-tosyl-3-pyrrolin-2-one **553** as a single regioisomer. The tosyl group can then be removed by reduction of **553** with $NaBH_4$.

The Tietz research group discovered a novel route from pyrroles to 3-pyrrolin-2-ones; they treated a pyrrole-2-methanol with HCl and obtained the corresponding 3-pyrrolin-2-one derivative (1996S851). Couturier and coworkers investigated interesting reactions involving the conversion of 2,3-halopyrrole derivatives into spirocyclic 3-pyrrolin-2-ones under acidic conditions (2012BMCL6292).

Finally, the most well-studied transformation of pyrroles into 3-pyrrolin-2-ones involves the vinylogous aldol addition of 2-silyloxypyrroles (e.g., **62**). This transformation, developed by the group of Casiraghi and Rassu (1992JOC3760), has been reviewed extensively (1995S607, 1997JMC168, 1999SL1333, 2000CR1929, 2000CSR109, 2009SL1525, 2011CR3076, 2014T2595). Pettus and coworkers have developed a diastereoselective aldol reaction involving silyloxypyrrole intermediates (Scheme 158; 2014OL4384). Treatment of methyl tetramate **555** with TMSOTf and $Et_3N$ generated silyloxypyrrole **556,** which upon treatment with isobutyraldehyde **557** and $SnCl_4$ gives the aldol adduct **558** with high *syn* stereoselectivity.

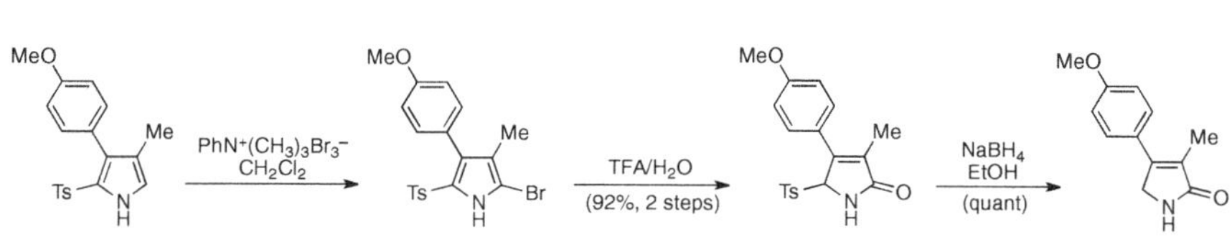


Scheme 157

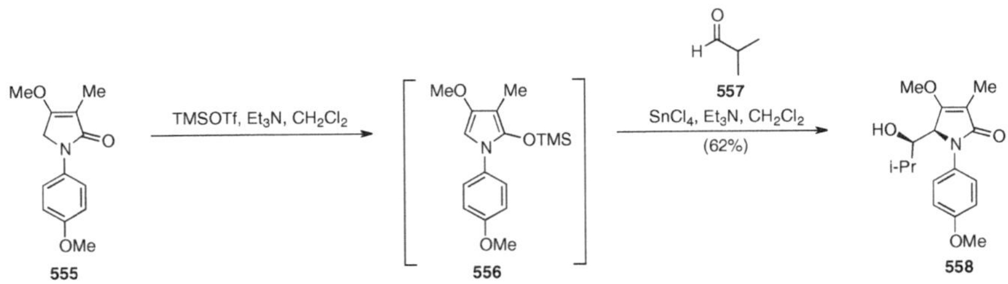


Scheme 158

## 5.7 Synthesis from Other Nitrogen Heterocycles

Other nitrogen heterocycles can serve as precursors to 3-pyrrolin-2-ones. Dejaegher and De Kimpe transformed azetidinones into 3-pyrrolin-2-ones (Scheme 159; 2004JOC5974). They treated the 1-allyl-2-azetidinone **559** with sodium methoxide in methanol to obtain a mixture of cyclized (**561**) and ring-opened (**560**) products. After flash chromatography only 3-pyrrolin-2-one **561** was present.

In 2013 Kise and colleagues reported a reductive coupling between aliphatic cyclic imides, including succinimides, with benzophenones to produce 3-pyrrolin-2-ones along with other cyclic and acyclic products (2013TL6944). Kise employed low-valent titanium, generated from zinc titanium tetrachloride, followed by reflux in benzene-containing catalytic *para*-toluenesulfonic acid to afford the corresponding 5-alkylidene-3-pyrrolin-2-ones. Wijnberg and coworkers demonstrated similar reactivity of succinimides through a regioselective reduction of 3,3-disubstituted *N*-methyl succinimides using sodium borohydride (1974TL4073). The bulkier the substituents at C-3, the more reduction occurred at C-2 than C-5, thus providing one major product, that is, the 3,3-disubstituted 5-hydroxy-3-pyrrolin-2-one.

In an attempt to synthesize a radiolabelled neurochemical compound from pyrrolidines, Vader and colleagues employed bromination using bromine in acetic acid to afford a mixture of compounds including 3-pyrrolin-2-ones (1994MI1).

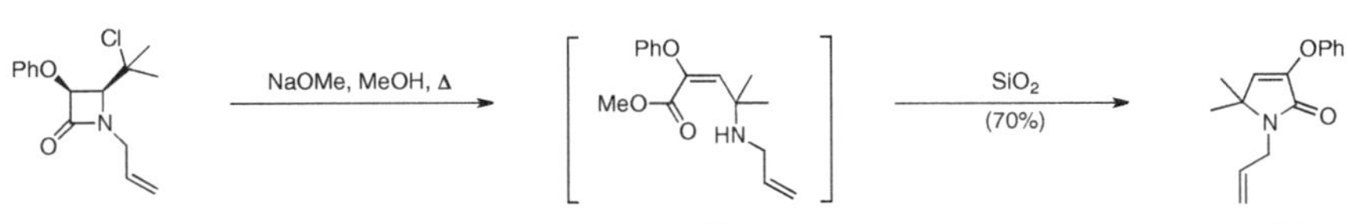


Scheme 159

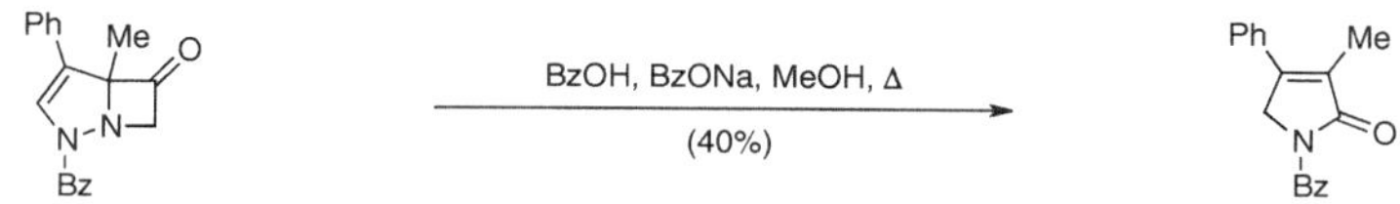

562 563

Scheme 160

Moore and coworkers demonstrated that azabicyclic systems can also be converted to 3-pyrrolin-2-ones (e.g., **562** → **563**), using methanol and benzoic acid with sodium benzoate (Scheme 160; 1962JA3022).

Diketopiperazines can undergo ring contractions to give various five-membered nitrogen heterocycles including 4-hydroxy-3-pyrrolin-2-ones (tetramic acids). Farran and colleagues treated Boc-activated 2,5-diketopiperazines **564** with sodium *tert*-butoxide in THF to produce tetramic acid **565** via stereocontrolled transannular rearrangement, similar to an aza-version of the Chan rearrangement (Scheme 161; 2008OBC3989). Three more steps allowed access to the synthetically challenging *spiro*-derivatives **566**. Farran demonstrated the application of this methodology to the synthesis of biologically-active compounds including dipeptide mimetics, statine derivatives, and $^{15}$N-labeled compounds (2008TL906). Lorenz and coworkers approached a similar synthesis using hermidin quinone with ethanol or methanol, water, and benzilic acid (2014HCA1606). The quinone underwent a spontaneous rearrangement and esterification to afford 5-hydroxy-4-methoxy-5-(alkoxycarbonyl)-1-methyl-3-pyrrolin-2-one.

Selective reduction and ring contraction of pyridazin-3-ones provide access to 3-pyrrolin-2-ones. Brown and coworkers explored the reduction of pyridazin-3-ones with zinc dust and acetic acid to give reduced products which underwent ring contractions, thus offering 3-pyrrolin-2-ones and 2-pyrrolidones (1984CC1373). Gewald, Hain, and Abdelrazek and coworkers employed a similar approach to 3,4,5-trisubstituted 3-pyrrolin-2-ones (1984S62, 2001T1813).

Oxazolium salts can undergo rearrangement in basic conditions to access 3-pyrrolin-2-ones. In 1997, Fisyuk and Vorontsova reported that treating 2-benzyl-3-ethyl-5-methyloxazolium tetrafluoroborate **567** with potassium

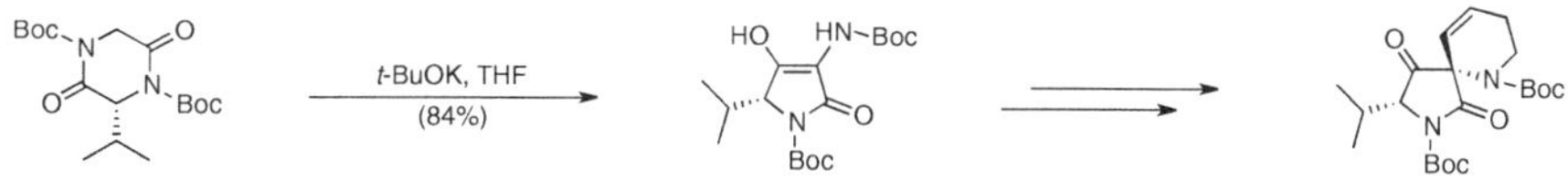

564 565 (>99% ee) 566 (>95% de)

Scheme 161

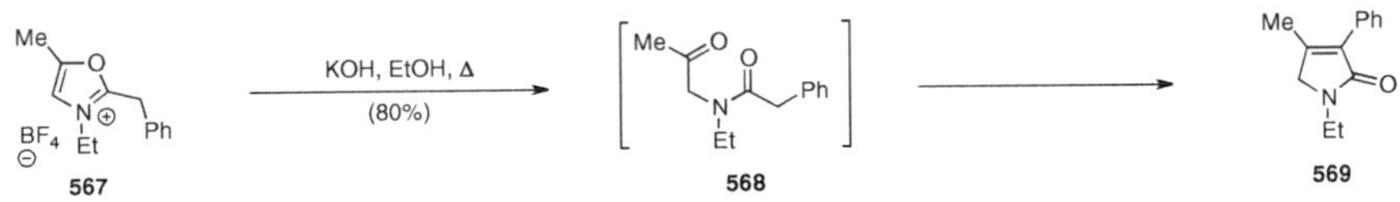

Scheme 162

hydroxide gives the corresponding 3-pyrrolin-2-one **569** via ring-opened intermediate **568** (Scheme 162; 1997MI1). Dillard and Easton demonstrated a comparable reaction with alkyl-substituted **567** (1966JOC2580). Poschenrieder and coworkers employed this rearrangement en route to tetramic acid isomers, which have shown potential as *N*-methyl-D-aspartate receptor ligands (1998MI1).

3-Pyrrolin-2-ones can be reached via cycloadditions of oxazoles. Bonne and colleagues employed a one-pot multicomponent cycloaddition using morpholine, cyclohexanecarboxaldehyde, and methyl *p*-nitrophenylisocyanoacetate to provide 5-methoxyoxazole which upon further heating in toluene and triethylamine with 3-phenylprop-2-ynoyl afforded the corresponding 3-pyrrolin-2-one (2007ACIE2485).

Isoxazoles have been converted to 3-pyrrolin-2-one derivatives by refluxing with sodium methoxide in methanol as shown by Coskun and Cetin (2009T648). Through imine formation followed by nucleophilic attack with methoxide, the isoxazole N—O bond is cleaved and cyclization affords the new 3-pyrrolin-2-one structure.

Cyclic sulphinylation of 2-arylated *N*-alkoxy-2-vinylglycolamides provides a range of 3-arylated 3-pyrrolin-2-one products in good yield. Geffken, Haerting, and Frobose carried out this reaction with freshly prepared 2,2-dipyridyl sulfite in dichloromethane (1997MI2). The 2-arylated *N*-alkoxy-2-vinylglycoamides quickly form cyclic sulfurous ester amide intermediates, which rearrange at room temperature expelling sulfur dioxide to afford 1-alkoxy-3-aryl-3-pyrrolin-2-ones.

Finally, Jones and coworkers reported that fused isoxazoles are useful precursors to 4-amino-3-pyrrolin-2-ones (Scheme 163; 2014CC1588). Hydrogenation of pyrrolo[3,4-*c*]isoxazole **570** gives 3-pyrrolin-2-one **571**.

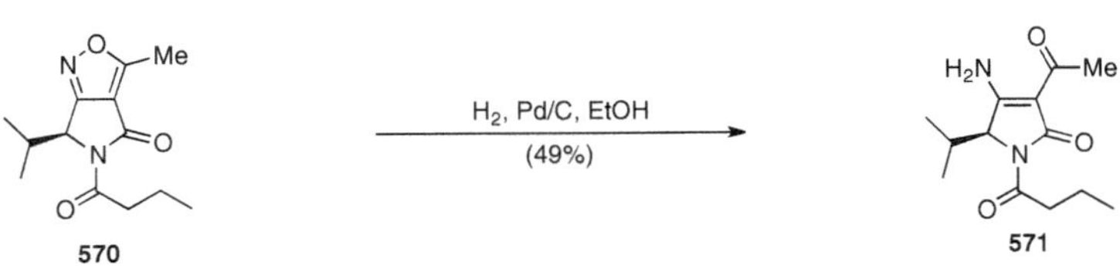

Scheme 163

## 5.8 Synthesis from Oxygen Heterocycles

Dialkoxydihydrofurans (e.g., **10**) can undergo addition and elimination reactions with primary amines to give 3-pyrrolin-2-ones. Baussanne and coworkers found that condensing dimethoxydihydrofuran **10** with (*R*)-(−)-phenylglycinol **572** afforded **574** when treated with HCl at room temperature and **573** when heated under reflux with acetic acid (Scheme 164; 1994TL3931, 1998TA797). This strategy has been used to prepare a variety of *N*-substituted 3-pyrrolin-2-ones (2002ARK(i)61, 2011TL1253, 2011ACIE5682). In 2000, Bella and colleagues treated 2,5-dimethoxy-3-phenylselenyl-(2,5*H*)-furan with benzylamine to give products similar to **574** (2000TL3669). Alves has shown that condensation works with benzylamine and chiral amines (2007MI2). In a review, Baussanne showed this transformation to be rather versatile as it can tolerate numerous chiral amines. This reaction has also been used for the preparation of complex polycyclic structures including an enantioselective synthesis of (−)-cephalotaxine (2006ARK(vii)57). Under acidic conditions, Saygili and coworkers formed analogs of **574** by condensing amino esters with **10** to give the corresponding pyrrolinones in good yield (Scheme 164; 2006MI1). The condensation of **10** with amines or amino esters has shown application in the preparation of complex natural products and RGD peptidomimetics. Halie and colleagues condensed **10** with racemic allylglycine under acidic conditions to afford the precursor for peptidomimetic products whose properties as cell adhesion inhibitors were evaluated in vitro (2009T1402). In 2010, Perard-Viret and coworkers developed a convergent, three-step synthesis of (±)-jamtine via the condensation of 2-(3,4-dimethoxyphenyl)ethylamine with dimethyldihydrofuran methyl ester (2010TL96).

In a similar fashion, 5-alkoxy-butyrolactones can undergo addition and elimination to give substituted 3-pyrrolin-2-ones. In 1973, Farina and coworkers found that 5-methoxyfuran-2-one **575** treated with aqueous ammonia gives 5-hydroxy-3-pyrrolin-2-one **576** (Scheme 165; 1973S167). Ten years later, Farina and coworkers showed that primary amines, such as benzylamine, give 5-hydroxy-3-pyrrolin-2-ones when stirred with 2-chloro-5-methoxyfuran-2-one (1983H(20)1761). The transformation

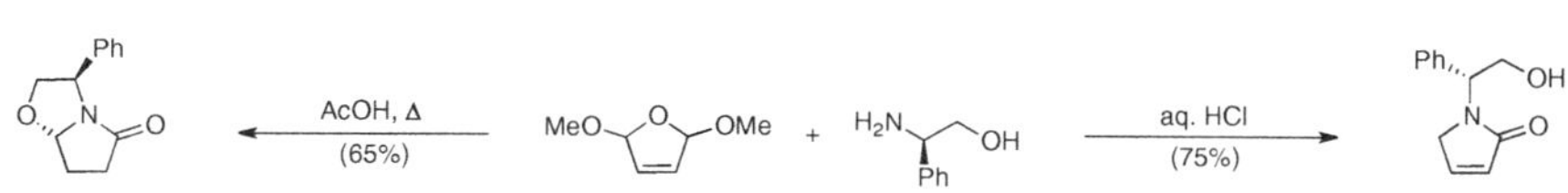


**Scheme 164**

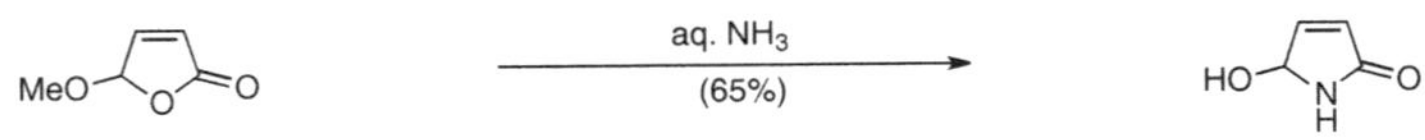


Scheme 165

proceeded with various 4-substituted-5-methoxyfuran-2-ones, with 4-methyl giving the highest yield (1984H(22)1733). Kosugi and Hamaguchi employed Farina's methodology with 5-ethoxyfuran-2-ones, which then underwent 1,3-dipolar cycloadditions with azides to give the corresponding triazolines (1984H(22)2363). This type of ammonolysis can work with 4-methoxybenzylamine, as shown by Inomata and coworkers. They used this step in the preparation of the A- and D-rings of phycocyanobilin (1995CL713). Moore and coworkers also investigated this chemistry en route to fused azetidinones (1976JA3728, 1981JA1769) and Cuiper and colleagues have examined the chiral applications of this chemistry. The addition of ammonia to racemic 5-methoxyfuran-2-one provides the corresponding racemic 5-hydroxy-3-pyrrolin-2-one, which can then be converted to the (+)-enantiomer in >99% ee using an enzymatic transesterification with a lipase, *Candida antarctica* (1999JOC2567).

5-Hydroxyfuran-2-ones can also undergo addition–elimination reactions with amines. Shiraki and coworkers employed this addition reaction in their total synthesis of a natural platelet aggregation-inhibiting γ-lactam PI-091 (**39**) (Scheme 166; 1996JOC2845). By treating **577** with ammonia and catalytic sodium cyanide in a sealed tube they obtained **578**.

Zhang and colleagues demonstrated a reductive amination using various amines and anilines with mucochloric acid **579** to give N-substituted 3,4-dichloro-3-pyrrolin-2-ones (e.g., **580**; Scheme 167; 2003OL553). Zhang employed aryl, alkyl, and chiral amines. Das Sarma and coworkers showed that chiral amino acid esters, amides, and short peptides also worked in the same reductive amination reaction (2006EJOC3730). From this development, after a simple dehalogenation/hydrogenation, they were able to synthesize the antiepileptic drug keppra®.

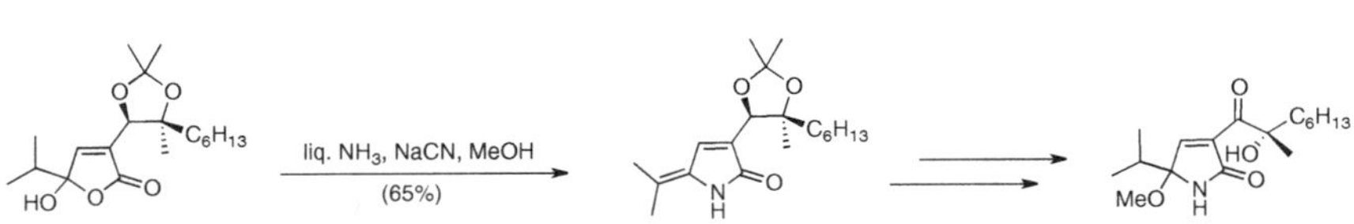


Scheme 166

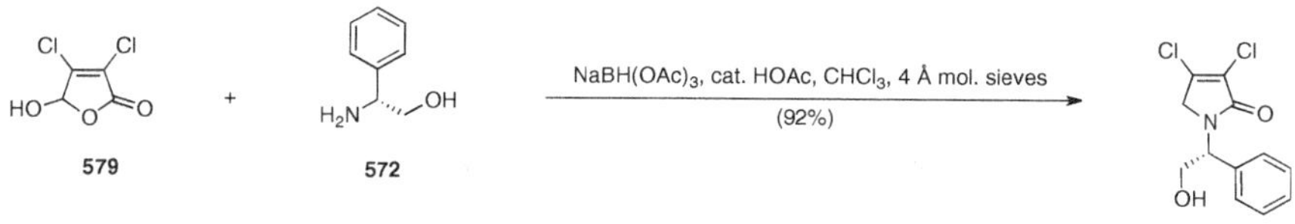


579 572 580

Scheme 167

In 1977, Wooldridge and Lightner synthesized alkyl-substituted 5-hydroxy-3-pyrrolin-2-ones from 5-chlorofuran-2-ones and ammonia (1977JHC1283). Liquid ammonia was stirred with **581** at room temperature in a sealed reaction bomb to give gram quantities of product **582** (Scheme 168). Duczek and colleagues investigated a similar transformation using a more complex primary amine as part of a chiral β-lactam to afford racemic 3-pyrrolin-2-one (1992LAC781).

Addition and elimination of 5-unsubstituted furan-2-ones has been shown with primary amines. In 1960, while developing a synthesis of amino derivatives of strophanthidin, Bembry and coworkers speculated that 4-phenylfuran-2-one treated with propylamine would give the corresponding N-substituted 3-pyrrolin-2-one (1960JOC1175). Vezouet and colleagues later showed that a similar furan-2-one **583** mixed with methylamine and hydrochloric acid produced the 3-pyrrolin-2-one **584** (Scheme 169; 2006T12252).

In exploring reactions with 3,4-diphenylfuran-2-ones, Rio and Serkiz synthesized 3,4-diphenyl-3-pyrrolin-2-ones from 3,4-diphenylfuran-2-one and ammonium acetate in benzene within a sealed tube (1976MI1). Egorova and coworkers showed a similar transformation with 5-alkyl(aryl)-3*H*-furan-2-ones and primary arylamines in xylene at 120 °C. Using this chemistry,

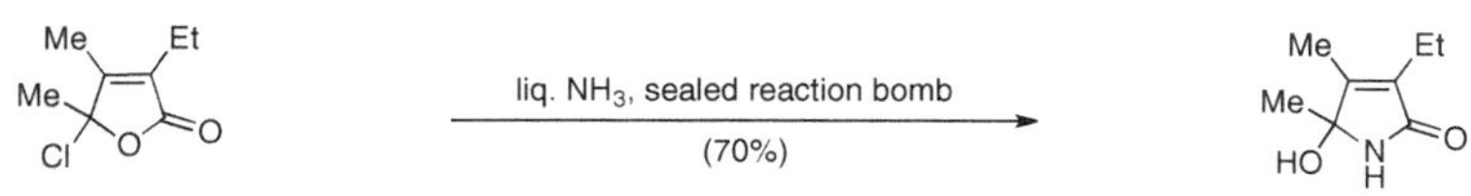


581 582

Scheme 168

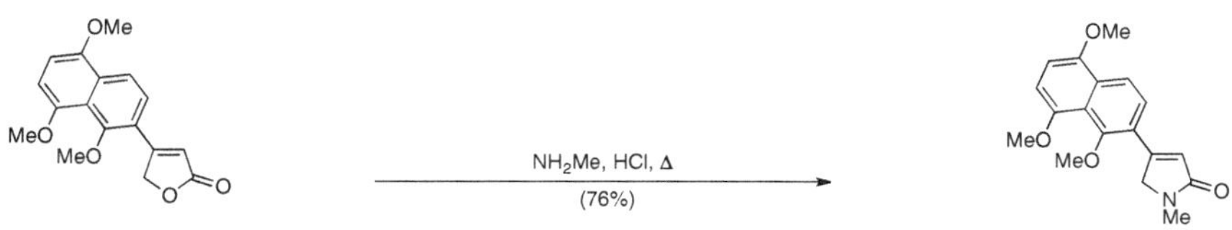


583 584

Scheme 169

Egorova produced mixtures of 1-aryl-5-alkyl(aryl)-*3H*-pyrrolin-2-one and 1-aryl-5-alkyl(aryl)-5*H*-pyrrolin-2-one isomers with the 3*H* isomer as the major product (2001MI4). A few 4-oxoalkanamide intermediates were isolated, and it was shown that the ratio of products was substrate dependent.

HIV-1 integrase inhibitors stop the insertion of viral DNA into the genome of healthy host cells. Melamed and colleagues realized the value of these integrase inhibitors and focused on preparing 4-substituted pyrrolidinones, which have shown activity against viral replication (2008BMCL5307). Furan-2-ones were treated with methylamine in methanol at 70 °C to give racemic aminopyrrolidinones, which were heated in TFA for 2 h to produce 1-methyl-4-aryl-3*H*- and 1-methyl-4-aryl-5*H*-pyrrolin-2-ones.

Linear tetrapyrroles are of interest due to their structural similarities with biliproteins, which are naturally-occurring chromophores. Tetrapyrroles are often found in various types of algae and act as light-harvesting proteins in photosynthesis. Thus, Jacobi and coworkers developed a new strategy for the synthesis of tetrapyrroles, known as the "B, C + D + A" approach (2000JOC205). Final steps involve the addition–elimination of 5-alkylidene-substituted furan-2-one as part of an oligopyrrole. Upon the addition of liquid ammonia at −33 °C followed by acid-catalyzed cyclodehydration, **585** afforded the 3-pyrrolin-2-one **586** (Scheme 170).

Haase and Langer employed a similar approach in the formal synthesis of pukeleimide A (2005SL453). 5-Alkylidene-substituted furan-2-one **587** stirred with ammonium acetate and acetic acid provides the corresponding 3-pyrrolin-2-ones **588** and **589** as a separable mixture of *E*/*Z* isomers (Scheme 171). In 2009, Haase and Langer showed more examples of this chemistry using 5-alkylidene-substituted furan-2-ones and alkyl- or arylamines. The ratio of *E*/*Z*-isomers is dependent on N-substitution (2009T4530).

Goh and colleagues converted fimbrolides to 1,5-dihydropyrrol-2-ones via addition of arylamines and dehydration using *p*-TsOH (2007TL2287). The resulting *Z*-isomers of 3-pyrrolin-2-ones demonstrated potent antagonist activity on Gram-negative bacteria. Felluga and coworkers employed

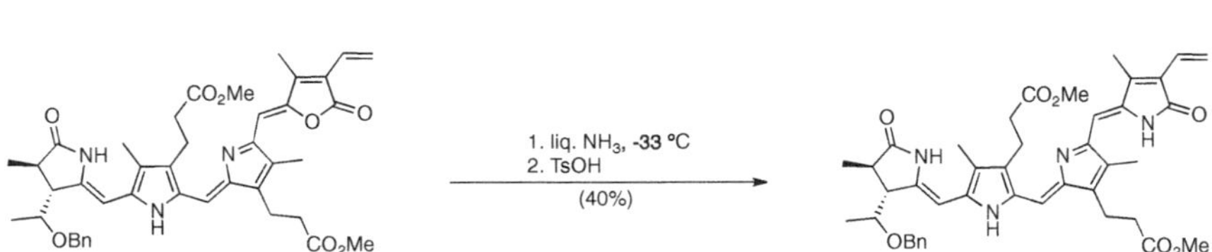


**Scheme 170**

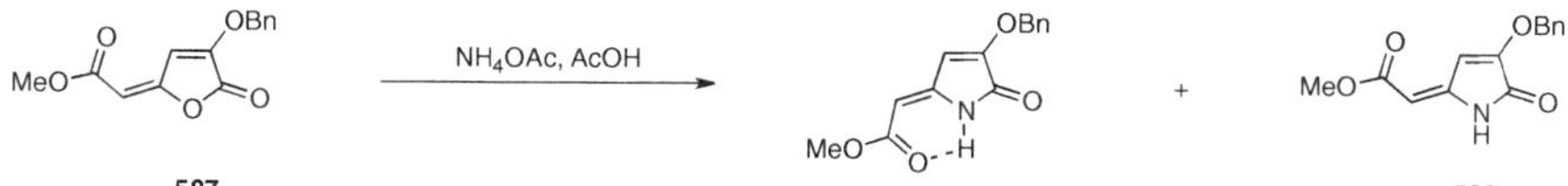


587

588
(55%)

589
(22%)

**Scheme 171**

this strategy in their efficient five-stage synthesis of pulchellalactam, a protein tyrosine phosphatase inhibitor. Pulchellalactone was treated with aqueous ammonia followed by addition of 10% hydrochloric acid to give (*Z*)-pulchellalactam (2007S1882).

Gupton and coworkers developed a six-step stereoselective synthesis of *N*-substituted 5-benzylidene-4-arylpyrrol-2-ones (2010T9113). Within this sequence, Gupton treated a variety of (*Z*)-5-benzylidene-4-arylfuran-2-ones with alkyl- and arylamines in methylene chloride at room temperature to afford corresponding 5-benzyl-5-hydroxy-4-arylpyrrol-2-ones.

Chatzimpaloglou and colleagues employed this approach for a total synthesis of phaeosphaeride A, a natural antitumor compound (2012JOC9659). They stirred 5-alkylidene-substituted furan-2-one with methoxyamine hydrochloide and LiHMDS in dry THF at −78 °C for 2 h to give the resulting 3-pyrrolin-2-one.

In a new route to antimalarial compounds, Kanishchev and coworkers used a ring-opening—ring-closure process to transform γ-lactones to 3-pyrrolin-2-ones (2013BMCL6167). 5-Alkylidene-substituted 3-(2,2,2-trifluoroethyl)furan-2-ones were stirred with aminoquinoline derivatives in THF and MeOH at room temperature affording the corresponding 3-pyrrolin-2-ones. These structures were then tested for antimalarial activity. Cornut and colleagues also explored synthetic routes to antimalarial compounds (2013JMC73). Using a similar approach as Kanishchev, they showed success with a wide range of amines.

Starting from 3,4-disubstituted maleic anhydride, Klopfleisch and coworkers synthesized the bilirubin oxidation end-product BOX A and isomers in four or six steps (2013OL4608). Within this synthesis they transformed 5-alkylidene-furan-2-ones to the corresponding substituted 3-pyrrolin-2-ones using ammonium acetate and glacial acetic acid. Seidel and coworkers employed a similar approach to BOX B analogs (2014TL6526).

Pereira and colleagues found that isobutylamine and highly substituted 5-alkylidene-furan-2-ones in methylene chloride at 0 °C provided *N*-isobutyl-3-pyrrolin-2-ones (2014EJMC127). Some of these products inhibited planktonic growth and biofilm formation better than previously shown in the literature.

Perfluorinated heterocyclic compounds are of interest due to their synthetic uses and bioactivity. Bouillon and coworkers transformed fluorinated furans into substituted pyrroles and 3-pyrrolin-2-ones (2004S711). 2-Perfluoroalkyl furans were treated with benzylamine in ether at room temperature for 16 h to provide a mixture of *N*-benzyl-3-pyrrolin-2-one diastereomers.

3-Pyrrolin-2-ones can be formed from rearrangements of 2-aminofurans. Ito and Yakushijin were among the first to show this type of rearrangement mediated by photocycloaddition of oxygen (1978H(9)1603). Yakushijin used 3,4-diphenyl-2-furalcarbamoyls in benzene at room temperature to provide 3,4-diphenyl-5-hydroxy-3-pyrrolin-2-ones (1980CPB2178). Yakushijin also applied this reaction to methyl-substituted 2-aminofurans (1980H(14)1073) and later examined the mechanism (1986CPB2049). During these studies of ring transformations of 2-methylfurylcarbamates, the revised structure of jatropham (**38**), an antitumor alkaloid, was elucidated. In 1981, Yakushijin published a synthesis of jatropham (**38**) using an photooxidation of 3-methyl-2-furylcarbamate **590** followed by Boc-removal from the product **591** using TFA (Scheme 172; 1981H(16)1157). Farina and colleagues have reported an alternative route to jatropham (**38**) that employs an oxidation of 3-methyl-2-furoic acid in methanol followed by addition of ammonia (1984H(22)1733). Furukawa and coworkers have applied the same photocycloaddition mechanism to the synthesis of nucleosides from glycosylfurans (1987JOC2368).

Padwa and coworkers realized the synthesis of 2,4-disubstituted pyrroles using rearrangements of 2-furanyl carbamates via 5-methoxy-3-pyrrolin-2-one intermediates. 2-Furanyl carbamate **593** was treated with iodine and sodium bicarbonate followed by methyl iodide and silver oxide to give the *N*-substituted 5-methoxy-3-pyrrolin-2-one **595** via the iodinated intermediate **594** (Scheme 173; 2009OL1233). Padwa employed this method to synthesize 5-methoxypyrrol-2-ones, which were then treated with alkyl lithiates to give tricyclic products (2009T6720).

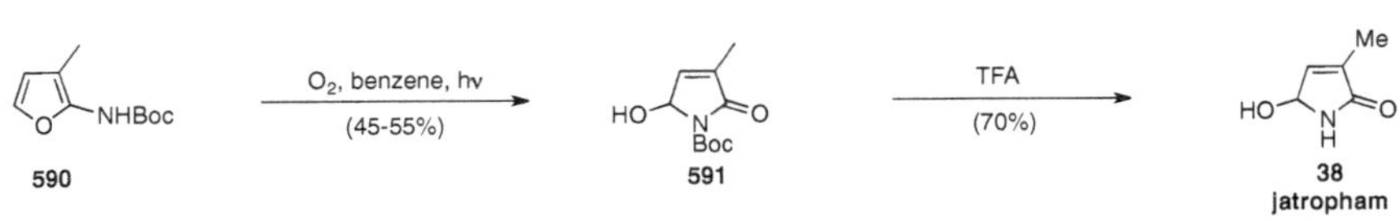


**Scheme 172**

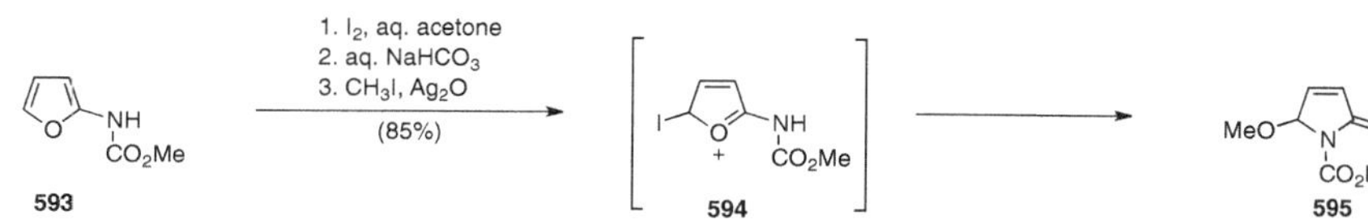

Scheme 173

Buehrdel has demonstrated that a rearrangement is possible from 2-iminolactone to form 5-alkylidene-substituted 3-pyrrolin-2-ones upon treatment with TFA (2008JHC845).

## ACKNOWLEDGMENTS

E.T.P. acknowledges the support of our research program directed at the synthesis and reactions of 3-pyrrolin-2-ones by the National Science Foundation (RUI: 1362183).

## REFERENCES


| | |
|---|---|
| 1932MI1 | H. Fischer, T. Yoskioka, and P. Hartmann, *Z. Physiol. Chem.*, **212**, 146 (1932). |
| 1935JA178 | J.T. Bashour and H.G. Lindwall, *J. Am. Chem. Soc.*, **57**, 178 (1935). |
| 1937LAC265 | H. Fischer and W. Lautsch, *Liebigs Ann. Chem.*, **528**, 265 (1937). |
| 1942MI1 | H. Fischer and H. Plieninger, *Z. Physiol. Chem.*, **274**, 231 (1942). |
| 1943LAC144 | W. Siedel, *Liebigs Ann. Chem.*, **554**, 144 (1943). |
| 1947MI1 | A. Hüni and F. Frank, *Z. Physiol. Chem.*, **282**, 96 (1947). |
| 1949HCA2010 | C.A. Grob and P. Ankli, *Helv. Chim. Acta*, **32**, 2010 (1949). |
| 1949JA2532 | P.L. Southwick and L.L. Seivard, *J. Am. Chem. Soc.*, **71**, 2532 (1949). |
| 1950JA1236 | J.A. King and F.H. McMillan, *J. Am. Chem. Soc.*, **72**, 1236 (1950). |
| 1951CB526 | W. Langenbeck and H. Boser, *Chem. Ber.*, **84**, 526 (1951). |
| 1952JA6304 | W.D. Celmer, F.W. Tanner, M. Harfenist, T.M. Lees, and I.A. Solomons, *J. Am. Chem. Soc.*, **74**, 6304 (1952). |
| 1953JA3413 | P.L. Southwick and R.T. Crouch, *J. Am. Chem. Soc.*, **75**, 3413 (1953). |
| 1954JCS850 | R.N. Lacey, *J. Chem. Soc.*, 850 (1954). |
| 1955JA2861 | W.D. Celmer and I.A. Solomons, *J. Am. Chem. Soc.*, **77**, 2861 (1955). |
| 1956JOC1087 | P.L. Southwick, E.P. Previc, J. Casanova, and E.H. Carlson, *J. Org. Chem.*, **21**, 1087 (1956). |
| 1956LAC198 | H. Plieninger and M. Decker, *Liebigs Ann. Chem.*, **598**, 198 (1956). |
| 1957JAB195 | H. Yamaguchi, *J. Antibiot.*, **10**, 195 (1957). |
| 1959HCA563 | v. L. Ettlinger, E. Gäumann, R. Hütter, W. Keller-Schierlein, F. Kradolfer, L. Neipp, V. Prelog, and H. Zahner, *Helv. Chim. Acta*, **42**, 563 (1959). |
| 1959JA4355 | E.G. Howard, R.V. Lindsey, and C.W. Theobald, *J. Am. Chem. Soc.*, **81**, 4355 (1959). |
| 1960JOC1175 | T.H. Bembry, R.C. Elderfield, and G.L. Krueger, *J. Org. Chem.*, **25**, 1175 (1960). |
| 1962ACIE265 | U. Schmidt and F. Geiger, *Angew. Chem. Int. Ed.*, **1**, 265 (1962). |
| 1962CI1576 | P. de Mayo and S.T. Reid, *Chem. Ind.*, 1576 (1962). |
| 1962JA3022 | J.A. Moore, F.J. Marascia, R.W. Medeiros, and E. Wyss, *J. Am. Chem. Soc.*, **84**, 3022 (1962). |

1962JOC3075 P.L. Southwick and J.A. Vida, *J. Org. Chem.*, **27**, 3075 (1962).
1962LAC165 H. Plieninger, H. Bauer, A.R. Katritzky, and U. Lerch, *Liebigs Ann. Chem.*, **654**, 165 (1962).
1963JA647 G. Büchi and G. Lukas, *J. Am. Chem. Soc.*, **85**, 647 (1963).
1963JA1430 D.J. Cram, O. Theander, H. Jager, and M.K. Stanfield, *J. Am. Chem. Soc.*, **85**, 1430 (1963).
1963LAC168 U. Schmidt and F. Geiger, *Liebigs Ann. Chem.*, **664**, 168 (1963).
1964JA5654 G. Büchi and G. Lukas, *J. Am. Chem. Soc.*, **86**, 5654 (1964).
1964JCS5999 J.H. Atkinson, R.S. Atkinson, and A.W. Johnson, *J. Chem. Soc.*, 5999 (1964).
1964LAC60 H. Plieninger and J. Kurze, *Liebigs Ann. Chem.*, **680**, 60 (1964).
1965JOC3824 J. Bordner and H. Rapoport, *J. Org. Chem.*, **30**, 3824 (1965).
1966JOC2580 R.D. Dillard and N.R. Easton, *J. Org. Chem.*, **31**, 2580 (1966).
1967JOC3330 E.B. Smith and H.B. Jensen, *J. Org. Chem.*, **32**, 3330 (1967).
1967T737 V. Bocchi, L. Chierici, and G.P. Gardini, *Tetrahedron*, **23**, 737 (1967).
1967TL3041 E. Höft, A.R. Katritzky, and M.R. Nesbit, *Tetrahedron Lett.*, **32**, 3041 (1967).
1969OPPI271 V. Bocchi and G.P. Gardini, *Org. Prep. Proc. Int.*, **1**, 271 (1969).
1970CC445 G. Stork and R. Matthews, *J. Chem. Soc. Chem. Commun.*, 445 (1970).
1970JAB271 H. Fujimoto, T. Kinoshita, H. Suzuki, and H. Umezawa, *J. Antibiot.*, **23**, 271 (1970).
1970T4073 V. Bocchi, L. Chierici, G.P. Gardini, and R. Mondelli, *Tetrahedron*, **26**, 4073 (1970).
1971CC1099 G.B. Quistad and D.A. Lightner, *Chem. Commun.*, 1099 (1971).
1971JOC31 R.W. Frank and J. Auerbach, *J. Org. Chem.*, **36**, 31 (1971).
1971MI1 R. Mondelli, V. Bocchi, G.P. Gardini, and L. Chierici, *Org. Magn. Reson.*, **3**, 7 (1971).
1971TL4417 G.B. Quistad and D.A. Lightner, *Tetrahedron Lett.*, **36**, 4417 (1971).
1972BSCF4598 R. Guy and D. Masure, *Bull. Soc. Chim. Fr.*, 4598 (1972).
1972JHC167 D.A. Lightner and L.K. Low, *J. Heterocycl. Chem.*, **9**, 167 (1972).
1973JHC273 D.A. Lightner and G.B. Quistad, *J. Heterocycl. Chem.*, **10**, 273 (1973).
1973MI1 R.M. Wiedhopf, E.R. Trumbull, and J.R. Cole, *J. Pharm. Sci.*, **62**, 1206 (1973).
1973S167 F. Fariña, M.V. Martín, and M.C. Paredes, *Synthesis*, 167 (1973).
1974BCSJ1484 K. Hagio and N. Yoneda, *Bull. Chem. Soc. Jpn.*, **47**, 1484 (1974).
1974JAB897 H. Sakakibara, H. Naganawa, M. Ohno, M. Maeda, and H. Umezawa, *J. Antibiot.*, **27**, 897 (1974).
1974JHC1097 D.A. Lightner, D.I. Kirk, and R.D. Norris, *J. Heterocycl. Chem.*, **11**, 1097 (1974).
1974TL4073 J.B.P.A. Wijnberg, W.N. Speckamp, and H.E. Schoemaker, *Tetrahedron Lett.*, **15**, 4073 (1974).
1975CC121 B.W. Bycroft and R. Pinchin, *J. Chem. Soc. Chem. Commun.*, 121 (1975).
1975JAB286 H.A. Kirst, E.F. Szymanski, D.E. Dorman, J.L. Occolowitz, N.D. Jones, M.O. Chaney, R.L. Hamill, and M.M. Hoehn, *J. Antibiot.*, **28**, 286 (1975).
1975JOC2724 D.A. Lightner and C.-S. Pak, *J. Org. Chem.*, **40**, 2724 (1975).
1975T1437 J.C. Hubert, J.B.P.A. Wunberg, and W.N. Speckamp, *Tetrahedron*, **31**, 1437 (1975).
1976BCSJ3341 J. Ide and Y. Yura, *Bull. Chem. Soc. Jpn.*, **49**, 3341 (1976).
1976JA802 D.A. Lightner, G.S. Bisacchi, and R.D. Norris, *J. Am. Chem. Soc.*, **98**, 802 (1976).

1976JA3728 H.W. Moore, L. Hernandez, and A. Sing, *J. Am. Chem. Soc.*, **98**, 3728 (1976).

1976JOC3587 P.A. Zoretic and P. Soja, *J. Org. Chem.*, **41**, 3587 (1976).

1976MI1 G. Rio and B. Serkiz, *Bull. Soc. Chim. Fr.*, 1491 (1976).

1976S335 D.A. Lightner, G.B. Quistad, and C.S. Pak, *Synthesis*, 335 (1976).

1977HCA660 V.W. Hofheinz and W.E. Oberhänsli, *Helv. Chim. Acta*, **60**, 660 (1977).

1977JAB275 S. Omura, Y. Iwai, A. Hirano, A. Nakagawa, J. Awaya, H. Tsuchiya, Y. Takahashi, and R. Masuma, *J. Antibiot.*, **30**, 275 (1977).

1977JAB334 K. Okamura, K. Soga, Y. Shimauchi, T. Ishikura, and J. Lein, *J. Antibiot.*, **30**, 334 (1977).

1977JHC681 P.A. Zoretic and P. Soja, *J. Heterocycl. Chem.*, **14**, 681 (1977).

1977JHC1283 T.A. Wooldridge and D.A. Lightner, *J. Heterocycl. Chem.*, **14**, 1283 (1977).

1977JOC2219 M.W. Moon, *J. Org. Chem.*, **42**, 2219 (1977).

1977JOC2891 J.E. Ellis, J.H. Fried, I.T. Harrison, E. Rapp, and C.H. Ross, *J. Org. Chem.*, **42**, 2891 (1977).

1978CB486 A. Gossauer and J.-P. Weller, *Chem. Ber.*, **111**, 486 (1978).

1978H(9)1603 K. Ito and K. Yakushijin, *Heterocycles*, **9**, 1603 (1978).

1978HCA977 T.W. Güntert, H.H.A. Linde, M.S. Ragab, and S. Spengel, *Helv. Chim. Acta*, **61**, 977 (1978).

1978JA4225 V.J. Lee, A.R. Branfinan, T.R. Herrin, and K.L. Rinehart, *J. Am. Chem. Soc.*, **100**, 4225 (1978).

1978JA5928 A. Gossauer and J.-P. Weller, *J. Am. Chem. Soc.*, **100**, 5928 (1978).

1978JOC283 A. Gossauer and R.-P. Hinze, *J. Org. Chem.*, **43**, 283 (1978).

1978MI1 M. Pinza and G. Pifferi, *Farm. Ed. Sci.*, **33**, 130 (1978).

1978PT(1)795 L. Vasvári-Debreczy, I. Hermecz, Z. Mészáros, A. Horváth, and P. Simon-Párkányi, *J. Chem. Soc. Perkin Trans. 1*, 795 (1978).

1978S614 R. Pellegata, M. Pinza, and G. Pifferi, *Synthesis*, 614 (1978).

1978T179 J.B.P.A. Wunberg, H.E. Schoemaker, and W.N. Speckamp, *Tetrahedron*, **34**, 179 (1978).

1979HCA872 W. Kunz and H. Kristinsson, *Helv. Chim. Acta*, **62**, 872 (1979).

1979JA240 J.H. Cardellina II, F.-J. Marner, and R.E. Moore, *J. Am. Chem. Soc.*, **101**, 240 (1979).

1979JAB549 M. Kenig and C. Reading, *J. Antibiot.*, **32**, 549 (1979).

1979JOC2798 J.T. Baker and S. Sifniades, *J. Org. Chem.*, **44**, 2798 (1979).

1979T463 D.A. Lightner and Y.-T. Park, *Tetrahedron*, **35**, 463 (1979).

1979TL2003 C.J. Simmons, F.-J. Marner, J.H. Cardellina II, R.E. Moore, and K. Seff, *Tetrahedron Lett.*, **20**, 2003 (1979).

1979TL2007 J.H. Cardellina II and R.E. Moore, *Tetrahedron Lett.*, **20**, 2007 (1979).

1980CPB2178 K. Yakushijin, M. Kozuka, and H. Furukawa, *Chem. Pharm. Bull.*, **28**, 2178 (1980).

1980H(14)1073 K. Yakushijin, M. Kozuka, Y. Ito, R. Suzuki, and H. Furukawa, *Heterocycles*, **14**, 1073 (1980).

1980HCA121 T. Schmidlin and C. Tamm, *Helv. Chim. Acta*, **63**, 121 (1980).

1980PT(1)227 L. Vasvári-Debreczy, I. Hermecz, Z. Mészáros, and P. Dvortsák, *J. Chem. Soc. Perkin Trans. 1*, 227 (1980).

1981AJC2641 R.D. Allan and H. Tran, *Aust. J. Chem.*, **34**, 2641 (1981).

1981H(16)795 K. Tabei, H. Ito, and T. Takada, *Heterocycles*, **16**, 795 (1981).

1981H(16)1157 K. Yakushijin, R. Suzuki, R. Hattori, and H. Furukawa, *Heterocycles*, **16**, 1157 (1981).

1981H(16)1987 T. Nagasaka, S. Esumi, N. Ozawa, Y. Kosugi, and F. Hamaguchi, *Heterocycles*, **16**, 1987 (1981).

1981JA1769 H.W. Moore, L. Hernandez, D.M. Kunert, F. Mercer, and A. Sing, *J. Am. Chem. Soc.*, **103**, 1769 (1981).

1981MI1 K. Spirkova, J. Kovac, I. Horsak, and M. Dandarova, *Collect. Czech. Chem. Commun.*, **46**, 1513 (1981).

1981MI2 W. Onkenhout, J.A. van Koevergine, and J. Lugtenburg, *Recueil*, **100**, 106 (1981).

1981PT(1)322 R. Bonnett, D.G. Buckley, and D. Hamzetash, *J. Chem. Soc. Perkin Trans. 1*, 322 (1981).

1981TL4301 Y. Tamura, H. Maeda, S. Akai, K. Ishiyama, and H. Ishibashi, *Tetrahedron Lett.*, **22**, 4301 (1981).

1981S200 J. Barluenga, F. Palacios, S. Fustero, and V. Gotor, *Synthesis*, 200 (1981).

1982ACIE620 G. Himbert and L. Henn, *Angew. Chem. Int. Ed.*, **21**, 620 (1982).

1982JA3511 Y. Ohfune and M. Tomita, *J. Am. Chem. Soc.*, **104**, 3511 (1982).

1982JAB635 B. Kunze, H. Reichenbach, H. Augustiniak, and G. Höfle, *J. Antibiot.*, **35**, 635 (1982).

1982JOC1534 E. Vedejs, J.B. Campbell, R.C. Gadwood, J.D. Rodgers, K.L. Spear, and Y. Watanabe, *J. Org. Chem.*, **47**, 1534 (1982).

1982MI1 J.-C. Chalchat, R.-P. Garry, and A. Michet, *C. R. l'Acad. Sci. Ser. II Univ.*, **295**, 871 (1982).

1982MI2 J.A. de Groot and J. Lugtenburg, *Recueil*, **10**, 333 (1982).

1983H(20)469 B. Sarstedt and E. Winterfeldt, *Heterocycles*, **20**, 469 (1983).

1983H(20)1761 F. Fariña, M.V. Martín, F. Sánchez, M.C. Maestro, and M.R. Martín, *Heterocycles*, **20**, 1761 (1983).

1983JA5510 G. Stork and E. Nakamura, *J. Am. Chem. Soc.*, **105**, 5510 (1983).

1983JHC687 C. Anselmi, A. Camparini, and M. Scotton, *J. Heterocycl. Chem.*, **20**, 687 (1983).

1983JOC4058 M. Mori, Y. Washioka, T. Urayama, K. Yoshiura, K. Chiba, and Y. Ban, *J. Org. Chem.*, **48**, 4058 (1983).

1983TL1441 I. Hughes and R.A. Raphael, *Tetrahedron Lett.*, **24**, 1441 (1983).

1983TL4755 R.C.F. Jones and G.E. Peterson, *Tetrahedron Lett.*, **24**, 4755 (1983).

1984CC122 W.C.A. Gelderblom, W.F.O. Marasas, P.S. Steyn, P.G. Thiel, P.H. Van der Merwe, P.H. Van Rooyen, R. Vleggaar, and P.L. Wessels, *J. Chem. Soc. Chem. Commun.*, 122 (1984).

1984CC1373 G.R. Brown, A.J. Foubister, and B. Wright, *J. Chem. Soc. Chem. Commun.*, 1373 (1984).

1984CPB4197 K. Matsuo, M. Kimura, T. Kinuta, N. Takai, and K. Tanaka, *Chem. Pharm. Bull.*, **32**, 4197 (1984).

1984H(22)1733 F. Fariña, M.V. Martín, M.C. Paredes, M.C. Ortega, and A. Tito, *Heterocycles*, **22**, 1733 (1984).

1984H(22)2363 Y. Kosugi and F. Hamaguchi, *Heterocycles*, **22**, 2363 (1984).

1984HCA1783 H. Köhler and H. Gerlach, *Helv. Chim. Acta*, **67**, 1783 (1984).

1984JA2645 R.W. Schoenleber, Y. Kim, and H. Rapoport, *J. Am. Chem. Soc.*, **106**, 2645 (1984).

1984JOMCC61 R.J.P. Corriu, V. Huynh, J. Iqbal, and J.J.E. Moreau, *J. Organomet. Chem*, **276**, C61 (1984).

1984S62 K. Gewald and U. Hain, *Synthesis*, **62** (1984).

1984TL1871 K.S. Kochhar, H.J. Carson, K.A. Clouser, J.W. Elling, L.A. Gramens, J.L. Parry, H.L. Sherman, K. Braat, and H.W. Pinnick, *Tetrahedron Lett.*, **25**, 1871 (1984).

1984TL2009 K. Inami and T. Shiba, *Tetrahedron Lett.*, **25**, 2009 (1984).
1984TL3483 S.H. Goh, C. Wei, and A.R.M. Ali, *Tetrahedron Lett.*, **25**, 3483 (1984).
1985BCSJ352 K. Inami and T. Shiba, *Bull. Chem. Soc. Jpn.*, **38**, 352 (1985).
1985JA1777 R.H. Schlessinger, G.R. Bebernitz, P. Lin, and A.J. Poss, *J. Am. Chem. Soc.*, **107**, 1777 (1985).
1985JHC1599 O. Igglessi-Markopoulou and C. Sandris, *J. Heterocycl. Chem.*, **22**, 1599 (1985).
1985JOC425 R.J. Sundberg and B.C. Pearce, *J. Org. Chem.*, **50**, 425 (1985).
1985JOC1344 R.H. Schlessinger and G.R. Bebernitz, *J. Org. Chem.*, **50**, 1344 (1985).
1985JOC4771 R.J. Capon and D.J. Faulkner, *J. Org. Chem.*, **50**, 4771 (1985).
1985LAC1095 R. Schmierer and H. Mildenberger, *Liebigs Ann. Chem*, 1095 (1985).
1985T5241 J.E. Baldwin, J.K. Cha, and L.I. Kruse, *Tetrahedron*, **41**, 5241 (1985).
1985T5465 M. Mori, N. Kanda, I. Oda, and Y. Ban, *Tetrahedron*, **41**, 5465 (1985).
1985TL85 C.-g. Shin, Y. Yonezawa, and E. Watanabe, *Tetrahedron Lett.*, **26**, 85 (1985).
1985TL3617 G.D. James, G. Pattenden, and S.D. Mills, *Tetrahedron Lett.*, **26**, 3617 (1985).
1986CPB2049 K. Yakushijin, R. Suzuki, N. Kawaguchi, Y. Tsuboi, and H. Furukawa, *Chem. Pharm. Bull.*, **34**, 2049 (1986).
1986JOC3656 J.C. Gilbert and B.K. Biackburn, *J. Org. Chem.*, **51**, 3656 (1986).
1986TL2501 S.H. Goh and A.R.M. Ali, *Tetrahedron Lett.*, **27**, 2501 (1986).
1986TL2691 G.A. Krafft, E.A. Garcia, A. Guram, B. O'Shaughnessy, and X. Xu, *Tetrahedron Lett.*, **27**, 2691 (1986).
1986TL5285 R.C.F. Jones and A.D. Bates, *Tetrahedron Lett.*, **27**, 5285 (1986).
1986TL6339 F. Henin, J. Muzart, and J.-P. Pete, *Tetrahedron Lett.*, **27**, 6339 (1986).
1987CC386 R. Tanikaga, J. Takeuchi, M. Takyu, and A. Kaji, *J. Chem. Soc. Chem. Commun.*, 386 (1987).
1987CPB2140 N. Ikota and A. Hanaki, *Chem. Pharm. Bull.*, **35**, 2140 (1987).
1987CPB4368 K. Hori, M. Arai, K. Nomura, and E. Yoshii, *Chem. Pharm. Bull.*, **35**, 4368 (1987).
1987IJC(B)63 P.R. Babu and T.R. Balasubramanian, *Indian J. Chem. Sect. B*, **26B**, 63 (1987).
1987JMC743 P. Berthelot, C. Vaccher, A. Musadad, N. Flouquet, M. Debaert, and M. Luyckx, *J. Med. Chem.*, **30**, 743–746 (1987).
1987JOC1177 R.P. Joyce, J.A. Gainor, and S.M. Weinreb, *J. Org. Chem.*, **52**, 1177 (1987).
1987JOC2368 I. Maeba, O. Hara, M. Suzuki, and H. Furukawa, *J. Org. Chem.*, **52**, 2368 (1987).
1987PH582 H. Shimomura, Y. Sashida, Y. Mimaki, and Y. Minegishi, *Phytochemistry*, **26**, 582 (1987).
1987PT(1)1177 P. Jouin, B. Castro, and D. Nisato, *J. Chem. Soc. Perkin Trans. 1*, 1177 (1987).
1987PT(1)2027 A.R. Battersby, M.G. Baker, H.A. Broadbent, C.J.R. Fookes, and F.J. Leeper, *J. Chem. Soc. Perkin Trans. 1*, 2027 (1987).
1987TL2963 R.J.P. Corriu, J.J.E. Moreau, and C. Vernhet, *Tetrahedron Lett.*, **28**, 2963 (1987).
1987TL6675 B. Tarnchompoo, C. Thebtaranonth, and Y. Thebtaranonth, *Tetrahedron Lett.*, **28**, 6675 (1987).
1988H(27)2843 W.R. Ewing and M.M. Joullié, *Heterocycles*, **27**, 2843 (1988).
1988JA4017 F. Gaviña, A.M. Costero, M.R. Andreu, M. Carda, and S.V. Luis, *J. Am. Chem. Soc.*, **110**, 4017 (1988).

1988JOC976 R.J. Sundberg, G.S. Hamilton, and J.P. Laurino, *J. Org. Chem.*, **53**, 976 (1988).

1988PH3101 R.E. Moore and M. Entzeroth, *Phytochemistry*, **27**, 3101 (1988).

1988PT(1)1557 A.R. Battersby, C.J. Dutton, C.J.R. Fookes, and S.P.D. Turner, *J. Chem. Soc. Perkin Trans. 1*, 1557 (1988).

1988PT(1)1569 A.R. Battersby, C.J. Dutton, and C.J.R. Fookes, *J. Chem. Soc. Perkin Trans. 1*, 1569 (1988).

1988TL4859 J. Barluenga, F.J. Fañanás, F. Foubelo, and M. Yus, *Tetrahedron Lett.*, **29**, 4859 (1988).

1988TL5829 S.V. Ley, S.C. Smith, and P.R. Woodward, *Tetrahedron Lett.*, **29**, 5829 (1988).

1989HCA518 A. Gossauer, F. Nydegger, E. Benedikt, and H.-P. Köst, *Helv. Chim. Acta*, **72**, 518 (1989).

1989JA1524 Y. Hamada, A. Kawai, Y. Kohno, O. Hara, and T. Shioiri, *J. Am. Chem. Soc.*, **111**, 1524 (1989).

1989JA5346 B. Capon and F.-C. Kwok, *J. Am. Chem. Soc.*, **111**, 5346 (1989).

1989JHC1713 R.S. Brinkmeyer and N.H. Terando, *J. Heterocycl. Chem.*, **26**, 1713 (1989).

1989JNP891 A. Maxwell and D. Rampersad, *J. Nat. Prod.*, **52**, 891 (1989).

1989JOC6005 G.R. Pettit, Y. Kamano, C. Dufresne, R.L. Cerny, C.L. Herald, and J.M. Schmidt, *J. Org. Chem.*, **54**, 6005 (1989).

1989MI1 H. Falk, *The Chemistry of Linear Oligopyrroles and Bile Pigments*, Springer-Verlag: Wien (1989).

1989T2417 A.P. Craven, H.J. Dyke, and E.J. Thomas, *Tetrahedron*, **45**, 2417 (1989).

1989T7899 S.H. Goh, A.R.M. Ali, Razak, and W.H. Wong, *Tetrahedron*, **45**, 7899 (1989).

1989TL7469 J.M. Clough, G. Pattendon, and P.G. Wight, *Tetrahedron Lett.*, **30**, 7469 (1989).

1990ABC813 Y. Kimura, H. Nakajima, T. Hamasaki, T. Matsumoto, Y. Matsuda, and A. Tsuneda, *Agric. Biol. Chem.*, **54**, 813 (1990).

1990HCA122 J. Ackermann, M. Matthes, and C. Tamm, *Helv. Chim. Acta*, **73**, 122 (1990).

1990JOC3880 J. Mathew and B. Alink, *J. Org. Chem.*, **55**, 3880 (1990).

1990JOMC375 G. Consiglio, L. Kollár, and R. Kölliker, *J. Organomet. Chem.*, **396**, 375 (1990).

1990LAC525 R. Grote, A. Zeeck, J. Stümpfel, and H. Zähner, *Liebigs Ann. Chem.*, 525 (1990).

1990MI1 A. Kawashima, Y. Yoshimura, S. Yuko, S. Noriyshi, K. Kamigoori, T. Mizutani, and S. Omura, γ-Lactam PI-091 from *Paecilomyces* species, *Jpn. Kokai Tokkyo Koho JP 02062859* [*Chem. Abstr.*, **113**, 113856(1990)].

1990PT(1)611 J. Poncet, P. Jouin, B. Castro, L. Nicholas, M. Boutar, and A. Gaudemer, *J. Chem. Soc. Perkin Trans. 1*, 611 (1990).

1990PT(1)2475 I. Hughes, W.P. Nolan, and R.A. Raphael, *J. Chem. Soc. Perkin Trans. 1*, 2475 (1990).

1990S1062 D.P. Shrout and D.A. Lightner, *Synthesis*, 1062 (1990).

1990SL63 T.J. Hagen, *Synlett*, **63** (1990).

1991CL2135 Y. Arai, T. Kontani, and T. Koizumi, *Chem. Lett.*, 2135 (1991).

1991EJMC395 P. Barthelot, C. Vaccher, N. Flouqet, M. Luyckx, C. Brunet, T. Boulanger, J.P. Frippiat, D.P. Vercauteren, M. Debaert, G. Evrard, and F. Durant, *Eur. J. Med. Chem.*, **26**, 395 (1991).

1991JA6692 G.R. Pettit, D.L. Herald, S.B. Singh, T.J. Thornton, and J.T. Mullaney, *J. Am. Chem. Soc.*, **113**, 6692 (1991).

1991JA8024 J.E. Bishop, J.F. O'Connell, and H. Rapoport, *J. Am. Chem. Soc.*, **113**, 8024 (1991).

1991JMC1011 B.L. Mylari, T.A. Beyer, and T.W. Siegel, *J. Med. Chem.*, **34**, 1011 (1991).

1991JMC1721 S.D. Larsen, C.H. Spilman, F.P. Bell, D.M. Dinh, E. Martinborough, and G.J. Wilson, *J. Med. Chem.*, **34**, 1721 (1991).

1991JOC2024 I. Ojima, A. Korda, and W.R. Shay, *J. Org. Chem.*, **56**, 2024 (1991).

1991JOC6027 Y.-S. Byun and D.A. Lightner, *J. Org. Chem.*, **56**, 6027 (1991).

1991LAC709 F.-P. Montforts and U.M. Schwartz, *Liebigs Ann. Chem.*, 709 (1991).

1991MI1 Y.-S. Hon, Y.-C. Chang, and M.-L. Gong, *Bull. Inst. Chem. Acad. Sin.*, **38**, 23 (1991).

1991T2087 S. Carmeli, R.E. Moore, and G.M.L. Patterson, *Tetrahedron*, **47**, 2087 (1991).

1991TL283 P. Somfai, H.M. He, and D. Tanner, *Tetrahedron Lett.*, **32**, 283 (1991).

1991TL1379 J.E. Baldwin, M.G. Moloney, and S.B. Shim, *Tetrahedron Lett.*, **32**, 1379 (1991).

1991TL3057 N. Langlois and R.Z. Andriamialisoa, *Tetrahedron Lett.*, **32**, 3057 (1991).

1991TL5983 Y. Hamada, Y. Tandada, F. Yokokawa, and T. Shioiri, *Tetrahedron Lett.*, **32**, 5983 (1991).

1991TL6949 K.-C. Woo and K. Jones, *Tetrahedron Lett.*, **32**, 6949 (1991).

1992AJC1903 L.T. Byrne, B.Q. Guevara, W.C. Patalinghug, B.V. Recio, C.R. Ualat, and A.H. White, *Aust. J. Chem.*, **45**, 1903 (1992).

1992CPB1670 Y. Arai, A. Fujii, T. Ohno, and T. Koizumi, *Chem. Pharm. Bull.*, **40**, 1670 (1992).

1992CPB1925 N. Ikota, *Chem. Pharm. Bull.*, **40**, 1925 (1992).

1992JNP613 F. Koehn, R.E. Longley, and J.K. Reed, *J. Nat. Prod.*, **55**, 613 (1992).

1992JOC1059 W.-J. Koot, H. Hiemstra, and W.N. Speckamp, *J. Org. Chem.*, **57**, 1059 (1992).

1992JOC3760 G. Casiraghi, G. Rassu, P. Spanu, and L. Pinna, *J. Org. Chem.*, **57**, 3760 (1992).

1992LAC473 H.D. Stachel, J. Nienaber, and T. Zoukas, *Leibigs Ann. Chem.*, 473 (1992).

1992LAC781 W. Duczek, K. Jähnisch, A. Kunath, G. Reck, G. Winter, and B. Schulz, *Liebigs Ann. Chem.*, 781 (1992).

1992MI1 S. Kagabu and C. Ito, *Biosci. Biotechnol. Biochem.*, **56**, 1164 (1992).

1992PH1084 E. Eisenreichova, M. Haladova, A. Buckova, J. Tomko, D. Ohrin, and K. Ubik, *Phytochemistry*, **31**, 1084 (1992).

1992PH2767 K. Ori, Y. Mimaki, K. Mito, Y. Sashida, T. Nikaido, T. Ohmoto, and A. Masuko, *Phytochemistry*, **31**, 2767 (1992).

1992S391 L. Duc, J.F. McGarrity, T. Meul, and A. Warm, *Synthesis*, 391 (1992).

1992T3313 H. Yoda, H. Kitayama, T. Katagiri, and K. Takabe, *Tetrahedron*, **48**, 3313 (1992).

1992T4115 N. Patino, E. Frérot, N. Galeotti, J. Poncet, J. Coste, M.-N. Dufour, and P. Jouin, *Tetrahedron*, **48**, 4115 (1992).

1992TA535 K. Arai, T. Kontani, and T. Koizumi, *Tetrahedron: Asymmetry*, **3**, 535 (1992).

1992TL1147 A.T. Pham, J.R. Carney, W.Y. Yoshida, and P.J. Scheuer, *Tetrahedron Lett.*, **33**, 1147 (1992).

1993CL1437 H. Kinoshita, Y. Hayashi, Y. Murata, and K. Inomata, *Chem. Lett.*, **22**, 1437 (1993).

1993JA9856 G.C. Fu, S.T. Nguyen, and R.H. Grubbs, *J. Am. Chem. Soc.*, **115**, 9856 (1993).

1993JAB1834 H. Shiozawa, T. Kagasaki, T. Kinoshita, H. Haruyama, H. Domon, Y. Utsui, K. Kodama, and S. Takahashi, *J. Antibiot.*, **46**, 1834 (1993).

1993JMC1041 J.W. Lampe, Y.-L. Chou, R.G. Hanna, S.V. Di Meo, P.W. Erhardt, A.A. Hagedorn, W.R. Ingebretsen, and E. Cantor, *J. Med. Chem.*, **36**, 1041 (1993).

1993JOC36 A.I. Meyers and L. Snyder, *J. Org. Chem.*, **58**, 36 (1993).

1993JOC4010 M. Bänziger, J.F. McGarrity, and T. Meul, *J. Org. Chem.*, **58**, 4010 (1993).

1993JOC6336 M.D. Unson, C.B. Rose, D.J. Faulkner, L.S. Brinen, J.R. Steiner, and J. Clardy, *J. Org. Chem.*, **58**, 6336 (1993).

1993PH1159 M.G. Nonato, M.J. Garson, R.J.W. Truscott, and J.A. Carver, *Phytochemistry*, **34**, 1159 (1993).

1993PT(1)1545 G. Graziano, I. N'Diaye, G. Chiasera, and F. Pietra, *J. Chem. Soc. Perkin Trans. 1*, 1545 (1993).

1993PT(1)2567 G.B. Gill, G.D. James, K.V. Oates, and G. Pattenden, *J. Chem. Soc. Perkin Trans. 1*, 2567 (1993).

1993PT(1)2581 G.D. James, S.D. Mills, and G. Pattenden, *J. Chem. Soc. Perkin Trans. 1*, 2581 (1993).

1993S216 U. Schmidt, B. Riedl, G. Haas, H. Griesser, A. Vetter, and S. Weinbrenner, *Synthesis*, 216 (1993).

1993SC1111 G.A. Revelli and E.G. Gros, *Synth. Commun.*, **23**, 1111 (1993).

1993T2185 M. Xie and D.A. Lightner, *Tetrahedron*, **49**, 2185 (1993).

1993T4603 R.J.P. Corriu, G. Bolin, J. Iqbal, J.J.E. Moreau, and C. Vernhet, *Tetrahedron*, **49**, 4603 (1993).

1993T9151 G.R. Pettit, Y. Kamano, C.L. Herald, Y. Fujii, H. Kizu, M.R. Boyd, F.E. Boettner, D.L. Doubek, J.M. Schmidt, J.-C. Chapuis, and C. Michel, *Tetrahedron*, **49**, 9151 (1993).

1993TL6705 J. Jiang, W.-R. Li, R.M. Przeslawski, and M.M. Joullié, *Tetrahedron Lett.*, **34**, 6705 (1993).

1993TL6821 M.A. Rachita and G.A. Slough, *Tetrahedron Lett.*, **34**, 6821 (1993).

1993TL6825 G.A. Slough, *Tetrahedron Lett.*, **34**, 6825 (1993).

1993TL7777 J. Barluenga, R. Gonzalez, and F.J. Fañanás, *Tetrahedron Lett.*, **48**, 7777 (1993).

1993TL8361 W. Harris, C.H. Hill, E. Keech, and P. Malsher, *Tetrahedron Lett.*, **34**, 8361 (1993).

1994BCSJ3088 K. Kohori, M. Hashimoto, H. Kinoshita, and K. Inomata, *Bull. Chem. Soc. Jpn.*, **67**, 3088 (1994).

1994BCSJ3320 H. Ngwe, H. Kinoshita, and K. Inomata, *Bull. Chem. Soc. Jpn.*, **67**, 3320 (1994).

1994BMCL1947 F. Roux, N. Galeotti, J. Poncet, P. Jouin, S. Cros, and G. Zenke, *Bioorg. Med. Chem. Lett.*, **16**, 1947 (1994).

1994H(38)1839 T. Iida, K. Hori, K. Nomura, and E. Yoshii, *Heterocycles*, **38**, 1839 (1994).

1994JA2139 H. Uno, J.E. Baldwin, and A.T. Russell, *J. Am. Chem. Soc.*, **116**, 2139 (1994).

1994JA8116 S. Paik, S. Carmeli, J. Cullingham, R.E. Moore, G.M.L. Patterson, and M.A. Tius, *J. Am. Chem. Soc.*, **116**, 8116 (1994).

1994JAB851 H. Shiozawa and S. Takahashi, *J. Antibiot.*, **47**, 851 (1994).

1994JHC1145 S. Hamilakis, D. Kontonassions, and C. Sandris, *J. Heterocycl. Chem.*, **31**, 1145 (1994).

1994JOC2447 D.S. Brown, M.C. Elliott, C.J. Moody, T.J. Mowlem, J.P. Marino, and A. Padwa, *J. Org. Chem.*, **59**, 2447 (1994).

1994MI1 J. Vader, F. Kaspersen, E. Sperling, I. Schlachter, A. Terpstra, P. Hilberink, and G. Wagenaars, *J. Label. Comp. Radiopharm.*, **34**, 845 (1994).

1994PT(1)15 K. Arai, T. Kontani, and T. Koizumi, *J. Chem. Soc. Perkin Trans. 1*, 15 (1994).

1994S25 J. Brüning, T. Hache, and E. Winterfeldt, *Synthesis*, **25** (1994).

1994T12097 G.R. Pettit, T.J. Thornton, J.T. Mullaney, M.R. Boyd, D.L. Herald, S.-B. Singh, and E.J. Flahive, *Tetrahedron*, **50**, 12097 (1994).

1994TL119 D. Griffart-Brunet and N. Langlois, *Tetrahedron Lett.*, **35**, 119 (1994).

1994TL3931 I. Baussanne, A. Chiaroni, H.-P. Husson, C. Riche, and J. Royer, *Tetrahedron Lett.*, **35**, 3931 (1994).

1994TL5555 G. Xie and J.W. Lown, *Tetrahedron Lett.*, **35**, 5555 (1994).

1994TL7159 A. Padwa, D.J. Austin, and A.T. Price, *Tetrahedron Lett.*, **35**, 7159 (1994).

1994TL7417 A.M.P. Koskinen and P.M. Pihko, *Tetrahedron Lett.*, **35**, 7417 (1994).

1994TL9135 J.T. Link and S.J. Danishefsky, *Tetrahedron Lett.*, **35**, 9135 (1994).

1995CL713 H. Ngwe, E. Nakayama, T. Higashi, H. Kinoshita, and K. Inomata, *Chem. Lett.*, 713 (1995).

1995CR1981 B.J.L. Royles, *Chem. Rev.*, **95**, 1981 (1995).

1995JA552 J.T. Link, S. Raghavan, and S.J. Danishefsky, *J. Am. Chem. Soc.*, **117**, 552 (1995).

1995JA3422 C. Copéret, T. Sugihara, G. Wu, I. Shimoyama, and E.-i. Negishi, *J. Am. Chem. Soc.*, **117**, 3422 (1995).

1995JAB535 S. Omura, Y. Sasaki, Y. Iwai, and H. Takeshima, *J. Antibiot.*, **48**, 535 (1995).

1995JAB1413 K. Shiomi, H. Yang, Q. Xu, N. Arai, M. Namiki, M. Hayashi, J. Inokoshi, H. Takeshima, R. Masuma, K. Komiyama, and S. Omura, *J. Antibiot.*, **48**, 1413 (1995).

1995JAB733 H. Kakeya, I. Takahashi, G. Okada, K. Isono, and H. Osada, *J. Antibiot.*, **48**, 733 (1995).

1995JAB907 H. Shiozawa, T. Kagasaki, A. Torikata, N. Tanaka, K. Fujimoto, T. Hata, Y. Furukawa, and S. Takahashi, *J. Antibiot.*, **48**, 907 (1995).

1995JOC7040 S.B. Singh, M.A. Goetz, E.T. Jones, G.F. Bills, R.A. Giacobbe, L. Herranz, S. Stevens-Miles, and D.L. Williams, *J. Org. Chem.*, **60**, 7040 (1995).

1995JOMC127 C. Bouancheau, M. Rudler, E. Chelain, H. Rudler, J. Vaissermann, and J.-C. Daran, *J. Organomet. Chem.*, **496**, 127 (1995).

1995MI1 S.-C. Yang and F.-R. Shea, *J. Chin. Chem. Soc.*, **42**, 969 (1995).

1995S607 G. Casiraghi and G. Rassu, *Synthesis*, 607 (1995).

1995S1511 M.M. Faul, K.A. Sullivan, and L.L. Winneroski, *Synthesis*, 1511 (1995).

1995SC1265 B. Li and M.B. Smith, *Synth. Commun.*, **25**, 1265 (1995).

1995T2129 G. Reginato, A. Capperucci, A. Degl'Innocenti, A. Mordini, and S. Pecchi, *Tetrahedron*, **51**, 2129 (1995).

1995T2763 P. Bonete and C. Nájera, *Tetrahedron*, **51**, 2763 (1995).

1995T5585 G.T. Crisp and A.G. Meyer, *Tetrahedron*, **51**, 5585 (1995).

1995T10107 J. Ezquerra, C. Pedregal, I. Collado, B. Yruretagoyena, and A. Rubio, *Tetrahedron*, **51**, 10107 (1995).

| | |
|---|---|
| 1995TA2081 | M. Yus, F. Foubelo, and L.R. Falvello, *Tetrahedron: Asymmetry*, **6**, 2081 (1995). |
| 1995TL291 | A. Murray and G.R. Proctor, *Tetrahedron Lett.*, **36**, 291 (1995). |
| 1995TL1771 | C. Copéret, T. Sugihara, and E-i. Negishi, *Tetrahedron Lett.*, **36**, 1771 (1995). |
| 1995TL4197 | J.P. Dittami, F. Xu, H. Qi, and M.W. Martin, *Tetrahedron Lett.*, **36**, 4197 (1995). |
| 1995TL4201 | J.P. Dittami, F. Xu, H. Qi, and M.V. Martin, *Tetrahedron Lett.*, **36**, 4201 (1995). |
| 1995TL5551 | R. Shiraki, A. Sumino, K-i. Tadano, and S. Ogawa, *Tetrahedron Lett.*, **36**, 5551 (1995). |
| 1995TL5851 | S.L. Miller, W.F. Tinto, J.-P. Yang, S. McLean, and W.F. Reynolds, *Tetrahedron Lett.*, **36**, 5851 (1995). |
| 1995TL7705 | M.J. Martín and F. Bermejo, *Tetrahedron Lett.*, **36**, 7705 (1995). |
| 1995TL7889 | R. Rico and F. Bermejo, *Tetrahedron Lett.*, **36**, 7889 (1995). |
| 1995TL8383 | T.B. Lowinger, J. Chu, and P.L. Spence, *Tetrahedron Lett.*, **36**, 8383 (1995). |
| 1996BMCL1543 | P.L. Toogood, J.J. Hollenbeck, H.M. Lam, and L. Li, *Bioorg. Med. Chem. Lett.*, **6**, 1543 (1996). |
| 1996CC1323 | A. Detsi, J. Markopoulos, and O. Igglessi-Markopoulou, *Chem. Commun.*, 1323 (1996). |
| 1996H(42)75 | H. Ogawa, T. Aoyama, and T. Shioiri, *Heterocycles*, **42**, 75 (1996). |
| 1996H(43)685 | M. Matsuda, T. Kobayashi, S. Nagao, T. Ohta, and S. Nozoe, *Heterocycles*, **43**, 685 (1996). |
| 1996JA2825 | J.T. Link, S. Raghavan, M. Gallant, S.J. Danishefsky, T.C. Chou, and L.M. Ballas, *J. Am. Chem. Soc.*, **118**, 2825 (1996). |
| 1996JA10656 | J.L. Wood, B.M. Stoltz, and S.N. Goodman, *J. Am. Chem. Soc.*, **118**, 10656 (1996). |
| 1996JAB967 | R.R. West, J. Van Ness, A.-M. Varming, B. Rassing, S. Biggs, S. Gasper, P.A. McKernan, and J. Piggott, *J. Antibiot.*, **49**, 967 (1996). |
| 1996JHC825 | S. Hamilakis, D. Kontonassions, and C. Sandris, *J. Heterocycl. Chem.*, **33**, 825 (1996). |
| 1996JOC2179 | M.P. Doyle and A.V. Kalinin, *J. Org. Chem.*, **61**, 2179 (1996). |
| 1996JOC2845 | R. Shiraki, A. Sumino, K-i. Tadano, and S. Ogawa, *J. Org. Chem.*, **61**, 2845 (1996). |
| 1996JOC3534 | C.P. Decicco and P. Grover, *J. Org. Chem.*, **61**, 3534 (1996). |
| 1996JOC4542 | G.P. Lutz, H. Hu, D.J. Gallagher, and P. Beak, *J. Org. Chem.*, **61**, 4542 (1996). |
| 1996MI1 | D. Geffken and J. Froböse, *Die Pharmazie*, **51**, 671 (1996). |
| 1996PH1225 | T. Satou, Y. Mimaki, M. Kuroda, Y. Sashida, and Y. Hatakeyama, *Phytochemistry*, **41**, 1225 (1996). |
| 1996PT(1)1839 | M. Yoshimatsu, K. Machida, T. Fuseya, H. Shimizu, and T. Kataoka, *J. Chem. Soc. Perkin Trans. 1*, 1839 (1996). |
| 1996PT(1)2799 | J. Löffler and R. Schobert, *J. Chem. Soc. Perkin Trans. 1*, 2799 (1996). |
| 1996S851 | L.F. Tietz, G. Kettschau, and K. Heitmann, *Synthesis*, 851 (1996). |
| 1996SL67 | C.M. Huwe, O.C. Kiehl, and S. Blechert, *Synlett*, **67** (1996). |
| 1996SL513 | R. D'Alessio and A. Rossi, *Synlett*, 513 (1996). |
| 1996T869 | Y. Aoyagi, T. Manabe, A. Ohta, T. Kurihara, G.-L. Pang, and T. Yuhara, *Tetrahedron*, **52**, 869 (1996). |
| 1996T3247 | A. Padwa, D.J. Austin, A.T. Price, and M.D. Weingarten, *Tetrahedron*, **52**, 3247 (1996). |
| 1996T3757 | A. Murray, G.R. Proctor, and P.J. Murray, *Tetrahedron*, **52**, 3757 (1996). |

1996T11529 C. Copéret, S. Ma, T. Sugihara, and E.-i. Negishi, *Tetrahedron*, **52**, 11529 (1996).

1996TA2365 D. Ma, J. Ma, W. Ding, and L. Dai, *Tetrahedron: Asymmetry*, **7**, 2365 (1996).

1996TL291 R.-H. Mattern, *Tetrahedron Lett.*, **37**, 291 (1996).

1996TL5809 R. Rico and F. Bermejo, *Tetrahedron Lett.*, **37**, 5809 (1996).

1996TL6263 D.G. Nagle, V.J. Paul, and M.A. Roberts, *Tetrahedron Lett.*, **37**, 6263 (1996).

1996TL7841 I. Manteca, N. Sotomayor, M.-J. Villa, and E. Lete, *Tetrahedron Lett.*, **37**, 7841 (1996).

1997ACIE2454 B.A. Kulkarni and A. Ganesan, *Angew. Chem. Int. Ed.*, **36**, 2454 (1997).

1997JA9652 J.L. Wood, B.M. Stoltz, S.N. Goodman, and K. Onwueme, *J. Am. Chem. Soc.*, **119**, 9652 (1997).

1997JAB449 H. Shiozawa, A. Shimada, and S. Takahashi, *J. Antibiot.*, **50**, 449 (1997).

1997JMC168 G. Rassu, F. Zanardi, L. Battistini, E. Gaetani, and G. Casiraghi, *J. Med. Chem.*, **40**, 168 (1997).

1997JMC391 H. Kakeya, C. Onozawa, M. Sato, K. Arai, and H. Osada, *J. Med. Chem.*, **40**, 391 (1997).

1997JOC1918 N. Iwasawa and K. Maeyama, *J. Org. Chem.*, **62**, 1918 (1997).

1997JOC3254 H. Kang and W. Fenical, *J. Org. Chem.*, **62**, 3254 (1997).

1997JOC5542 M.B. Andrus, W. Li, and R.F. Keyes, *J. Org. Chem.*, **62**, 5542 (1997).

1997MI1 A.S. Fisyuk and M.A. Vorontsova, *Chem. Heterocycl. Comp.*, **33**, 857 (1997).

1997MI2 D. Geffken, M. Haerting, and J. Froböse, *Heterocycl. Commun.*, **3**, 127 (1997).

1997MI3 L.-J. Chen, S.S. Hecht, and L.A. Peterson, *Chem. Res. Toxicol.*, **10**, 866 (1997).

1997OM2726 H. Arzoumanian, M. Jean, D. Nuel, J.L. Garcia, and N. Rosas, *Organometallics*, **16**, 2726 (1997).

1997PT(1)3543 M. Petroliagi and O. Igglessi-Markopoulou, *J. Chem. Soc. Perkin Trans. 1*, 3543 (1997).

1997S305 M.D. Andrews, A.G. Brewster, J. Chuhan, A.J. Ibbett, M.G. Moloney, K. Prout, and D. Watkin, *Synthesis*, 305 (1997).

1997S559 A. Diaz, J.G. Siro, J.L. García-Navío, J.J. Vaquero, and J. Alvarez-Builla, *Synthesis*, 559 (1997).

1997SL1179 M. Arisawa, E. Takezawa, A. Nishida, M. Mori, and M. Nakagawa, *Synlett*, 1179 (1997).

1997TA515 G.B. Giovenzana, M. Sisti, and G. Palmisano, *Tetrahedron: Asymmetry*, **8**, 515 (1997).

1997TL5503 J.-N. Denis, S. Tchertchian, A. Tomassini, and Y. Vallée, *Tetrahedron Lett.*, **38**, 5503 (1997).

1997TL6849 Y. Gao, M. Shirai, and F. Sato, *Tetrahedron Lett.*, **38**, 6849 (1997).

1998BMC1273 D.L. Whitehouse, K.H. Nelso, S.N. Savinov, R.S. Löwe, and D.J. Austin, *Bioorg. Med. Chem.*, **6**, 1273 (1998).

1998CL455 T. Masukawa, H. Kato, T. Kakiuchi, K.P. Yayasundera, H. Kinoshita, and K. Inomata, *Chem. Lett.*, **27**, 455 (1998).

1998JAB515 K.A. Alvi, A. Casey, and B.G. Nair, *J. Antibiot.*, **51**, 515 (1998).

1998JOC92 P.M. Pihko and A.M.P. Koskinen, *J. Org. Chem.*, **63**, 92 (1998).

1998JOC3164 N. Iwasawa, T. Ochiai, and K. Maeyama, *J. Org. Chem.*, **63**, 3164 (1998).

1998JOC4808 J. Matthews and R.A. Rivero, *J. Org. Chem.*, **63**, 4808 (1998).

1998JOC5031 Z. Xu and X. Lu, *J. Org. Chem.*, **63**, 5031 (1998).

1998JOC6053 M.M. Faul, L.L. Winneroski, and C.A. Krumrich, *J. Org. Chem.*, **63**, 6053 (1998).

1998JOC6220 S.E. Boiadjiev and D.A. Lightner, *J. Org. Chem.*, **63**, 6220 (1998).

1998MI1 H. Poschenrieder, G. Höfner, and H.-D. Stachel, *Arch. Pharm.*, **331**, 389 (1998).

1998PT(1)667 T.J. Donohoe, P.M. Guyo, R.L. Beddoes, and M. Helliwell, *J. Chem. Soc. Perkin Trans. 1*, 667 (1998).

1998PT(1)2443 A. Detsi, M. Micha-Screttas, and O. Igglessi-Markopoulou, *J. Chem. Soc. Perkin Trans. 1*, 2443 (1998).

1998SL507 J. Cossy, M. Cases, and D.G. Pardo, *Synlett*, 507 (1998).

1998TA797 I. Baussanne, C. Travers, and J. Royer, *Tetrahedron: Asymmetry*, **9**, 797 (1998).

1998T6909 E.M. Beccalli, M.L. Gelmi, and A. Marchesini, *Tetrahedron*, **54**, 6909 (1998).

1998T9447 G. Guillena, B. Mancheño, C. Nájera, J. Ezquerra, and C. Pedregal, *Tetrahedron*, **54**, 9447 (1998).

1998T10227 G. Reginato, A. Mordini, A. Degl'Innocenti, S. Manganiello, A. Capperucci, and G. Poli, *Tetrahedron*, **54**, 10227 (1998).

1998T12361 I. Manteca, B. Etxarri, A. Ardeo, S. Arrasate, I. Osante, N. Sotomayor, and E. Lete, *Tetrahedron*, **54**, 12361 (1998).

1998TL4369 B.A. Kulkarni and A. Ganesan, *Tetrahedron Lett.*, **39**, 4369 (1998).

1999EJOC3315 H. Rudler, A. Parlier, M. Ousmer, and J. Vaissermann, *Eur. J. Org. Chem.*, 3315 (1999).

1999H(50)463 R. Bossio, S. Marcaccini, R. Pepino, and T. Torroba, *Heterocycles*, **50**, 463 (1999).

1999JA1758 T. Morimoto, N. Chatani, and S. Murai, *J. Am. Chem. Soc.*, **121**, 1758 (1999).

1999JA6501 Y. Kobayashi, T. Fujimoto, and T. Fukuyama, *J. Am. Chem. Soc.*, **121**, 6501 (1999).

1999JA9253 Q. Chen, M.T. Huggins, D.A. Lightner, W. Norona, and A.F. McDonagh, *J. Am. Chem. Soc.*, **121**, 9253 (1999).

1999JCC223 P.C. Miller, T.J. Owen, J.M. Molyneaux, J.M. Curtis, and C.R. Jones, *J. Comb. Chem.*, **1**, 223 (1999).

1999JOC405 K. Akaji, Y. Hayashi, Y. Kiso, and N. Kuriyama, *J. Org. Chem.*, **64**, 405 (1999).

1999JOC2567 A.D. Cuiper, M. Brzostowska, J.K. Gawronski, W.J.J. Smeets, A.L. Spek, H. Hiemstra, R.M. Kellogg, and B.L. Feringa, *J. Org. Chem.*, **64**, 2567 (1999).

1999JOC6005 A.G.M. Barrett, J. Head, M.L. Smith, N.C. Stock, A.J.P. White, and D.J. Williams, *J. Org. Chem.*, **64**, 6005 (1999).

1999JOC8275 A. Fürstner, J. Grabowski, and C.W. Lehmann, *J. Org. Chem.*, **64**, 8275 (1999).

1999JOMC(579)177 N. Chatani, T. Morimoto, A. Kamitani, Y. Fukumoto, and S. Murai, *J. Organomet. Chem.*, **579**, 177 (1999).

1999MI1 R. Shiraki and K-i. Tadano, *Rev. Heteroatom Chem.*, **20**, 283 (1999).

1999MI2 A.K. Tipton and D.A. Lightner, *Monatsh. Chem.*, **130**, 425 (1999).

1999OL67 B.M. Trost and G.J. Roth, *Org. Lett.*, **1**, 67 (1999).

1999OPRD177 S.P. Raillard, G. Ji, A.D. Mann, and T.A. Baer, *Org. Proc. Res. Dev.*, **3**, 177 (1999).

1999S275 S. Eils and E. Winterfeldt, *Synthesis*, 275 (1999).

| | |
|---|---|
| 1999SL602 | C. Dagoneau, J.-N. Denis, and Y. Vallée, *Synlett*, 602 (1999). |
| 1999SL901 | T. Kakiuchi, H. Kinoshita, and K. Inomata, *Synlett*, 901 (1999). |
| 1999SL1333 | G. Rassu, F. Zanardi, L. Battistini, and G. Casiraghi, *Synlett*, 1333 (1999). |
| 1999T5839 | F. Ghelfi, F. Bellesia, L. Forti, G. Ghirardini, R. Grandi, E. Libertini, M.C. Montemaggi, U.M. Pagnoni, A. Pinetti, L. De Buyck, and A.F. Parsons, *Tetrahedron*, **55**, 5839 (1999). |
| 1999TA1873 | M. Petroliagi and O. Igglessi-Markopoulou, *Tetrahedron: Asymmetry*, **10**, 1873 (1999). |
| 1999TA3827 | P.Q. Huang, Q.F. Chen, C.L. Chen, and H.K. Zhang, *Tetrahedron: Asymmetry*, **10**, 3827 (1999). |
| 1999TA4469 | N. Mase, T. Nishi, Y. Takamori, H. Yoda, and K. Takabe, *Tetrahedron: Asymmetry*, **10**, 4469 (1999). |
| 1999TL8595 | F. Ghelfi, G. Ghirardini, E. Libertini, L. Forti, and U.M. Pagnoni, *Tetrahedron Lett.*, **40**, 8595 (1999). |
| 2000BCSJ497 | K.P. Jayasundera, H. Kinoshita, and K. Inomate, *Bull. Chem. Soc. Jpn.*, **73**, 497 (2000). |
| 2000BMCL1745 | J. Bosch, T. Roca, J.-L. Catena, O. Llorens, J.-J. Pérez, C. Lagunas, A.G. Fernández, I. Miquel, A. Fernández-Serrat, and C. Farrerons, *Bioorg. Med. Chem. Lett.*, **10**, 1745 (2000). |
| 2000CPB1593 | M. Arisawa, M. Takahashi, E. Takezawa, T. Yamaguchi, Y. Torisawa, A. Nishida, and M. Nakagawa, *Chem. Pharm. Bull.*, **48**, 1593 (2000). |
| 2000CR1929 | G. Casiraghi, F. Zanardi, G. Appendino, and G. Rassu, *Chem. Rev*, **1000**, 1929 (2000). |
| 2000CSR109 | G. Rassu, F. Zanardi, L. Battistini, and G. Casiraghi, *Chem. Soc. Rev.*, **29**, 109 (2000). |
| 2000EJOC3961 | K. Rück-Braun and P. Amrhein, *Eur. J. Org. Chem.*, 3961 (2000). |
| 2000H(53)2119 | G.E. Burtin, D.J. Madge, and D.L. Selwood, *Heterocycles*, **53**, 2119 (2000). |
| 2000JA8168 | S.B. Garber, J.S. Kingsburgy, B.L. Gray, and A.H. Hoveyda, *J. Am. Chem. Soc.*, **122**, 8168 (2000). |
| 2000JAB1247 | N. Kakinuma, H. Iwai, S. Takahashi, K. Hamano, T. Yanagisawa, K. Nagai, K. Tanaka, K. Suzuki, F. Kirikae, T. Kirikae, and A. Nakagawa, *J. Antibiot.*, **53**, 1247 (2000). |
| 2000JAB1252 | S. Takahashi, N. Kakinuma, H. Iwai, T. Yanagisawa, K. Nagai, A. Suzuki, T. Tokunaga, and A. Nakagawa, *J. Antibiot.*, **53**, 1252 (2000). |
| 2000JHC143 | S.E. Ghammarti, B. Rigo, H. Mejdl, J.-P. Henichart, and D. Couturier, *J. Heterocycl. Chem.*, **37**, 143 (2000). |
| 2000JNP768 | S. Suzuki, T. Hosoe, K. Nozawa, K-i. Kawai, Y. Yaguchi, and S-i. Udagawa, *J. Nat. Prod.*, **63**, 768 (2000). |
| 2000JNP965 | K.E. Milligan, B. Márquez, R.T. Williamson, M. Davies-Coleman, and W.H. Gerwick, *J. Nat. Prod.*, **63**, 965 (2000). |
| 2000JNP1576 | J-i. Kobayashi, M. Sekiguchi, S. Shimamoto, H. Shigemori, and A. Ohsaki, *J. Nat. Prod.*, **63**, 1576 (2000). |
| 2000JOC205 | P.A. Jacobi, L.D. Coutts, J. Guo, S.I. Hauck, and S.H. Leung, *J. Org. Chem.*, **65**, 203 (2000). |
| 2000JOC2204 | A. Fürstner, O.R. Thiel, L. Ackermann, H.-J. Schanz, and S.P. Nolan, *J. Org. Chem.*, **65**, 2204 (2000). |
| 2000JOC6412 | C. Osterhage, R. Kaminsky, G.M. König, and A.D. Wright, *J. Org. Chem.*, **65**, 6412 (2000). |
| 2000JOC8478 | P.A. Jacobi, R.W. DeSimone, I. Ghosh, J. Guo, S.H. Leung, and D. Pippin, *J. Org. Chem.*, **65**, 8478 (2000). |
| 2000MI1 | M. Watabe, K. Machida, and H. Osada, *Cancer Res.*, **60**, 5214 (2000). |

2000OL3801 S. Ma and H. Xie, *Org. Lett.*, **2**, 3801 (2000).

2000OM3527 P. Amrhein, D. Schollmeyer, and K. Rück-Braun, *Organometallics*, **19**, 3527 (2000).

2000S925 N. Chatani, M. Tobisu, T. Asaumi, and M. Murai, *Synthesis*, 925 (2000).

2000T7869 J.O. Brower, D.A. Lightner, and A.F. McDonagh, *Tetrahedron*, **56**, 7869 (2000).

2000TL2419 Y. Nagao, S. Sano, K. Morimoto, H. Kakegawa, T. Takatani, M. Shiro, and N. Katunuma, *Tetrahedron Lett.*, **41**, 2419 (2000).

2000TL2825 C. Pichon-Santander and A.I. Scott, *Tetrahedron Lett.*, **41**, 2825 (2000).

2000TL2939 J-i. Kobayashi, M. Sekiguchi, H. Shigemori, and O. Ayumi, *Tetrahedron Lett.*, **41**, 2939 (2000).

2000TL3669 M. Bella, G. Piancatelli, A. Squarcia, and C. Trolli, *Tetrahedron Lett.*, **41**, 3669 (2000).

2000TL9477 C. Peschko and W. Steglich, *Tetrahedron Lett.*, **41**, 9477 (2000).

2000TL9859 K. Takabe, M. Suzuki, T. Nishi, M. Hiyoshi, Y. Takamori, H. Yoda, and N. Mase, *Tetrahedron Lett.*, **41**, 9859 (2000).

2001ACIE4212 D. Witthaut, R. Fröhlich, and H.J. Schäfer, *Angew. Chem., Int. Ed*, **40**, 4212 (2001).

2001ACIE4251 J.S. Kingsburgy, S.B. Garber, J.M. Giftos, B.L. Gray, M.M. Okamoto, R.A. Rarrer, J.T. Fourkas, and A.H. Hoveyda, *Angew. Chem. Int. Ed.*, **40**, 4251 (2001).

2001CBC60 A. Fürstner, J. Grabowsi, C.W. Lehmann, T. Kataoka, and K. Nagai, *ChemBioChem*, **2**, 60 (2001).

2001CL590 S. Takeda, K.P. Yayasundera, T. Kakiuchi, H. Kinoshita, and K. Inomata, *Chem. Lett.*, **30**, 590 (2001).

2001CL738 V. Nair, J.S. Mathen, A.U. Vinod, and R.L. Varma, *Chem. Lett.*, **30**, 738 (2001).

2001EJOC1673 L. Bhat, A.G. Steinig, R. Appelbe, and A. de Meijere, *Eur. J. Org. Chem.*, 1673 (2001).

2001H(55)2099 B.A. Frieman, C.W. Bock, and K.L. Bhat, *Heterocycles*, **55**, 2099 (2001).

2001JA10214 B.G. Van den Hoven and H. Alper, *J. Am. Chem. Soc.*, **123**, 10214 (2001).

2001JA11312 J. Louie, C.W. Bielawski, and R.H. Grubbs, *J. Am. Chem. Soc.*, **123**, 11312 (2001).

2001JAB109 K. Tatsuta, H. Misawa, and K. Chikauchi, *J. Antibiot.*, **54**, 109 (2001).

2001JAB831 W.-G. Kim, H.K. Song, and I.D. Yoo, *J. Antibiot.*, **54**, 831 (2001).

2001JHC527 P. Bobál and D.A. Lightner, *J. Heterocycl. Chem.*, **38**, 527 (2001).

2001JHC591 R.L. Hudkins and N.W. Johnson, *J. Heterocycl. Chem.*, **38**, 591 (2001).

2001JHC1203 G. Athanasellis, E. Gavrielatos, and O. Igglessi-Markopoulou, *J. Heteerocycl. Chem.*, **38**, 1203 (2001).

2001JMC3074 S. Hanessian, D.B. MacKay, and N. Moitessier, *J. Med. Chem.*, **44**, 3074 (2001).

2001JNP1486 G.H. Goetz, G.G. Harrigan, and J. Likos, *J. Nat. Prod.*, **64**, 1486 (2001).

2001JOC8287 F.A. Bermejo, R. Rico-Ferreira, S. Bamidele-Sanni, and S. García-Granda, *J. Org. Chem.*, **66**, 8287 (2001).

2001MI1 B. Oliva, A. O'Neill, J.M. Wilson, P.J. O'Hanlon, and I. Chopra, *Antimicrob. Agents Chemother.*, **45**, 532 (2001).

2001MI2 A.P. Bai, Z.-R. Guo, W.H. Hu, F. Shen, and G.F. Cheng, *Chin. Chem. Lett.*, **12**, 775 (2001).

2001MI3 D.-W. Ma, Y.-D. Zhang, X.-R. Zhang, S.-H. Wu, and F.-G. Tao, *Chin. J. Chem.*, **19**, 489 (2001).

2001MI4 A.Y. Egorova, V.A. Sedavkina, and Z.Y. Timofeeva, *Chem. Heterocycl. Comp.*, **37**, 694 (2001).

2001MI5 A.J. Clark, C.P. Dell, and J.P. McDonagh, *C. R. Acad. Sci. Paris Chim.*, **4**, 575 (2001).

2001MI6 E.-C. Wang, K.-S. Huang, G.-W. Lin, J.-R. Lin, and M.-K. Hsu, *J. Chin. Chem. Soc.*, **48**, 83 (2001).

2001MI7 J.O. Brower and D.A. Lightner, *Monatsh. Chem.*, **132**, 1572 (2001).

2001OL2851 S.-K. Kang, K.-J. Kim, C.-M. Yu, J.-W. Hwang, and Y.-K. Do, *Org. Lett.*, **3**, 2851 (2001).

2001PT(1)1421 J. Courcambeck, F. Bihel, C. De Michelis, G. Quéléver, and J.L. Kraus, *J. Chem. Soc. Perkin Trans. 1*, 1421 (2001).

2001SC1481 C. Vaccher, *Synth. Commun.*, **31**, 1481 (2001).

2011SL37 G. V-Thanh, V. Boucard, H. Sauriat-Dorizon, and F. Gulbe, *Synlett*, **37** (2001).

2001SL1437 P. Langer and M. Döring, *Synlett*, 1437 (2001).

2001SL1653 G. Athanasellis, E. Gavrielatos, and O. Igglessi-Markopoulou, *Synlett*, 1653 (2001).

2001T1813 F.M. Abdelrazek, A.M. Salah El-Din, and A.E. Mekky, *Tetrahedron*, **57**, 1813 (2001).

2001T6353 C.M. Acevedo, E.F. Kogut, and M.A. Lipton, *Tetrahedron*, **57**, 6353 (2001).

2001T7813 J.O. Brower, D.A. Lightner, and A.F. McDonagh, *Tetrahedron*, **57**, 7813 (2001).

2001TL573 A. Chihab-Eddine, A. Daïch, A. Jilale, and B. Decroix, *Tetrahedron Lett.*, **42**, 573 (2001).

2001TL1999 A.J. Clark, G.M. Battle, and A. Bridge, *Tetrahedron Lett.*, **42**, 1999 (2001).

2001TL3831 C. Grison, S. Genéve, and P. Coutrot, *Tetrahedron Lett.*, **42**, 3831 (2001).

2001TL4573 F. Bellesia, L. De Buyck, M.V. Colucci, F. Ghelfi, I. Laureyn, E. Libertini, A. Mucci, U.M. Pagnoni, A. Pinetti, T.M. Rogge, and C.V. Stevens, *Tetrahedron Lett.*, **42**, 4573 (2001).

2001TL7907 R.-Y. Kuo, F.-R. Chang, and Y.-C. Wu, *Tetrahedron Lett.*, **42**, 7907 (2001).

2001TL8523 M.F. Greaney, W.B. Mortherwell, and D.A. Tocher, *Tetrahedron Lett.*, **42**, 8523 (2001).

2002ACIE2602 T. Akiyama and S. Kobayashi, *Angew. Chem. Int. Ed.*, **41**, 2602 (2002).

2002ARK(i)61 R. Halim, M.A. Brimble, and P.D. Woodgate, *Arkivoc*, **61** (2002).

2002BMCL561 P. Zarantonello, C.P. Leslie, R. Ferritto, and W.M. Kazmierski, *Bioorg. Med. Chem. Lett.*, **12**, 561 (2002).

2002BMCL3337 S. Tsymbalov, T.J. Hagen, W.M. Moore, G.M. Jerone, J.R. Connor, P.T. Manning, B.S. Pitzele, and E.A. Hallinan, *Bioorg. Med. Chem. Lett.*, **12**, 3337 (2002).

2002H(57)881 M. Ito, H. Okui, H. Nakagawa, S. Mio, T. Iwasaki, and J. Iwabuchi, *Heterocycles*, **57**, 881 (2002).

2002JA4716 S. Hanessian, M. Bayrakdarian, and X. Luo, *J. Am. Chem. Soc.*, **124**, 4716 (2002).

2002JA4954 J.J. Van Veldhuizen, S.B. Garber, J.S. Kingsburgy, and A.H. Hoveyda, *J. Am. Chem. Soc.*, **124**, 4954 (2002).

2002JAB696 L. Lamari, A. Zitouni, H. Boudjella, B. Badji, N. Sabaou, A. Lebrihi, G. Lefebvre, E. Seguin, and F. Tillequin, *J. Antibiot.*, **55**, 696 (2002).

2002JBC31243 K. Machida, Y. Hayashi, and H. Osada, *J. Biol. Chem.*, **277**, 31243 (2002).

2002JNP1206 N. Borbone, S. De Marino, M. Iorizzi, F. Zollo, C. Debitus, G. Esposito, and T. Iuvone, *J. Nat. Prod.*, **65**, 1206 (2002).

2002JNP1307 Y. Liu, J. Hong, C.-O. Lee, K.S. Im, N.D. Kim, J.S. Choi, and J.H. Jung, *J. Nat. Prod.*, **65**, 1307 (2002).

2002JOC847 R.K. Dieter and K. Lu, *J. Org. Chem.*, **67**, 847 (2002).

2002JOC4702 W.-R. Li, S.T. Lin, N.-M. Hsu, and M.-S. Chern, *J. Org. Chem.*, **67**, 4702 (2002).

2002JOC5445 D.A. DeGoey, H.-J. Chen, W.J. Flosi, D.J. Grampovnik, C.M. Yeung, L.L. Klein, and D.J. Kempf, *J. Org. Chem.*, **67**, 5445 (2002).

2002JOC7319 R. Di Fabio, G. Alvaro, B. Bertani, D. Donati, S. Giacobbe, C. Marchioro, C. Palma, and S.M. Lynn, *J. Org. Chem.*, **67**, 7319 (2002).

2002MI1 F. Shen, A.P. Bai, Z.-R. Guo, and G.F. Cheng, *Acta Pharm. Sin.*, **23**, 762 (2002).

2002MI2 J. Westman and K. Orrling, *Comb. Chem. High Throughput Screening*, **5**, 571 (2002).

2002OL119 J.P.N. Papillon and R.J.K. Taylor, *Org. Lett.*, **4**, 119 (2002).

2002OL2005 B.M. Trost, M.J. Krische, V. Berl, and E.M. Grenzer, *Org. Lett.*, **4**, 2005 (2002).

2002OL2633 A. Namsa-aid and S. Ruchirawat, *Org. Lett.*, **4**, 2633 (2002).

2002OPRD416 Z. Tian, M. Rasmussen, and S.J. Wittenberger, *Org. Proc. Res. Dev.*, **6**, 416–418 (2002).

2002PT(1)707 N. Mase, T. Nishi, M. Hiyoshi, K. Ichihara, J. Bessho, H. Yoda, and K. Takabe, *Perkin Trans. 1*, 707 (2002).

2002S869 F.F. Paintner, M. Metz, and G. Bauschke, *Synthesis*, 869 (2002).

2002SC1031 Q. Chen, T. Wang, Y. Zhang, Q. Wang, and J. Ma, *Synth. Commun.*, **32**, 1031 (2002).

2002SC2527 I. Yavari and M. Bayat, *Synth. Commun.*, **32**, 2527 (2002).

2002SL593 I. Gonzalez-Temprano, N. Sotomayor, and E. Lete, *Synlett*, 593 (2002).

2002SL947 V.R. Pattabiraman, S. Padakanti, V.R. Veeramaneni, M. Pal, and K.R. Yeleswarapu, *Synlett*, 947 (2002).

2002T7959 H. Luesch, W.Y. Yoshida, R.E. Moore, and V.J. Paul, *Tetrahedron*, **58**, 7959 (2002).

2002T8113 V. Nair, J.S. Mathen, S. Viji, R. Srinivas, M.V. Nandakumar, and L. Varma, *Tetrahedron*, **58**, 8113 (2002).

2002T9239 R. Akué-Gédu, S.A.A. Ebrik, A. Witczak-Legrand, D. Fasseur, S. El Ghammarti, D. Couturier, B. Decroix, M. Othman, M. Debacker, and B. Rigo, *Tetrahedron*, **58**, 9239 (2002).

2002TA601 A.S. Demir, F. Aydogan, and I.M. Akhmedov, *Tetrahedron: Asymmetry*, **13**, 601 (2002).

2002TL1569 N.A. Nedolya, N.I. Schlyakhtina, V.P. Sinov′eva, A.I. Albanov, and L. Brandsma, *Tetrahedron Lett.*, **43**, 1569 (2002).

2002TL1633 H. He, R. Bigelis, E.H. Solum, M. Greenstein, and G.T. Carter, *Tetrahedron Lett.*, **43**, 1633 (2002).

| | |
|---|---|
| 2002TL2649 | M.P. Green, J.C. Prodger, and C.J. Hayes, *Tetrahedron Lett.*, **43**, 2649 (2002). |
| 2002TL4609 | R. Detterbeck and M. Hesse, *Tetrahedron Lett.*, **43**, 4609 (2002). |
| 2002TL5057 | S.J.F. Macdonald, G.G.A. Inglis, D. Bentley, and M.D. Dowle, *Tetrahedron Lett.*, **43**, 5057 (2002). |
| 2002TL5131 | S.-H. Li, H.-J. Zhang, S.-X. Qiu, X.-M. Niu, B.D. Santarsiero, A.D. Mesecar, H.H.S. Fong, N.R. Farnsworth, and H.-D. Sun, *Tetrahedron Lett.*, **43**, 5131 (2002). |
| 2002TL6609 | M.P. Green, J.C. Prodger, and C.J. Hayes, *Tetrahedron Lett.*, **43**, 6609 (2002). |
| 2003ACIE3582 | A. Fürstner, *Angew. Chem. Int. Ed.*, **43**, 3582 (2003). |
| 2003ARK112 | H. Inagaki, K. Sugita, R.N. Miyauchi, S. Miyauchi, T. Takeda, M. Itoh, H. Takahashi, and M. Takemura, *Arkivoc*, 112 (2003). |
| 2003BMCL2505 | Z. Liu, X. Ruan, and X. Huang, *Bioorg. Med. Chem. Lett.*, **13**, 2505 (2003). |
| 2003CC776 | M. Oba, N. Nishiyama, and K. Nishiyama, *Chem. Commun.*, 776 (2003). |
| 2003EJOC47 | T.R. Appel, N.A.M. Yehia, U. Baumeister, H. Hartung, R. Kluge, D. Ströhl, and E. Fanghänel, *Eur. J. Org. Chem.*, **47** (2003). |
| 2003EJOC4593 | A. Detsi, A. Afantis, G. Athanasellis, J. Markopoulos, O. Igglessi-Markopoulou, and C.-K. Skylaris, *Eur. J. Org. Chem.*, 4593 (2003). |
| 2003JA4028 | M. Lautens, W. Han, and J.H.-C. Liu, *J. Am. Chem. Soc.*, **125**, 4028 (2003). |
| 2003JHC135 | R.L. Hudkins and C.H. Park, *J. Heterocycl. Chem.*, **40**, 135 (2003). |
| 2003JMC1005 | H. Inagaki, S. Miyauchi, R.N. Miyauchi, H.C. Kawato, H. Ohki, N. Matsuhashi, K. Kawakami, H. Takahashi, and M. Takemura, *J. Med. Chem.*, **46**, 1005 (2003). |
| 2003JOC1552 | R. Andrukiewicz, R. Loska, V. Prisyahnyuk, and K. Stalinksi, *J. Org. Chem.*, **68**, 1552 (2003). |
| 2003JOC4523 | X.-R. Zhang, W. Jiang, and Z. Sui, *J. Org. Chem.*, **68**, 4523 (2003). |
| 2003MI1 | A.Y. Egorova and Z.Y. Timofeeva, *Izbrannye Metody Sinteza i Modifikatsii*, 181 (2003) [*Chem. Abstr.*, **141**, 331971(2003)]. |
| 2003OL553 | J. Zhang, P.G. Blazecka, and J.G. Davidson, *Org. Lett.*, **5**, 553 (2003). |
| 2003OL2063 | A.J. Clark, C.P. Dell, A.F. McDonagh, J. Geden, and P. Mawdsley, *Org. Lett.*, **5**, 2063 (2003). |
| 2003OL2615 | F. Nishino, K. Miki, Y. Kato, K. Ohe, and S. Uemura, *Org. Lett.*, **5**, 2615 (2003). |
| 2003OL4341 | P.-Q. Huang, T.-J. Wu, and Y.-P. Ruan, *Org. Lett.*, **5**, 4341 (2003). |
| 2003PH1173 | P.K. Cheplogoi and D.A. Mulholland, *Phytochemistry*, **62**, 1173 (2003). |
| 2003S497 | G.M. Reddy, S.-Y. Chen, and B.-J. Uang, *Synthesis*, 497 (2003). |
| 2003SC223 | E.S. Greenwood, P.J. Parsons, and M.J. Young, *Synth. Commun.*, **33**, 223 (2003). |
| 2003SL1007 | M.A. Bonache, G. Gerona-Navarro, M. Martin-Martinez, M.T. Garcia-Lopez, P. Lopez, C. Cativiela, and R. Gonzalez-Muniz, *Synlett*, 1007 (2003). |
| 2003SL2013 | G. Verniest and N. De Kimpe, *Synlett*, 2013 (2003). |
| 2003T217 | C. Herdeis and B. Kelm, *Tetrahedron*, **59**, 217 (2003). |
| 2003T1147 | F. Ghelfi, C.V. Stevens, I. Laureyn, E. Van Meenen, T.M. Rogge, L. De Buyck, K.V. Nikitin, R. Grandi, E. Libertini, U.M. Pagnoni, and L. Schenetti, *Tetrahedron*, **59**, 1147 (2003). |
| 2003T6221 | J.S. Bryans, N.E.A. Chessum, N. Huther, A.F. Parsons, and F. Ghelfi, *Tetrahedron*, **59**, 6221 (2003). |

2003TL4853 M.K. Gurjar, R.A. Joshi, S.R. Chaudhuri, S.V. Joshi, A.R. Barde, L.K. Gediya, P.V. Ranade, S.M. Kadam, and S.J. Naik, *Tetrahedron Lett.*, **44**, 4853 (2003).

2003TL7417 T. Komoda, Y. Sugiyama, N. Abe, M. Imachi, H. Hirota, H. Koshino, and A. Hirota, *Tetrahedron Lett.*, **44**, 7417 (2003).

2003TL7485 J.A. Grzyb and R.A. Batey, *Tetrahedron Lett.*, **44**, 7485 (2003).

2003TL7803 C. Gill, D.A. Greenhalgh, and N.S. Simpkins, *Tetrahedron Lett.*, **44**, 7803 (2003).

2004CC1422 P.G. Turner, T.J. Donohoe, and R.P.C. Cousins, *Chem. Commun.*, 1422 (2004).

2004EJOC1897 J.T. Anders, H. Görls, and P. Langer, *Eur. J. Org. Chem.*, 1897 (2004).

2004H(63)699 H. Inagaki, T. Takeda, R.N. Miyauchi, K. Kawakami, H. Takahashi, and M. Takemura, *Heterocycles*, **63**, 699 (2004).

2004H(63)1009 R. Naitoh, Y. Nakamura, E. Katano, Y. Nakamura, E. Okada, and M. Asaoka, *Heterocycles*, **63**, 1009 (2004).

2004H(63)1013 J.-i. Bessho, Y. Shimotsu, S. Mizumoto, N. Mase, H. Yoda, and K. Takabe, *Heterocycles*, **63**, 1013 (2004).

2004IJC(B)2467 K. Bhandari and V.L. Sharma, *Indian J. Chem. Sect. B*, **43B**, 2467 (2004).

2004JCC684 Y. Chen, H. Zhang, and F. Nan, *J. Comb. Chem.*, **6**, 684 (2004).

2004JMC3180 W.A. Carroll, K.A. Agrios, R.J. Altenbach, S.A. Buckner, Y. Chen, M.J. Coghlan, A.V. Daza, I. Drizin, M. Gopalakrishnan, R.F. Henry, M.E. Kort, P.R. Kym, I. Milicic, J.C. Smith, R. Tang, S.C. Turner, K.L. Whiteaker, H. Zhang, and J.P. Sullivan, *J. Med. Chem.*, **47**, 3180 (2004).

2004JNP1532 J. Hiort, K. Maksimenka, M. Reichert, S. Perovic-Ottstadt, W.H. Lin, V. Wray, K. Steube, K. Schaumann, H. Weber, P. Proksch, R. Ebel, W.E.G. Müller, and G. Bringmann, *J. Nat. Prod.*, **67**, 1532 (2004).

2004JOC6105 R.T. Watson, V.K. Gore, K.R. Chandupatla, R.K. Dieter, and J.P. Snyder, *J. Org. Chem.*, **69**, 6105 (2004).

2004JOC6305 S. Ma, B. Ni, and Z. Liang, *J. Org. Chem.*, **69**, 6305 (2004).

2004JOC5974 Y. Dejaegher and N. De Kimpe, *J. Org. Chem.*, **69**, 5974 (2004).

2004JOC8952 B.B. Snider and B.J. Neubert, *J. Org. Chem.*, **69**, 8952 (2004).

2004MI1 A.Y. Egorova and Z.Y. Timofeeva, *Chem. Heterocycl. Comp.*, **40**, 1243 (2004).

2004MI2 X.-H. Chen, J.-Y. Bai, F. Shen, A.P. Bai, Z.-R. Guo, and G.F. Cheng, *Acta Pharm. Sin.*, **25**, 927 (2004).

2004MI3 K.V. Nikitin and N.P. Andryukhova, *Chem. Heterocycl. Comp.*, **40**, 561 (2004).

2004OBC3524 R. Schobert, C. Jagusch, C. Melanophy, and G. Mullen, *Org. Biomol. Chem.*, **2**, 3524 (2004).

2004OL39 B. Beck, A. Picard, E. Herdtweck, and A. Dumling, *Org. Lett.*, **6**, 39 (2004).

2004OL577 R.S. Coleman and P.-H. Liu, *Org. Lett.*, **6**, 577 (2004).

2004OL609 J. Clayden, R. Turnbull, and I. Pinto, *Org. Lett.*, **6**, 609 (2004).

2004OL3229 Y. Tang and C. Li, *Org. Lett.*, **6**, 3229 (2004).

2004S711 J.-P. Bouillon, B. Tinant, J.-M. Nuzillard, and C. Portella, *Synthesis*, 711 (2004).

2004S1560 S. Mangaleswaran and N.P. Argade, *Synthesis*, 1560 (2004).

2004SL247 P.-Q. Huang and J. Deng, *Synlett*, 247 (2004).

| | |
|---|---|
| 2004SL1059 | G. Verniest, S. Boterberg, F. Bombeke, C.V. Stevens, and N. De Kimpe, *Synlett*, 1059 (2004). |
| 2004SL2779 | J.T. Anders, V.T.H. Nguyen, and P. Langer, *Synlett*, 2779 (2004). |
| 2004T3987 | M. Pal, N.K. Swamy, P.S. Hamseed, S. Padakanti, and K.R. Yeleswarapu, *Tetrahedron*, **60**, 3987 (2004). |
| 2004T9017 | B. Tu, B. Ghosh, and D.A. Lightner, *Tetrahedron*, **60**, 9017 (2004). |
| 2004T11493 | C. Danieli, F. Ghelfi, A. Mucci, U.M. Pagnoni, A.F. Parsons, M. Pattarozzi, and L. Schenetti, *Tetrahedron*, **60**, 11493 (2004). |
| 2004TL57 | M. Marhold, A. Buer, H. Hiemstra, J.H. van Maarseveen, and G. Haufe, *Tetrahedron Lett.*, **45**, 57 (2004). |
| 2004TL719 | R. Badorrey, C. Cativiela, M. Díaz-de-Villegas, R. Díaz, and J.A. Gálvez, *Tetrahedron Lett.*, **45**, 719 (2004). |
| 2004TL959 | V. De Matteis, F.L. van Delft, R. de Gelder, J. Tiebes, and F.P.J.T. Rutjes, *Tetrahedron Lett.*, **45**, 959 (2004). |
| 2004TL1413 | M. Quai, S. Frattini, U. Vendrame, M. Mondoni, S. Dossena, and E. Cereda, *Tetrahedron Lett.*, **45**, 1413 (2004). |
| 2004TL2063 | K. Cherry, J. Thibonnet, A. Duchêne, J.-L. Parrain, and M. Abarbri, *Tetrahedron Lett.*, **45**, 2063 (2004). |
| 2004TL8793 | S.-J. Park, K.-N. Cho, W.-G. Kim, and K.-I. Lee, *Tetrahedron Lett.*, **45**, 8793 (2004). |
| 2004TL8931 | A.S. Kyei, K. Tchabanenko, J.E. Baldwin, and R.M. Adlington, *Tetrahedron Lett.*, **45**, 8931 (2004). |
| 2004TL9197 | R. Akué-Gédu, J.-P. Hénichart, D. Couturier, and B. Rigo, *Tetrahedron Lett.*, **45**, 9197 (2004). |
| 2004TL9353 | G. Tsolomiti and A. Tsolomiti, *Tetrahedron Lett.*, **45**, 9353 (2004). |
| 2005ACIE2777 | A. Fürstner, K. Radkowski, and H. Peters, *Angew. Chem. Int. Ed.*, **44**, 2777 (2005). |
| 2005ACIE3110 | J. Yamaguchi, H. Kakeya, T. Uno, M. Shoji, H. Osada, and Y. Hayashi, *Angew. Chem. Int. Ed.*, **44**, 3110 (2005). |
| 2005AJC882 | R. Adhikari, D.A. Jones, A.J. Liepa, and R.H. Nearn, *Aust. J. Chem.*, **58**, 882 (2005). |
| 2005BMCL4043 | A.K. Mandal, J. Hines, K. Kuramochi, and C.M. Crews, *Bioorg. Med. Chem. Lett.*, **15**, 4043 (2005). |
| 2005CEJ4227 | T.J. Donohoe, H.O. Sintim, L. Sisangia, K.W. Ace, P.M. Guyo, A. Cowley, and J.D. Harling, *Chem. Eur. J.*, **11**, 4227 (2005). |
| 2005CL800 | H. Kinoshita, M.A.S. Hammam, and K. Inomata, *Chem. Lett.*, **34**, 800 (2005). |
| 2005H(65)1481 | Y. Yoshida, S. Ichikawa, Y. Shinozuka, M. Satoh, K. Mohri, and K. Isobe, *Heterocycles*, **65**, 1481 (2005). |
| 2005H(65)2451 | K. Matsuo, J. Kawanishi, M. Kobayashi, and S. Ueno, *Heterocycles*, **65**, 2451 (2005). |
| 2005JA4510 | J.S. Kingsburgy and A.H. Hoveyda, *J. Am. Chem. Soc.*, **127**, 4510 (2005). |
| 2005JA8298 | A. Endo and S.J. Danishefsky, *J. Am. Chem. Soc.*, **127**, 8298 (2005). |
| 2005JA12230 | X. Zhang and R.C. Larock, *J. Am. Chem. Soc.*, **127**, 12230 (2005). |
| 2005JA16038 | R.S. Coleman, M.C. Walczak, and E.L. Campbell, *J. Am. Chem. Soc.*, **127**, 16038 (2005). |
| 2005JHC731 | Z.R. Woydziak, B.J. Frost, and D.A. Lightner, *J. Heterocycl. Chem.*, **42**, 731 (2005). |
| 2005JMC5932 | S.M. Jones, J.E. Urch, M. Kaiser, R. Brun, J.L. Harwood, C. Berry, and I.H. Gilbert, *J. Med. Chem.*, **48**, 5932 (2005). |
| 2005JMC8009 | R. Dayam, T. Sanchez, and N. Neamati, *J. Med. Chem.*, **48**, 8009 (2005). |

2005JOC2109 R.K. Dieter, N. Chen, H. Yu, L.E. Nice, and V.K. Gore, *J. Org. Chem.*, **70**, 2109 (2005).

2005JOC3826 X.-L. Qiu and F.-L. Qing, *J. Org. Chem.*, **70**, 3826 (2005).

2005JOC6196 P.G. Williams, G.O. Buchanan, R.H. Feling, C.A. Kauffman, R.J. Jensen, and W. Fenical, *J. Org. Chem.*, **70**, 6196 (2005).

2005JOC8417 Z.R. Woydziak, S.E. Boiadjiev, W.S. Norona, A.F. McDonagh, and D.A. Lightner, *J. Org. Chem.*, **70**, 8417 (2005).

2005MI1 S. Yamato, T. Fusaka, and Y. Tanaka, *J. Pestic. Sci.*, **30**, 384 (2005).

2005OL31 K.L. Burgess, N.J. Lajkiewicz, A. Sanyal, W. Yan, and J.K. Snyder, *Org. Lett.*, **7**, 31 (2005).

2005OL5521 D.M. Fitch, K.A. Evans, D. Chai, and K.J. Duffy, *Org. Lett.*, **7**, 5521 (2005).

2005OM4893 N. Rosas, P. Sharma, I. Arellano, M. Ramírez, D. Pérez, S. Hernández, and A. Cabrera, *Organometallics*, **24**, 4893 (2005).

2005SL83 A. Stepien, R. Loska, P. Cmoch, and K. Stalinksi, *Synlett*, **83** (2005).

2005SL453 C. Haase and P. Langer, *Synlett*, 453 (2005).

2005SL935 B. Gabriele, P. Plastina, G. Salerno, and M. Costa, *Synlett*, 935 (2005).

2005SL1734 R. Badorrey, C. Cativiela, M.D. Diaz-de-Villegas, R. Diez, and J.A. Galvez, *Synlett*, 1734 (2005).

2005SL2763 K.C. Prousis, A. Detsi, and O. Igglessi-Markopoulou, *Synlett*, 2763 (2005).

2005T251 S. Ma and H. Xie, *Tetrahedron*, **61**, 251 (2005).

2005T2301 R. Schobert and C. Jagusch, *Tetrahedron*, **61**, 2301 (2005).

2005T5713 J. Clayden, B. Read, and K. Hebditch, *Tetrahedron*, **61**, 5713 (2005).

2005T8456 M. Oba, N. Nishiyama, and K. Nishiyaya, *Tetrahedron*, **61**, 8456 (2005).

2005TA2235 J. Clayden, R. Turnbull, and I. Pinto, *Tetrahedron: Asymmetry*, **16**, 2235 (2005).

2005TL407 J-i. Matsuo and Y. Aizawa, *Tetrahedron Lett.*, **46**, 407 (2005).

2005TL2475 A. Sanyal, Q. Yuan, and J.K. Snyder, *Tetrahedron Lett.*, **46**, 2475 (2005).

2005TL3123 A.J. Pearson and X. Wang, *Tetrahedron Lett.*, **46**, 3123 (2005).

2005TL7893 Y.-C. Shen, Y.-S. Lin, Y.-H. Kuo, and Y.-B. Cheng, *Tetrahedron Lett.*, **46**, 7893 (2005).

2006ARK(vii)57 I. Baussanne, B. Dudot, J. Pérard-Viret, L. Planas, and J. Royer, *Arkivoc*, **57** (2006).

2006ARK(iii)163 M.G. Banwell, D.A.S. Beck, and A.C. Willis, *Arkivoc*, 163 (2006).

2006BCSJ1561 M.A.S. Hammam, H. Nakamura, Y. Hirata, H. Khawn, Y. Murata, H. Kinoshita, and K. Inomata, *Bull. Chem. Soc. Jpn.*, **79**, 1561 (2006).

2006BMCL2205 K.A. Evans, D. Chai, T.L. Graybill, G. Burton, R.T. Sarisky, J. Lin-Goerke, V.K. Johnston, and R.A. Rivero, *Bioorg. Med. Chem. Lett.*, **16**, 2205 (2006).

2006BMCL3906 F. Micheli, A. Paquarello, G. Tedesco, D. Hamprecht, G. Bonanomi, A. Checcia, A. Jaxa-Chamiec, F. Damiani, S. Davalli, D. Donati, C. Gallotti, M. Petrone, M. Rinaldi, G. Riley, S. Terreni, and M. Wood, *Bioorg. Med. Chem. Lett.*, **16**, 3906 (2006).

2006CEJ524 M. Shindo, T. Yoshikawa, Y. Itou, S. Mori, T. Nishii, and K. Shishido, *Chem. Eur. J.*, **12**, 524 (2006).

2006EJOC1166 V. De Matteis, F.L. Van Delft, J. Tiebes, and F.P.J.T. Rutjes, *Eur. J. Org. Chem.*, 1166 (2006).

2006EJOC2843 F. Palacios, J. Vicario, and D. Aparicio, *Eur. J. Org. Chem.*, 2843 (2006).

| | |
|---|---|
| 2006EJOC3730 | K.D. Sharma, J. Zhang, Y. Huang, and J.G. Davidson, *Eur. J. Org. Chem.*, 3730 (2006). |
| 2006EJOC4861 | K.Y. Koltunov, G.K.S. Prakash, G. Rasul, and G.A. Olah, *Eur. J. Org. Chem.*, 4861 (2006). |
| 2006JA1464 | K. Tonogaki, K. Itami, and J-i. Yoshida, *J. Am. Chem. Soc.*, **128**, 1464 (2006). |
| 2006JAB418 | T. Sasaki, S. Takahashi, K. Uchida, S. Funayama, M. Kainosho, and A. Nakagawa, *J. Antibiot.*, **59**, 418 (2006). |
| 2006JCC480 | G. Larbig and B. Schmidt, *J. Comb. Chem.*, **8**, 480 (2006). |
| 2006JHC949 | Y.-C. Wu, L. Liu, D. Wang, and Y.-J. Chen, *J. Heterocycl. Chem.*, **43**, 949 (2006). |
| 2006JNP580 | Y.C. Park, S.P. Gunasekera, J.V. Lopez, P.J. McCarthy, and A.E. Wright, *J. Nat. Prod.*, **69**, 580 (2006). |
| 2006JNP1212 | A. Grube and M. Köck, *J. Nat. Prod.*, **69**, 1212 (2006). |
| 2006JOC1220 | N. Fukuda, K. Sasaki, T.V.R.S. Sastry, and M. Kanai, *J. Org. Chem.*, **71**, 1220 (2006). |
| 2006JOC2173 | A. Gheorghe, M. Schulte, and O. Reiser, *J. Org. Chem.*, **71**, 2173 (2006). |
| 2006JOC4703 | F. Issa, J. Fischer, P. Turner, and M.J. Coster, *J. Org. Chem.*, **71**, 4703 (2006). |
| 2006JOC6678 | A.R. Coffin, M.A. Roussell, E. Tserlin, and E.T. Pelkey, *J. Org. Chem.*, **71**, 6678 (2006). |
| 2006JOMC5336 | C. Spino, L. Boisvert, J. Douville, S. Roy, S. Lauzon, J. Minville, D. Gagnon, F. Beaumier, and C. Chabot, *J. Organomet. Chem.*, **691**, 5336 (2006). |
| 2006MI1 | N. Saygili, A. Altunbas, and A. Yesilada, *Turk. J. Chem.*, **30**, 125 (2006). |
| 2006NPR1007 | C. Sánchez, C. Méndez, and J.A. Salas, *Nat. Prod. Rep.*, **23**, 1007 (2006). |
| 2006OBC193 | C.J. Hayes, A.E. Sherlock, and M.D. Selby, *Org. Biomol. Chem.*, **4**, 193 (2006). |
| 2006OBC1512 | B. Clark, R.J. Capon, E. Lacey, S. Tennant, and J.H. Gill, *Org. Biomol. Chem.*, **4**, 1512 (2006). |
| 2006OBC1868 | J.F. Bower, S. Chakthong, J. Svenda, A.J. Williams, R.M. Lawrence, P. Szeto, and T. Gallagher, *Org. Biomol. Chem.*, **4**, 1868 (2006). |
| 2006OL83 | C. Mukai, T. Yoshida, M. Sorimachi, and A. Odani, *Org. Lett.*, **8**, 83 (2006). |
| 2006OL939 | S. Roy and C. Spino, *Org. Lett.*, **8**, 939 (2006). |
| 2006OL2103 | M. Hosseini, H. Kringelum, A. Murray, and J.E. Tønder, *Org. Lett.*, **8**, 2103 (2006). |
| 2006OL2711 | Y. Kobayashi, H. Kamisaki, R. Yanada, and Y. Takemoto, *Org. Lett.*, **8**, 2711 (2006). |
| 2006OL4067 | K.N. Maloney, W. Hao, J. Xu, J. Gibbons, J. Hucul, D. Roll, S.F. Brady, F.C. Schroeder, and J. Clardy, *Org. Lett.*, **8**, 4067 (2006). |
| 2006OL5191 | T.R. Hoye, V. Dvornikovs, and E. Sizova, *Org. Lett.*, **8**, 5191 (2006). |
| 2006OPPI347 | S.E. Boiadjiev and D.A. Lightner, *Org. Prep. Proced. Int.*, **38**, 347 (2006). |
| 2006S2019 | A. Arcadi, F. Marinelli, L. Rossi, and M. Verdecchia, *Synthesis*, 2019 (2006). |
| 2006S3431 | A. Alizadeh, F. Movahedi, H. Masrouri, and L.-G. Zhu, *Synthesis*, 3431 (2006). |

| | |
|---|---|
| 2006S3490 | S. Lebrun, A. Couture, E. Deniau, and P. Grandclaudon, *Synthesis*, 3490 (2006). |
| 2006S3902 | R. Schobert, M. Dietrich, G. Mullen, and J.-M. Urbina-Gonzalez, *Synthesis*, 3902 (2006). |
| 2006SC2253 | J.-P. Hermet, V. Caubert, and N. Langlois, *Synth. Commun.*, **36**, 2253 (2006). |
| 2006SL2807 | J. Toueg and J. Prunet, *Synlett*, 2807 (2006). |
| 2006T746 | F. Bellesia, C. Danielli, L. De Buyck, R. Galeazzi, F. Ghelfi, A. Mucci, M. Orena, U.M. Pagnoni, A.F. Parsons, and F. Roncaglia, *Tetrahedron*, **62**, 746 (2006). |
| 2006T6018 | B. Metten, M. Kostermans, G. Van Baelen, M. Smet, and W. Dehaen, *Tetrahedron*, **62**, 6018 (2006). |
| 2006T7043 | S.E. Boiadjiev, Z.R. Woydziak, A.F. McDonagh, and D.A. Lightner, *Tetrahedron*, **62**, 7043 (2006). |
| 2006T12252 | R. Le Vézouët, A.J.P. White, J.N. Burrows, and A.G.M. Barrett, *Tetrahedron*, **62**, 12252 (2006). |
| 2006TL3387 | L.T. Simmons, K.L. McPhail, E. Ortega-Barria, S.L. Mooberry, and W.H. Gerwick, *Tetrahedron Lett.*, **47**, 3387 (2006). |
| 2006TL6263 | H. Ishibashi, S. Haruki, M. Uchiyama, O. Tamura, and J-i. Matsuo, *Tetrahedron Lett.*, **47**, 6263 (2006). |
| 2007ACIE2485 | D. Bonne, M. Dekhand, and J. Zhu, *Angew. Chem. Int. Ed.*, **46**, 2485 (2007). |
| 2007ACIE3912 | M. McLaughlin, M. Takahashi, and G.C. Micalizio, *Angew. Chem. Int. Ed.*, **46**, 3912 (2007). |
| 2007ACIE4930 | S. Ogoshi, H. Ikeda, and H. Kurosawa, *Angew. Chem. Int. Ed.*, **46**, 4930 (2007). |
| 2007ARK206 | I. Osante, M.N. Abdullah, S. Arrasate, N. Sotomayor, and E. Lete, *Arkivoc*, 206 (2007). |
| 2007ASC1066 | J. Lim, S.S. Lee, S.N. Riduan, and J.Y. Ying, *Adv. Synth. Cat.*, **349**, 1066 (2007). |
| 2007BMC4601 | B. Li, M.P.A. Lyle, G. Chen, J. Li, K. Hu, L. Tang, M.A. Alaoui-Jamali, and J. Webster, *Bioorg. Med. Chem.*, **15**, 4601 (2007). |
| 2007BMC5487 | T. Kawasuji, M. Fuji, T. Yoshinaga, A. Sato, T. Fujiwara, and R. Kiyama, *Bioorg. Med. Chem.*, **15**, 5487 (2007). |
| 2007BMCL1908 | D.S. Dodd, S. Sheriff, C.Y.J. Chang, D.K. Stetsko, L.M. Phillips, Y. Zhang, M. Launany, D. Potin, W. Vaccaro, M.A. Poss, M. McKinnon, J.C. Barrish, S.J. Suchard, and T.G.M. Dhar, *Bioorg. Med. Chem. Lett.*, **17**, 1908 (2007). |
| 2007BMCL3426 | H.-X. Liu, G.-R. Huang, H.-M. Zhang, J.-R. Wu, and Z.-J. Yao, *Bioorg. Med. Chem. Lett.*, **17**, 3426 (2007). |
| 2007BMCL6155 | R. Dayam, L.Q. Al-Mawsawi, and N. Neamati, *Bioorg. Med. Chem. Lett.*, **17**, 6155 (2007). |
| 2007CAJ20 | M. Shibasaki, M. Kanai, and N. Fukuda, *Chem. Asian J.*, **2**, 20 (2007). |
| 2007CAJ114 | S. Suntorchaswej, K. Suwanborirux, K. Koga, and M. Isobe, *Chem. Asian J.*, **2**, 114 (2007). |
| 2007CC728 | J.F. Bower, T. Riis-Johannessen, P. Szeto, A.J. Whitehead, and T. Gallagher, *Chem. Commun.*, 728 (2007). |
| 2007CEJ6985 | Y.-L. Yang, C.-P. Lu, M.-Y. Chen, K.-Y. Chen, Y.-C. Wu, and S.-H. Wu, *Chem. Eur. J.*, **13**, 6985 (2007). |
| 2007EJOC1287 | Y. Makhynya, Z. Hussain, T. Bauschlicher, P. Schwinte, F. Siebert, and W. Gärtner, *Eur. J. Org. Chem.*, 1287 (2007). |

2007EJOC1294 P.B. Shrestha-Dawadi and J. Lugtenburg, *Eur. J. Org. Chem.*, 1294 (2007).

2007EJOC2430 A. Arcadi, A. Inesi, F. Marinelli, L. Rossi, and M. Verdecchia, *Eur. J. Org. Chem.*, 2430 (2007).

2007JMC433 R.L. Hudkins, N.W. Johnson, T.S. Angeles, G.W. Gessner, and J.P. Mallamo, *J. Med. Chem.*, **50**, 433 (2007).

2007JMC1528 J. Regourd, A.A.-S. Ali, and A. Thompson, *J. Med. Chem.*, **50**, 1528 (2007).

2007JNP1180 G. Schlingmann, T. Taniguchi, H. He, R. Bigelis, H.Y. Yang, F.E. Koehn, G.T. Carter, and N. Berova, *J. Nat. Prod.*, **70**, 1180 (2007).

2007JNP1640 T.N. Duong, R. Edrada, W. Rainer, V. Wray, W. Frank, A.T. Duong, W.H. Lin, and P. Proksch, *J. Nat. Prod.*, **70**, 1640 (2007).

2007JNP1741 R. Liu, Y. Liu, Y.-D. Zhou, and D.G. Nagle, *J. Nat. Prod.*, **70**, 1741 (2007).

2007JNP1812 G. Carr, M. Raszek, R. Van Soest, T. Matainaho, M. Shopik, C.F.B. Holmes, and R.J. Andersen, *J. Nat. Prod.*, **70**, 1812 (2007).

2007JNP1961 Y.-C. Shen, Y.-B. Cheng, J.-i. Kobayashi, T. Kubota, Y. Takahashi, Y. Mikami, J. Ito, and Y.-S. Lin, *J. Nat. Prod.*, **70**, 1961 (2007).

2007JOC2008 Y. Wada, H. Nagasaka, M. Tokuda, and K. Orito, *J. Org. Chem.*, **72**, 2008 (2007).

2007JOC2547 A.J. Pearson, H. Sun, and X. Wang, *J. Org. Chem.*, **72**, 2547 (2007).

2007JOC3961 R. Shen and X. Huang, *J. Org. Chem.*, **72**, 3961 (2007).

2007JOC7190 D. Mori, Y. Kimura, S. Kitamura, Y. Sakagami, Y. Yoshioka, T. Shintani, T. Okamoto, and M. Ojika, *J. Org. Chem.*, **72**, 7190 (2007).

2007JOC7443 J.H. Lee, S. Shin, J. Kang, and S.-g. Lee, *J. Org. Chem.*, **72**, 7443 (2007).

2007JOC10108 J.-C. Legeay and N. Langlois, *J. Org. Chem.*, **72**, 10108 (2007).

2007JOMC579 M. Murai, S. Kawai, K. Miki, and K. Ohe, *J. Organomet. Chem.*, **692**, 579 (2007).

2007MI1 R. Schobert, *Naturwiss.*, **94**, 1 (2007).

2007MI2 J.C.F. Alves, *J. Braz. Chem. Soc.*, **18**, 855 (2007).

2007MI3 S.K. Dey and D.A. Lightner, *Monatsh. Chem.*, **138**, 687 (2007).

2007OBC344 T. Hjelmgaard, M. Givskov, and J. Nielsen, *Org. Biomol. Chem.*, **5**, 344 (2007).

2007OBC2207 M. Hosseini, J.S. Grau, K.K. Sørensen, I. Søtofte, D. Tanner, A. Murray, and J.E. Tønder, *Org. Biomol. Chem.*, **5**, 2207 (2007).

2007OBC2357 S. Spiess, C. Berthold, R. Weihofen, and G. Helmchen, *Org. Biomol. Chem.*, **5**, 2357 (2007).

2007OBC2636 J.F. Bower, A.J. Williams, H.L. Woodward, P. Szeto, R.M. Lawrence, and T. Gallagher, *Org. Biomol. Chem.*, **5**, 2636 (2007).

2007OBC3486 M. Hosseini, D. Tanner, A. Murray, and J.E. Tønder, *Org. Biomol. Chem.*, **5**, 3486 (2007).

2007OL1717 A. Hercouet, F. Berrée, C.H. Lin, L. Toupet, and B. Carboni, *Org. Lett.*, **9**, 1717 (2007).

2007OL1879 J.T. Reeves, *Org. Lett.*, **9**, 1879 (2007).

2007OL3037 G. Yao and L.C. Chang, *Org. Lett.*, **9**, 3037 (2007).

2007OL4939 C. Jiang and A.J. Frontier, *Org. Lett.*, **9**, 4939 (2007).

2007S663 M.S. Majik, J. Shet, S.G. Tilve, and P.S. Parameswaran, *Synthesis*, 663 (2007).

2007S1499 R. Schobert and A. Wicklein, *Synthesis*, 1499 (2007).

| | |
|---|---|
| 2007S1882 | F. Felluga, F. Ghelfi, U.M. Pagnoni, A.F. Parsons, M. Pattarozzi, F. Roncaglia, and E. Valentin, *Synthesis*, 1882 (2007). |
| 2007S2317 | K.M. Dorward, N.J. Guthrie, and E.T. Pelkey, *Synthesis*, 2317 (2007). |
| 2007SC813 | J.Y. Mun and M.B. Smith, *Synth. Commun.*, **37**, 813 (2007). |
| 2007SC1503 | S.P. Chavan, A.B. Pathak, A.N. Dhawane, and U.R. Kalkote, *Synth. Commun.*, **37**, 1503 (2007). |
| 2007SL1294 | M. Brasholz and H.-U. Reissig, *Synlett*, 1294 (2007). |
| 2007SL1521 | M.B.T. Thuong, S. Sottocornola, G. Prestat, G. Broggini, D. Madec, and G. Poli, *Synlett*, 1521 (2007). |
| 2007SL1557 | D. Albrecht and T. Bach, *Synlett*, 1557 (2007). |
| 2007T3217 | S. Suntornchashwej, K. Suwanborirux, and M. Isobe, *Tetrahedron*, **63**, 3217 (2007). |
| 2007T8987 | M.-J. Fan, B. Qian, L.-B. Zhao, and Y.-M. Liang, *Tetrahedron*, **63**, 8987 (2007). |
| 2007T11030 | S.D. Roth, T. Shkindel, and D.A. Lightner, *Tetrahedron*, **63**, 11030 (2007). |
| 2007TL381 | V. Caubert, J. Massé, P. Retailleau, and N. Langlois, *Tetrahedron Lett.*, **48**, 381 (2007). |
| 2007TL827 | S.J.P. Yoon-Miller, S.M. Opalka, and E.T. Pelkey, *Tetrahedron Lett.*, **48**, 827 (2007). |
| 2007TL1571 | T. Quiroz, D. Corona, A. Covarruvias, J.G. Avila-Zárraga, and M. Romero-Ortega, *Tetrahedron Lett.*, **48**, 1571 (2007). |
| 2007TL2287 | W.K. Goh, G. Iskander, D. StC. Black, and N. Kumar, *Tetrahedron Lett.*, **48**, 2287 (2007). |
| 2007TL2819 | R. Hunter, S.C.M. Rees-Jones, and H. Su, *Tetrahedron Lett.*, **48**, 2819 (2007). |
| 2007TL7259 | M. Anwar and M.G. Moloney, *Tetrahedron Lett.*, **48**, 7259 (2007). |
| 2007TL7399 | L.J. Wilson, C. Yang, and W.V. Murray, *Tetrahedron Lett.*, **48**, 7399 (2007). |
| 2007TL8056 | M. Adib, M. Mahdavi, M.A. Noghani, and H.R. Bijanzadeh, *Tetrahedron Lett.*, **48**, 8056 (2007). |
| 2007TL8956 | T. Muramatsu, S. Tamashita, Y. Nakamura, M. Suzuki, N. Mase, H. Yoda, and K. Takabe, *Tetrahedron Lett.*, **48**, 8956 (2007). |
| 2007TL9199 | K.-i. Kanno, S. Ren, Y. Li, K. Nakak'jima, and T. Takahashi, *Tetrahedron Lett.*, **48**, 9199 (2007). |
| 2008ARK(xiv)7 | T. Taniguchi, R. Kawajiri, and H. Ishibashi, *Arkivoc*, **xiv**, 7 (2008). |
| 2008BMC4419 | E. Conchon, F. Anizon, B. Aboab, R.M. Golsteyn, S. Léonce, B. Pfeiffer, and M. Prudhomme, *Bioorg. Med. Chem.*, **16**, 4419 (2008). |
| 2008BMCL1623 | M.S. Coumar, J.-S. Wu, J.-S. Leou, U.-K. Tan, C.-Y. Chang, T.-Y. Chang, W.-H. Lin, J.T.A. Hsu, Y.-S. Chao, S.-Y. Wu, and H.-P. Hseih, *Bioorg. Med. Chem. Lett.*, **18**, 1623 (2008). |
| 2008BMCL1840 | S. Peukert, Y. Sun, R. Zhang, B. Hurley, M. Sabio, X. Shen, C. Gray, J. Dzink-Fox, J. Tao, R. Cebula, and S. Wattanasin, *Bioorg. Med. Chem. Lett.*, **18**, 1840 (2008). |
| 2008BMCL3865 | P. Pace, S.A.H. Spieser, and V. Summa, *Bioorg. Med. Chem. Lett.*, **18**, 3865 (2008). |
| 2008BMCL5307 | J.Y. Melamed, M.S. Egbertson, S. Varga, J.P. Vacca, G. Moyer, L. Gabryelski, P.J. Felock, K.A. Stillmock, M.V. Witmer, W. Schleif, D.J. Hazuda, Y. Leonard, L. Jin, J.D. Ellis, and S.D. Young, *Bioorg. Med. Chem. Lett.*, **18**, 5307 (2008). |
| 2008CC4312 | J. Lim, S.S. Lee, and J.Y. Ying, *Chem. Commun.*, **44**, 4312 (2008). |

2008CL198 H. Khawn, L.-Y. Chen, H. Kinoshita, and K. Inomata, *Chem. Lett.*, **37**, 198 (2008).

2008CSR789 A.S. Ivanov, *Chem. Soc. Rev.*, **37**, 789 (2008).

2008EJOC4350 X. Li, H. Deng, S. Luo, and J.-P. Cheng, *Eur. J. Org. Chem.*, 4350 (2008).

2008HCA1500 M.S. Majik, P.S. Parameswaran, and S.G. Tilve, *Helv. Chim. Acta*, **91**, 1500 (2008).

2008JA11303 U. Robben, U. Lindner, and W. Gärtner, *J. Am. Chem. Soc.*, **130**, 11303 (2008).

2008JHC845 G. Buehrdel, R. Beckert, B. Friedrich, and H. Goerls, *J. Heterocycl. Chem.*, **45**, 845 (2008).

2008JMC1487 R. Yendpally, J.G. Hurdle, E.I. Carson, R.B. Lee, and R.E. Lee, *J. Med. Chem.*, **51**, 1487 (2008).

2008JMC3814 C. Peifer, R. Selig, K. Kinkel, D. Ott, F. Totzke, C. Schächtele, R. Heidenreich, M. Röcken, D. Schollmeyer, and S. Laufer, *J. Med. Chem.*, **51**, 3814 (2008).

2008JMC5833 S.R. Donohue, J.H. Krushinki, V.W. Pike, E. Chernet, L. Phebus, A.K. Chesterfield, C.C. Felder, C. Halldin, and J.M. Schaus, *J. Med. Chem.*, **51**, 5833 (2008).

2008JOC2041 C.J. Hayes, A.E. Sherlock, M.P. Green, C. Wilson, A.J. Blake, M.D. Selby, and J.C. Prodger, *J. Org. Chem.*, **73**, 2041 (2008).

2008JOC2704 S.K. Dey and D.A. Lightner, *J. Org. Chem.*, **73**, 2704 (2008).

2008JOC3658 Q.-F. Yu, Y.-H. Zhang, Q. Yin, B.-X. Tang, R.-Y. Tang, P. Zhong, and J.-H. Li, *J. Org. Chem.*, **73**, 3658 (2008).

2008JOC6239 S. Lauzon, F. Tremblay, D. Gagnon, C. Godbout, C. Chabot, C. Mercier-Shanks, S. Perreault, H. DeSéve, and C. Spino, *J. Org. Chem.*, **73**, 6239 (2008).

2008JOC9008 B.-X. Tang, Q. Yin, R.-Y. Tang, and J.-H. Li, *J. Org. Chem.*, **73**, 9008 (2008).

2008JOC9174 M. Murai, K. Miki, and K. Ohe, *J. Org. Chem.*, **73**, 9174 (2008).

2008MI1 V.L. Gein, V.V. Yushkov, T.A. Splina, L.F. Gein, K.V. Yatsenko, and S.G. Shevtsova, *Pharm. Chem. J.*, **42**, 255 (2008).

2008MI2 R.N. Vydzhak and S.Y. Panchishin, *Russ. J. Gen. Chem.*, **78**, 1641 (2008).

2008MI3 V.L. Gein, E.P. Tsyplyakova, G.A. Stashina, and V.A. Bakulev, *Russ. J. Org. Chem.*, **44**, 478 (2008).

2008OBC3989 D. Farran, I. Parrot, L. Toupet, J. Martinez, and G. Dewynter, *Org. Biomol. Chem.*, **6**, 3989 (2008).

2008OL597 A. Mallagaray, G. Domínguez, A. Gradillas, and J. Pérez-Castells, *Org. Lett.*, **10**, 597 (2008).

2008OL1063 B.-X. Tang, D.-J. Tang, S. Tang, Q.-F. Yu, Y.-H. Zhang, Y. Liang, P. Zhong, and J.-H. Li, *Org. Lett.*, **10**, 1063 (2008).

2008OL4149 R. Kontnik and J. Clardy, *Org. Lett.*, **10**, 4149 (2008).

2008OL4947 M.D. Sadar, D.E. Williams, N.R. Mawji, B.O. Patrick, T. Wikanta, E. Chasanah, H.E. Irianto, R. Van Soest, and R.J. Andersen, *Org. Lett.*, **10**, 4947 (2008).

2008S3131 F. Ghelfi, M. Pattarozzi, F. Roncaglia, A.F. Parsons, F. Felluga, U.M. Pagnoni, E. Valentin, A. Mucci, and F. Bellesia, *Synthesis*, 3131 (2008).

2008SL2107 M.G. Moloney and M. Yaqoob, *Synlett*, 2107 (2008).

2008T4387 S. Kafka, A. Klasek, J. Polis, V. Rosenbreierová, C. Palík, V. Mrkvicka, and J. Kosmrlj, *Tetrahedron*, **64**, 4387 (2008).

2008TA2115 S. Yamashita, N. Mase, and K. Takabe, *Tetrahedron: Asymmetry*, **19**, 2115 (2008).

2008TL906 T. Coursindel, D. Farran, J. Martinez, and G. Dewynter, *Tetrahedron Lett.*, **49**, 906 (2008).

2008TL1824 S. Ashari, M. Tajbakhsh, and V. Taghipour, *Tetrahedron Lett.*, **49**, 1824 (2008).

2008TL3659 P. Haldar and J.K. Ray, *Tetrahedron Lett.*, **49**, 3659 (2008).

2008TL7334 Q. Yang, Y.-Y. Lai, W.-J. Xiao, and H. Alper, *Tetrahedron Lett.*, **49**, 7334 (2008).

2009BMCL2112 Z. Feng, X. Li, G. Zheng, and L. Huang, *Bioorg. Med. Chem. Lett.*, **19**, 2112 (2009).

2009BMCL2270 Z. Feng, F. Chu, Z. Guo, and P. Sun, *Bioorg. Med. Chem. Lett.*, **19**, 2270 (2009).

2009BMC6879 R. Miyauchi, K. Kawakami, M. Ito, N. Matsuhashi, H. Ohki, H. Inagaki, H. Takahashi, and M. Takemura, *Bioorg. Med. Chem.*, **17**, 6879 (2009).

2009CBC1634 K. Taori, Y. Liu, V.J. Paul, and H. Luesch, *ChemBioChem*, **10**, 1634 (2009).

2009CPB979 H. Miyazaki, T. Ogiku, H. Sai, Y. Moritani, A. Ohtani, and H. Ohmizu, *Chem. Pharm. Bull.*, **57**, 979 (2009).

2009EJOC3141 M. Toyofuku, E. Murase, H. Nagai, S-i. Fujiwara, T. Shin-ike, H. Kuniyasu, and N. Kambe, *Eur. J. Org. Chem.*, 3141 (2009).

2009HCA1665 Y. Lou, F. Zhao, Z. Wu, K.-F. Peng, X.-C. Wei, L.-X. Chen, and F. Qui, *Helv. Chim. Acta*, **92**, 1665 (2009).

2009JA14579 M.D. Clift and R.J. Thomson, *J. Am. Chem. Soc.*, **131**, 14579 (2009).

2009JCC685 Q. Zhu, H. Jiang, J. Li, S. Liu, C. Xia, and M. Zhang, *J. Comb. Chem.*, **11**, 685 (2009).

2009JCC851 N. Pudelová and V. Krchnák, *J. Comb. Chem.*, **11**, 851 (2009).

2009JHC447 S.J.P. Yoon-Miller, K.M. Dorward, K.P. White, and E.T. Pelkey, *J. Heterocycl. Chem.*, **46**, 447 (2009).

2009JMC6494 C.C. Cosner, J.T. Markiewicz, P. Bourbon, C.J. Mariani, O. Wiest, M. Rujoi, A.I. Rosenbaum, A.Y. Huang, F.R. Maxfield, and P. Helquist, *J. Med. Chem.*, **52**, 6494 (2009).

2009JNP14 R.G. Linington, B.R. Clark, E.E. Trimble, A. Almanza, L.-D. Urena, D.E. Kyle, and W.H. Gerwick, *J. Nat. Prod.*, **72**, 14 (2009).

2009JNP319 G. Yao, T.P. Kondratyuk, G.T. Tan, J.M. Pezzuto, and L.C. Chang, *J. Nat. Prod.*, **72**, 319 (2009).

2009JNP1691 D.E. Williams, S.O. Marques, E. Hajdu, S. Peixnho, R.J. Andersen, and R.G.S. Berlinck, *J. Nat. Prod.*, **72**, 1691 (2009).

2009JOC4149 J. Chen, X.-G. Fu, L. Zhou, J.-T. Zhang, X.-L. Qi, and X.-P. Cao, *J. Org. Chem.*, **74**, 4149 (2009).

2009JOC5032 M. Storgaard, F.Z. Dörwald, B. Peschke, and D. Tanner, *J. Org. Chem.*, **74**, 5032 (2009).

2009JOC7274 M.N. Alberti, G.C. Vougioukalakis, and M. Orfanopoulos, *J. Org. Chem.*, **74**, 7274 (2009).

2009JOC7457 J. Chen, P.-Q. Huang, and Y. Queneau, *J. Org. Chem.*, **74**, 7457 (2009).

2009JOC7566 M. Enomoto and S. Kuwahara, *J. Org. Chem.*, **74**, 7566 (2009).

2009MI1 M.F. Mohammet, Z. Shaameri, and A.S. Hamzah, *Molecules*, **14**, 250 (2009).

2009MI2 J.-L. Guo, B.-Z. Li, W.-M. Chen, P.-H. Sun, and Y. Wang, *Lett. Drug Des. Disc.*, **6**, 107 (2009).

2009MI3 S.K. Dey and D.A. Lightner, *Monatsh. Chem.*, **140**, 161 (2009).

2009OL1233 S. Kiren, X. Hong, C.A. Leverett, and A. Padwa, *Org. Lett.*, **11**, 1233 (2009).

2009OL2892 H. Nomura and C.J. Richards, *Org. Lett*, **11**, 2892 (2009).

2009OL4068 A. Basso, L. Banfi, A. Galatini, G. Guanti, F. Rastrelli, and R. Riva, *Org. Lett.*, **11**, 4068 (2009).

2009OL5522 X.-C. Wang, Z.-P. Zheng, X.-W. Gan, and L.-H. Hu, *Org. Lett.*, **11**, 5522 (2009).

2009S891 Z.-Q. Wang, B.-X. Tang, H.-P. Zhang, F. Wang, and J.-H. Li, *Synthesis*, 891 (2009).

2009S2407 P.R. Likhar, S.S. Racharlawar, M.V. Karkhelikar, M.S. Subhas, and B. Sridhar, *Synthesis*, 2407 (2009).

2009SL297 L.-J. Jiang, H.-Q. Lan, J.-F. Zheng, J.-L. Ye, and P.-Q. Huang, *Synlett*, 297 (2009).

2009SL1525 G. Casiraghi, F. Zanardi, L. Battistini, and G. Rassu, *Synlett*, 1525 (2009).

2009SL2487 Y.-C. Jeong and M.G. Moloney, *Synlett*, 2487 (2009).

2009SL2621 S. Lebrun, A. Couture, E. Deniau, and P. Grandclaudon, *Synlett*, 2621 (2009).

2009T648 N. Coskun and M. Cetin, *Tetrahedron*, **65**, 648 (2009).

2009T1402 D. Halie, J. Pérard-Viret, S. Dufour, and J. Royer, *Tetrahedron*, **65**, 1402 (2009).

2009T1481 F. Roncaglia, C.V. Stevens, F. Ghelfi, M. Van der Steen, M. Pattarozzi, and L. De Buyck, *Tetrahedron*, **65**, 1481 (2009).

2009T2905 S. De Marino, C. Festa, M.V. D'Auria, M.-L. Bourguet-Kondracki, S. Petek, C. Debitus, R.M. Andrés, M.C. Terencio, M. Payá, and A. Zampella, *Tetrahedron*, **65**, 2905 (2009).

2009T4530 C. Haase and P. Langer, *Tetrahedron*, **65**, 4530 (2009).

2009T6720 S. Kiren, X. Hong, C.A. Leverett, and A. Padwa, *Tetrahedron*, **65**, 6720 (2009).

2009TL1705 J.H. Spatz, S.J. Welsch, D.-E. Duhaut, N. Jäger, T. Boursier, M. Frederich, L. Allmendinger, G. Ross, J. Kolb, C. Burdack, and M. Umkehrer, *Tetrahedron Lett*, **1988**, 1705 (2009).

2009TL2402 S. Pinheiro, R.C. da Silva Júnior, A.S. de Souza, J.W.d. M. Carneiro, E.M.F. Muri, and O.A.C. Antunes, *Tetrahedron Lett.*, **50**, 2402 (2009).

2009TL5053 M. Oba, C. Ito, T. Hayashi, and K. Nishiyama, *Tetrahedron Lett.*, **50**, 5053 (2009).

2010ACIE9702 Z. Gu and A. Zakarian, *Angew. Chem. Int. Ed.*, **49**, 9702 (2010).

2010ASC893 S. Arai, Y. Koike, and A. Nishida, *Adv. Synth. Cat.*, **352**, 893 (2010).

2010BCSJ1309 K. Nishiyama, A. Kamiya, M.A.S. Hammam, H. Kinoshita, S. Fujinami, Y. Ukaji, and K. Inomata, *Bull. Chem. Soc. Jpn.*, **83**, 1309 (2010).

2010BMC4468 C. Alp, D. Ekinci, M.S. Gültekin, M. Sentürk, E. Sahin, and O.I. Küfruvioglu, *Bioorg. Med. Chem.*, **18**, 4468 (2010).

2010BMCL1924 S. Hanessian, Z. Shao, C. Betschart, J.-M. Rondeau, U. Neumann, and M. Tintelnot-Blomley, *Bioorg. Med. Chem. Lett.*, **20**, 1924 (2010).

2010CC806 J. Lim, S.S. Lee, and J.Y. Ying, *Chem. Commun.*, **46**, 806 (2010).

2010CC5319 H.-Q. Lan, Y.-P. Ruan, and P.-Q. Huang, *Chem. Commun.*, **46**, 5319 (2010).

2010CC8055 T. Ozawa, H. Horie, T. Kurahashi, and S. Matsubara, *Chem. Commun.*, **46**, 8055 (2010).

2010CC9064 G.B. Bajracharya, P.S. Koranne, R.N. Nadaf, R.K.M. Gabr, K. Takenaka, S. Takizawa, and H. Sasai, *Chem. Commun.*, **46**, 9064 (2010).

2010EJOC1142 A. Mallinger, B. Nadal, N. Chopin, and T. Le Gall, *Eur. J. Org. Chem.*, 1142 (2010).

2010EJOC5989 D. Matiadis and O. Igglessi-Markopoulou, *Eur. J. Org. Chem.*, 5989 (2010).

2010JCC84 A. Pendri, T.L. Troyer, M.J. Sofia, M.A. Walker, B.N. Naidu, J. Banville, N.A. Meanwell, I. Dicker, Z. Lin, M. Krystal, and S.W. Gerritz, *J. Comb. Chem.*, **12**, 84 (2010).

2010JMC5108 P.Y. Zhang, I.L.K. Wong, C.S.W. Yan, X.Y. Zhang, T. Jiang, L.M.C. Chow, and S.B. Wan, *J. Med. Chem.*, **53**, 5108 (2010).

2010JNP1164 R. Merrouche, N. Bouras, Y. Coppel, F. Mathieu, M.-C. Monje, N. Sabaou, and A. Lebrihi, *J. Nat. Prod.*, **73**, 1164 (2010).

2010JOC582 T.-j. Meng, Y. Hu, and S. Wang, *J. Org. Chem.*, **75**, 582 (2010).

2010JOC7435 M. Aginagalde, T. Bello, C. Masdeu, Y. Vara, A. Arrieta, and F.P. Cossio, *J. Org. Chem.*, **75**, 7435 (2010).

2010MI1 V.L. Gein, N.N. Kasimova, Z.G. Aliev, and M.I. Vakhrin, *Russ. J. Org. Chem.*, **46**, 875 (2010).

2010OL1100 Y. Jin, Y. Liu, Z. Wang, S. Kwong, Z. Xu, and T. Ye, *Org. Lett.*, **12**, 1100 (2010).

2010OL4560 A.P. Marcus and R. Sarpong, *Org. Lett.*, **12**, 4560 (2010).

2010OL5576 T. Conroy, J.T. Guo, N.H. Hunt, and R.J. Payne, *Org. Lett.*, **12**, 5576 (2010).

2010S2816 M. Passiniemi and A.M.P. Koskinen, *Synthesis*, 2816 (2010).

2010SC766 A. Arfaoui, F. Béji, and H. Amri, *Synth. Commun.*, **40**, 766 (2010).

2010SC1040 F. Ghelfi, M. Pattarozzi, F. Roncaglia, V. Giangiordano, and A.F. Parsons, *Synth. Commun.*, **40**, 1040 (2010).

2010SC1350 M. Anary-Abbasinejad, M. Mirhossaini, A. Parhami, and E. Pourhassan, *Synth. Commun.*, **40**, 1350 (2010).

2010SL2156 K. Hirai, Y. Tamura, I. Sato, and M. Hirama, *Synlett*, 2156 (2010).

2010SL2944 T. Sengoku, J. Wierzejska, M. Takahashi, and H. Yoda, *Synlett*, 2944 (2010).

2010T172 L.-J. Jiang, B. Teng, J.-F. Zheng, J.-L. Ye, and P.-Q. Huang, *Tetrahedron*, **66**, 172 (2010).

2010T1274 F. Mo, F. Li, D. Qiu, and J. Wang, *Tetrahedron*, **66**, 1274 (2010).

2010T1357 M. Pattarozzi, F. Roncaglia, L. Accorsi, A.F. Parsons, and F. Ghelfi, *Tetrahedron*, **66**, 1357 (2010).

2010T2422 P. Basabe, A. Blanco, O. Bodero, M. Martín, I.S. Marcos, D. Díez, F. Mollinedo, and J.G. Urones, *Tetrahedron*, **66**, 2422 (2010).

2010T6526 P.J. Parsons, J. Board, D. Faggiani, P.B. Hitchcock, L. Preece, and A.J. Waters, *Tetrahedron*, **66**, 6526 (2010).

2010T9113 J.T. Gupton, N. Telang, E.J. Banner, E.J. Kluball, K.E. Hall, K.L. Finzel, X. Jia, S.R. Bates, R.S. Welden, B.C. Giglio, J.E. Eaton, P.J. Barelli, L.T. Firich, J.A. Stafford, M.B. Coppock, E.F. Worrall, R.P.F. Kanters, K. Keertikar, and R. Osterman, *Tetrahedron*, **66**, 9113 (2010).

2010TA895 Z.-B. Ye, J. Chen, W.-H. Meng, and P.-Q. Huang, *Tetrahedron Asymmetry*, **21**, 895 (2010).

2010TA1758 M. Anwar, A.R. Cowley, and M.G. Moloney, *Tetrahedron: Asymmetry*, **21**, 1758 (2010).

2010TL96 J. Pérard-Viret, F. Souquet, M.-L. Manisse, and J. Royer, *Tetrahedron Lett.*, **51**, 96 (2010).

2010TL4317 W.-H. Huang, Q.-R. Li, L.-M. Chao, X.-S. Lei, and B.-G. Wei, *Tetrahedron Lett.*, **51**, 4317 (2010).

2011ACIE5188 E. Ascic, J.F. Jensen, and T.E. Nielsen, *Angew. Chem. Int. Ed.*, **50**, 5188 (2011).

2011ACIE5682 Y. Xie, Y. Zhao, B. Qian, L. Yang, C. Xia, and H. Huang, *Angew. Chem. Int. Ed.*, **50**, 5682 (2011).

2011ASC2966 W. Hess and J.W. Burton, *Adv. Synth. Cat.*, **353**, 2966 (2011).

2011BMC4903 A. Zall, D. Kieser, N. Höttecke, E.C. Naumann, B. Thomaszewski, K. Schneider, D.T. Steinbacher, R. Schubenel, S. Masur, K. Baumann, and B. Schmidt, *Bioorg. Med. Chem.*, **19**, 4903 (2011).

2011BMCL1218 E. Choi, C. Lee, J.E. Park, J.J. Seo, M. Cho, J.S. Kang, H.M. Kim, S.-K. Park, K. Lee, and G. Han, *Bioorg. Med. Chem. Lett.*, **21**, 1218 (2011).

2011CAJ560 H. Tokuyama, K. Okano, H. Fujiwara, T. Noji, and T. Fukuyama, *Chem. Asian J.*, **6**, 560 (2011).

2011CAJ2260 E.A. Ilardi and A. Zakarian, *Chem. Asian J.*, **6**, 2260 (2011).

2011CC3219 S. Tekkam, M.A. Alam, S.C. Jonnalagadda, and V.R. Mereddy, *Chem. Commun.*, **47**, 3219 (2011).

2011CEJ958 H.-Q. Lan, J.-L. Ye, A.-E. Wang, Y.-P. Ruan, and P.-Q. Huang, *Chem. Eur. J.*, **17**, 958 (2011).

2011CEJ13544 T. Conroy, J.T. Guo, R.G. Linington, N.H. Hunt, and R.J. Payne, *Chem. Eur. J.*, **17**, 13544 (2011).

2011CR3076 G. Casiraghi, L. Battistini, C. Curti, G. Rassu, and F. Zanardi, *Chem. Rev.*, **111**, 3076 (2011).

2011CS1487 R. Leon, A. Jawalekar, T. Redert, and M.J. Gaunt, *Chem. Sci.*, **2**, 1487 (2011).

2011EJOC2981 J. Sun, Q. Wu, E.-Y. Xia, and C.-G. Yan, *Eur. J. Org. Chem.*, 2981 (2011).

2011H(82)1157 R. Sakata, R. Iwamoto, S. Fujinami, Y. Ukaji, and K. Inomata, *Heterocycles*, **82**, 1157 (2011).

2011JA740 B.M. Trost, J.-P. Lumb, and J.M. Azzarelli, *J. Am. Chem. Soc.*, **133**, 740 (2011).

2011JA12350 A. Raghuraman, E. Ko, L.M. Perez, T.R. Ioerger, and K. Burgess, *J. Am. Chem. Soc.*, **133**, 12350 (2011).

2011JOC1342 Y.-C. Jeong and M.G. Moloney, *J. Org. Chem.*, **76**, 1342 (2011).

2011JOC8203 J.G. Greger, S.J.P. Yoon-Miller, N.R. Bechtold, S.A. Flewelling, J.P. MacDonald, C.R. Downey, E.A. Cohen, and E.T. Pelkey, *J. Org. Chem.*, **76**, 8203 (2011).

2011JOC9472 F. Palacios, A.M. Ochoa de Retana, and A.V. del Burgo, *J. Org. Chem.*, **76**, 9472 (2011).

2011MI2 I.W. Wu, S.B. Kang, G. Keum, and Y. Kim, *Bull. Korean Chem. Soc.*, **32**, 3167 (2011).

2011OBC2535 M. Blangetti, A. Deagostino, G. Gervasio, D. Marabello, C. Prandi, and P. Venturello, *Org. Biomol. Chem.*, **9**, 2535 (2011).

2011OBC4353 J.C. Ball, R. Gleave, and S. Jones, *Org. Biomol. Chem.*, **9**, 4353 (2011).

2011OBC5825 K. Kobayashi, I. Okamato, N. Morita, T. Kiyotani, and O. Tamura, *Org. Biomol. Chem.*, **9**, 5825 (2011).

2011OBC6761 M. Daly, K. Gill, M. Sime, G.L. Simpson, and A. Sutherland, *Org. Biomol. Chem.*, **9**, 6761 (2011).

2011OL161 T. Terauchi, T. Kamikawai, M.G. Vinogradov, E.V. Starodubtseva, M. Takeda, and M. Kainosho, *Org. Lett.*, **13**, 161 (2011).

2011OL5468 J.W. Tucker and C.R.J. Stephenson, *Org. Lett.*, **13**, 5468 (2011).

2011S1267 M. Bregoli, F. Felluga, V. Frenna, F. Ghelfi, U.M. Pagnoni, A.F. Parsons, G. Petrillo, and D. Spinelli, *Synthesis*, **43**, 1267 (2011).

2011S1831 M.-Y. Yu, K. Li, N. Wang, T. He, and X.-Q. Yu, *Synthesis*, **43**, 1831 (2011).

2011SL1631 A. Duris, A. Daïch, and D. Berkes, *Synlett*, 1631 (2011).

2011SL2657 K.C. Majumdar, T. Ghosh, and P.K. Shyam, *Synlett*, 2657 (2011).

2011T408 R. Casolari, F. Felluga, V. Frenna, F. Ghelfi, U.M. Pagnoni, A.F. Parsons, and D. Spinelli, *Tetrahedron*, **67**, 408 (2011).

2011T4318 R.D. Grigg, J.M. Schomaker, and V. Timokhin, *Tetrahedron*, **67**, 4318 (2011).

2011T10026 R. Bastin, J.W. Dale, M.G. Edwards, J.P.N. Papillon, M.R. Webb, and R.J.K. Taylor, *Tetrahedron*, **67**, 10026 (2011).

2011TL1253 V. André, A. Vidal, J. Ollivier, S. Robin, and D.J. Aitken, *Tetrahedron Lett.*, **52**, 1253 (2011).

2012BMC1240 Y. Zhao, Q. Wang, Q. Meng, D. Ding, H. Yang, G. Gao, D. Li, W. Zhu, and H. Zhou, *Bioorg. Med. Chem.*, **20**, 1240 (2012).

2012BMCL4189 C. Lee, E. Choi, M. Cho, B. Lee, S.J. Oh, S.-K. Park, K. Lee, H.M. Kim, and G. Han, *Bioorg. Med. Chem. Lett.*, **22**, 4189 (2012).

2012BMCL4765 H. Uchiro, N. Shionozaki, Y. Kobayakawa, H. Nakagawa, and K. Makino, *Bioorg. Med. Chem. Lett.*, **22**, 4765 (2012).

2012BMCL6292 C. Couturier, A. Bauer, A. Rey, C. Schroif-Dufour, and M. Broenstrup, *Bioorg. Med. Chem. Lett.*, **22**, 6292 (2012).

2012CAJ504 G.-Y. Chen, H. Huang, J.-L. Ye, A.-E. Wang, H.-Y. Huang, H.-K. Zhang, and P.-Q. Huang, *Chem. Asian J.*, **7**, 504 (2012).

2012CC6363 M.-Q. Jia and S.-L. You, *Chem. Commun.*, **48**, 6363 (2012).

2012CEJ13274 H.-J. Yu, C. Shao, Z. Cui, C.-G. Feng, and G.-Q. Lin, *Chem. Eur. J.*, **18**, 13274 (2012).

2012EJMC232 Q. Zhu, L. Gao, Z. Chen, S. Zheng, H. Shu, J. Li, H. Jiang, and S. Liu, *Eur. J. Med. Chem.*, **54**, 232 (2012).

2012EJOC5324 F. Pan, J.-M. Chen, T.-Y. Qin, S.X.-A. Zhang, and W.-W. Liao, *Eur. J. Org. Chem.*, 5324 (2012).

2012JMC9630 C. Zhuang, Z. Miao, L. Zhu, G. Dong, Z. Guo, S. Wang, Y. Zhang, Y. Wu, J. Yao, C. Sheng, and W. Zhang, *J. Med. Chem.*, **55**, 9630 (2012).

2012JMC10766 E. Choi, C. Lee, M. Cho, J.J. Seo, J.S. Yang, S.J. Oh, K. Lee, S.-K. Park, H.M. Kim, H.J. Kwon, and G. Han, *J. Med. Chem.*, **55**, 10766 (2012).

2012JOC2837 B.-X. Tang, Y.-H. Zhang, R.-J. Song, D.-J. Tang, G.-B. Deng, Z.-Q. Wang, Y.-X. Xie, Y.-Z. Xia, and J.-H. Li, *J. Org. Chem.*, **77**, 2837 (2012).

2012JOC4391 T. Sengoku, Y. Nagae, Y. Ujihara, M. Takahashi, and H. Yoda, *J. Org. Chem.*, **77**, 4391 (2012).

2012JOC5808 H. Zali-Boeini, M. Mobin, K. Hajibabaei, and M. Ghani, *J. Org. Chem.*, **77**, 5808 (2012).

2012JOC6778 A.J. Clark, A.E.C. Collis, D.J. Fox, L.L. Halliwell, N. James, R.K. O'Reilly, H. Parekh, A. Ross, A.B. Sellars, H. Willcock, and P. Wilson, *J. Org. Chem.*, **77**, 6778 (2012).

2012JOC8038 S.G. Parameswarappa and F.C. Pigge, *J. Org. Chem.*, **77**, 8038 (2012).

2012JOC9659 A. Chatzimpaloglou, M.P. Yavropoulou, K.E. Rooij, R. Biedermann, U. Mueller, S. Kaskel, and V. Sarli, *J. Org. Chem.*, **77**, 9659 (2012).

2012MI1 V.D. Bobade, P.C. Mhaske, K.S. Vadgaonkar, and S.H. Shelke, *Monatsch. Chem.*, **143**, 847 (2012).

2012OBC921 D. Fedoseyenko, A. Raghuraman, E. Ko, and K. Burgess, *Org. Biomol. Chem.*, **10**, 921 (2012).

2012OBC7313 A.R. Choudhury and S. Mukherjee, *Org. Biomol. Chem.*, **10**, 7313 (2012).

2012OL648 E.E. Schultz, B.G. Pujanauski, and R. Sarpong, *Org. Lett.*, **14**, 648 (2012).

2012OL3862 W.-J. Bai, S.K. Jackson, and T.R.R. Pettus, *Org. Lett.*, **14**, 3862 (2012).

2012OL5022 Q. Cai, X.-W. Liang, S.-G. Wang, J.-W. Zhang, X. Zhang, and S.-L. You, *Org. Lett.*, **14**, 5022 (2012).

2012OL5142 Y. Ujihara, K. Nakayama, T. Sengoku, M. Takahashi, and H. Yoda, *Org. Lett.*, **14**, 5142 (2012).

2012OL5566 J.K. Lam, Y. Schmidt, and C.D. Vanderwal, *Org. Lett.*, **14**, 5566 (2012).

2012OL5640 Y. Luo, X. Lu, Y. Ye, Y. Guo, H. Jiang, and W. Zeng, *Org. Lett.*, **14**, 5640 (2012).

2012PH190 Y.-x. Ou, Y.-y. Li, X.-m Qian, and Y.-m. Shen, *Phytochemistry*, **78**, 190 (2012).

2012PHL541 L. Hua, J. Chen, and K. Gao, *Phytochem. Lett.*, **5**, 541 (2012).

2012S2673 M.S. Majik and S.G. Tilve, *Synthesis*, **44**, 2673 (2012).

2012S3523 E. Benedetti, M. Lomazzi, F. Tibiletti, J.-P. Goddard, L. Fensterbank, M. Malacria, G. Palmisano, and A. Penoni, *Synthesis*, **44**, 3523 (2012).

2012T697 I. Loke, N. Park, K. Kempf, C. Jagusch, R. Schobert, and S. Laschat, *Tetrahedron*, **68**, 697 (2012).

2012T3336 J. Nilsson, R. Gidlöf, M. Johansson, and O. Sterner, *Tetrahedron*, **68**, 3336 (2012).

2012T4506 N. Castellucci, L. Gentilucci, and C. Tomasini, *Tetrahedron*, **68**, 4506 (2012).

2012T8539 Y. Han, Q. Wu, J. Sun, and C.-G. Yan, *Tetrahedron*, **68**, 8539 (2012).

2012TL837 Y. Maekawa, T. Sakaguchi, H. Tsuchikawa, and S. Katsumura, *Tetrahedron Lett.*, **53**, 837 (2012).

2012TL1891 Y. Kothapalli, R.K. Puthukanoori, and S.R. Alapati, *Tetrahedron Lett.*, **53**, 1891 (2012).

2012TL2622 A.T. Khan, A. Ghosh, and M.M. Khan, *Tetrahedron Lett.*, **53**, 2622 (2012).

2012TL4114 F. Leonetti, G. Muncipinto, A. Stefanachi, O. Nicolotti, S. Cellamare, M. Catto, L. Pisani, G. Pellegrino, and A. Carotti, *Tetrahedron Lett.*, **53**, 4114 (2012).

2012TL6300 R. Raju, O. Gromyko, V. Fedorenko, A. Luzhetskyy, and R. Müller, *Tetrahedron Lett.*, **53**, 6300 (2012).

2012TL6861 X. Yu, A. Yang, W. Lin, and S.-M. Li, *Tetrahedron Lett.*, **53**, 6861 (2012).

2013BMC5983 T. Hasui, T. Ohra, N. Ohyabu, K. Asano, H. Matsui, A. Mizukami, N. Habuka, S. Sogabe, S. Endo, C.S. Siedem, T.P. Tang, C. Gauthier, L.A. De Meese, S.A. Boyd, and S. Fukumoto, *Bioorg. Med. Chem.*, **21**, 5983 (2013).

2013BMCL6167 O.S. Kanishchev, A. Lavoignat, S. Picot, M. Médebielle, and J.-P. Bouillon, *Bioorg. Med. Chem. Lett.*, **23**, 6167 (2013).

2013CAJ1243 N. Shinozaki, N. Iwamura, R. Tanaka, K. Makino, and H. Uchiro, *Chem. Asian J.*, **8**, 1243 (2013).

2013CEJ16746 M. Jaekel, J. Qu, T. Schnitzer, and G. Helmchen, *Chem. Eur. J.*, **49**, 16746 (2013).

2013CMC272 M. Cho, E. Choi, J.S. Yang, C. Lee, J.J. Seo, B.S. Kim, S.J. Oh, H.M. Kim, K. Lee, S.-K. Park, H.J. Kwon, and G. Han, *ChemMedChem*, **8**, 272 (2013).

2013CMC1554 Y. Gao, S. Samanta, T. Cui, and Y. Lam, *ChemMedChem*, **8**, 1554 (2013).

2013CS2262 E.C. Izgu and T.R. Hoye, *Chem. Sci.*, **4**, 2262 (2013).

2013HCA1564 P. Radha Krishna and B. Karunakar Reddy, *Helv. Chim. Acta*, **96**, 1564 (2013).

2013HCA1991 A. Alizadeh, V. Saberi, and J. Mokhtari, *Helv. Chim. Acta*, **96**, 1991 (2013).

2013JA1006 T. Fukuyama, N. Nakashima, T. Okada, and I. Ryu, *J. Am. Chem. Soc.*, **135**, 106 (2013).

2013JA8201 R. Liu, G.N. Winston-McPherson, Z.-Y. Yang, X. Zhou, W. Song, I.A. Guzei, X. Xu, and W. Tang, *J. Am. Chem. Soc.*, **135**, 8201 (2013).

2013JA19127 Z. Xu, Q. Wang, and J. Zhu, *J. Am. Chem. Soc.*, **135**, 19127 (2013).

2013JMC73 D. Cornut, H. Lemoine, O. Kanishchev, E. Okada, F. Albrieux, A.H. Beavogui, A.-L. Bienvenu, S. Picot, J.-P. Bouillon, and M. Médebielle, *J. Med. Chem.*, **56**, 73 (2013).

2013JNP957 C.-Y. Gan, Y.-Y. Low, N.F. Thomas, and T.-S. Kam, *J. Nat. Prod.*, **76**, 957 (2013).

2013JOC2455 K. Kempf, A. Raja, F. Sasse, and R. Schobert, *J. Org. Chem.*, **78**, 2455 (2013).

2013JOC2661 C.-W. Lee, R. Lira, J. Dutra, K. Ogilvie, B.T. O'Neill, M. Brodney, C. Helal, J. Young, E. Lachapelle, S. Sakya, and J.C. Murray, *J. Org. Chem.*, **78**, 2661 (2013).

2013JOC4823 A. Raghuraman, D. Xin, L.M. Perez, and K. Burgess, *J. Org. Chem.*, **78**, 4823 (2013).

2013MAC3895 S. Lu, M. Drees, Y. Yao, D. Boudinet, H. Yan, H. Pan, J. Wang, Y. Li, H. Usta, and A. Facchetti, *Macromolecules*, **46**, 3895 (2013).

2013MI1 L. Wei and G.-Y. Lu, *Adv. Mat. Res.*, **652-654**, 689 (2013).

2013OBC3834 S. Rastogi, E. Marchal, I. Uddin, B. Groves, J. Colpitts, S.A. McFarland, J.T. Davis, and A. Thompson, *Org. Biomol. Chem.*, **11**, 3834 (2013).

2013OBC4695 W. Balansa, R. Islam, F. Fontaine, A.M. Piggott, H. Zhang, X. Xiao, T.I. Webb, D.F. Gilbert, J.W. Lynch, and R.J. Capon, *Org. Biomol. Chem.*, **11**, 4695 (2013).

2013OL1714 J. K Howard, C.J.T. Hyland, J. Just, and J.A. Smith, *Org. Lett.*, **15**, 1714 (2013).

2013OL1814 W. Hou, B. Zhou, Y. Yang, H. Feng, and Y. Li, *Org. Lett.*, **15**, 1814 (2013).

2013OL1986 M.T. Petersen and T.E. Nielsen, *Org. Lett.*, **15**, 1986 (2013).

2013OL2734 C.-X. Zhou, L.-S. Gan, and Y.-L. Cao, *Org. Lett.*, **15**, 2734 (2013).

2013OL4608 M. Klopfeisch, R.A. Siedel, H. Görls, H. Richter, R. Beckert, W. Imhof, M. Reiher, G. Pohnert, and M. Westerhausen, *Org. Lett.*, **15**, 4608 (2013).

2013OPRD985 K.E. Henegar, R. Lira, H. Kim, and J. Gonzalez-Hernandez, *Org. Proc. Res. Dev.*, **17**, 985 (2013).

2013PHL162 M. Isaka, P. Chinthanom, P. Rachtawee, W. Somyong, J.J. Luangsa-ard, and N.L. Hywel-Jones, *Phytochem. Lett.*, **6**, 162 (2013).

2013PHL183 L.A. Shaala, D.T.A. Youssef, K.L. McPhail, and M. Elbandy, *Phytochem. Lett.*, **6**, 183 (2013).

2013PHL350 T. Damsud, S. Akisakwattana, and P. Phuwapraisirisan, *Phytochem. Lett.*, **6**, 350 (2013).

2013S2888 R.U. Batwal and N.P. Argade, *Synthesis*, 2888 (2013).

2013SL1861 C.E. Shaner, G.M. Ferrence, and T.A. Mitchell, *Synlett*, 1861 (2013).

2013T589 H. Gao, J. Sun, and C.-G. Yan, *Tetrahedron*, **69**, 589 (2013).

2013T3054 M. Ghandi, A.-T. Ghomi, and M. Kubicki, *Tetrahedron*, **69**, 3054 (2013).

2013TL506 H. Uchiro, N. Shionozaki, R. Tanaka, H. Kitano, N. Iwamura, and K. Makino, *Tetrahedron Lett.*, **54**, 506 (2013).

2013TL1344 L. Low-Beinart, X. Sun, E. Sidman, and T. Kesharwani, *Tetrahedron Lett.*, **54**, 1344 (2013).

2013TL1371 S. Rana, M. Brown, A. Dutta, A. Bhaumik, and C. Mukhopadhyay, *Tetrahedron Lett.*, **54**, 1371 (2013).

2013TL6067 L. Pawar and F.C. Pigge, *Tetrahedron Lett.*, **54**, 6067 (2013).

2013TL6944 N. Kise, S. Kinameri, and T. Sakurai, *Tetrahedron Lett.*, **54**, 6944 (2013).

2014ACIE1346 E.J. Choi, E. Kim, Y. Lee, A. Jo, and S.B. Park, *Angew. Chem. Int. Ed.*, **53**, 1346 (2014).

2014ACIE9334 P.N. Carlsen, T.J. Mann, A.H. Hoveyda, and A.J. Frontier, *Angew. Chem. Int. Ed.*, **53**, 9334 (2014).

2014ACIE10970 K.C. Nicolaou, L. Shi, M. Lu, M.R. Pattanayak, A.A. Shah, H.A. Ioannidou, and M. Lamani, *Angew. Chem. Int. Ed.*, **53**, 10970 (2014).

2014ACSC2374 D. Guerin, A.-C. Gaumont, I. Dez, M. Mauduit, S. Couve-Bonnaire, and X. Pannecoucke, *ACS Catal.*, **4**, 2374 (2014).

2014ARK(iii)256 I. Kholod, O. Vallat, A.-M. Buciumas, and R. Neier, *Arkivoc*, 256 (2014).

2014BMCL1052 U.A. Pereira, L.C.A. Barbosa, C.R.A. Maltha, A.J. Demuner, M.A. Masood, and A.L. Pimenta, *Bioorg. Med. Chem. Lett.*, **24**, 1052 (2014).

2014BMCL5727 Y.H. Seo, J.-K. Kim, and J.-G. Jun, *Bioorg. Med. Chem. Lett.*, **24**, 5727 (2014).

2014CC1588 R.C.F. Jones, J.P. Bullous, C.C.M. Law, and M.R.J. Elsegood, *Chem. Commun.*, **50**, 1388 (2014).

2014CC13998 L.-J. Wang, A.-Q. Wang, Y. Xia, X.-X. Wu, X.-Y. Liu, and Y.-M. Liang, *Chem. Commun.*, **50**, 13998 (2014).

2014CEJ2154 J. Liu, Z. Liu, N. Wu, P. Liao, and X. Bi, *Chem. Eur. J.*, **20**, 2154 (2014).

2014CEJ4105 Y. Hoshimoto, T. Ohata, M. Ohashi, and S. Ogoshi, *Chem. Eur. J.*, **20**, 4105 (2014).

2014CMC649 M. Cho, E. Choi, J.H. Kim, H. Kim, H.M. Kim, J.I. Lee, K.-C. Hwang, H.-J. Kim, and G. Han, *ChemMedChem*, **9**, 649 (2014).

2014EJMC127 U.A. Pereira, L.C.A. Barbosa, C.R.A. Maltha, A.J. Demuner, M.A. Masood, and A.L. Pimenta, *Eur. J. Med. Chem.*, **82**, 127 (2014).

2014EJOC3170 J.M. Herrmann, M. Untergehrer, G. Jürgenliemk, J. Heilmann, and B. König, *Eur. J. Org. Chem.*, 3170 (2014).

2014EJOC5540 Y. Liu, J. Ma, Y. Wang, P.O. Donkor, Q. Li, S. Gao, Y. Hou, Y. Xu, J. Cui, L. Ding, F. Zhao, N. Kang, L. Chen, and F. Qui, *Eur. J. Org. Chem.*, 5540 (2014).

2014EJOC5777 M. Marhold, C. Stillig, R. Froehlich, and G. Haufe, *Eur. J. Org. Chem.*, 5777 (2014).

2014EJOC7865 I. Kholod, O. Vallet, A.-M. Buciumas, A. Neels, and R. Neier, *Eur. J. Org. Chem.*, 7865 (2014).

2014HCA1606 P. Lorenz, J. Conrad, S. Duckstein, D.R. Kammerer, and F.C. Stintzing, *Helv. Chim. Acta*, **97**, 1606 (2014).

2014JHCE25 Y.-X. Liu, H.-P. Zhao, H.-B. Song, Y.-C. Gu, and Q.-M. Wang, *J. Heterocycl. Chem.*, **51**, E25 (2014).

2014JA15877 Y. Hoshimoto, T. Ohata, Y. Sasaoka, M. Ohashi, and S. Ogoshi, *J. Am. Chem. Soc.*, **136**, 15877 (2014).

2014JMC2334 S.S. Zimmerman, A. Khatri, E.C. Garnier-Amblard, P. Mullasseril, N.L. Kurtkaya, S. Gyoneva, K.B. Hansen, S.F. Traynelis, and D.C. Liotta, *J. Med. Chem.*, **57**, 2334 (2014).

2014JMC10557 T. Conroy, J.T. Guo, N. Elias, K.M. Cergol, J. Gut, J. Legac, L. Khatoon, Y. Liu, S. McGowan, P.J. Rosenthal, N.H. Hunt, and R.J. Payne, *J. Med. Chem.*, **57**, 10557 (2014).

2014JNP327 Y.-Y. Low, F.-J. Hong, K.-H. Lim, N.F. Thomas, and T.-S. Kam, *J. Nat. Prod.*, **77**, 327 (2014).

2014JNP983 G.-Y. Zhu, G. Chen, L. Liu, L.-P. Bai, and Z.-H. Jiang, *J. Nat. Prod.*, **77**, 983 (2014).

2014JNP1396 C.-K. Kim, I.-H. Song, H.Y. Park, Y.-J. Lee, H.-S. Lee, C.J. Sim, D.-C. Oh, K.-B. Oh, and J. Shin, *J. Nat. Prod.*, **77**, 1396 (2014).

2014JNP2537 R.C. Jadulco, M. Koch, T.B. Kakule, E.W. Schmidt, A. Orendt, H. He, J.E. Janso, G.T. Carter, E.C. Larson, C. Pond, T.K. Matainaho, and L.R. Barrows, *J. Nat. Prod.*, **77**, 2537 (2014).

2014JOC4043 A. Chatzimpaloglou, M. Kolosov, T.K. Eckols, D.J. Tweardy, and V. Sarli, *J. Org. Chem.*, **79**, 4043 (2014).

2014JOC4582 X.-H. Ouyang, R.-J. Song, Y. Li, B. Liu, and J.-H. Li, *J. Org. Chem.*, **79**, 4582 (2014).

2014JOC8049 A.A. van Loon, M.K. Holton, C.R. Downey, T.M. White, C.E. Rolph, S.R. Bruening, G. Li, K.M. Delaney, S.J. Pelkey, and E.T. Pelkey, *J. Org. Chem.*, **79**, 8049 (2014).

2014MI1 P.M.G. Sabido and D.A. Lightner, *Monatsh. Chem.*, **145**, 775 (2014).

2014NPR905 B. Li, W.J. Wever, C.T. Walsh, and A.A. Bowers, *Nat. Prod. Rep.*, **31**, 905 (2014).

2014OBC1090 A.I. Lansakara, D.P. Farrell, and F.C. Pigge, *Org. Biomol. Chem.*, **12**, 1090 (2014).

2014OBC1833 L.-W. Ye, C. Shu, and F. Gagosz, *Org. Biomol. Chem.*, **12**, 1833 (2014).

2014OBC4132 E. Marchal, D.A. Smithen, M.I. Uddin, A.W. Robertson, D.L. Jakeman, V. Mollard, C.D. Goodman, K.S. MacDougall, S.A. McFarland, G.I. McFadden, and A. Thompson, *Org. Biomol. Chem.*, **12**, 4132 (2014).

2014OL1294 M.G. Weaver, W.-J. Bai, S.K. Jackson, and T.R.R. Pettus, *Org. Lett.*, **16**, 1294 (2014).

2014OL3704 R.-Q. Ran, J. He, S.-D. Xiu, K.-B. Wang, and C.-Y. Li, *Org. Lett.*, **16**, 3704 (2014).

2014OL3780 B.J. Marsh, H. Adams, M.D. Barker, I.U. Kutama, and S. Jones, *Org. Lett.*, **16**, 3780 (2014).

2014OL4384 J.G. David, W.-J. Bai, M.G. Weaver, and T.R.R. Pettus, *Org. Lett.*, **16**, 4384 (2014).

2014OL5604 S. Jia, S. Su, C. Li, X. Jia, and J. Li, *Org. Lett.*, **16**, 5604 (2014).

| | |
|---|---|
| 2014S1539 | C. Joie, K. Deckers, G. Raabe, and D. Enders, *Synthesis*, **46**, 1539 (2014). |
| 2014T2595 | X. Jusseau, L. Chabaud, and C. Guillou, *Tetrahedron*, **70**, 2565 (2014). |
| 2014T5993 | M. Ergun, C. Dengiz, M.S. Özer, E. Sahin, and M. Balci, *Tetrahedron*, **70**, 5993 (2014). |
| 2014T8768 | C. Han, W. Meng, H. Liu, Y. Liu, and J. Tao, *Tetrahedron*, **70**, 8768 (2014). |
| 2014TL2142 | M.I. García-Aranda, M.T. García-López, M.J.P. de Vega, and R. González-Muñiz, *Tetrahedron Lett.*, **55**, 2142 (2014). |
| 2014TL4716 | D. Hahn, J. Chin, H. Kim, I. Yang, D.H. Won, M. Ekins, H. Choi, S.-J. Nam, and H. Kang, *Tetrahedron Lett.*, **55**, 4716 (2014). |
| 2014TL4753 | G. Saidachary, K.V. Prasad, M. Sairam, and B.C. Raju, *Tetrahedron Lett.*, **55**, 4753 (2014). |
| 2014TL6455 | J. Feng, Y. Wang, Q. Li, R. Jiang, and Y. Tang, *Tetrahedron Lett.*, **55**, 6455 (2014). |
| 2014TL6526 | R.A. Seidel, B. Schowtka, M. Klopfleisch, T. Kühl, A. Weiland, A. Koch, H. Görls, D. Imhof, G. Pohnert, and M. Westerhausen, *Tetrahedron Lett.*, **55**, 6526 (2014). |

CHAPTER FOUR

# Recent Advances in the Chemistry of Acridines

**Andreas Schmidt*, Ming Liu**
Institute of Organic Chemistry, Clausthal University of Technology, Clausthal-Zellerfeld, Germany
*Corresponding author: E-mail: schmidt@ioc.tu-clausthal.de

## Contents



*Advances in Heterocyclic Chemistry*, Volume 115
ISSN 0065-2725
http://dx.doi.org/10.1016/bs.aihch.2015.04.004

## Abstract

This review summarizes advances in the chemistry of acridines concerning properties, isolation of natural products possessing the acridine ring as a partial structure, new structures of biologically active acridines, and applications of acridines as fluorescent, chemiluminescent, or sensor materials. New synthetic methods for the preparation of acridines mainly published in the last decade as well as new reactions are also reviewed.

**Keywords:** Acridane; Acridine dyes; Acridine orange; Acridine sensors; Acridine synthesis; Acridine yellow; Acridinium; Acridone; Acridone alkaloids; Acriflavine; Acronycine; Amsacrine; Arborinine; Asulacrine; BRACO 19; Chemiluminescent acridines; Euflavine; Fluorescent acridines; Ledacrine; Mepacrin; Neutroflavine; Oriciacridones; Proflavine; Quinacrine; Rivanol; Tacrine; Toddaliopsin; Trypaflavine

## 1. INTRODUCTION

A stable, crystalline, and basic substance was identified in 1870 by Graebe and Caro as an impurity of crude anthracene that was isolated from coal tar. As the authors realized, due to an irritating effect on skin, the name acridine was given to that substance (lat. acer = sharp) (1870MI746; 1871MI265). Acridine, the structure and numbering of which is shown in Scheme 1, can also be called dibenzo[*b*,*e*]pyridine, benzo[*b*]quinoline, 2,3,5,6-dibenzopyridine, or 10-azaanthracene. The development of "Orange Dye," patented in 1895 (1895USP) and the introduction of acriflavine (vide infra) in 1912 by Ehrlich and Benda (1912MI1787) undoubtedly are milestones in the early history of acridines. These were widely applied as antiseptics and antibacterials during the world wars, until they were almost replaced by penicillins and other antibiotics. Nevertheless, considerable interest has been focused on syntheses and applications of acridines due to a large number of their fascinating properties. Among those, anticancer, antibacterial, and antiprion activities can be mentioned, in addition to their important roles as dyes or sensors for biological or technical purposes.

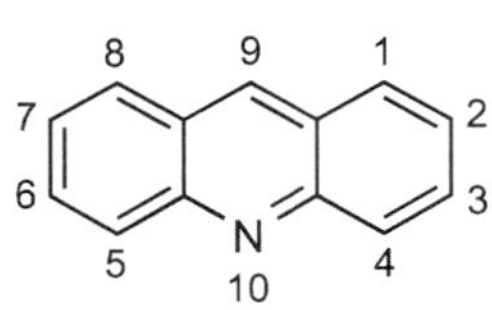


**Scheme 1**

As a consequence, the chemistry, physics, and applications of acridines have been the subject of numerous books, monograph chapters, and reviews. The syntheses of acridines have been summarized in 2005 (2005SOS(15)987) as an update of several earlier reviews (1992HOU(E7b)115, 1987MI, 1973MI, 1966MI, 1956MI). The reactivity of the acridine ring (2004S313), syntheses of 9-substituted acridines (1977H(6)987), and kinetic data and mechanisms of the reactions occurring at position 9 or 10 of 9-substituted acridines have also been reviewed (1980H(14)985). As acridines constitute a class of compounds of considerable pharmaceutical interest, their biological activities have been summarized in several overviews that cover portions of that field. Recent reviews deal with natural and synthetic acridines and acridones as antitumor agents (2011MI305, 2004MI55), as DNA intercalators (2009MOL1725, 2005CMC127), antimalarials (2011MI11), or antibacterials (2001MI1). Genetic effects due to acridines have been summarized (1979MR261), and their role in anticancer therapy (2013MMCIAE1256, 2008MI1211, 2007MI139) has been discussed. Some other reviews describe more specialized aspects of biologically active acridines. Thus, the potential of acridine carboxamide platinum complexes as anticancer agents (2014MI695) and structure—activity studies of peptide—acridine/acridone conjugates were summarized (2007CMC3079). The significance of acridines has been reflected in a large number of patents, and patent overviews were presented (2014MI647). Additional references, including some review articles, are cited in the following sections.

This review summarizes advances in the chemistry of acridines published mainly in the last decade. As more than 1000 publications on acridines appear every year, this review cannot be comprehensive. It presents recent, selected examples of acridines as natural products, as biologically active compounds, some applications of technically or biotechnically interesting acridines, examples of recently described syntheses of acridines, and some reactions on the acridine ring.

## 2. PHYSICAL PROPERTIES OF ACRIDINE

A summary of physical and spectroscopic properties of acridine can be found in the literature (1992HOU(E7b)115). Acridine is known to crystallize in six different polymorphs, and the influence of solvents as well as isotopes on the crystallization has recently been studied (2012MI5966). The acridine ring has been the subject of numerous calculations during the last decade. The characteristics of the electron density at the ring critical point (RCP), which is defined as the point of the minimum electron density within the ring surface and a maximum on the ring line (2000MI), have been discussed as possible descriptors for the aromaticity of several heteroaromatics, including acridine. Calculated values of HOMA, NICS(1), PDI, ATI, FLU, and $\Delta_6^{\pi}$ indices for the three individual rings of acridine have been presented. The HOMA and NICS(1) values calculated 1 Å above the RCP of the A, B, and C ring of acridine (cf. Scheme 2) are 0.6275, 0.7287, 0.6271, and −16.0218, −22.4097, −16.0185, respectively (2008CPL378). The role of benzo-annulation on the local aromaticity of a series of conjugated heterocycles, an NICS(1) calculated 1 Å above the ring center of the central ring of acridine logically gave a different value, calculated to be −13.35 (2014JCP(A)11591). The standard molar heat capacity $C^0_{p,m}$ of acridine in the crystalline state at 0.1 MPa and 298.15 K was calculated to be 203.5 $J \cdot K^{-1} \cdot mol^{-1}$ (2012MI330) and measured to be 201.5 $J \cdot K^{-1} \cdot mol^{-1}$ (1989MI81). Molecular size, polarizability, and geometry significantly influence the proton affinities of nitrogen-containing heteroaromatics. Thus, the gas-phase basicity ($GB_{298}$), the proton affinity (PA), and the protonation entropy $S_{prot}$ of acridine were determined to be 224.4 kcal $mol^{-1}$, 232.5 kcal $mol^{-1}$, and 0.7 kcal $mol^{-1}$, respectively, and these values have been compared with calculated values. The experimental proton affinities increase with molecular size and change from pyridine (222.2 kcal $mol^{-1}$) to quinoline (227.8 kcal $mol^{-1}$) and acridine (see above) (2015MI118). The calculated polarizability and the dipole moment of acridine are 22.2 $Å^3$ and 2.01 D, respectively, and these values have been correlated with the calculated theoretical proton affinities of a series of heteroaromatics. The calculated atomic charges of the non-protonated acridine are presented in Scheme 2; they have been compared with the values of the corresponding N-protonated form, that is, the acridinium ion. On protonation, the charge of C-9 of acridine changes by +0.10 to −0.01, whereas the charge differences of the other atoms are smaller on quaternization (2015JPC(A)127).

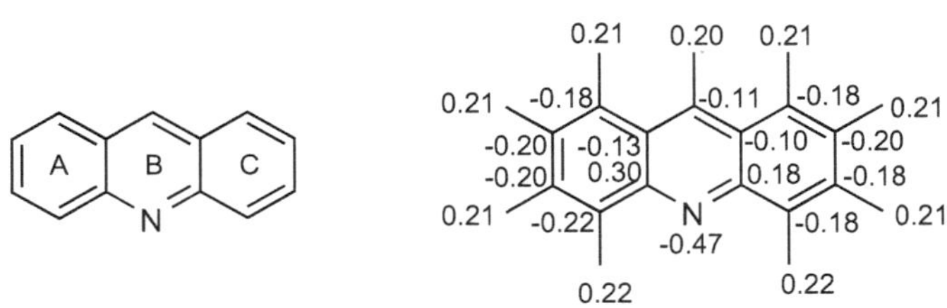


Scheme 2

## 3. NATURAL PRODUCTS POSSESSING THE ACRIDINE CORE

Acridines are found in nature as acridone alkaloids or as a partial structure of tetracyclic, pentacyclic, or even larger ring systems. The first acridone alkaloids were isolated from the relatively small plant family of Rutaceae species, and have been found to be widely distributed across that species, among other alkaloids such as indoles, phenanthridines, and aporphines. The class of Rutaceous alkaloids served as an inspiration for the design of novel antitumor compounds summarized in a recent review (2007MI65). Information to 2006 on acridine alkaloids can be found in natural product reports (2008NPR166) and some monographs (1998MI). Selected acridine alkaloids proved to be active as antiviral and antimalarials, and details are given in a review (2011MI11). In addition, they are described as antileishmanial, antifungal, cytotoxic, and photosensitive agents (2007PLM41). Two main groups have been distinguished. The first mentioned class comprises acridine alkaloids which are oxygenated in ring A or in ring B. The second group possesses acyclic or cyclic C5 side chains in ring A (1978MI).

Acronycine alkoids form a subgroup which is formally derived from 1,3-dihydroxyacridone and one or more isoprene units (1978MI). The well-known parent, the alkaloid acronycin, which is 6-methoxy-3,3,12-trimethyl-3*H*-pyrano[2,3-*c*]acridin-7(12*H*)-one, is an example of a tetracyclic natural system possessing acridine partial structures. Acronycine proved to be a potent antitumor alkaloid and reached clinical trials (2004CMC83). New derivatives of acronycine have been developed as antimalarials (2005BMC1653) and as cytotoxic agents (2005CPB1540). The cytotoxicity of the tetracyclic ring system arborinine and a series of furanoacridones such as evoxanthine, isogravacridone chlorine, rutacridone, and others isolated from *Ruta graveolens* L. was tested (2007PLM41).

Pyridoacridines form an important class of pentacyclic alkaloids possessing the acridine ring. They have been isolated from *Cystodytes violatinctus* (2013JNP1801), from the sponge *Xestospongia* sp. (2010MI1769, 2007JOC8501), from the western Mediterranean ascidian *Cystodytes dellechiajei* (2010JNP1044), and from the Thai marine sponge *Petrosia n.* sp. (2008BMC6560). Methanolic extracts of the sponge *Oceanapia* sp. contain the pyridoacridine alkaloid petrosamine B (2005JNP804). The new alkaloids isolissoclinotoxin B, diplamine B, and lissoclinidine B have been isolated from *Lissoclinum* cf. *badium* (2008BMC10022). A review on biological activities of pyridoacridines appeared in 2004 (2004NPR731), and as a consequence of the broad spectrum of biological activities of pyridoacridines, modifications of that ring system have been performed (2009BMC4836). Other pentacyclic alkaloids possess the dipyrano[2,3-*c*:2′,3′-*h*]acridin-7(2*H*)-one and the pyrano[2,3-*c*]acridin-7(12*H*)-one ring. As an example the alkaloids macranthan and 7-hydroxynoracronycine from dried stem barks of *Glycosmis macrantha* (Rutaceae) growing in the India-Malayan region (2001MOL4401) are mentioned.

Recent findings concerning acridine alkaloids without additional annulated rings, which are in fact acridone alkaloids, are as follows. The natural products **1** and **2** were isolated from *Boenninghausenia japonica* NAKAI (Table 1). They resemble glycitrine-II **3** from citrus plants (2004MI1312). Four toddaliopsins A—D **4—7** have been isolated from the leaves of *Toddaliopsis bremekampii* (Rutaceae) from South Africa, three of them possess substituted N-methyl groups, a very rare substitution pattern in these natural products. Toddaliopsin C displayed moderate anti-inflammatory activities, and the role of the hydroxyl group at C-1 on the biological activity was discussed (2005P1724).

Eleven acridone alkaloids were identified from methanol extracts of the stem bark of *Swinglea glutinosa* (Rutaceae), among them seven acridones **12—18** (Table 2) and four tetracyclic ring systems possessing acridine rings. Eight were identified as inhibitors of cathepsin V (2011BMC1477). Three new acridone alkaloids, the pyrano[3,2-*b*]acridin-6(11*H*)-ones cycloatalaphylline-A, *N*-methylcycloatalaphylline-A, and the furo[3,2-*b*]acridin-5(10*H*)-one *N*-methylbuxifoliadine-E, in addition to the known structures citrusinine-I **17**, buxifoliadine-A **19**, *N*-methylatalaphylline **20**, atalaphylline (which is the N10-demethylated derivative of **20**), and junosine (**21**) were isolated from the roots of the South-Asian *Atalantia monophylla* by extraction with dichloromethane and acetone. Antiallergic activities were tested (2008P2616).

Two new acridone dimers, glycobismines-F **22** and G **23**, were isolated from the roots of *Glycosmis citrifolia* collected in Taiwan (2004CPB362) and the oriciacridones A—F from the stem bark of *Oriciopsis glaberrima* (Rutaceae). The oriciacridones A **24** and B **25** are shown in Scheme 3 (2006CPB292, 2006P475).

## 4. BIOLOGICALLY ACTIVE ACRIDINES

A number of acridine derivatives serve as chemotherapeutic agents, especially in the field of antitumor DNA-binding agents (2004MI55).

**Table 1** Acridine alkaloids: One

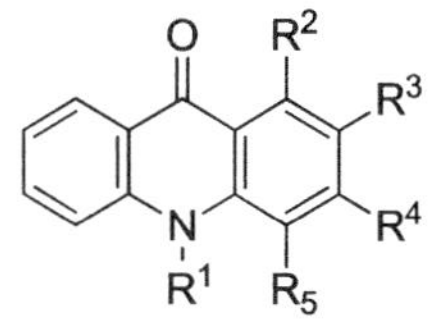

| No. | $R^1$ | $R^2$ | $R^3$ | $R^4$ | $R^5$ |
|---|---|---|---|---|---|
| **1** | Me | OH | H | OH | 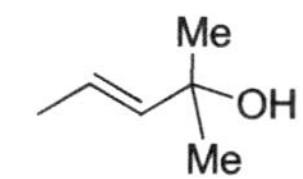 |
| **2** | Me | OH | H | OH | 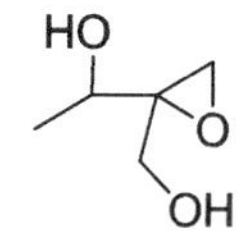 |
| **3** | Me | OH | H | OH | 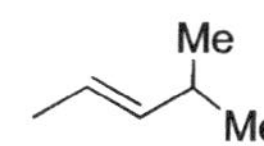 |
| **4** | H | OMe | OMe | OMe | H |
| **5** | 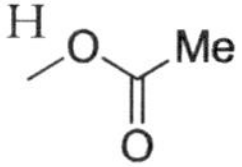 | OMe | OMe | OMe | H |
| **6** | 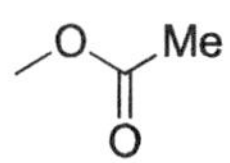 | OH | OMe | OMe | H |
| **7** | $-CH_2OMe$ | OH | OMe | OMe | H |
| **8** | Me | OH | OMe | OMe | H |

**Table 2** Acridine alkaloids: Two

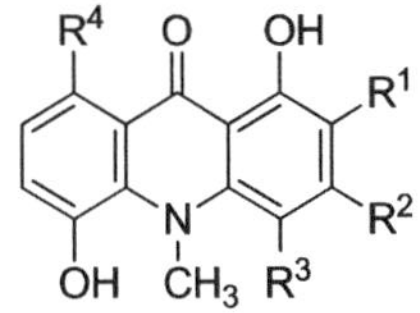


| No. | R¹ | R² | R³ | R⁴ |
|---|---|---|---|---|
| **12** | 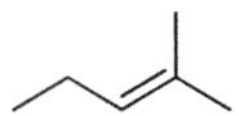 | OH | OMe | 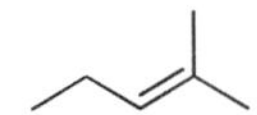 |
| **13** | 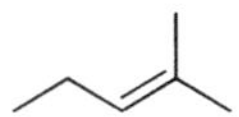 | OH | OH | 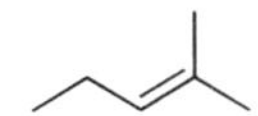 |
| **14** | 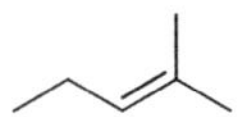 | OH | OMe | H |
| **15** | H | OH | 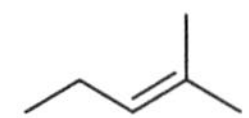 | H |
| **16** | H | OH | OMe | H |
| **17** | H | OMe | OMe | H |
| **18** | OMe | OMe | OMe | H |
| **19** | 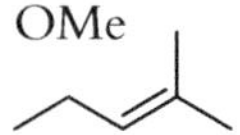 | OMe | 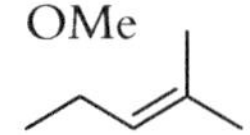 | H |
| **20** | 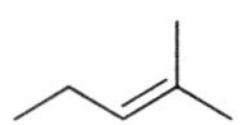 | OH | 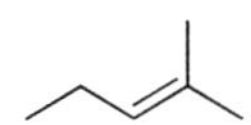 | H |
| **21** | 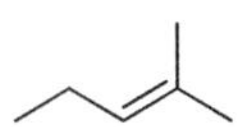 | OH | H | H |

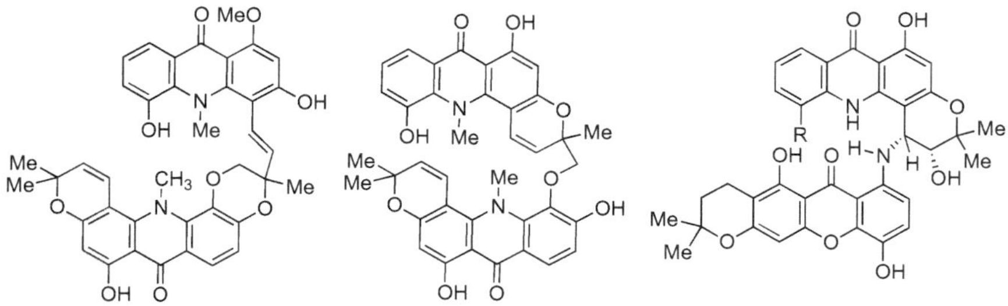


Scheme 3

## 4.1 Some Established Acridines as Biologically Active Compounds

Amsacrine is *N*-(4-(acridin-9-ylamino)-3-methoxyphenyl)methanesulfonamide (synonyms: m-AMSA, acridinyl anisidide), which is an antineoplastic agent applied against acute leukemia (Scheme 4). The antitumor agent N1,N1-dimethyl-N3-(1-nitroacridin-9-yl)propane-1,3-diamine is called Ledakrin (synonyms: Ledacrine, Nitracrina, Nitracrine, Nitracrinum). Acriflavine (Acriflavinium chloride, Trypaflavine, Neutroflavine, Euflavine), one of the milestones in the history of acridines and developed in 1912 (see above), is the name of a mixture of 3,6-diamino-10-methylacridinium chloride and 3,6-diaminoacridinium hydrochloride. It has been applied as a topical antiseptic, as antibacterial, and as anticancer agent and

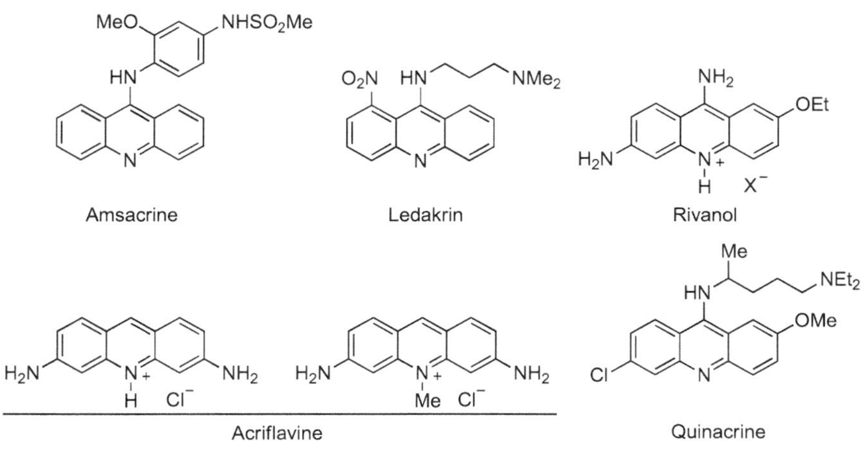


Scheme 4

has also been recently shown to exhibit antimalarial activities in vitro and in vivo (2014MI2366). Rivanol (ethacridine monolactate monohydrate, Acrinol) is the trade name of 2-ethoxy-6,9-diaminoacridinium monolactate monohydrate (X = lactate = 2-hydroxypropionate), used as antiseptic and as abortifacient. The racemic N4-(6-chloro-2-methoxyacridin-9-yl)-N1,N1-diethylpentane-1,4-diamine is known as Quinacrine (synonyms: Mepacrin, Atebrin, Atabrine), used for therapy and prophylaxis of malaria. Recently, the effect of the expression of Quinacrine on antioxidant, apoptotic, and anti-apoptotic genes was evaluated (2014MI8093). A possible explanation of the discrepancy between the ability of Quinacrine to reduce effectively the incidence of prion protein in cell culture on the one hand and its weak prion-binding affinity on the other was discussed. Based on the finding that several 9-aminoacridine derivatives cleave off the acridine ring in the presence of primary aliphatic amines especially in the course of solid-phase syntheses, investigations showed that the α-amino group in peptides is a weak acridine acceptor, whereas the ε-amino group of lysine is a moderate acridine acceptor (2006MI605).

1,2,3,4-Tetrahydroacridin-9-amine, Scheme 5, is called Tacrine (synonym: Cognex). It has been applied as a cholinesterase inhibitor for the treatment of Alzheimer's disease and as parasympathomimetic. A succinic acid bridged Tacrine–Silibinin codrug developed recently has shown neuro- and hepatoprotective effects (2012JMC5231). Proflavine is the well-known name of 3,6-diaminoacridine having a number of pharmaceutical uses

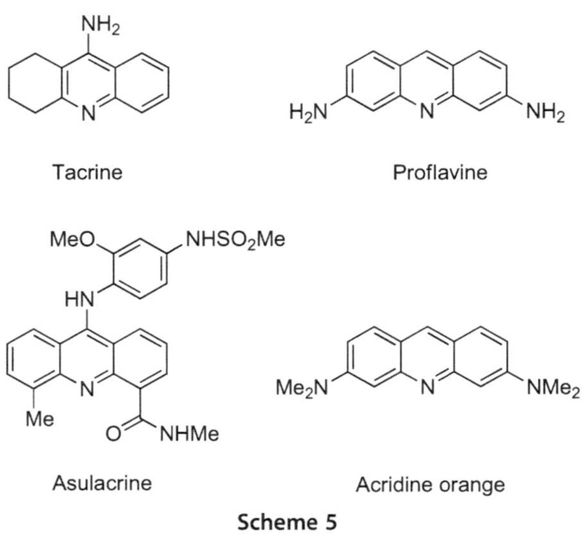


**Scheme 5**

(2002CMC655, 2007MI139) and is commercially available in non-protonated form, as a hydrochloride, or as a hemisulfate hydrate. The mutagenic effect is due to intercalation between nucleic acid base pairs in double-stranded DNA which hinders DNA replication in rapidly growing tumor cells (1995B13682). Telomeric G-quadruplex DNA (GQ-DNA) is an alternative structure motif of DNA of interest as a target for anticancer drug design. Steady-state and time-resolved fluorescence spectroscopy in combination with circular dichroism (CD) and isothermal titration calorimetry (ITC) demonstrated that proflavine acts as a stabilizer of GQ-DNA due to the presence of a heteroaromatic capable of $\pi$ interactions to nucleobases as well as a cationic center in the protonated form for Coulomb interactions with the phosphate backbone of DNA (2014JPC(B)11090, 2012MI1109, 2008JPC(B)9243). Proflavine is also in use as disinfectant active against Gram-positive bacteria and can be applied as a dihydrochloride or a hemisulfate salt as topical antiseptic. The adsorption of proflavine, 9-aminoacridine hydrochloride hydrate (9AA-HCl, vide infra), acridine yellow (AY; 2,7-dimethylacridin-3,6-diamine, vide infra), and acridine orange (AO, vide infra) on citrate-stabilized gold nanoparticle surface was examined (UV–vis, FT-IR, TEM). The gold nanoparticle-conjugated 9AA-HCl and acridine orange had better antibacterial efficacy than the bare drug against various strains of Gram-positive and Gram-negative bacteria (2014MI636). A mixture of 9-aminoacridine and 4-hexylresorcinol is called Acrisorcin, a fungicide. Asulacrine is 9-[4-(methanesulfonamido)-2-methoxyanilino]-N,5-dimethylacridine-4-carboxamide (synonyms: ASL, Ci-921, UNII-S8P50T62B6, CCRIS 1030, NSC 343499, Amsalog), and therefore an analog of amsacrine is an inhibitor of topoisomerase II (1988MI1783) and has been examined in the development of an anticancer liposomal formulation (2014MI528). Its activity against breast and lung cancer in clinical trials phase I and II has proven to be better than Amsacrine (2001MI333, 1992MI309). A disadvantage is the danger of phlebitis on intravenous infusion. Therefore, efforts have been directed toward the development of a stable long-circulating liposome system with a high Asulacrine content for intravenous infusions (2014MI1). Acridine (AO) orange is N,N,N′,N′-tetramethylacridine-3,6-diamine (synonyms: Euchrysine, Rhoduline Orange), first extracted from coal tar, and displaying various biological activities, proved to be a useful fluorescent dye specific for DNA and RNA, and is an antitumor and antimalarial drug, photosensitizer with biological as well as technical applications, and a detector of bacteria and parasites. Moreover, it is a pH indicator and accumulates in musculoskeletal sarcomae which enables selective cytocidal effects on irradiation. A recent review summarizes results concerning acridine orange as a potential anticancer agent (2007MI205). Fluorescence spectroscopic techniques applying acridine orange differentiate between dead (orange fluorescence) and live cancer cells (green fluorescence) (2013MI1921). They were used for the rapid, precise, and accurate enumeration of parasitemia in rodents, and this method was discussed in malaria research (2007MI662). Acridine orange was used to functionalize magnetic nanoparticles for nucleus labeling (2014MI150) and fluorescent hollow microcapsules via congregation of poly(lysine) interceded by silica nanoparticles which can detect DNA in 100 ng $ml^{-1}$ concentration ranges (2014PEA320). Another work deals with the adsorption behavior of acridine orange onto nanozeolites in aqueous media (2013MI909). A (poly(acridine orange))-GR (graphene)/carbon/ionic liquid paste electrode showed good selectivity, stability, and reproducibility toward the detection of rutin (2013 MI298). Acridine orange was also described as a DNA-binding agent for the detection of DNA hybridization and the discrimination of mutations (2012JA14099). Degradation mechanisms have also been investigated dealing with the photocatalyzed degradation of acridine orange in aqueous suspensions of titanium dioxide under a variety of conditions (2007MI233).

## 4.2 Acridines with 9-Amino, 9-Alkylamino, and 9-Imino Substituents

In the last decade, publications have dealt with the syntheses of 9-amino, 9-alkylamino, and 9-anilino derivatives of acridine, often in combination with the investigation of their biological activities. The simplest derivative of a monosubstituted acridine with N9 substitution pattern, 9-aminoacridine **26** (9AA) displayed a selective toxicity for infected leukemic cells (Scheme 6). In combination with other chemotherapeutics or the monoclonal antibody Campath-1H 9-aminoacridine resulted in increased efficacy and selective killing of tumor cells (2014MI43). A clear structure—activity relationship between the N-alkyl chain length of a series of 9-alkylaminoacridines **27** and their acridin-3-yl(4-methylpiperazin-1-yl)methanone derivatives and antibacterial activity against methicillin-resistant and methicillin-sensitive *Staphylococcus aureus* strains was found. The 9-alkylaminoacridines are believed to function as amphiphilic membrane-active disruptors, similar to quaternary ammonium antimicrobials (2014BMC3014). Intercalation and groove binding of an acridine—spermine conjugate **28** on two 16-mer oligonucleotides possessing alternating guanine—cytosine or adenine—thymine sequences

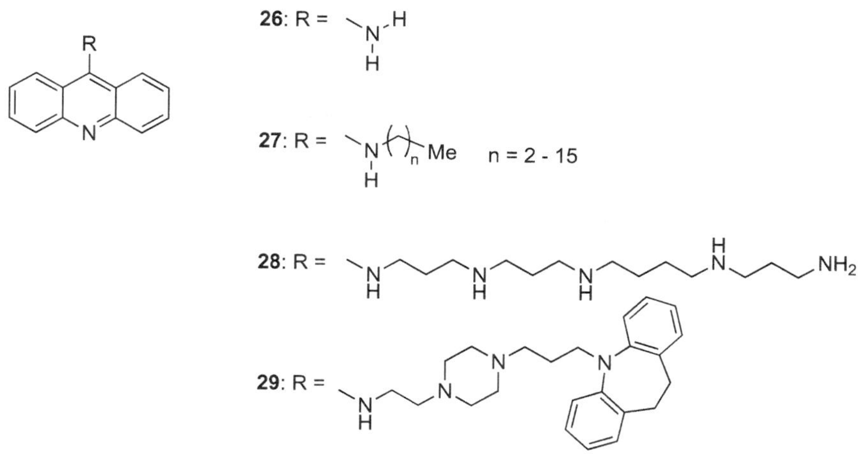


Scheme 6

were studied by FT–Raman and UV–vis absorptions. Each oligonucleotide chain is able to host only one molecule of **28**. Raman spectra showed wave-number shifts for skeletal and base vibrations on intercalation and the data suggested some sequence selectivity induced by the spermine chain (2005JMS699). The antiprion-active acridine **29**, Scheme 6, is an additional example of an N9-substituted acridine with biological activities that combines structure elements of quinacrine and imipramine (2006JMC6591). Some other acridines of that substitution pattern have been described, among those acridine possessing 9-alkylamino chains as spacer between the hetero-aromatic and a reactive EDTA-derived moiety which displayed antiviral properties (2004EJMC975). A gene transfer agent was prepared by combining a melittin fusogenic peptide with a polyacridine peptide of the general formula (Acr-X)$_n$-Cys, where Acr stands for a lysine which is modified on its $\varepsilon$-amine with an acridine ring, and X represents the amino acids arginine, leucine, or lysine with $n = 2$, 3, or 4 (2010MI74).

A prototype of a class of cationic DNA-targeted hybrid agents is [PtCl(en)(ACRAMTU-S)]$(NO_3)_2$ **30** ("PT-ACRAMTU"), where "en" stands for ethane-1,2-diamine, and "ACRAMTU" for 1-[2-(acridin-9-ylamino)-ethyl]-1,3-dimethylthiourea (Scheme 7). It cleaves DNA by platinum binding to guanine or adenine and intercalation of the acridine ring onto the base pair near the site of platination (2001JMC4492, 2002BCP191, 2003JA9629, 2005JA1160). An exchange of the thiourea moiety to an amidine group leads to a platinum-acridine agent with activity in non-small-cell lung cancer in vitro and in vivo (2008JMC7574). A screening

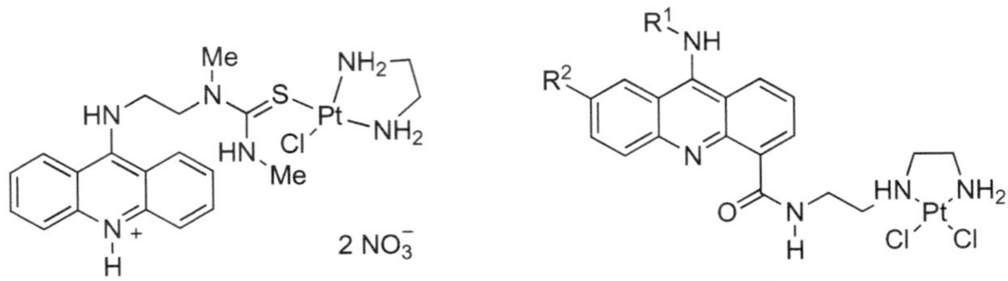

30

31a: $R^1 = R^2 = H$
31b: $R^1 = H$, $R^2 = F$
31c: $R^1 = H$, $R^2 = OMe$
31d: $R^1 = -CH_2CH_2OH$, $R^2 = H$

Scheme 7

method was developed for related hybrids based on metal-mediated amine-to-nitrile addition that yielded a library of 60 platinum–acridines (2012JMC10198). A comparison of the DNA sequence specificity showed that the DNA-targeted 9-aminoacridine complexes **31a–d** had greatly increased levels of DNA damage at CG and GA dinucleotides as compared with cisplatin (2014MI997). Furthermore, gold(I) analogs of a platinum-acridine antitumor agent are only moderately cytotoxic but display activity against *Mycobacterium tuberculosis* (2009JMC6519).

2,6,9-Trisubstituted biologically active acridines with N9 substituents are shown in part in Scheme 8. A library of 175 9-aminoacridine derivatives related to quinacrine starting from 9-chloroacridine was designed and screened against two strains of *Plasmodium falciparum.* The derivatives possess methoxy or chloride substituents in position 6 and trifluoromethyl groups or methoxy groups in position 2, among others, in addition to varying substituents at N9 (2006BMC334). A similar substitution pattern was used to examine structure–activity relationships of 9-aminoacridines in scrapie-infected neuroblastoma cells (2006BMCL4913). An oligonucleotide **32** with (*R*)-3-*O*-[4-(9-aminoacridin-2-ylethynyl)benzyl]glycerol moiety was prepared, via Sonogashira coupling of iodoacridine, which led to an increased stability of parallel triplexes in comparison to unmodified triplexes. Molecular modeling suggests that the acridine moiety is stacking between the bases of the duplex, while the phenyl ring is located between the bases of the triplex-forming oligonucleotide (2006MI950). Compound **33** displayed a good activity against breast cancer cells, and others from the series against lung, liver, and colon cancer cells (2014MI2). Antimalarial activities of 9-aminoacridines such as **34** and artemisinin–acridine hybrid **35**, prepared under microwave irradiation from the amines and 2-bromo-(10β-dihydroartemisinoxy)ethane,

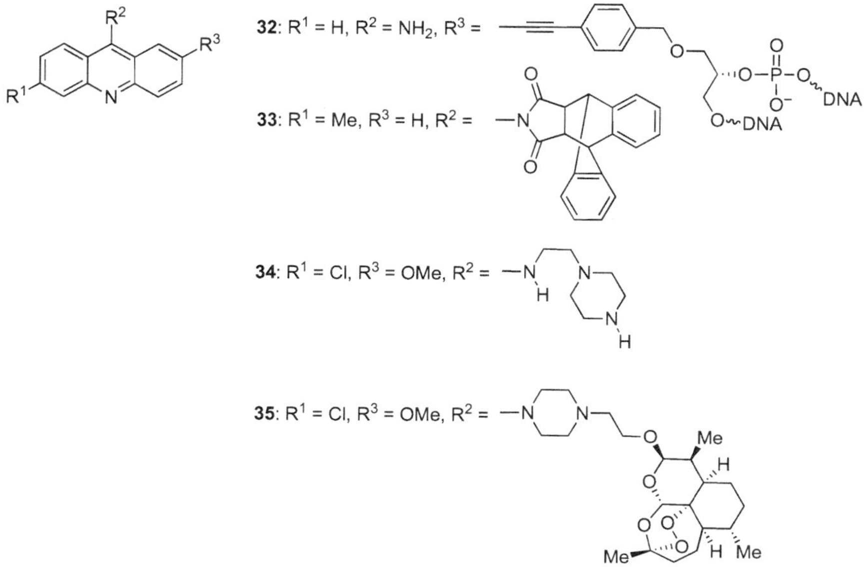

Scheme 8

were studied, and the in vitro cytotoxicity and anticancer activity against HeLa cell lines were elucidated (2014MI16).

Apart from 2,6,9-trisubstituted acridines described above, 2,4,9-trisubstituted acridines have been prepared and examined (cf. Scheme 9). The syntheses of

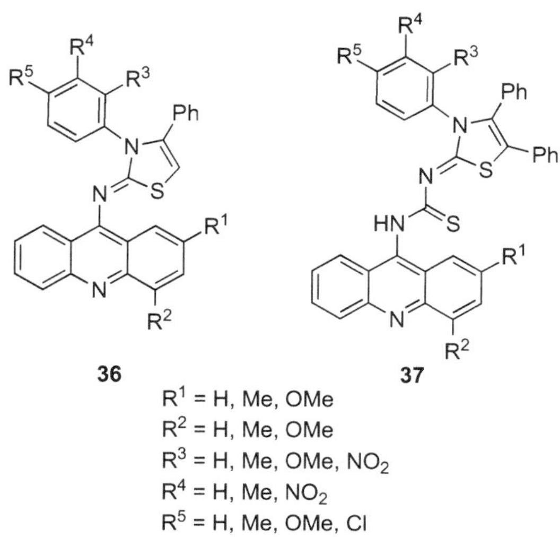

Scheme 9

acridinyl-thiazolino derivatives **36** and **37** and their evaluation for anti-inflammatory, analgesic, and kinase inhibition activities were published (2005 BMC4291).

1-Nitro-9-hydroxyethylamino acridine **38a** (C-857), a member of the class of 1-nitroacridines and a representative of a 1,4,7,9-tetrasubstituted acridine, is a well-documented anticancer compound (Scheme 10). Derivatives such as **38b** and **c** have been prepared in an effort to enhance antitumor efficacy and to lower the toxicity. When $R^3$ = Me in C-857 (=C-1748), a significantly higher therapeutic efficacy was observed. A study was undertaken to correlate the mutagenicity of these derivatives. Due to a relatively low mutagenic potential in combination with a higher antitumor effect in prostate cancer, C-1748 seemed to be a promising chemotherapeutic agent for clinical development (2005MI2312). The 3,7,9,10-tetrasubstitution patterns of biologically active acridines is realized in **39**, tested against *Plasmodium falciparum* (2007BMC3278).

## 4.3 Acridines with 9-Anilino Substituents

Novel acridines possessing 9-anilino substituents are DNA-binding and apoptosis-inducing antitumor (2013BMC4170) and anti-prion agents (2006 EJMC1124); their quantitative structure–activity relationship (QSAR) studies have been performed (2004ARK36). The DNA-binding properties of a series including 9-anilinoacridines have been reviewed and discussed using 27 QSAR models, which reveal determinants for the activity such as Hammett electronic ($\sigma$ and $\sigma^+$), hydrophobic, molar refractivity, and Sterimol width parameters (2008MI88).

Among the N9-monosubstituted acridines from the 9-anilinoacridine series, the derivative **40**, Scheme 11, possesses the bis(chloroethyl)amine substituent (a nitrogen mustard derivative) and induces monofunctional

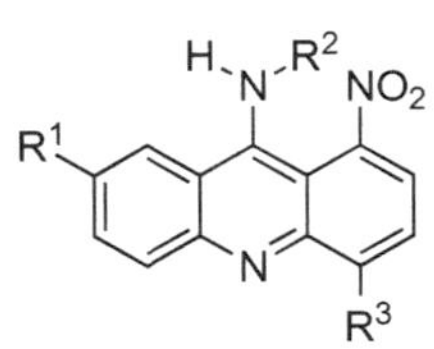


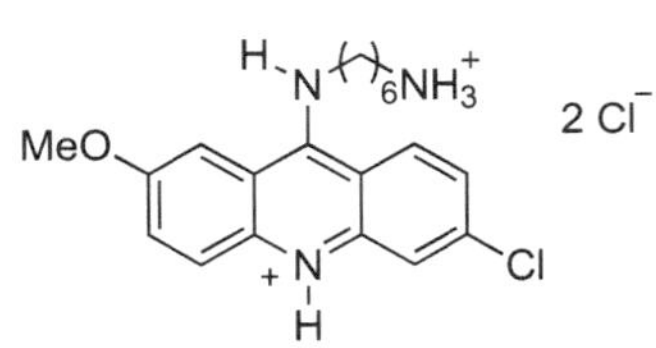


**38a**: $R^1 = R^3 = H$, $R^2 = -(CH_2)_2OH$
**38b**: $R^1 = H$, $R^2 = -(CH_2)_2OH$, $R^3 = Me$
**38c**: $R^1 = -OMe$, $R^2 = -(CH_2)_2OH$, $R^3 = Me$

**39**

**Scheme 10**

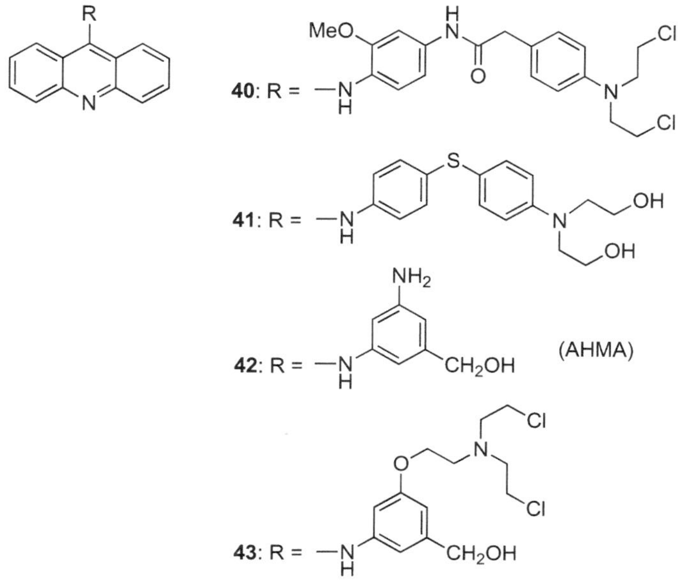


Scheme 11

adduct formation with N3 of adenine (1999MI1166). Therefore, it is a DNA minor-groove-modifying agent. The sulfide **41** proved to be highly active in vivo against P388 leukemia, whereas the corresponding sulfoxide is considerably less cytotoxic than **41** and was discussed as potential bio-reductive anticancer prodrug (2005MI118). A series of 9-anilinoacridine derivatives also possessing the alkylating N-mustard residue at the anilino ring such as **43** was synthesized. An evaluation of cytotoxicity against human lymphoblastic leukemic cells (CCRF-CEM) showed that all exhibited potent cytotoxicity similar to AHMA **42** or were more effective (2004BMCL4719).

We continue this summary with a series of 4,5,9-trisubstituted acridines, also described as biologically active compounds. The 9-anilinoacridines **44** (Scheme 12) possessing the alkylating N-mustard residue at C4 displayed potent antitumor activity in nude mice bearing the human breast carcinoma MX-1 xenograft (2006JMC3710). The syntheses and antitumor activities of 5-(9-acridinylamino)anisidines such as **45** and **46** (and isomers with respect to OMe and $NH_2$) were described as additional examples (2005BMC6513). The compound Asulacrine (cf. Scheme 5) belongs to this series of substituted acridines.

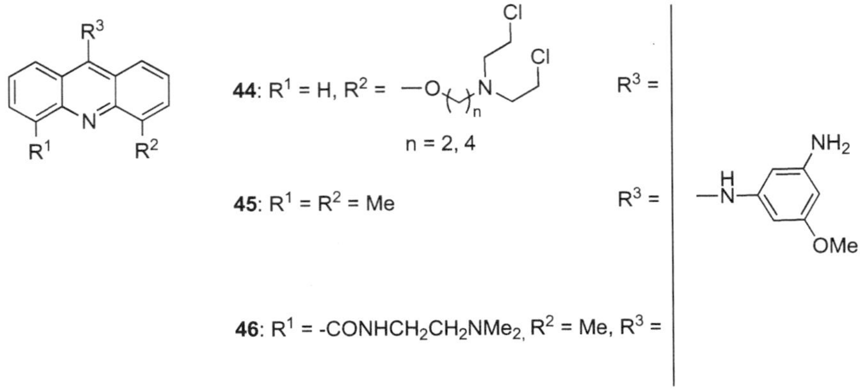


**Scheme 12**

The aminoacridine **47**, a 2,6,9-trisubstituted derivative, Scheme 13, binds to a GAAA tetraloop (2007JMC4096) and **48** displayed antitubercular activity (2004MI713).

BRACO 19 is the name of *N,N′*-(9-((4-(dimethylamino)phenyl)amino) acridine-3,6-diyl)bis(3-(pyrrolidin-1-yl)propanamide) which is a representative of a 3,6,9-trisubstituted acridine (Scheme 14). It is commercially available as a trihydrochloride, displays potent G quadruplex stabilization abilities that were evaluated by telomerase repeat amplification protocol (TRAP) assays (2001PNA4844, 2006JMC582, 2005MI1489). BRACO 19 is a cancer cell proliferation inhibitor (2007MI679). A series of 3,6,9-trisubstituted acridines related to BRACO 19, **49a–d**, was synthesized and studied with respect to its stabilizing effects on the human telomeric DNA quadruplex (2014MI13).

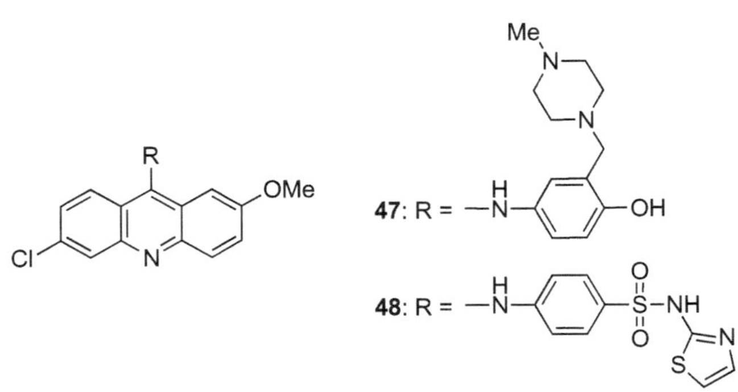


**Scheme 13**

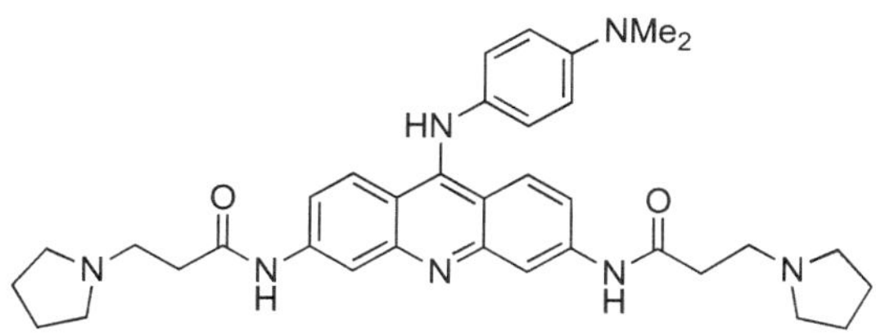


BRACO19

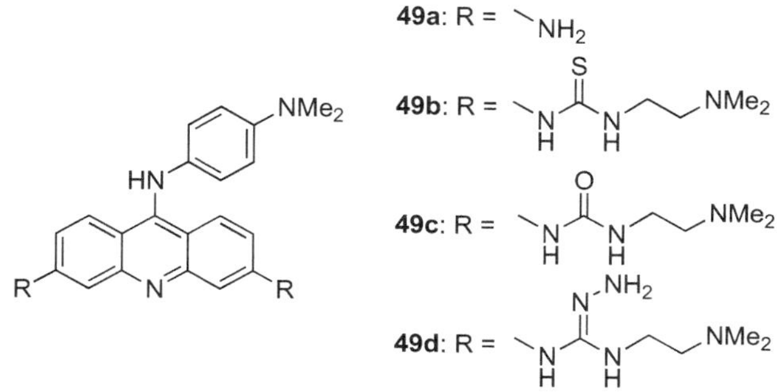


Scheme 14

## 4.4 Acridines Having N-Heterocycles at C-9

The preparation of substituted acridinyl pyrazoline derivatives **50** and their evaluation for anti-inflammatory activity (2010EJMC1772) and also acridinyl benzotriazoles as potential antibacterials have been described (2011 MI193; Scheme 15).

## 4.5 Acridines Bearing 9-Alkyl, 9-Carbonyl, and 9-Aryl Substituents

In addition to N9-substituted acridines, several C9-substituted acridines have been described as biologically active compounds. Examples are shown

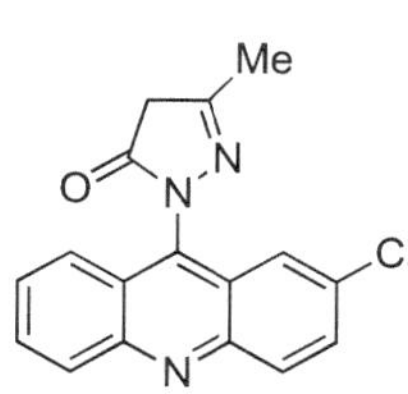


**50**

Scheme 15

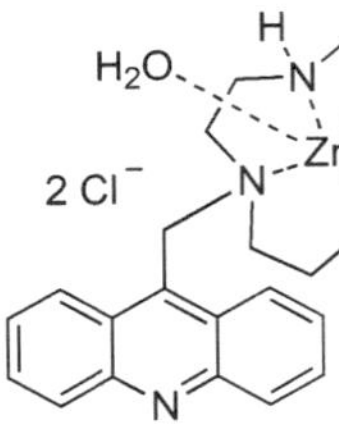


**51**

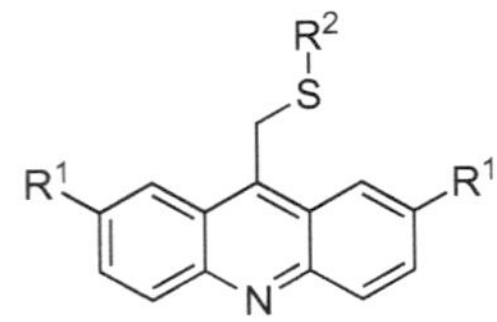


**52a-c**

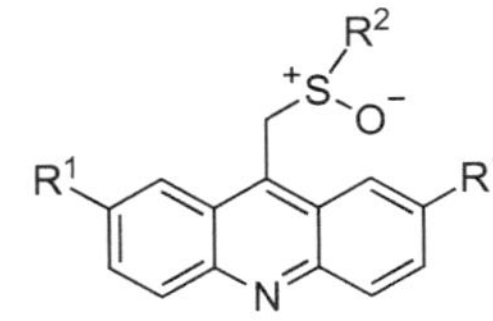


**53a-c**

**a**: $R^1$ = H, $R^2$ = $C_2H_4Cl$
**b**: $R^1$ = OMe, $R^2$ = $CH_2C_6H_4NO_2$
**c**: $R^1$ = OMe, $R^2$ = $C_2H_4NEt_2$

**Scheme 16**

in Scheme 16. The $Zn^{2+}$ complex containing an acridine pendent, Zn(ACR) **51**, binds tightly to pentanucleotides with both tandem and spaced thymines, as well as indiscriminately binds to both H-telo and duplex DNA (2014IC11540). Additional examples are the acridine thioethers **52a–c** that have been prepared, among them sulfur mustards and epoxy sulfides, and converted into their corresponding sulfoxides **53** by treatment with hydrogen peroxide and $(NH_4)_6Mo_7O_{24}$ tetrahydrate. Some sulfones within that series have also been synthesized. Results of in vitro studies against human cancer cell lines are reported and show that the activity is increased five to ten times from sulfides to sulfoxides (2004 EJMC1029).

Scheme 17 shows examples of 9-carbonyl-substituted biologically active acridines. UV spectrophotometric measurements of 1-[6-(acridine-9-carbonyloxy)hexyl]pyridinium chloride **54** reveal that the pyridinium salt is a reversible and competitive inhibitor of acetylcholinesterase. NMR investigations found evidence that the pyridinium salt adopts a conformation that allows hydrophobic interactions between the aliphatic chain and the deep and narrow gorge of AchE on the one hand, and interactions between the acridine group and the catalytic subsites situated at the bottom of the gorge on the other (2006JPOC148). Compound **55**, an additional example of a 9-carbonyl substituted acridine, was tested against *Plasmodium falciparum* (2007BMC3278). The oligopyrrole carboxamides **56** with a flexible spacer at the N-terminal end (combilexins) are attached to an acridine as intercalating system. The cytotoxicity was examined (2007EJMC752). The dependence of the structure of aminoacetylhydrazones of 9-formylacridine on the DNA affinity was studied by inhibition of the polymerase chain reaction (2005MI183).

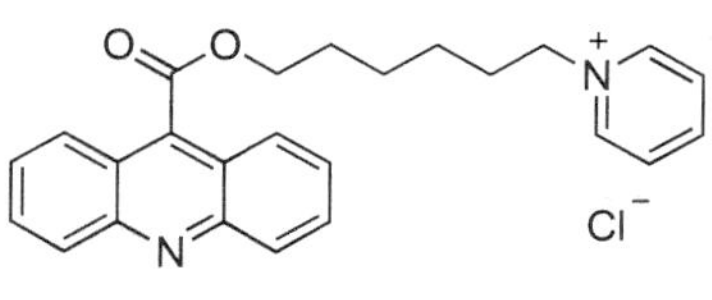

**54**

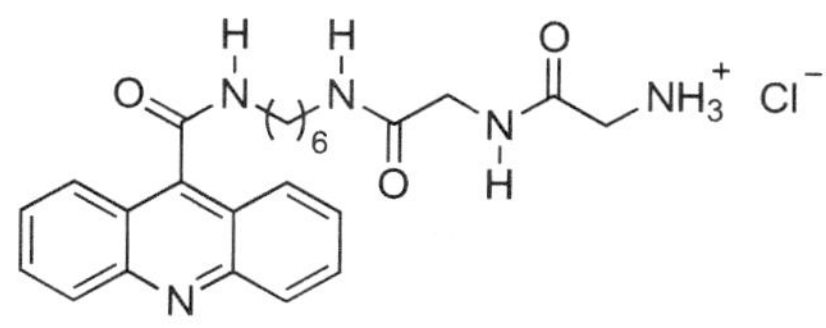

**55**

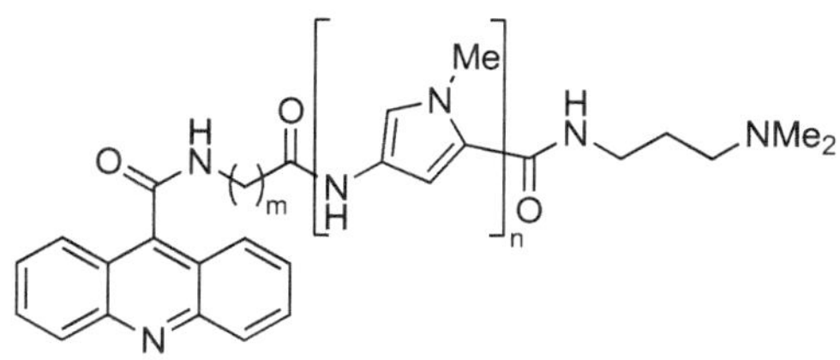

**56a**: m = 2, n = 1
**56b**: m = 0, n = 2

Scheme 17

Compounds **57** and **58**, shown in Scheme 18, possess aromatic and heteroaromatic rings at C-9 of the acridine. Microorganisms were grown in the presence of chromogenic aminopeptidase substrates such as **57**, whereupon colored colonies were produced in a potential method for the detection of aminopeptidase activity and for the identification of microorganisms (2007BMCL1418). Biophysical studies on the binding of 9-phenyl acridine to DNA elucidate their chemotherapeutic potential (2010MI1243). Ruthenium complexes such as bis(1,10-phenanthroline)(11-(9-acridinyl)

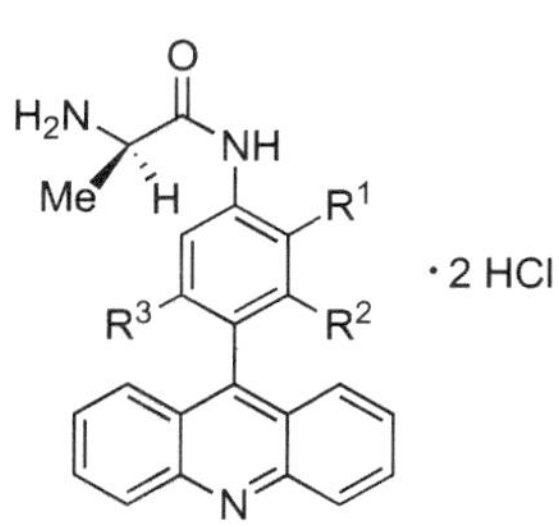

**57a**: $R^1 = R^2 = R^3 = H$
**57b**: $R^1 = OMe$, $R^2 = R^3 = H$
**57c**: $R^1 = H$, $R^2 = R^3 = OMe$

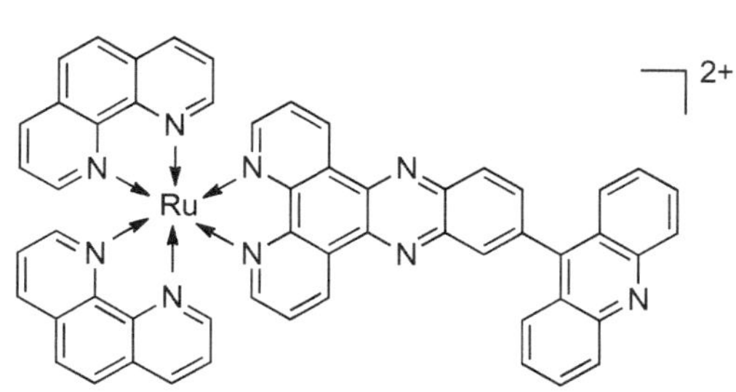

**58**

Scheme 18

dipyridon[3,2-*a*:2′,3′-*c*]phenazine)ruthenium(II) chloride **58** have been examined as potential anticancer agents (2015MI1292).

## 4.6 Biologically Active Acridines with Sulfur Substituents in Position 9

In comparison to acridines with an N9-substituent, those possessing S9-substituents form a considerably smaller group of biologically active compounds. Nevertheless, more than 1000 substances are known and approximately one third of them have been subjected to biological testing. Results have been summarized (2013MI1220). A recent example is **59a** (Scheme 19) which was identified as a potent haspin kinase inhibitor, and **59b** as a moderately potent DYRK2 inhibitor (2010BMCL3491).

## 4.7 Biologically Active Acridines Possessing a Hydrogen in Position 9

The trifunctional metal complex **60** (Scheme 20), designed for receptor-specific nuclear targeting, consists of the peptide bombesine (BBN) which is coupled to a 2,4-dipyc ligand. BBN binds to the GRP receptor which is over-expressed on numerous tumor cell lines and which therefore can serve for cancer imaging and therapy. The preformed 2,4-dipyc modified BBN with $[Re(OH_2)_3(CO)_3]^+$ in water gave a rhenium complex which was treated with an acridinium ligand to give **60**. Specific cell targeting was examined (2007CEJ3842).

A series of tetra-acridine derivatives **61** were examined (Scheme 21), among those acridines with H9 (**61a**), N9 (**61b**), and Cl9 (**61c**) groups, as dual inhibitors of topoisomerase II and the human proteasome (2007MI1863).

Scheme 22 shows acridines with carbonyl groups in position 4. The acridine-4-carboxamides **62**, either unsubstituted as shown (n = 2−6) or bearing one methoxy group in position 1, 2, 5, 6, or 7, respectively, have been examined as antitumor agents. DACA (n = 2) has been under clinical

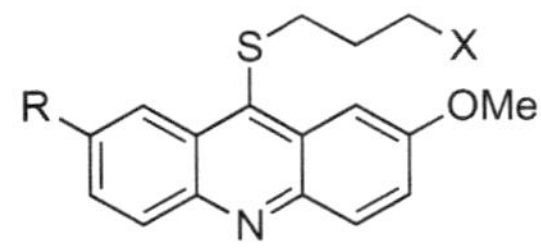


**59a**: R = Cl, X = $NH_2$
**59b**: R = OMe, X = Phth

**Scheme 19**

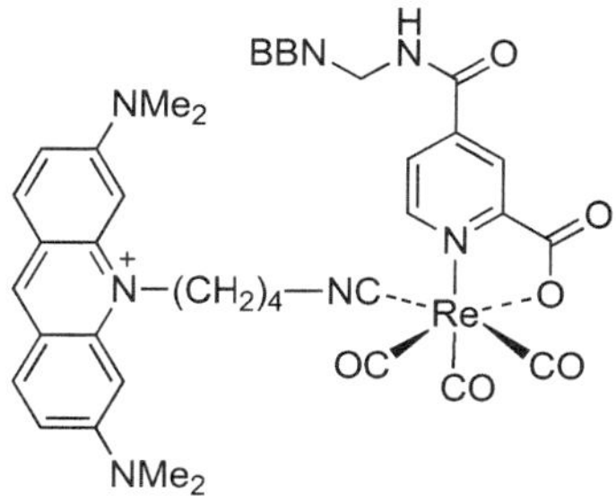


60

BBN = bombesine

Scheme 20

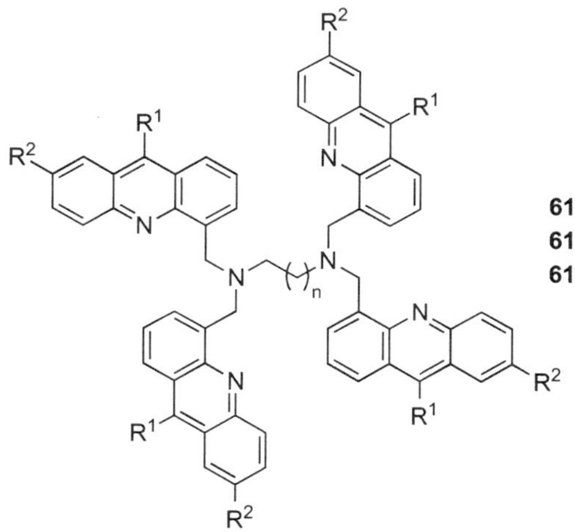


**61a**: n = 1; $R^1 = R^2 = H$
**61b**: n = 1; $R^1 = NH_2$, $R^2 = H$
**61c**: n = 3; $R^1 = Cl$, $R^2 = Br$

Scheme 21

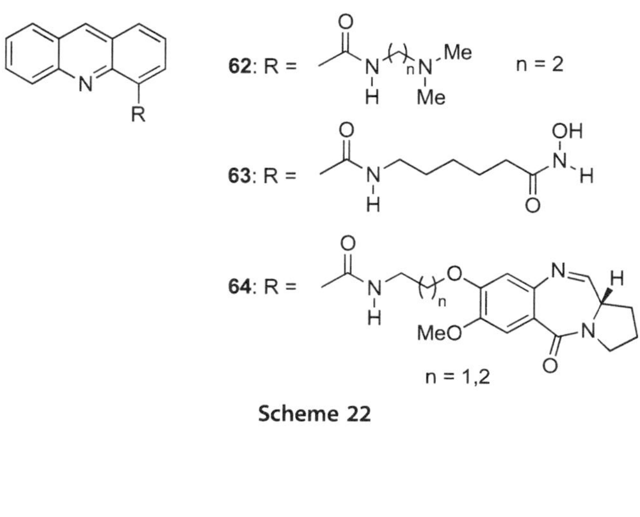


Scheme 22

evaluation since 2001. Results have been summarized in a review (2001PAC1421). A hybrid **63** of the histone deacetylase (HDAC) inhibitor vorinostat (*N*-hydroxy-*N*′-phenyloctanediamide) and the topoisomerase inhibitor DACA [(*N*-[2-(dimethylamino)ethyl]acridine-4-carboxamide] exhibited anticancer activity against human hormone-refractory metastatic prostate cancer (2014MI320). Hybrids **64** between the biologically active pyrrolobenzodiazepine ring and acridines have been prepared for DNA-binding affinity and in vitro anticancer studies (2004BMCL4107).

In vitro antileishmanial activities of 4,5-di-substituted acridines **65** and **66** shown in Scheme 23 have been tested. In contrast to 4-mono-substituted homologues, which exhibited a weak specificity for *Leishmania* parasites, these showed interesting amastigote-specific activities. It is thought that the mechanism of action does not involve intercalation to DNA (2005BMC5560). Some other 4,5-bis(dialkylaminoalkyl)-substituted acridines have been prepared as potent telomeric G-quadruplex ligands (2009EJMC3880).

Tuneable photonucleases **67** possess a central 3,6-acridinediamine chromophore (Scheme 24) substituted by metal-coordinating imidazole rings that cause an enhanced photocleavage of DNA in the presence of metal salts of Hg(II), Fe(III), Cd(II), Zn(II), V(V), or Pb(II). DNA photocleavage can also be modulated by modifying the buffer type and pH (2005IC6159). 3,6-Bis((1-alkyl-5-oxo-imidazolidin-2-yliden)imino)acridine hydrochlorides (AcrDIMs) **68a–c** (2011BMC1790) serve as potential photosensitizers for photodynamic antitumor therapy (2014BMC4684).The syntheses, interactions with DNA, and the cytotoxic activity of dithiazolidinone-substituted proflavines **69a–c** were reported (2007MI93). Triazole-linked acridines have been designed on the basis of structure-based modeling methods with selectivity for human telomeric quadruplex DNAs. The selectivity was tested on two promoter quadruplexes. The lead compounds also displayed inhibitory effects on the proliferation of cancer cell lines (2010JA12263).

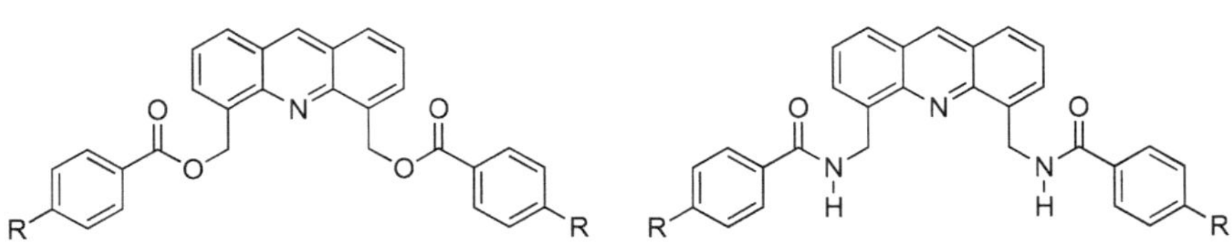

**65**
R = H, Cl, F, OMe, $NMe_2$

**66**
R = H, Cl, F, OMe, $NMe_2$

**Scheme 23**

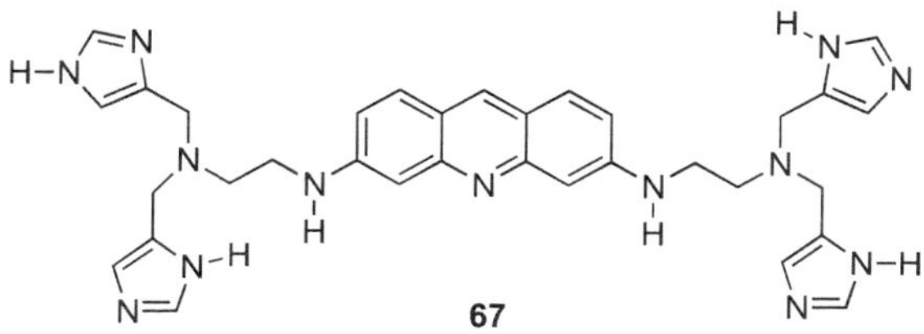

67

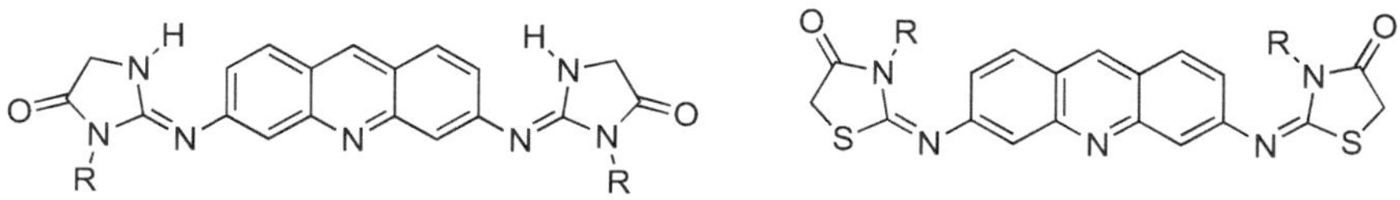

**68a**: R = Et
**68b**: R = nPr
**68c**: R = nBu

**69a**: R = Et
**69b**: R = nPr
**69c**: R = n-hexyl

Scheme 24

The cell distribution of 3-amino-4-hydroxymethyl acridine derivatives **70** (Scheme 25), which has the N3—C4—I6 substitution pattern, was studied by real-time fluorescence microscopy and SIMS (*structured illumination microscopy*). The latter method required the introduction of an iodine atom at position 6 of the acridine which influences the lipophilicity but does not modify significantly the biological properties such as IC50 and subcellular localization (2009EJMC4758). A co-polymer **71** consisting of water-soluble maleic anhydride-containing poly[maleic anhydride-*alt*-acrylic acid] (poly(MA-alt-AA) or MAAA) copolymer was modified with acriflavine (AF) which displayed antibacterial activity on EHEC and *Staphylococcus aureus* (2014MI2903).

## 5. APPLICATIONS OF ACRIDINES

In addition to their biological activities and their uses as drugs, antiseptics, or disinfectants, acridines are widely applied as dyes, fluorescent

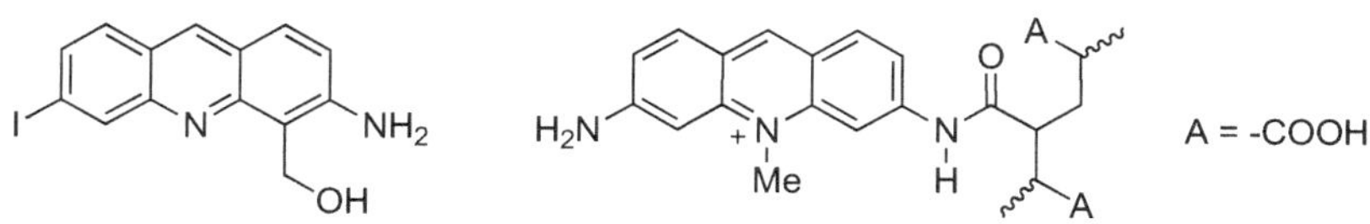

70

71

Scheme 25

materials, and, as a consequence, as sensors for a variety of organic as well as inorganic substrates. With respect to this, an interesting area of overlap between the two topics "biologically active acridines" and "applications of acridines" exists. Again, this section cannot be comprehensive due to the huge amount of publications in the various fields of applications of acridines. However, some representative examples published in the last decade are given here. Thus, the acridinium-containing cyanine dyes **72** (Scheme 26) were presented and their DNA-binding properties were examined (2010JOC204). Chemiluminogenic properties of 10-methyl-9-(phenoxycarbonyl)acridinium cations were also examined (2010JPC(A)10550), while the arylethynylacridines **73** displayed electrogenerated chemiluminescence (2004OBC3113). The second-order NLO response of the acridine-benzothiazolylamine NLO chromophore **74** is enhanced by the presence of the tautomeric minor form of the corresponding benzo[*d*]thiazol-2(3*H*)-ylidene derivative (2005T9075). Acridines as building block of fluorescent polymers have been published (2010MI614), while the oxadisiole-fused 9-aminoacridines **75** show potential as blue emitters for OLED applications (2014EJO4170). 10-Ethyl-acridine-3-sulfonyl chloride has been applied as pre-column labeling reagent for the determination of aliphatic amines (2009MI1351). Acridinium esters can be triggered by basic hydrogen

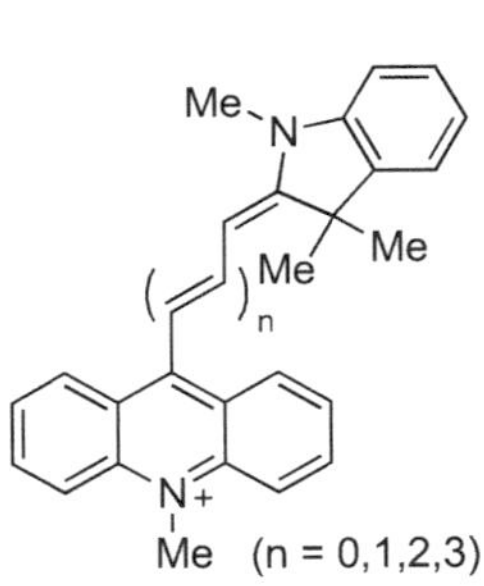


**72**

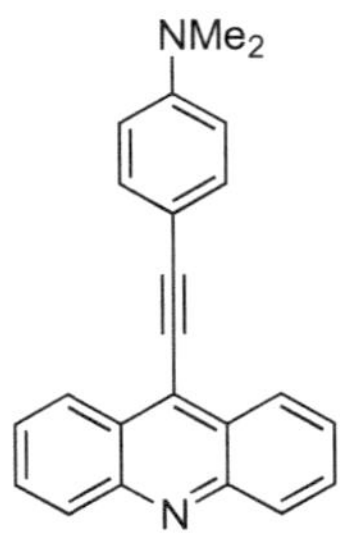


**73**

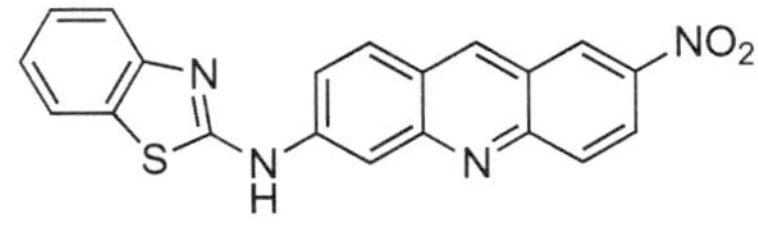


**74**

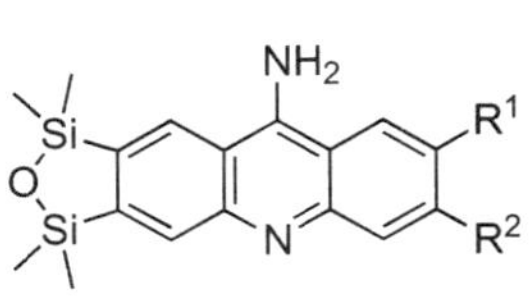


**75a-g**

**a**: $R^1 = OCH_3$, $R^2 = OCH_3$
**b**: $R^1 = H$, $R^2 = OCH_3$
**c**: $R^1 = H$, $R^2 = CH_3$
**d**: $R^1 = R^2 = H$
**e**: $R^1 = H$, $R^2 = Cl$
**f**: $R^1 = Br$, $R^2 = H$
**g**: $R^1 = F$, $R^2 = H$

**Scheme 26**

peroxide and triggering also occurred with emission of chemiluminescence by zinc metal or reduced forms of iron and copper salts (2011MI2945). They were also employed as a chemiluminescent emitter in nucleic acid hybridization assays (2009OBC386).

As a result of their fluorescent properties, acridines have been widely applied as sensors for metal ions, anions, and for organic and inorganic substrates. The monosubstituted acridine 9-acrylamidoacridine displayed a selective fluorescence enhancement on titration with $Hg^{2+}$ (2012MI2736). The novel fluorescent probe **76** in Scheme 27 was rationally designed and synthesized. The fluorescent probe exhibited good selectivity and sensitivity toward $Cr^{3+}$ in methanol, and the **76**-$Cr^{3+}$ solution also displayed excellent selectivity to $PO_4^{3-}$ (2014MI388). Two novel biotinylated fluorophores (2-biotinyloxyethyl) acridine-9-carboxylate (**77a**) and (2-biotinylamidoethyl) acridine-9-carboxamide (**77b**) were synthesized as fluorescent materials and labels in bioanalytical applications (2005MI215). Highly sensitive solid-phase fluorescence-based immunoassays that may find applications in proteomics are biotinylated acridinium derivatives such as the 9-(2-biotinyl-oxyethyl)-carboxylate-10-methyl-acridinium triflates **78a** and **b** (2005MI159).

The nanomolar-level determination of hydrogen peroxide concentrations in natural water by chemiluminescence-based measurements required

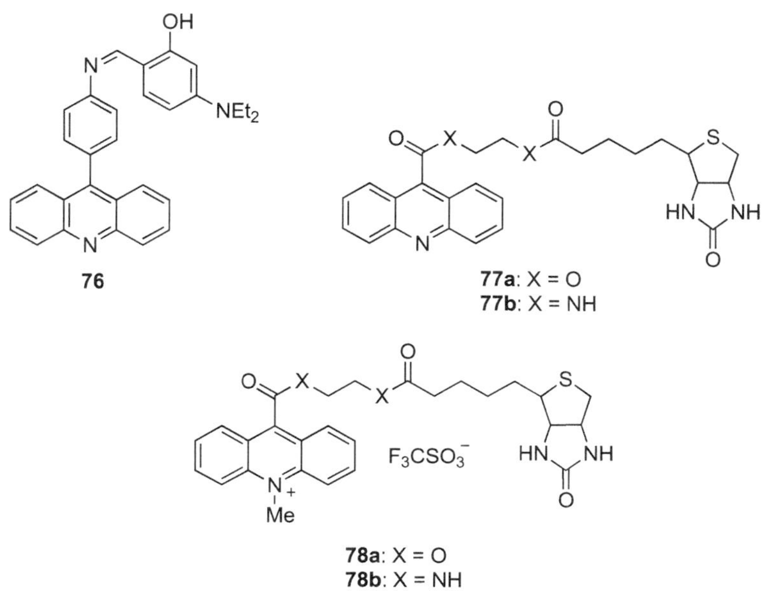


**Scheme 27**

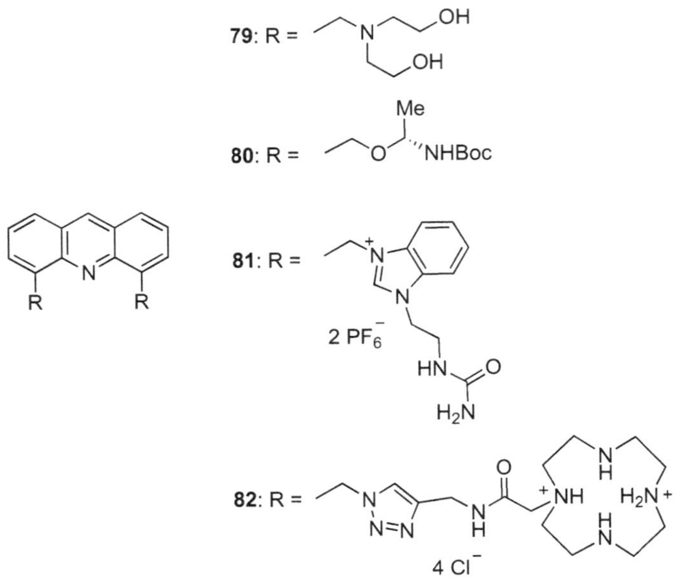


Scheme 28

a highly sensitive compound and 10-methyl-9-(*p*-formylphenyl)acridinium carboxylate trifluoromethanesulfonate shown in Scheme 28 fulfilled this requirement (2007MI4169). 4,5-Bis(*N*,*N*-di(2-hydroxyethyl)iminomethyl) acridine **79** is a new acridine that can be applied as a fluorescent sensor toward metal ions in buffered aqueous media and is a reversible chemosensor for $Cd^{2+}$ (2011MI126). 4,5-Bis(hydroxymethyl) acridine was discovered as simple-structured $Pd^{2+}$ fluorescent probe (2015MI33). The acridine **80** was examined as sensor for malate anions (2014MI169, 2014NJC1004), and the tweezer-like fluorescent probe **81** can serve as sensor for dihydrogenphosphate and hydrogensulfate anions (2013T10457). Macrocyclic compounds possessing a pentamine chain attached to the 2,7 positions of an acridine moiety were studied with respect to its Cu(II) complexation by means of potentiometric, UV–vis and EPR measurements in aqueous solutions (2009JCS(D)1223). In addition, the fluorescent bis-cyclen tweezer receptor **82** was described as inositol (1,4,5)-trisphosphate sensor (2011T3803).

DNA-binding studies through spectroscopic investigations, viscosity, and thermal denaturation temperature measurements indicated that water-soluble systems such as **83** and **84** (Scheme 29) interact with DNA preferentially through intercalation of the acridinium moiety. DNA association

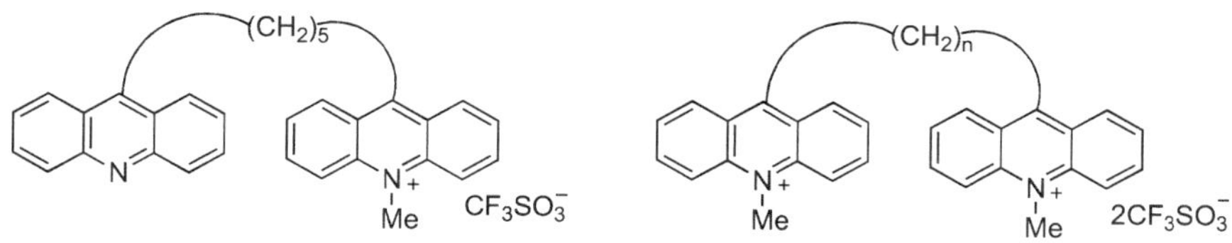


**83**

**84a**: n = 5
**84b**: n = 10

**Scheme 29**

constants were determined, and mono-and bis-intercalation properties were examined (2005JPC(B)21997).

Enantiopure dioctyl-substituted acridino-18-crown-6 ether-type ligands such as **85**, shown in Scheme 30, are potentiometric sensor molecules when incorporated into plasticized PVC membrane electrodes (2009T(A)2795). Another article introduced two acridines **86** as fluorescent chemosensors for $Hg^{2+}$ and $Cd^{2+}$ in aqueous solution. These acridines displayed selective chelation-enhanced fluorescence effects with $Hg^{2+}$ or $Cd^{2+}$ among the metal ions examined. The practical use of these probes was demonstrated by their applications to the detection of $Hg^{2+}$ and $Cd^{2+}$ ions in mammalian cells (2008TL1261). The cationic cyclophane derivate **87** with bridging and

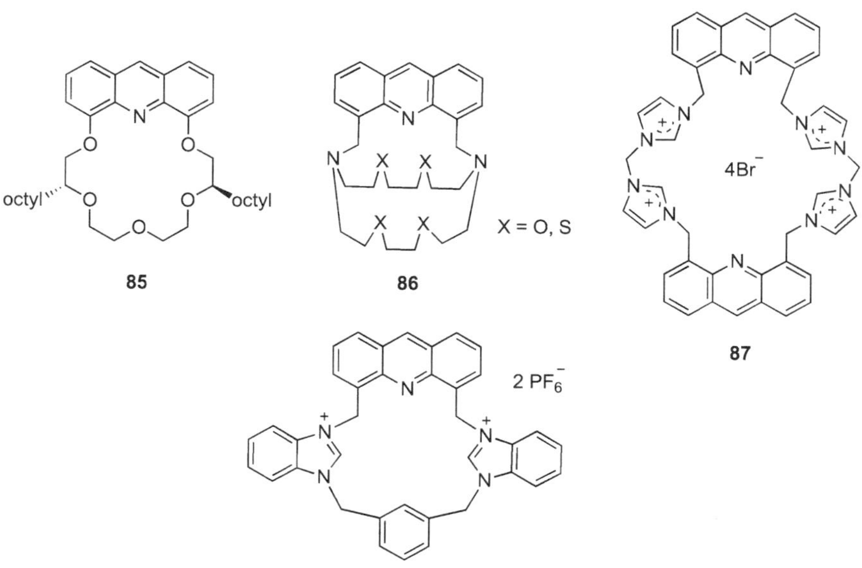


**85**

**86**

**87**

**88**

**Scheme 30**

spacer groups can undergo diverse intermolecular interactions due to a semi-rigid cavity. It possesses the acridine moiety as a fluorophore unit and imidazolium moieties as binding sites for GTP recognition that induces a considerable fluorescent quenching (2014OL2150). Macrocyclic acridine bisbenzimidazolium fluorescent sensors such as **88** have cavity sizes that varied by different linkers between the benzimidazolium rings. A bathochromic shift between 36 and 126 nm of the fluorescence emission was observed on addition of the dihydrogenphosphate anion depending on the structure of the macrocycle and, as a consequence, color changes from blue to orange red (2013OBC3375).

The C2-symmetric ligand 1,8-bis(3-*tert*-butyl-9-acridyl)naphthalene *N,N′*-dioxide **89** (Scheme 31) forms a highly fluorescent scandium complex, used as enantioselective sensor for chiral amino acids. This compound allows for the measurement of the total amount as well as the enantiomeric excess of a variety of amino alcohols at micromolecular concentrations (2006CC4242). The synthesis and an interesting ring-closure reaction of a very similar species are shown in Scheme 70 (see above).

## 6. SYNTHESES OF ACRIDINES

The most widely applied syntheses of acridines involve ring closures because substitution reactions often give mixtures of regioisomers (2004 S313). Often, aromatizations are subsequent steps. Several types of ring closures of suitable precursors to acridines can be differentiated as shown in Figure 1. Two types F and G supplement the classification system that was suggested earlier (2005SOS(15)987, 1992HOU(E7b)115). We sorted recent advances in the synthesis of acridines according to these types and took the architecture of the starting materials for classification. A different classification can be envisaged when the mechanisms leading to acridines are taken as the basis, since often intermediates are formed which suggest

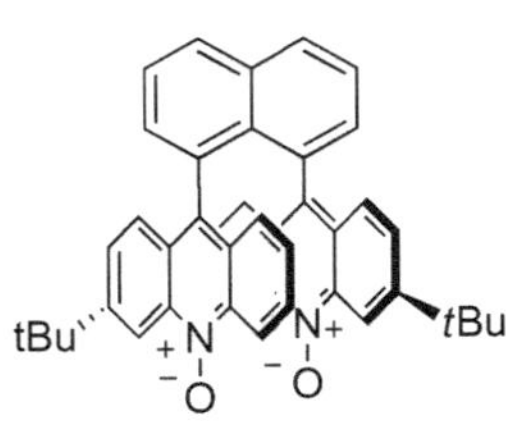


**89**

**Scheme 31**

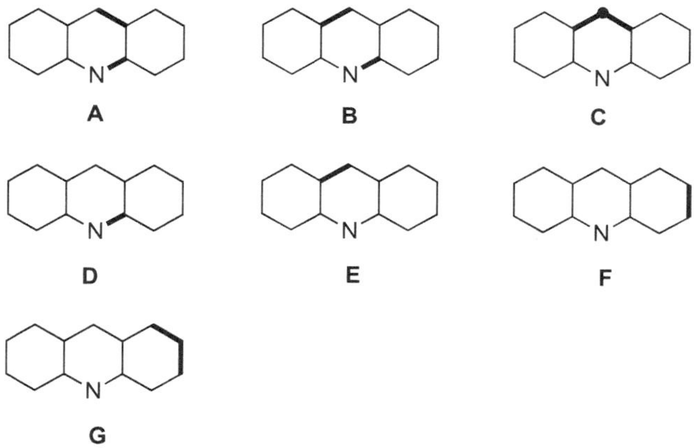


**Figure 1** Classification of ring-closure reactions to prepare acridines.

another type of ring closure. In addition to these ring closures, acridines can be synthesized by rearrangements, or by oxidation and reduction of acridanes or acridones, respectively.

## 6.1 Type A Ring-Closure Reactions

Type A of ring closure to acridines is represented by reactions of arynes with suitable partners. Recently, 2-aminoaryl ketones **90** with arynes generated from **91** to give acridines **92** have been developed and optimized (Scheme 32). The major by-products are the corresponding (2-(phenylamino)phenyl) ketones and result by protonation after attack of the amino group on the

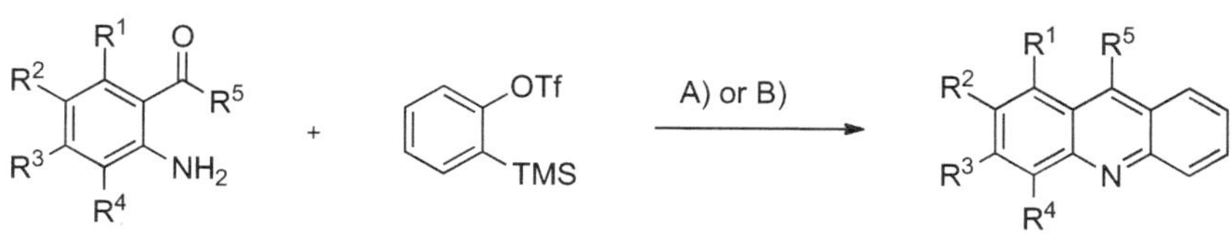


**90** **91** **92**

**a**: $R^1 = R^2 = R^3 = R^4 = H$, $R^5 = Me$ (70%; A)
**b**: $R^1 = R^4 = H$, $R^2 = R^3 = OMe$, $R^5 = Me$ (78%; A)
**c**: $R^1 = R^3 = H$, $R^2 = R^4 = Br$, $R^5 = Me$ (77%; A)
**d**: $R^1 = R^3 = R^4 = H$, $R^2 = Cl$, $R^5 = 2\text{-FPh}$ (72%; B)
**e**: $R^1 = R^3 = R^4 = H$, $R^2 = Cl$, $R^5 = H$ (30%; A)

A) CsF, MeCN, 65 °C, 24 h
B) TBAT, DME, rt, 24 h

**Scheme 32**

aryne and before the ring closure took place. The solvent plays a crucial role and acetonitrile proved to be the best choice when caesium fluoride was used as benzyne initiator. Water scavengers such as magnesium sulfate had no influence on the overall yield, although 1 equivalent of water is formed for every equivalent of acridine. Dimethoxymethane proved to be the best solvent when tetrabutylammonium triphenyldifluorosilicate (TBAT) was employed (2010JOC2289).

Arynes, starting from oxadisilole-fused benzene **93** also reacted with 2-aminoaryl ketones **94** to oxadisilole-fused acridines **95** and dioxatrisilole-fused acridines **96** that were separated by column chromatography (2014T5820). This synthetic procedure is shown in Scheme 33.

Cross-coupling reactions can also be regarded as examples of type A ring-closure reactions to acridines, and an example is presented in Scheme 34. Thus, starting from aryl halides **97**, N-tosylhydrazones such as **98** and $Pd_2(dba)_3$ as $Pd^0$ source were applied to prepare acridines such as **99a–e**. Among the phosphane ligands tested, RuPhos had the best performance when LiOtBu in dioxane was used as base in the presence of water. Not only a series of dihalo compounds underwent the acridine synthesis as shown, but also substituted hydrazones that allowed the variation of the substituent in position 9. Mechanistic studies revealed that **A** is the initially formed intermediate that undergoes a fast intramolecular C–N-cross-coupling (2012EJO6586). From a mechanistic viewpoint, this reaction could also be classified as a type D ring closure, as only one C–N bond is formed starting from the intermediate substituted aniline.

The 2-oxo-substituted aryl azides **100** in Scheme 35 reacted with benzene in the presence of trifluoroborane etherate to give the acridines **101** as main the products and 2-aminobenzenecarbonyls as by-products. Analogous reactions have been performed using aluminum chloride or a mixture

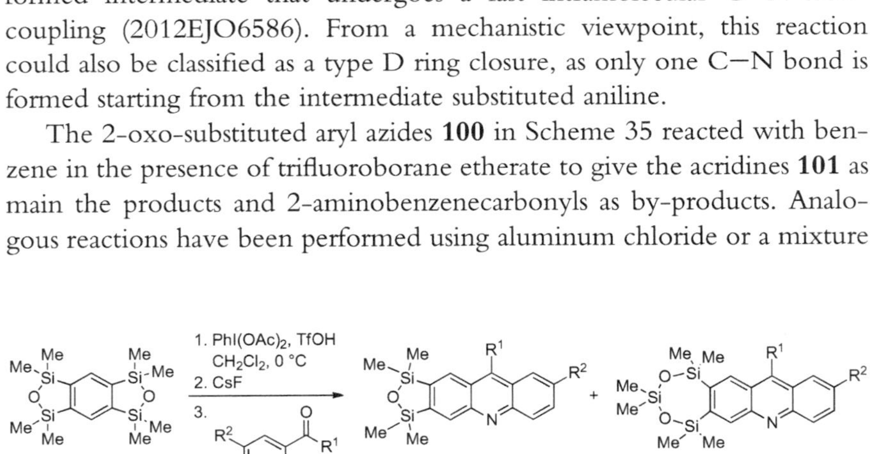


**93** **(94)** **95** **96**

**a**: $R^1$ = Ph, $R^2$ = Me (57%; 34:23)
**b**: $R^1$ = 2-FPh, $R^2$ = Cl (55%; 34:21)
**c**: $R^1$ = $R^2$ = H (30%; 17:13)
**d**: $R^1$ = Me, $R^2$ = H (89%, 60:29)

**Scheme 33**

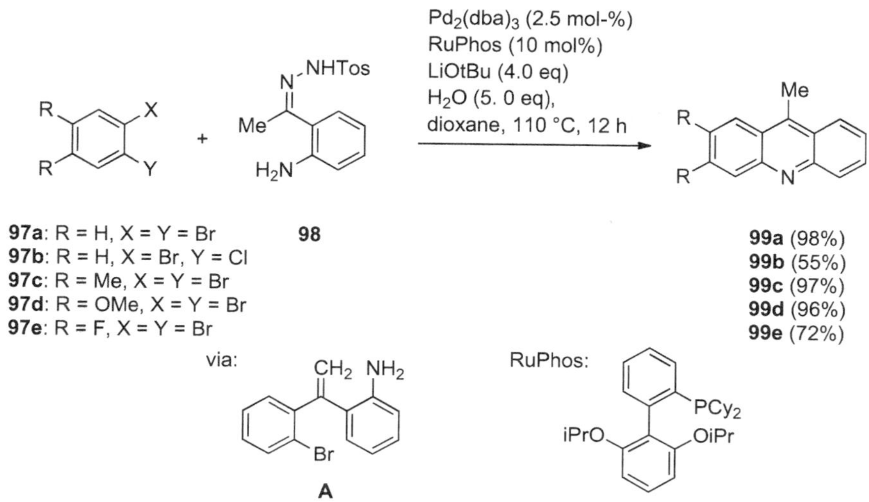


Scheme 34

of trifluoromethanesulfonic acid and triflic acid (TFSA/TFA) (2006 ARK111).

## 6.2 Type B Ring-Closure Reactions

A synthesis of acridinium salt **104** was developed which belongs to type B ring closure. It involves an aryne addition to hydrazones **102** to give ketimines that undergo a Friedel–Crafts addition (Scheme 36). Either the imines or the corresponding ketones that might be formed by imine hydrolysis under these conditions might be involved (2012JOC11232). Mechanistically, it can also be classified as an example of a type E ring closure, as imine or ketone **B** is the intermediate.

Type B of ring closure to acridines is exemplified by a Pd(II)-catalyzed one-pot, three-step route starting from **105** and **106** for the synthesis of unsymmetrical acridines **107**, Scheme 37. The synthetic scope has been

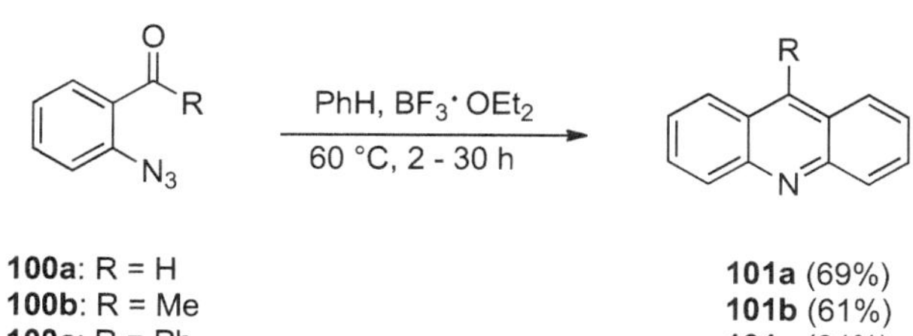


Scheme 35

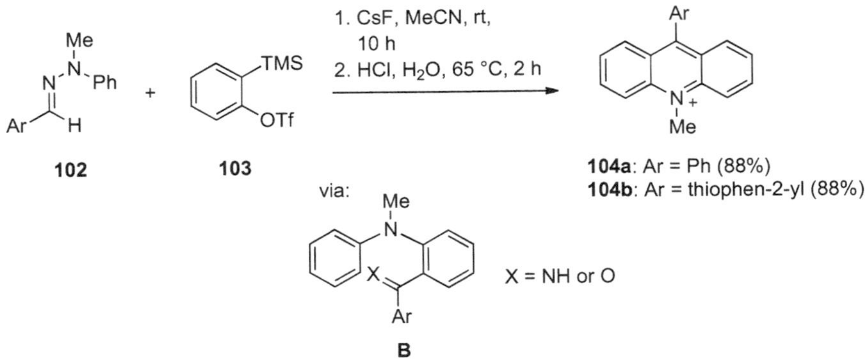


Scheme 36

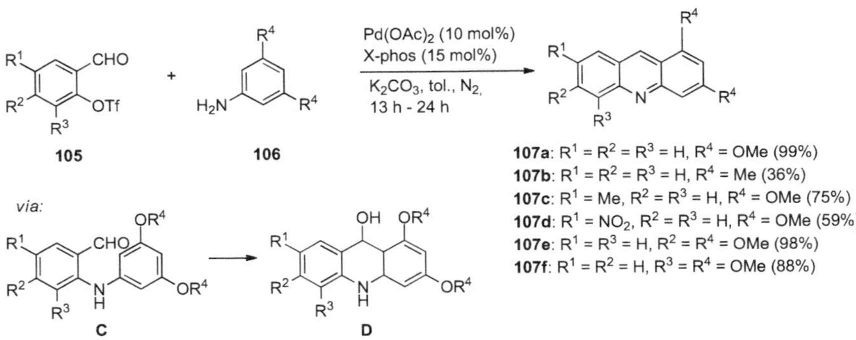


Scheme 37

elucidated by variations of the substitution pattern (2013OL5460). This synthesis can also be classified as type E ring closure from a mechanistic viewpoint, as **C** is the intermediate.

A one-pot synthesis of acridine **110** from **108** and **109** was developed employing a Buchwald reaction as the key step without purification of intermediates (Scheme 38). The sequence involved an ester hydrolysis that also avoids aqueous workup (eight examples) (2004SL2391). Barium hydroxide, which is soluble in methanol, is used in this procedure.

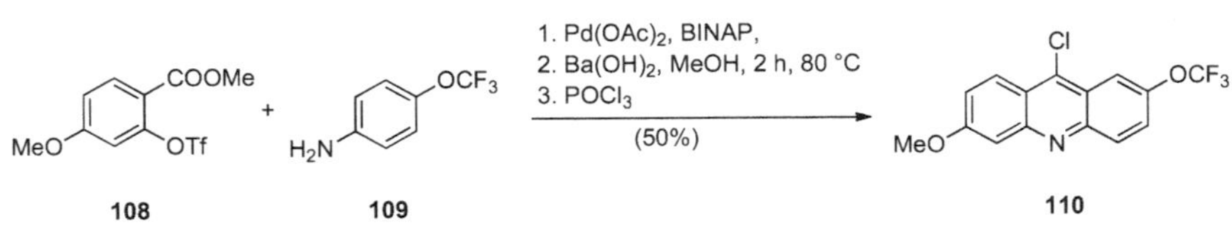


Scheme 38

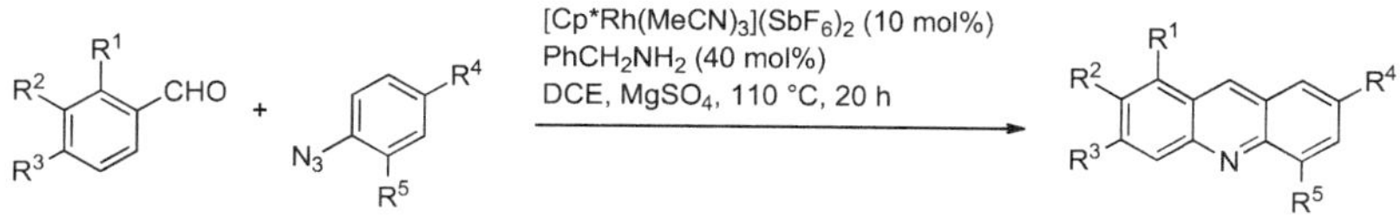


**111** **112**

**113a**: $R^1 = R^2 = R^4 = R^5 = H$, $R^3 = Cl$ (82%)
**113b**: $R^1 = R^2 = R^4 = R^5 = H$, $R^3 = COOMe$ (90%)
**113c**: $R^1 = F$, $R^2 = R^3 = R^4 = R^5 = H$ (83%)
**113d**: $R^1 = R^3 = R^4 = R^5 = H$, $R^2 = Me$ (79%)
**113e**: $R^1 = R^2 = R^3 = R^5 = H$, $R^4 = CF_3$ (74%)
**113f**: $R^1 = COOMe$, $R^2 = R^3 = R^4 = H$, $R^5 = Me$ (85%)

**Scheme 39**

Scheme 39 shows a formal [3 + 3] annulation of aromatic azides **112** with aromatic imines proceeding through a cascade process of Rh(III)-catalyzed amination followed by an intramolecular electrophilic aromatic substitution and aromatization to prepare the acridines **113** and can be classified as a type B ring closure. The reaction can be performed starting from aldehydes **111** which form imines in situ, and these direct the CH amination. Benzylamine proved to be the best amine, and magnesium sulfate was applied as a water scavenger. The substrate scope was examined, which is in part shown in Scheme 39 (2013JA12548).

It proved to be advantageous, however, to preform ketone-derived imines **114** instead of employing ketones and amines in the presence of water scavengers. Slightly modified conditions were applied to prepare acridines **116** from these ketone imines (Scheme 40; 2013JA12548).

The mechanistic proposal, Scheme 41, involves an *ortho*-directed CH bond activation to form a metallacycle followed by a complexation of the azide, nitrogen formation, and insertion to give a six-membered metallacycle. An electrophilic aromatic substitution gives an 9-amino-8a,9-dihydroacridinium ion that aromatizes under formation of the aromatic amine to give the target acridines (2013JA12548).

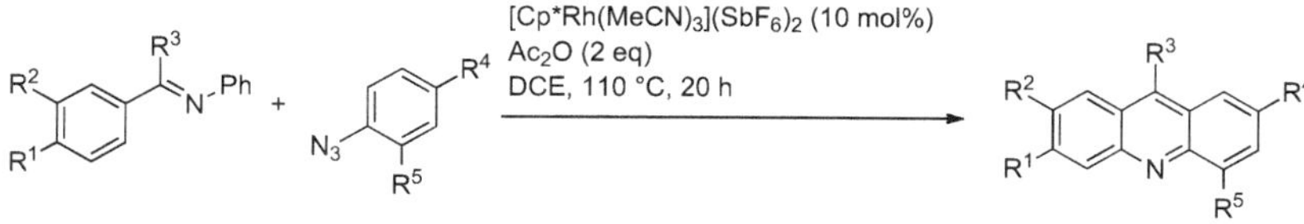


**114** **115**

**116a**: $R^1 = R^2 = R^4 = R^5 = H$, $R^3 = Me$ (87%)
**116b**: $R^1 = R^2 = R^4 = R^5 = H$, $R^3 = COOMe$ (79%)
**116c**: $R^1 = CF_3$, $R^2 = R^4 = R^5 = H$, $R^3 = Me$ (83%)
**116d**: $R^1 = R^3 = Me$, $R^2 = R^4 = R^5 = H$ (78%)
**116e**: $R^1 = R^2 = R^4 = H$, $R^3 = Me$, $R^5 = Br$ (87%)
**116f**: $R^1 = CF_3$, $R^2 = R^5 = H$, $R^3 = Me$, $R^4 = OMe$ (91%)

**Scheme 40**

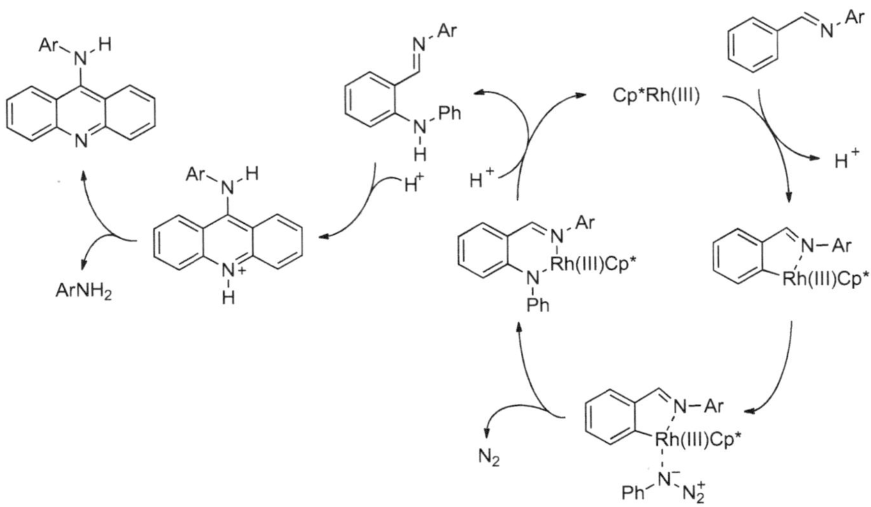


**Scheme 41**

Another pathway is displayed in Scheme 42. An acridine-9-carbonitrile (**119**) synthesis from carbanions and nitroarenes **117** followed by reductive transformation of the nitro group with silylating or acylating agents can be regarded as a recent example of a type B or D ring closure, if the starting

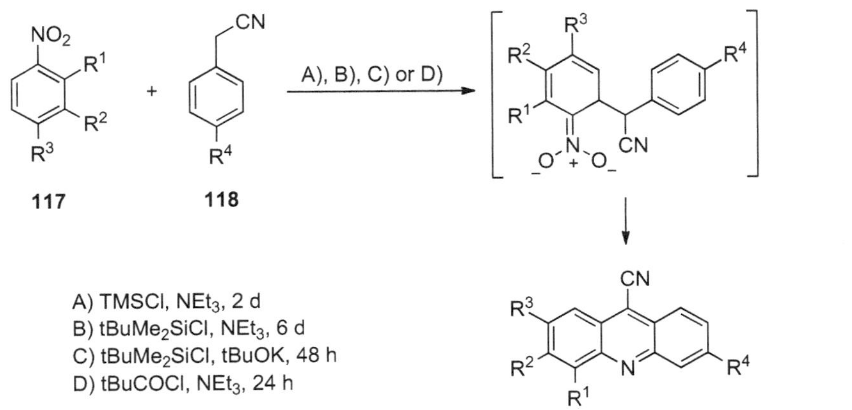


**119a**: $R^1 = R^2 = R^4 = H$, $R^3 = Cl$ (83%; C; 77%; D)
**119b**: $R^1 = R^2 = R^4 = H$, $R^3 = Br$ (70%; B)
**119c**: $R^1 = R^3 = Cl$, $R^2 = R^4 = H$ (24%; C)
**119d**: $R^1 = R^2 = H$, $R^3 = Cl$, $R^4 = Br$ (75%; A)

**Scheme 42**

materials or the intermediate adduct is taken to classify this reaction, respectively. Four different conditions that can be considered as nucleophilic substitution of hydrogen on nitroaromatics were tested (2007T11048).

Four major factors control the regioselectivity of the Pd-catalyzed transformation, Scheme 43. 2-Chloro-*N*-(2-vinylphenyl)anilines **122**, readily prepared from 2-bromostyrene **120** and 2-chloroanilines **121**, were converted into 5*H*-dibenzazepines, 9-methylacridines, or 1-vinylcarbazols. Factors influencing this reaction are (1) the ligand, (2) the base, (3) the solvent, and (4) the temperature. A regioselective 6-exo cyclization to 9-methylacridines was performed when tri-*tert*-butylphosphine (as salt of $HBF_4$) was used as ligand under the conditions shown. The usage of DavePhos yielded the carbazoles, whereas TrixiePhos gave azepines (2010JA14048).

## 6.3 Type C Ring-Closure Reactions

The Bernthsen reaction is an example of a type C ring closure. A short summary on the characteristic features of the Bernthsen reaction and recent applications have appeared recently (2011MI). The classical Bernthsen

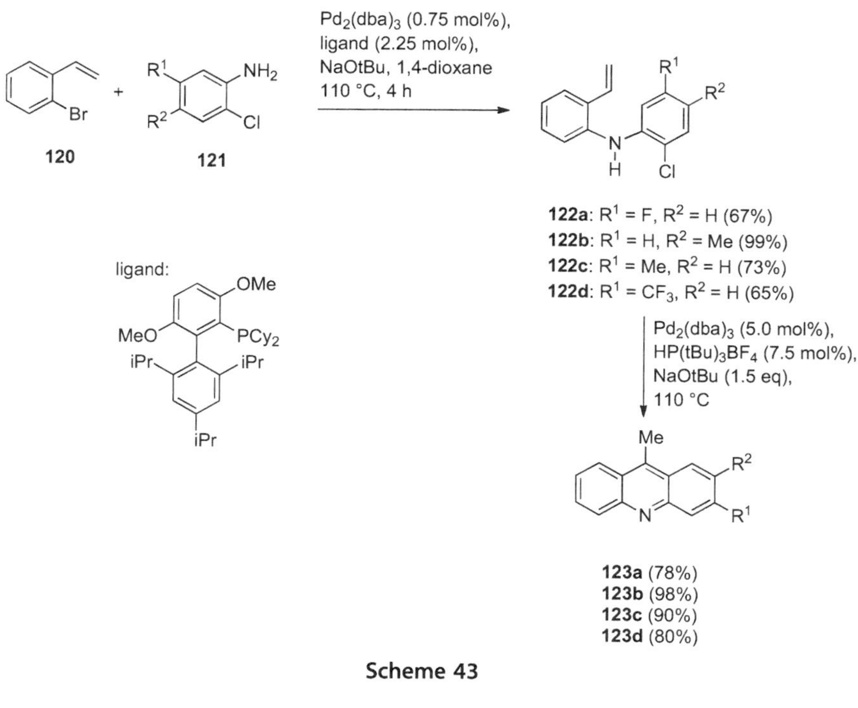


Scheme 43

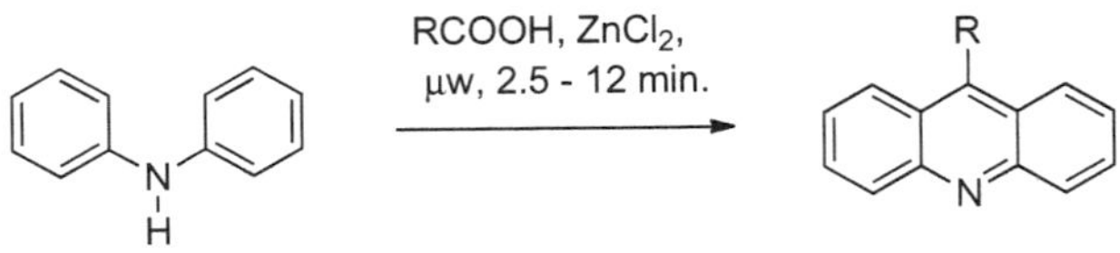

124

125a: R = Ph (98%)
125b: R = m-MePh (60%)
125c: R = p-PhPh (88%)
125d: R = n-hexyl (57%)

Scheme 44

reaction was first reported in 1884 (1884MI1) and described the coupling of an aromatic or aliphatic carboxylic acid and diphenylamine in the presence of excess zinc chloride at 200–270 °C over a period of 20 h (1883CB1802). This reaction is still widely applied, despite its vigorous reaction conditions. The Bernthsen reaction was used to prepare antimicrobial acridines (2010BMC6324). Formic acid and others can be applied under thermal (2004MI1230) and under microwave conditions (2002 H(57)1299).

A modification of the Bernthsen reaction uses polyphosphoric acid instead of zinc chloride; however, the conditions are also quite vigorous with respect to the temperature (200 °C; 15 min); in addition, the yields of 9-substituted acridines were low (1962JOC2658). In 2002, two synthetic procedures have been published, which use microwave conditions and zinc chloride as catalyst to prepare **125** from **124** (2002SC729; 2002MI390; Scheme 44). Reaction conditions for the synthesis of 9-substituted acridines have been optimized for microwave conditions in the presence of 2,2-dimethoxypropane as water scavenger very recently (2014H(89)1877).

Various other potential catalysts of the Bernthsen reaction have been tested under thermal as well as microwave conditions, among those are basic $Al_2O_3$ and ceric ammonium nitrate (CAN). Best results were achieved with *p*-toluenesulfonic acid under microwave conditions, Scheme 45. Other potential catalysts such as zirconium oxychloride octahydrate, potassium dichromate, or alumium chloride failed (2011MI131).

## 6.4 Type D Ring-Closure Reactions

This type is seldom used (2005SOS(15)987). Electrocyclic reactions of 6-benzylidenecyclohexa-2,4-dienimines, acid-induced ring-closure of 1-benzyl-2-nitrobenzenes, or the photolysis of the corresponding azides belong into that class of acridine syntheses (2005SOS(15)987).

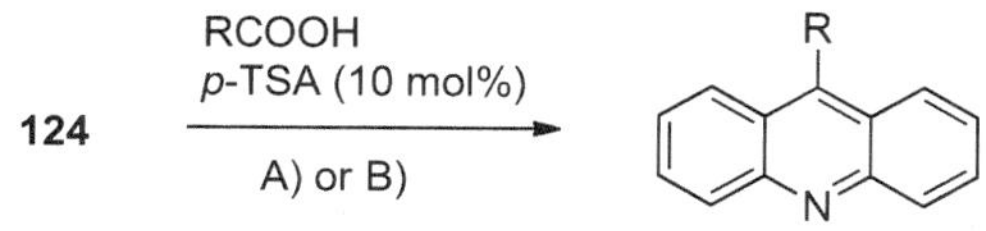


A) μw, 450 W - 600 W, 2 - 9 min.
B) Δ, 10 - 15 h

**125a**: R = Ph (A: 80%; B: 75 %)
**125e**: R = *o*-ClPh (A: 82%; B: 73%)
**125f**: R = *p*-$NH_2$Ph (A: 66%; B: 38%)
**125g**: R = -$(CH_2)_4$COOH (A: 88%; B: 71%)

**Scheme 45**

## 6.5 Type E Ring-Closure Reactions

Most ring closures of this type require diphenylamine-2-carbonyls as starting materials. *N*-phenylanthranilic acids can be prepared by the Ullmann–Jourdan reaction which usually requires high temperatures and often an excess of copper or copper salts (2002COC597, 2003OL2453, 2001OL2583, 2001OL3803). Alternatively, Buchwald–Hartwig Pd(0)-catalyzed syntheses have been developed (2004T5737). C–N bond formation was accomplished by $POCl_3$ to give 9-chloroacridines. Cyclization of diphenylamine-2-carbonyls can be performed in the presence of strong acids such as trifluoroacetic acid and sulfuric acid, and usually require high temperatures. A recent example is the ring closure of 1-(2-((2-nitrophenyl) amino)phenyl)ethanone to 9-methyl-4-nitroacridine in a mixture of sulfuric acid and acetic acid at 95–105 °C over a period of 30 min, which gave the corresponding acridine in 82% yield (2004T4283). As a comparison, indium triflate (5 mol%) in 1,2-dichloroethane at 115 °C over a period of 24 h was used to cyclize 2-(phenylamino)benzaldehyde to acridine in 97% yield (2011JOC7005). An intramolecular aryl-aldehyde $C_{sp}^2$–$C_{sp}^2$ bond formation starting from 2-(phenylamino)benzaldehydes **126** in the presence of the hypervalent iodine compound $PhI(OAc)_2$ was developed (Scheme 46). The solvent DMF was shown to give the best results; various additives have been tested, and yields improved considerably on addition of benzoyl peroxide. While the N-unproteced derivative gave acridine, the N-methyl and N-benzyl-substituted 2-(phenylamino)benzaldehydes gave the acridones **127** (2014JOC7451).

A copper-catalyzed aerobic C–H and C–C functionalization of 1-(2-(phenylamino)phenyl)ethanones **128** (Scheme 47) resulted in the formation of acridones **129**. A mechanism was suggested which involves acridines that are oxidized by copper and oxygen. Pyridine serves as additive, and

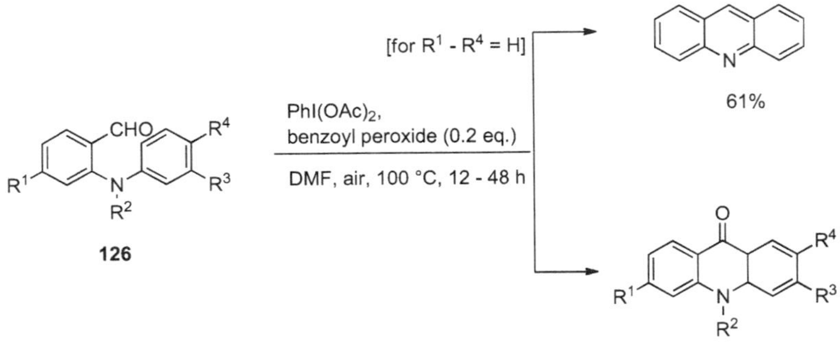


**127a**: $R^1 = R^3 = R^4 = H$, $R^2 = Me$ (81%)
**127b**: $R^1 = R^4 = H$, $R^2 = Me$, $R^3 = Br$ (70%)
**127c**: $R^1 = R^3 = H$, $R^2 = Me$, $R^4 = F$ (58%)
**127d**: $R^1 = Cl$, $R^3 = R^4 = H$, $R^2 = Me$ (55%)
**127e**: $R^1 = R^3 = R^4 = H$, $R^2 = Bn$ (74%)

**Scheme 46**

copper(II)trifluoroacetate as reagent. An aromatic C—H bond and an acetyl C—C bond are cleaved (2013CEJ4271).

Highly functionalized phenols **131** starting materials for an indium triflate-induced ring closure are available from cyclobutenones **130** which with activated methylene ketones give the acridines **132** (2012JOC5173), Scheme 48.

The Ruppert—Prakash reagent [(trifluoromethyl)trimethylsilane] was employed to prepare *N*-substituted *o*-trifluoroacetylanilines **133**, then subjected to a ring closure to give 9-trifluoromethylacridines **134** (2012 JFC272), Scheme 49. The trifluoromethyl-substituted acridines are of interest to medicinal chemists.

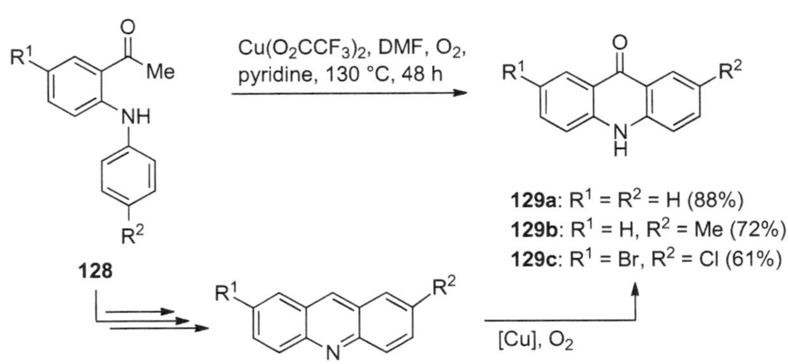


**Scheme 47**

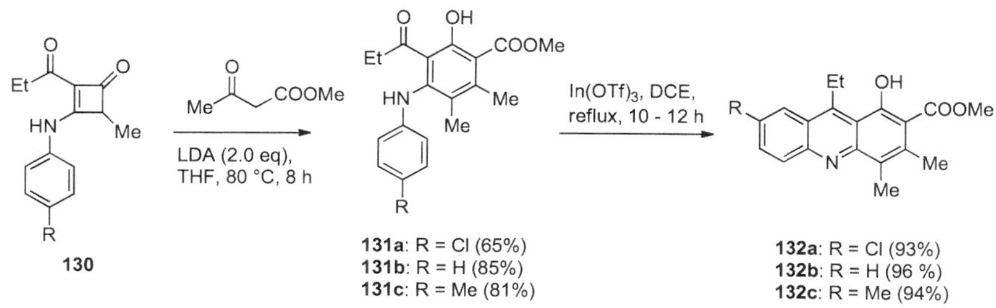


**Scheme 48**

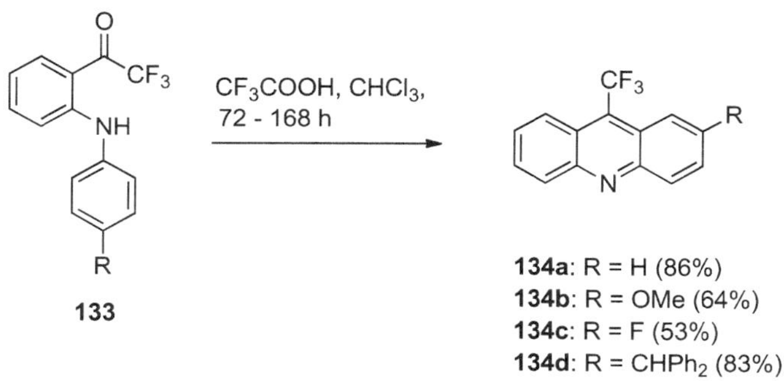


**Scheme 49**

Ring closure of *N*-phenylanthranilic acids **135** (Scheme 50) can be performed with phosphorus oxychloride under microwave conditions (2014THL3308, 2010JPO382) or under reflux temperature to prepare 9-chloroacridine **136** (2013EJMC469). The chloro-substituent in position 9 is extremely valuable as it can be replaced in a great variety of nucleophilic substitutions.

The syntheses of acridines by $ZnCl_2$-promoted intramolecular cyclization of *o*-arylaminophenyl Schiff bases **137** to **138** and some examples are shown in Scheme 51. Sterically demaning substituents failed to react as demonstrated by the attempt to synthesize **138c** (2014OL18).

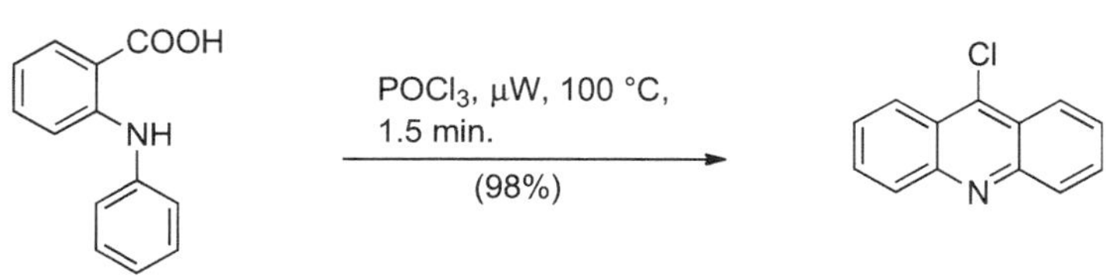


135 136

**Scheme 50**

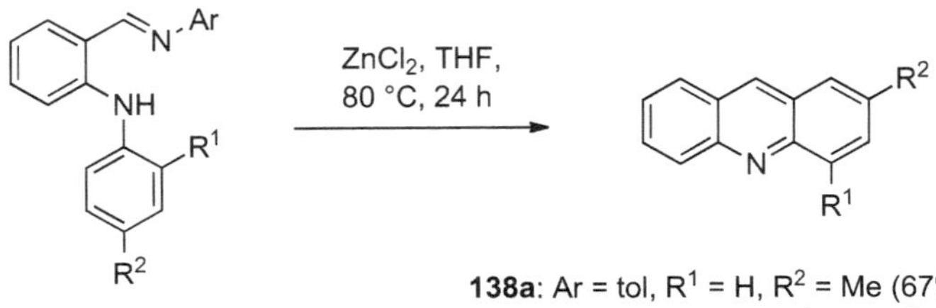


137

**138a**: Ar = tol, $R^1$ = H, $R^2$ = Me (67%)
**138b**: Ar = 2,6-$Me_2$Ph, $R^1$ = H, $R^2$ = Me (73%)
**138c**: Ar = 2,6-i$Pr_2$Ph, $R^1$ = H, $R^2$ = Me (0 %)
**138d**: Ar = 2-MeOPh, $R^1$ = OMe, $R^2$ = H (95 %)

**Scheme 51**

Depending on reaction time, substitution pattern, and temperature, *N*-alkyl-2-benzylanilines **139** cyclize to 2-phenylindole, 2,3-diphenylindole, acridine, and anthracene in varying yields and ratios (Scheme 52). Under the conditions shown, acridine is the main product (2008JHC837). Calcium oxide was also used to convert 2-benzylaniline, 2-benzoylaniline, or (2-aminophenyl)(phenyl)methanol into acridines. However, temperatures of 450–600 °C over periods of 40 min were applied. While 2-benzylaniline and (2-aminophenyl)(phenyl)methanol gave mainly acridine (58–75% and 12–50% yield, respectively), 2-benzoylanilines gave mixtures of acridine, 10,10a-dihydroacridin-9(8a*H*)-one, and 9*H*-carbazole (2004JHC1).

## 6.6 Type F Ring-Closure Reactions

Starting from 1-(2-ethynylquinolin-3-yl)ethanone **140**, available by subsequent Sonogashira reaction from 2-chloroquinoline, the Grignard reaction with MeMgBr to give a secondary alcohol, and Dess–Martin periodinane oxidation, a method was developed which provided 1-aminoacridines **141** (Scheme 53). On addition of secondary amines, enamines **E** formed under $TiCl_4$ catalysis (method A), in the presence of molecular sieves

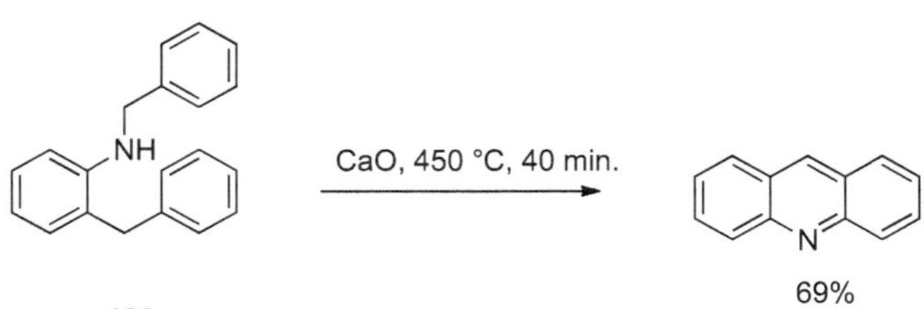


139

**Scheme 52**

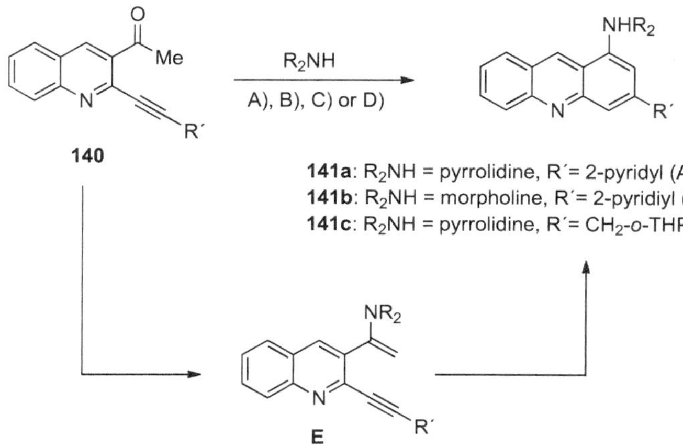


**141a**: $R_2NH$ = pyrrolidine, R´= 2-pyridyl (A: 46%; B: 78%; C: 85%)
**141b**: $R_2NH$ = morpholine, R´= 2-pyridiyl (D: 81%)
**141c**: $R_2NH$ = pyrrolidine, R´= $CH_2$-*o*-THP (C: 84%)

A) i ) hexane, 0-5 °C, $TiCl_4$, 20 min; ii) 25 min, rt; iii) $PtCl_2$, 17 h, reflux.
B) i) $CH_2Cl_2$, 4A molecular sieves, rt 3 - 5 h; ii) $PtCl_2$, 17 h, rt.
C) $CH_2Cl_2$, 4A molecular sieves, reflux, 3 - 4 h.
D) $CH_2Cl_2$, 4A molecular sieves, $Al_2O_3$, 15 h, reflux.

Scheme 53

with neutral aluminum oxide in dichloromethane (method B/D), or without aluminum oxide (method C). $PtCl_2$ is a catalyst for enyne cyclizations and was used in methods A and B. This catalyst is not necessary in methods C and D, which failed, however, in the case of piperidine (2005 OL1793).

A rhodium-catalyzed benzannulation of **142**, Scheme 54, supplements the synthetic procedure described above (2004TL2783) and can also be classified as an example of a type F ring closure to acridines.

## 6.7 Type G Ring-Closure Reactions

Several approaches to acridines **145** starting from 2-ethynylquinoline-3-carbaldehydes **144** represent type a G ring closure, Scheme 55, which at room

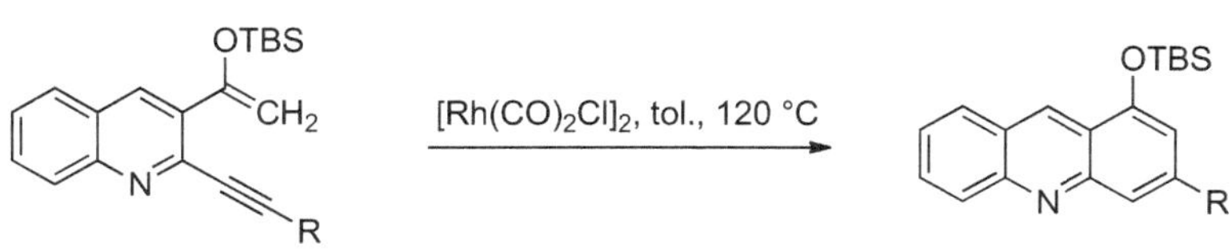


**142a**: R = $C_4H_9$
**142b**: R = Ph

**143a** (70%)
**143b** (60%)

Scheme 54

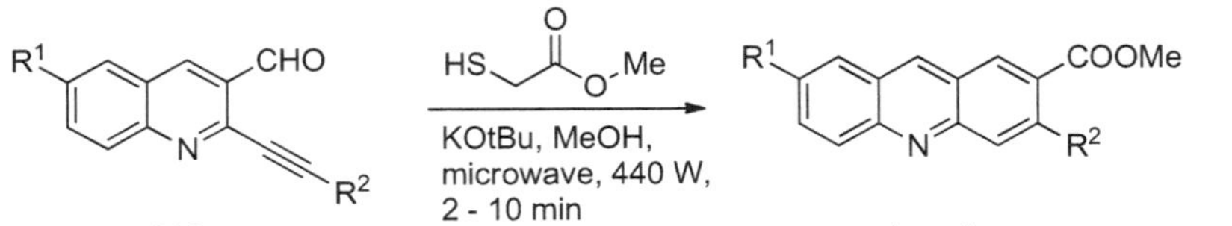


**144**

**145a**: $R^1 = R^2 = H$ (76%)
**145b**: $R^1 = H$, $R^2 = Ph$ (99%)
**145c**: $R^1 = OMe$, $R^2 = Ph$ (75%)
**145d**: $R^1 = OEt$, $R^2 = Ph$ (51%)

**Scheme 55**

temperature over a period of 2–4 h gave lower yields in comparison to those under microwave irradiation (2001T706).

## 6.8 Acridines by Rearrangements

9-Amino-substituted acridines are available by a thermal rearrangement of indazolium salts (2013OBC3558, 2014H(89)2356, 2015ZN(B)83), which were deprotonated with mild bases such as potassium phosphate whereupon a rearrangement to the acridines **147** took place (Scheme 56). A great variety of substitution patterns was realized.

According to a DFT calculation the mechanism, shown in Scheme 57, involves deprotonation of the indazolium salt such as **146a** to the N-heterocyclic carbene of indazole **F** which undergoes a ring cleavage of the pyrazolium ring to **G**, followed by trans–cis isomerization of the resulting phenylimine group, 6π-electrocyclic ring closure to **H** and subsequent tautomerization to the 9-aminoacridine **147a**. This ring-transformation can be regarded as a special example of a type E ring closure. The N-heterocyclic carbenes of pyrazole behave similarly and give quinolines by rearrangement

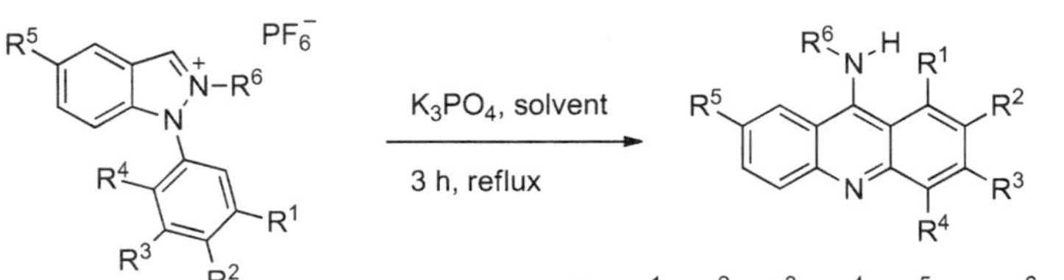


**146a-e**

**147a**: $R^1 = R^2 = R^3 = R^4 = R^5 = H$, $R^6 = Me$ (87%)
**147b**: $R^1 = R^3 = R^6 = Me$, $R^2 = R^4 = R^5 = H$ (58%)
**147c**: $R^1 = R^3 = R^4 = R^5 = H$, $R^2 = CF_3$, $R^6 = Me$ (51%)
**147d**: $R^1 = R^2 = R^3 = R^4 = R^5 = H$, $R^6 = Et$ (60%)
**147e**: $R^1 = R^2 = R^3 = R^5 = H$, $R^4 = Cl$, $R^6 = Me$ (83%)

**Scheme 56**

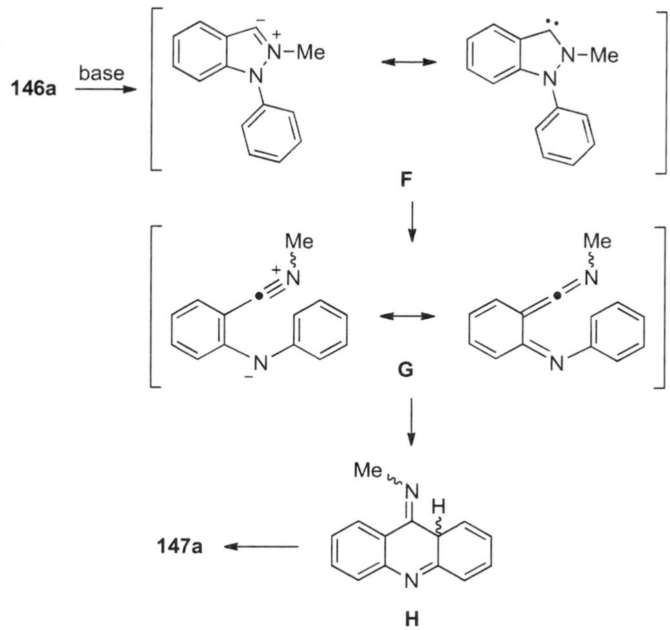


Scheme 57

(2010EJO4296, 2010AG2851, 2010AGE2790). A review appeared (2011 COC2897).

Heterocyclic synthesis with N-heterocyclic carbenes of indazoles are very versatile (2008EJO4313, 2008S2798), and results were summarized (2006S1882). The outcome of rearrangements can be determined by the conditions. Thus, deprotonation at lower temperatures results in the formation of spiro[indazole-3,2′-indolines] **148** as mixtures of *Z*- and *E*-isomers with respect to the exocyclic imine group (Scheme 58). A dimerization followed by ring-opening of one of the indazole rings took place. These spiro

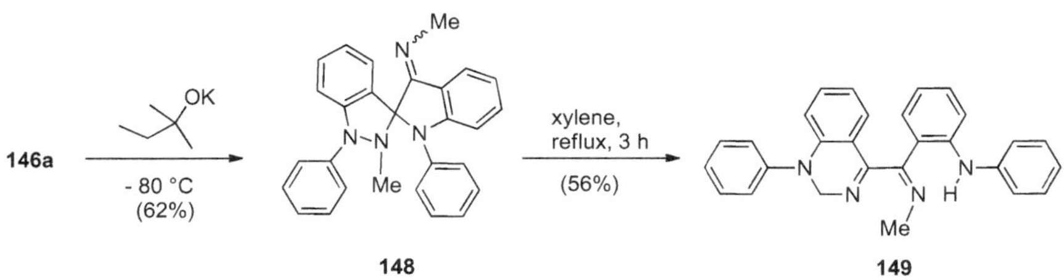


Scheme 58

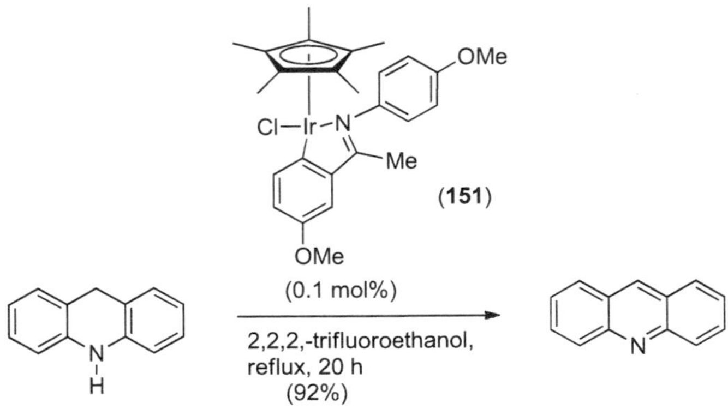


150

Scheme 59

compounds rearrange to the substituted quinazolines **149**. A mechanism via an ylide was proposed (2014BJOC832).

## 6.9 Acridines by Oxidation

Oxidative dehydrogenation with oxygen or air of N-heterocycles including dihydroacridine was developed using Pt nanowire catalyst and was performed under 1 bar of oxygen in methanol at 40 °C in quantitative yield (2013CCC2183).

An electron-rich dimethoxyacridane was oxidized in acetic acid under an oxygen atmosphere at reflux over a period of 5 h with 80% yield (2009T(A) 2795). An iridium catalyst **151** was able to dehydrogenate a series of nitrogen heterocycles, among those 9,10-dihydroacridine to acridine (2013AGE 6983; Scheme 59).

10-Methyl-9,10-dihydroacridine-9-carbonitrile in anhydrous ethanol gave the tribromide **153** which formed 2,7-dibromo-10-methyl-9(10*H*)-acridone on recrystallization (2007MI1067; Scheme 60).

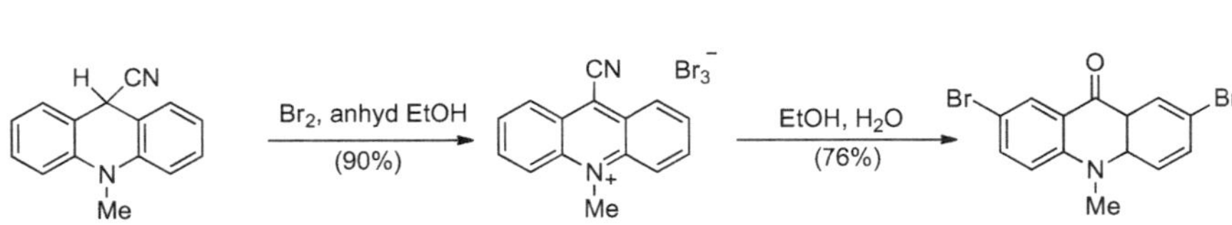


152 153 154

Scheme 60

## 6.10 Acridines by Reduction of Acridones

Numerous methods exist for the reduction of acridones to the corresponding acridines, among them hydrochloric acid in the presence of tin(II)chloride (2000OPPI573), hydrogen produced in situ by elemental sodium in propanol (2009T(A)2795), or sodium amalgam in the presence of sodium hydrogencarbonate in ethanol (1999SL641). Acridones can also be converted into acridines via acridanes by a borane tetrahydrofuran complex, Scheme 61 (2009TL6894).

# 7. REACTIONS OF ACRIDINES

The reactivity of acridine has been reviewed (1966MI, 1973MI), and more recent results have been summarized in an excellent overview in 2004 (2004S313). Methods to reduce the acridine ring to acridanes, 1,2,3,4,4a,9,9a,10-octahydroacridine, tetradecahydroacridine, 1,2,3,4,5,6,7,8-octahydroacridine, or 1,2,3,4-tetrahydroacridine have been described and reviewed (1973MI2; 2004S313). A report describing the catalytic hydrogenation of acridines over $Rh/Al_2O_3$ in a combined experimental and theoretical studies appeared (2013JMCC63).

## 7.1 Reactions at Position 1

Position 1 of the acridine ring can be activated toward electrophilic substitution by electron-donating groups at C-2. However, some new fused heteropolycycles have been synthesized via direct intramolecular Friedel–Crafts cycloalkylations, among them acridine. Thus, starting from 4-(acridin-9-yl)-2-methylbutan-2-ol **157** the tetracyclic acridine derivative 3,3-dimethyl-2,3-dihydro-1*H*-benzo[*kl*]acridine **158** was obtained in 80% yield (2010ARK338; Scheme 62). Following a related pathway, the syntheses of pyridoacridines have been described (1997SC2311).

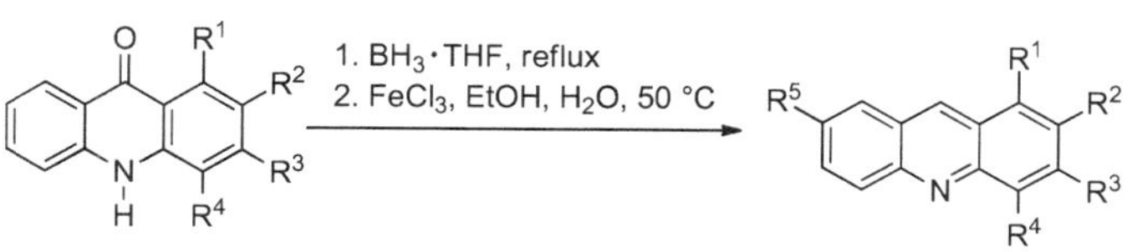


**155**

**156a**: $R^1 = R^3 = R^4 = R^5 = H$, $R^2 = Cl$ (47%)
**156b**: $R^1 = NO_2$, $R^2 = R^3 = R^4 = R^5 = H$ (50%)
**156c**: $R^1 = R^2 = R^4 = R^5 = H$, $R^3 = COOEt$ (73%)
**156d**: $R^1 = R^2 = R^3 = H$, $R^4 = COOMe$, $R^5 = Br$ (70%)

**Scheme 61**

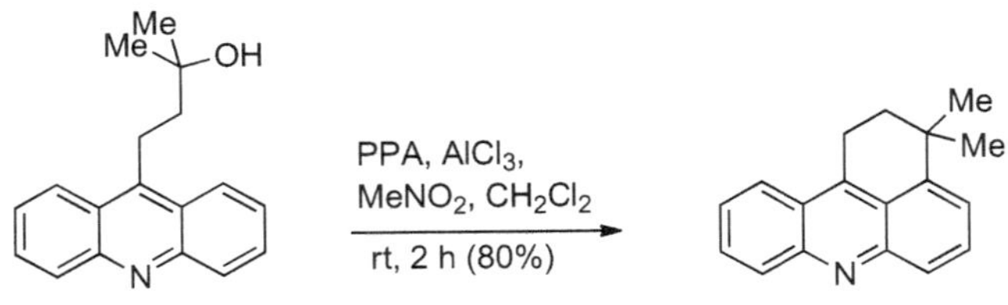


157 158

Scheme 62

## 7.2 Reactions at Position 4

The electrophilic substitution of C-4 of the acridine ring can be accomplished under Tscherniac-Einhorn conditions, used to prepare a library of potential anti-tumor agents (2003H(60)1653). Bromo(methoxy)methane or related compounds react with the 4-position of acridine or with the four- and 5-positions in the presence of sulfuric acid at 20 °C within several hours to give 4-(bromomethyl)acridines or 4,5-bis(bromomethyl)acridines (2003SL2349), then subjected to a variety of further transformations (2008H1943). As examples, **65** and **66** (Scheme 23; 2005BMC5560), **82** (Scheme 28; 2011T3803), **87** (Scheme 30; 2014OL2150), and the aforementioned telomeric G-quadruplex ligands (2009EJMC3880), among others, have been prepared. A catalyst consisting of Rh and the N-heterocyclic carbene 1,3-dimesityimidazol-2-ylidene (IMES) is able to directly phenylate the 4-position of acridine to give **159** (Scheme 63). A possible mechanistic pathway involves either a bimetallic intermediate **I** or a monometallic intermediate **J** (2011JA3780).

## 7.3 Reactions at Position 9

The reactivity of position 9 of acridines was the subject of a review (1977H(6)987). Position 9 is very susceptible toward nucleophilic attacks to give acridanes, which sometimes rearomatize spontaneously (2004 S313). As examples, azulene couples with acridine to a 1-dihydroacridine-azulene **160** or a 1,3-bis(dihydroacridine)azulene **161** (Scheme 64). The outcome depends on the stoichiometry of the partners. Aromatization to **162** and **163**, respectively, was then accomplished by potassium hydroxide in ethanol (2007TL1099, 2011EJO5311).

Sodium difluoroethylsulfinate (DFES-Na) can be used to introduce the 1,1-difluoroethyl group, which is a bioisoster to the methoxy group and

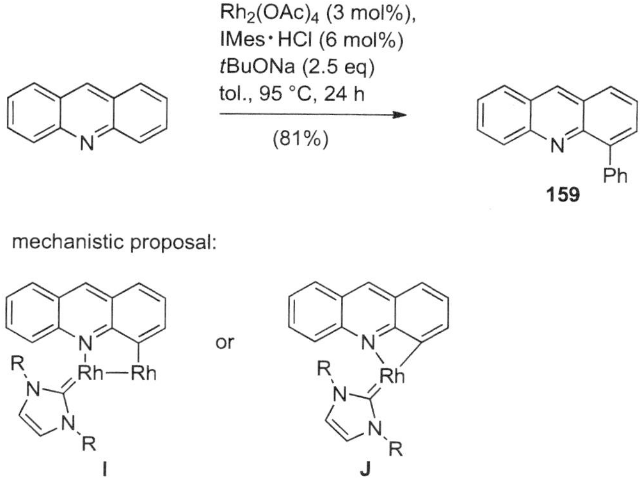


Scheme 63

therefore of considerable interest, into the position 9 of acridine (Scheme 65). Thus, **164** was formed in 72% yield (2013AGE3949).

Organozinc reagents were able to functionalize the CH bond in position 9 of acridines, and acridines **165** with diphenylzinc in the presence of various

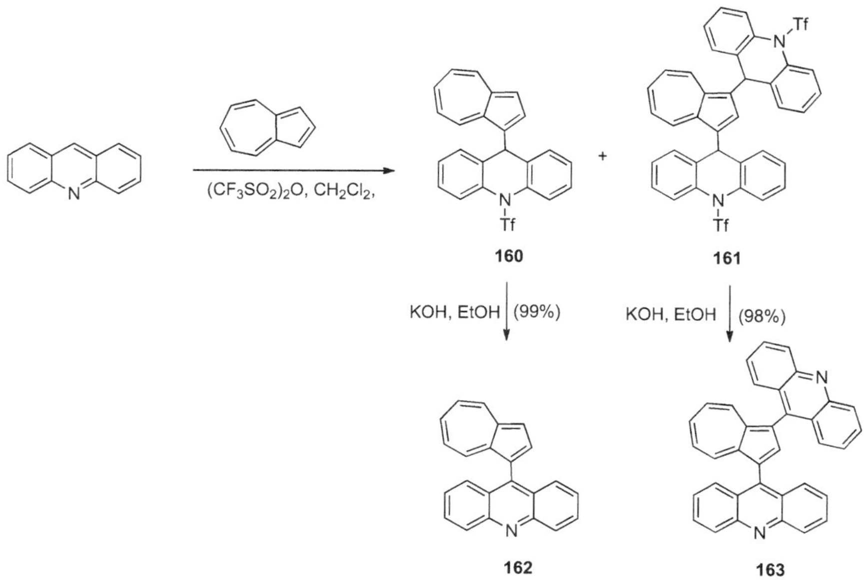


Scheme 64

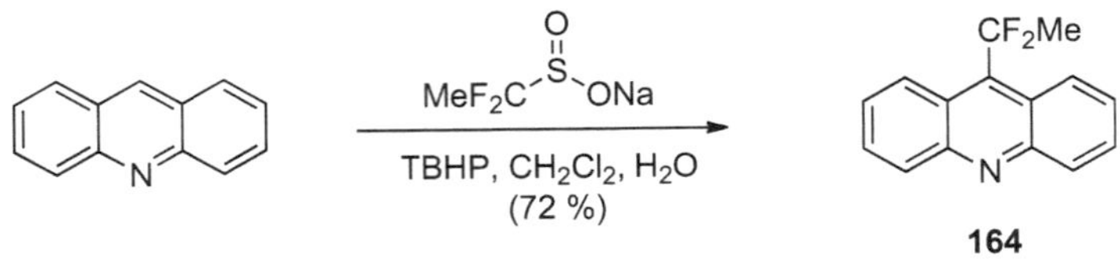


**Scheme 65**

metal catalysts ($Cu(OTf)_2$, $FeCl_3$, $InCl_3$, $Pd(OAc)_2$ etc.) have been tested. [$RhCl(cod)_2$] proved to be the best, as 82% of 9-phenylacridines over 72 h were obtained. Some examples (**166** and **167** from **165**) are presented in Scheme 66. Diisopropylzinc gave 9-isopropyl-9,10-dihydroacridine in quantitative yields even in the absence of a Rh(I) catalyst. Aromatization yielded the 9-alkylated acridines **167** (2012CC308).

9-Chloroacridines are very valuable starting materials for syntheses of a great variety of acridines. They can readily be prepared from acridone with $SOCl_2$ (2–4 h, 99%) (2009ARK(XII)31) or $POCl_3$ (12 h, 89%) (2014H(88)535), or starting from aminobenzoic acids with $POCl_3$ (2 h, 135–140 °C, 96%) (2010JA14006; 2013BMC4170). They react readily with amines and are used for the synthesis of sensor molecules (2014OBC6677, 2013EJMC469). Under microwave irradiation the 9-chloro substituent can also be replaced to 1,2-diaminophenyls, and this method was developed

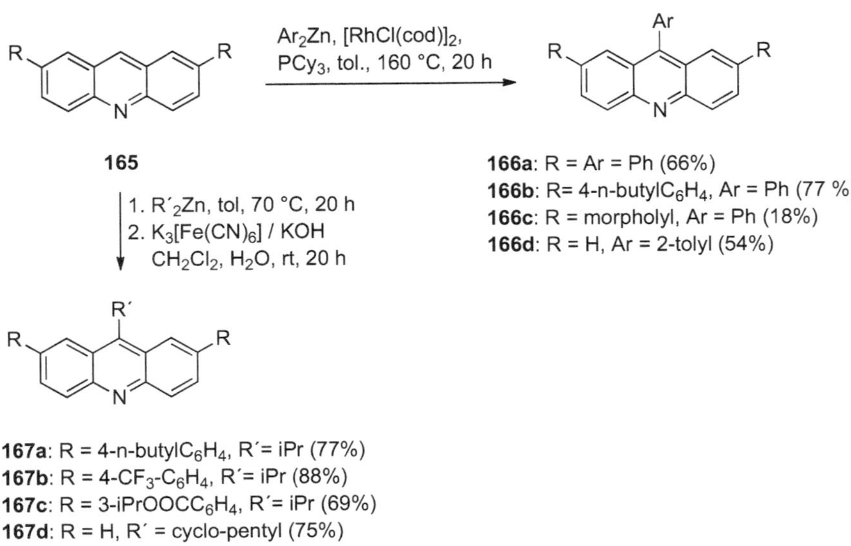


**Scheme 66**

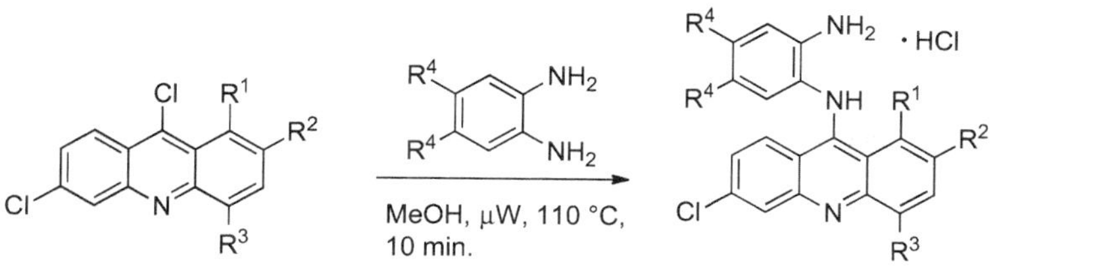


**168**

**169a**: $R^1 = R^3 = R^4 = H$, $R^2 = OMe$ (93 %)
**169b**: $R^1 = R^3 = OMe$, $R^2 = H$, $R^4 = Cl$ (84%)
**169c**: $R^1 = R^4 = H$, $R^2 = R^3 = OMe$ (69 %)

**Scheme 67**

as chromatography-free route to 9-amino-substituted acridines **169** from **168**, Scheme 67 (2014THL3308).

Recent examples of transition metal-catalyzed reactions of 9-chloroacridines are as follows:

Negishi cross-coupling of **170** to **171** were catalyzed by acenaphthoimidazolylidene palladium complexes (2013JOC7436), and a similar catalyst was able to induce a Suzuki–Miyaura reaction with the sterically hindered boronic acid, Scheme 68 (2012OL4250).

Other Suzuki–Miyaura reactions with 9-chloroacridine employed palladium(II)acetate in the presence of a carbazolyl-derived phosphine ligand

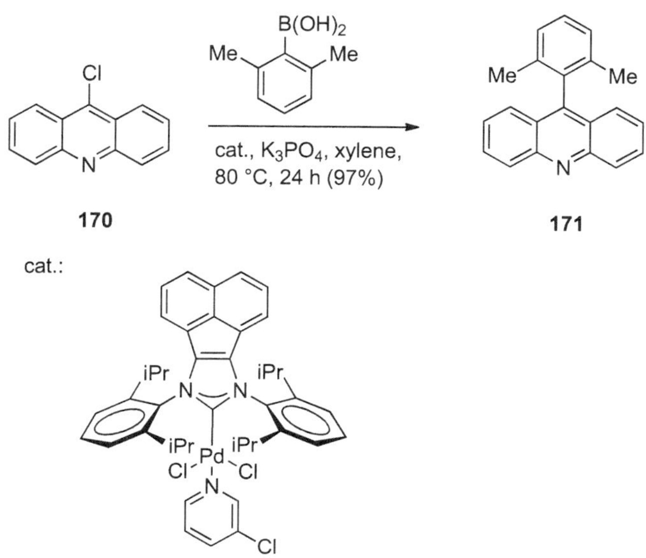


**Scheme 68**

(2011CC5079). Sonogashira reactions have also been described starting from 6-chloroacridines ($(PPh_3)_2PdCl_2$, CuI, $PPh_3$, 8–10 h reflux) to prepare phenylethynylacridines (2004OBC3113). A Stille reaction of 9-trimethylstannylacridine, prepared from 9-bromoacridine and 1,8-dibromonaphthalene (2003JA10651) was applied to prepare the dicationic acridinium derivative **174** upon treatment of **173** with excess methyltriflate (Scheme 69). The di-cation is no fluorophore, and an intramolecular charge shift from the acridinium to the naphthalene unit is given as a possible explanation for the fluorescence quenching. An X-ray structure revealed that the two acridinium rings are in a surprising proximity, the closest contact is between the two C-9 carbon atoms (292.7(7) pm). This is the condition for a two-electron reduction and C–C bond formation to **175** which occurs on treatment of **174** with Zn in THF/$NEt_3$ which was isolated in 95% yield as stable crystalline solid. Regeneration of the di-cation was observed during treatment of 2 equivalents of tris(*p*-bromophenyl)ammonium hexachloroantimonate in 85% yield (2004THL8289).

Another type of reaction is the formation of spiro compounds at position 9 of acridines. Thus, 1-(acridin-9-yl)methanamine **176** with isothiocyanates gave thioureas **177**, which are very unstable as they cyclized rapidly to the

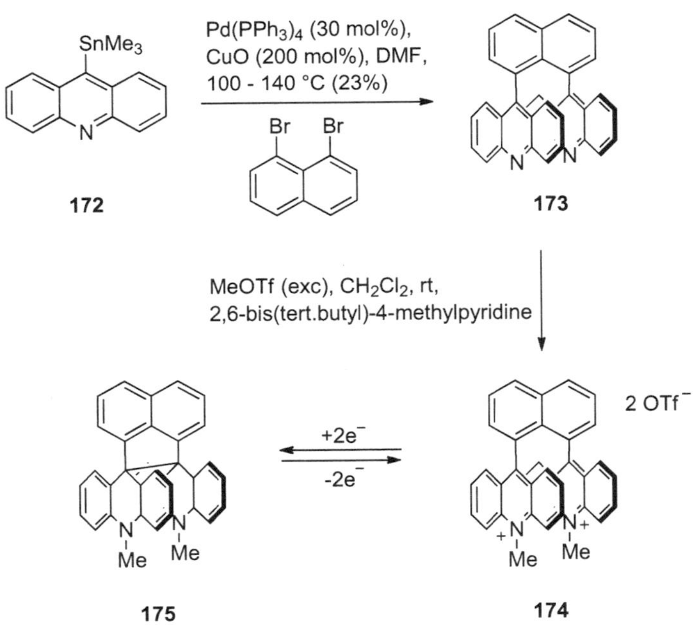


Scheme 69

corresponding spiro **178**. Mesitylnitrile oxide (MNO) converted those into the corresponding spiro[dihydroacridine-9′(10′*H*),5-imidazoline]-2-ones **179**. The equilibrium did not entirely favor the spiro form. Ring-opened products **180** were formed in varying amounts (Scheme 70; 2014T944).

## 7.4 Reactions at Position 10

The nitrogen atom of the acridine ring is the main nucleophilic center of the molecule (2004S313) and can be alkylated easily with methyl iodide in acetonitrile over a period of 6–8 h at reflux in quantitative yield (2004 MI1230) or in THF at slightly elevated temperatures (2010JOC204). Alternatively, trimethyloxonium tetrafluoroborate can be used to methylate 9-chloroacridine in dichloromethane (2006ASC1862). Benzylations were achieved by benzylbromide in the presence of sodium perchlorate (2007JPOC484) which induced an anion exchange. A direct phenylation of the nitrogen atom of acridine was also reported (2012CHE301). As shown in Scheme 71, a Mitsunobu reaction for the preparation of salts including N-methylacridinium salt **182** was described (2008TL3663).

Apart from alkylations and arylation of N10 of acridine, the formation of N–B bonds is possible. A borenium cation **184** was formed on addition of aluminum chloride to the acridine–trichloroborane adduct **183** formed under the conditions shown in Scheme 72 (2013OM6712). Depending on the conditions, different products were obtained.

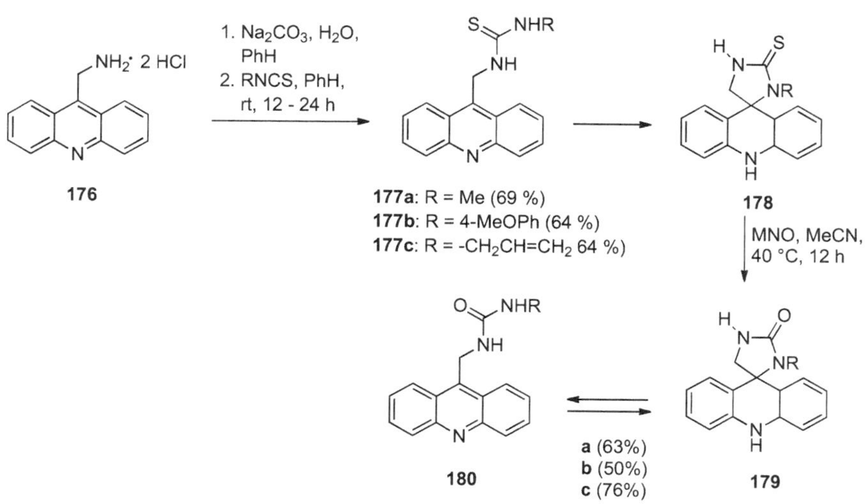


Scheme 70

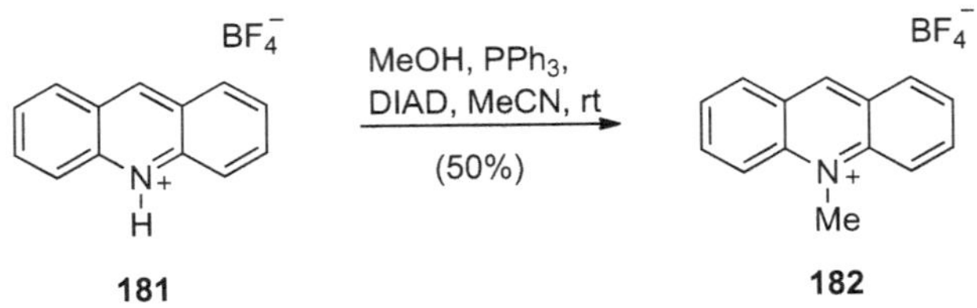


Scheme 71

As an application of acridinium–boron compounds, hydrophobically directed selective reductions of quaternized β-ketoamines with amine boranes such as (9-aminoacridinium-10-yl)trihydroborate **187** (Scheme 73) were studied (2006THL127).

## 7.5 Acridines in Catalytic Reactions

Some acridines are generally effective ligands in Pd(II)-catalyzed $C(sp^3)$–H and $C(sp^2)$–H alkylation of simple amides **188** with various alkyl iodides to give **189** (Scheme 74; 2014JA13194).

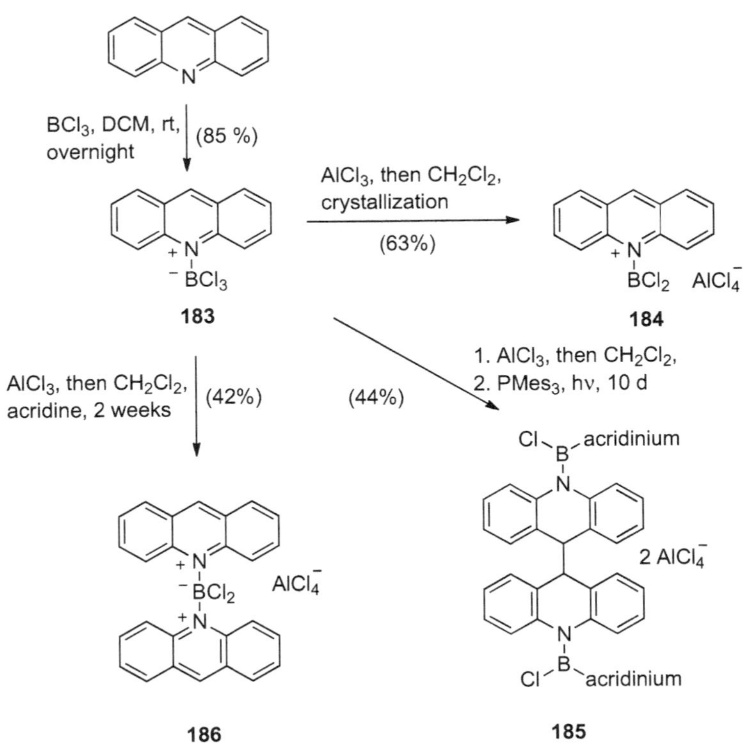


Scheme 72

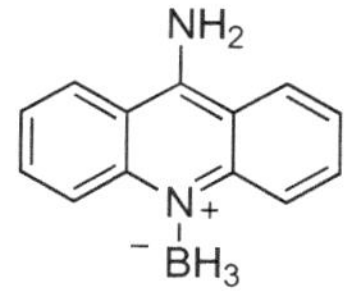


**187**

**Scheme 73**

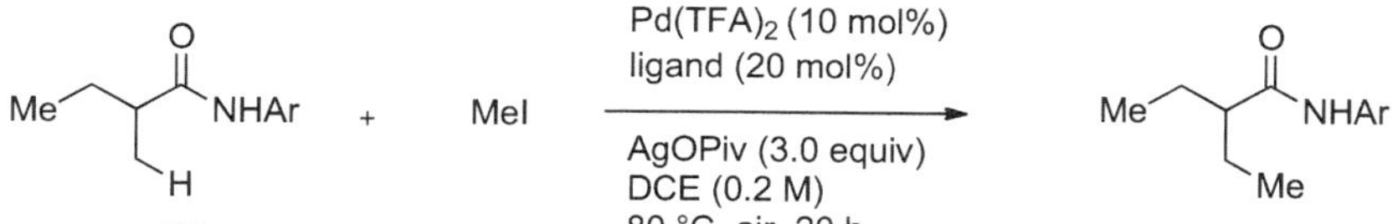


**188** **189**

ligand: acridine (77%)
9-Me-acridine (81%)
9-OiBu-acridine (79%)
9-Cl-acridine (69%)
acridine orange (54%)

**Scheme 74**

An acridinylidene palladium carbene complex, starting from 9-chloro-10-methylacridine and $Pd(PPh_3)_4$ by oxidative addition at 60 °C in toluene in 53% yield, was applied to catalyze Mizoroki–Heck reactions of **190** and **191** to **192** (Scheme 75). Increasing the catalyst concentration to 0.001 mol% with respect to the palladium, an almost quantitative conversion was measured by gas chromatography (2006ASC1862).

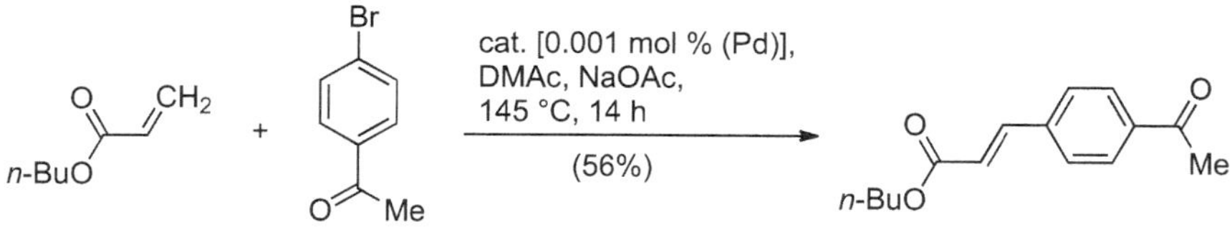


**190** **191** **192**

cat.:

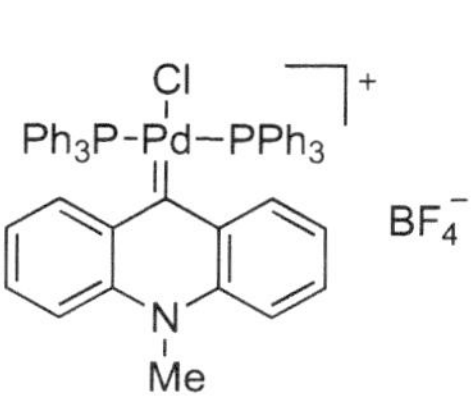


**Scheme 75**

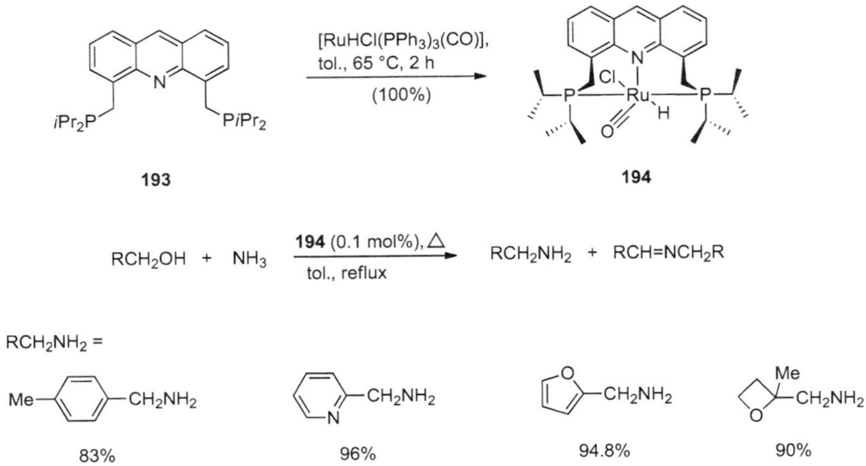


**Scheme 76**

An air-stable acridine pincer complex **194** was quantitatively prepared on treatment of the tridentate PNP ligand **193** with a ruthenium complex and was applied as a catalyst for the direct synthesis of primary amines from alcohols and ammonia, Scheme 76 (2008AGE8661, 2011ACR588).

New iron pincer complexes were also prepared. 4,5-Bis(diphenylphosphino)acridine **195** with iron(II)bromide in tetrahydrofuran gave the

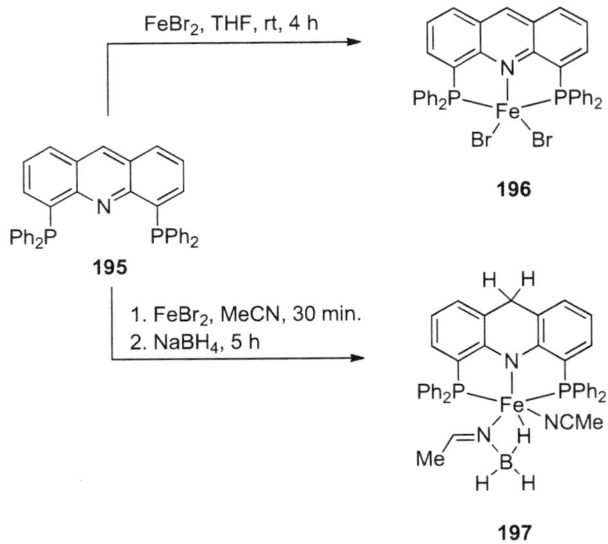


**Scheme 77**

paramagnetic **196**, whereas in acetonitrile followed by treatment with sodium borohydride it gave the imino borohydride **197**, Scheme 77. These two complexes were investigated as catalysts for the hydrogenation of alkynes. Complex **197** gave trans-stilbene from diphenylacetylene in 99% yield within 12 h at 90 °C. The scope of this hydrogenation was studied. Complex **197** showed excellent catalytic properties for a large variation of acetylenes (2013AGE14131).

Acridines are very versatile compounds. Their applications range from biologically highly active and widely applied compounds to sensor molecules, dyes, and catalysts for organic transformations. Perhaps, this brief overview will serve as a stimulus for further developments.

## REFERENCES

1870MI746 C. Graebe and H. Caro, *Ber. Dtsch. Chem. Ges.*, **3**, 746 (1870).
1871MI265 C. Graebe and H. Caro, *Ann. Chem.*, **158**, 265 (1871).
1883CB1802 A. Bernthsen and F. Bender, *Chem. Ber.*, **16**, 1802 (1883).
1884MI1 A. Bernthsen, *Ann. Chem.*, **224**, 1 (1884).
1895USP J. Schmid and K. Jedlicka, US 537723 A.
1912MI1787 L. Benda, *Ber. Dtsch. Chem. Ges.*, **45**, 1787 (1912).
1956MI R.M. Acheson, editor: *Acridines*, Interscience: New York (1956).
1962JOC2658 F.D. Popp, *J. Org. Chem.*, **22**, 2658 (1962).
1966MI A.A. Albert, In *The Acridines*, Edward Arnold: London (1966).
1973MI R.M. Acheson, editor: *Acridines*, 2nd ed., Interscience: New York (1973).
1973MI2 I.A. Selby, In *Acridines*, 2nd ed., John Wiley & Sons: New York (1973), p 433.
1977H(6)987 S. Skonieczny, *Heterocycles*, **6**, 987 (1977).
1978MI B.P. Swann and A. McKillop, In 2nd ed., In S. Coffey and M.F. Ansell, editors: *Rodd's Chem. Carbon Compd.*, **Vol. 4(G)**, (1978), p 257.
1979MR261 A. Nasim and T. Brychcy, *Mut. Res.*, **65**, 261 (1979).
1980H(14)985 S. Skonieczny, *Heterocycles*, **14**, 985 (1980).
1987MI J.D. Hepworth, In 2nd ed., Suppl. 1, In M.F. Ansell, editor: *Rodd's Chemistry of Carbon Compounds*, **Vol. 4G**, Elsevier: Amsterdam (1987), p 1.
1988MI1783 E. Schneider, S.J. Darkin, P.A. Lawson, L.-M. Ching, R.K. Ralph, and B.C. Baguley, *Eur. J. Cancer Clin. Oncol.*, **24**, 1783 (1988).
1989MI81 W.V. Steele, R.D. Chirico, I.A. Hossenlopp, A. Nguyen, N.K. Smith, and B.E. Gammon, *J. Chem. Thermodyn.*, **21**, 81 (1989).
1992HOU(E7b)115 U. Kuckländer, In R. Kreher, editor: *Houben-Weyl*, **Vol. E7b**, Thieme Verlag: Stuttgart (1992), p 115.
1992MI309 N.T. Sklarin, P.H. Wiernik, W.R. Grove, L. Benson, A. Mittelman, J.A. Maroun, J.A. Stewart, F. Robert, J.H. Doroshow, P.J. Rosen, J. Jolivet, J.C. Ruckdeschel, N.J. Robert, E. Vèlez-Garcìa, D.E. Bergsagel, L.C. Panasci, A.M. van der Merwe, J.J. Longueville, J. Leiby, and C.D. Kowal, *Inves. New Drugs*, **10**, 309 (1992).
1995B13682 J.M. Crenshaw, D.E. Graves, and W.A. Denny, *Biochemistry*, **34**, 13682 (1995).

1997SC2311 N. Fixler, M. Demeunynck, and J. Lhomme, *Synth. Commun.*, **27**, 2311 (1997).

1998MI J.P. Michael, In 2nd ed., In M. Sainsbury, editor: *Rodd's Chem. Carbon Compd.*, **Vol. 4, part F/part G**, (1998), p 483.

1999MI1166 J.Y. Fan, S.J. Ohms, M. Boyd, and W.A. Denny, *Chem. Res. Toxicol.*, **12**, 1166 (1999).

1999SL641 S. Ismaili, G. Boyer, and J.-P. Galy, *Synlett*, **641** (1999).

2000MI P. Popelier, *Atoms in Molecules, An Introduction*, Prentice Hall: Pearson Education Ltd (2000).

2000OPPI573 R.P. Dixon, J.S. Snyder, L. Bradley, and J. Linnenbrink, *Org. Prep. Proc. Int.*, **32**, 573 (2000).

2001JMC4492 E.T. Martins, H. Baruah, J. Kramarczyk, G. Saluta, C.S. Day, G.L. Kucera, and U. Bierbach, *J. Med. Chem.*, **44**, 4492 (2001).

2001MI1 M. Wainwright, *J. Antimicr. Chemotherapy*, **47**, 1 (2001).

2001MI333 D. Fyfe, C. Price, R.E. Langley, C. Pagonis, J. Houghton, L. Osborne, P.J. Woll, C. Gardner, B.C. Baguley, and J. Carmichael, *Cancer Chemother. Pharmacol.*, **47**, 333 (2001).

2001MOL4401 M.A. Yahayu, M. Rahmani, N.M. Hashim, M.A.M. Amin, G.C.L. Ee, M.A. Sukari, and A.M. Akim, *Molecules*, **16**, 4401 (2011).

2001OL2583 D. Ma and C. Xia, *Org. Lett.*, **3**, 2583 (2001).

2001OL3803 M. Wolter, A. Klapars, and S.L. Buchwald, *Org. Lett.*, **3**, 3803 (2001).

2001PAC1421 J. Konopa, *Pure Appl. Chem.*, **73**, 1421 (2001).

2001PNA4844 M. Read, R.J. Harrison, B. Romagnoli, F.A. Tanious, S.M. Gowan, A.P. Reszka, W.D. Wilson, L.R. Kelland, and S. Neidle, *Proc. Natl. Acad. Sci. U.S.A.*, **98**, 4844 (2001).

2001T706 I. Cikotiene, R. Buksnaitiene, and R. Sazinas, *Tetrahedron*, **67**, 706 (2001).

2002BCP191 H. Baruah, C.L. Rector, S.M. Monnier, and U. Bierbach, *Biochem. Pharmacol.*, **64**, 191 (2002).

2002CMC655 W.A. Denny, *Curr. Med. Chem.*, **9**, 1655 (2002).

2002COC597 J. Finet, A. Fedorov, S. Combes, and G. Boyer, *Curr. Org. Chem.*, **6**, 597 (2002).

2002H(57)1299 H. Koshima and K. Kutsunai, *Heterocycles*, **57**, 1299 (2002).

2002MI390 J.A. Seijas, M.P. Váuquez-Tato, M.M. Martínez, and J. Rodríguez-Parga, *Green Chem.*, **4**, 390 (2002).

2002SC729 E. Veverková, M. Nosková, and S. Toma, *Synth. Commun.*, **32**, 729 (2002).

2003H(60)1653 J. Chiron and J.P. Galy, *Heterocycles*, **60**, 1653 (2003).

2003JA10651 C. Wolf and X. Mei, *J. Am. Chem. Soc.*, **125**, 10651 (2003).

2003JA9629 C.G. Barry, H. Baruah, and U. Bierbach, *J. Am. Chem. Soc.*, **125**, 9629 (2003).

2003OL2453 D. Ma, Q. Cai, and H. Zhang, *Org. Lett.*, **5**, 2453 (2003).

2003SL2349 J. Chiron and J.P. Galy, *Synlett*, **15**, 2349 (2003).

2004ARK36 A. Thakura, M. Thakurb, N. Kakanic, A. Joshia, S. Thakurd, and A. Guptab, *ARKIVOC*, **xiv**, 36 (2004).

2004BMCL4107 A. Kamal, O. Srinivas, P. Ramulu, G. Ramesh, and P.P. Kumar, *Bioorg. Med. Chem. Lett.*, **14**, 4107 (2004).

2004BMCL4719 V.A. Bacherikov, T.-C. Chou, H.-J. Dong, C.-H. Chen, Y.-W. Lin, T.-J. Tsaia, and T.-L. Sua, *Bioorg. Med. Chem. Lett.*, **14**, 4719 (2004).

2004CMC83 M.H. David-Cordonnier, W. Laine, T. Gaslonde, S. Michel, F. Tillequin, M. Koch, S. Leonce, A. Pierre, and C. Bailly, *Curr. Med. Chem. Anticancer Agents*, **4**, 83 (2004).

| | |
|---|---|
| 2004CPB362 | N. Negi, Y. Jinguji, K. Ushijima, S. Ikeda, Y. Takemura, M. Ju-ichi, T.-S. Wu, C. Ito, and H. Furukawa, *Chem. Pharm. Bull.*, **52**, 362 (2004). |
| 2004EJMC1029 | C. Santelli-Rouvier, J.-M. Barret, C.M. Farrell, D. Sharples, B.T. Hill, and J. Barbe, *Eur. J. Med. Chem.*, **39**, 1029 (2004). |
| 2004EJMC975 | R. Csuk, A. Barthel, T. Brezesinski, and C. Raschke, *Eur. J. Med. Chem.*, **39**, 975 (2004). |
| 2004JHC1 | T. Horaguchi, T. Oyanagi, E. Cuevas Creencia, K. Tanemura, and T. Suzuki, *J. Heterocycl. Chem.*, **41**, 1 (2004). |
| 2004MI55 | M. Demeunynck, *Expert Opin. Ther. Patents*, **14**, 55 (2004). |
| 2004MI1230 | J. Joseph, E. Kuruvilla, A.T. Achuthan, D. Ramaiah, and G.B. Schuster, *Bioconj. Chem.*, **15**, 1230 (2004). |
| 2004MI1312 | N. Chaya, K. Terauchi, Y. Yamagata, J. Kinjo, and H. Okabe, *Biol. Pharm. Bull.*, **27**, 1312 (2004). |
| 2004MI713 | E.I. Aly and A.H. Abadi, *Arch. Pharm. Res.*, **27**, 713 (2004). |
| 2004NPR731 | K.M. Marshall and L.R. Barrows, *Nat. Prod. Rep.*, **21**, 731 (2004). |
| 2004OBC3113 | A. Elangovan, H.-H. Chiu, S.-W. Yang, and T.-I. Ho, *Org. Biomol. Chem.*, **2**, 3113 (2004). |
| 2004S313 | J. Chiron and J.-P. Galy, *Synthesis*, **313** (2004). |
| 2004SL2391 | M.O. Anderson, J. Moser, J. Sherrill, and R.K. Guy, *Synlett*, **2391** (2004). |
| 2004T4283 | Y. Kitahara, T. Mizuno, and A. Kubo, *Tetrahedron*, **60**, 4283 (2004). |
| 2004T5737 | R. Csuk, A. Barthel, and C. Raschke, *Tetrahedron*, **60**, 5737 (2004). |
| 2004THL8289 | H. Kawai, T. Takeda, K. Fujiwara, and T. Suzuki, *Tetrahedron Lett.*, **45**, 8289 (2004). |
| 2004TL2783 | P. Belmont, J.-C. Andrez, and C.S.M. Allan, *Tetrahedron Lett.*, **45**, 2783 (2004). |
| 2005BMC1653 | J.P. Dheyongera, W.J. Geldenhuys, T.G. Dekker, M.G. Matsabisad, and C.J. Van der Schyf, *Bioorg. Med. Chem.*, **13**, 1653 (2005). |
| 2005BMC4291 | S.M. Sondhi, N. Singh, A.M. Lahoti, K. Bajaj, A. Kumar, O. Lozach, and L. Meijer, *Bioorg. Med. Chem.*, **13**, 4291 (2005). |
| 2005BMC5560 | C. Di Giorgio Carole, M. De Méo, J. Chiron, F. Delmas, A. Nikoyan, S. Jean, G. Dumenil, P. Timon-David, and J.-P. Galy, *Bioorg. Med. Chem.*, **13**, 5560 (2005). |
| 2005BMC6513 | V.A. Bacherikov, J.-Y. Chang, Y.-W. Lin, C.-H. Chen, W.-Y. Pan, H. Dong, R.-Z. Lee, T.-C. Chou, and T.-L. Su, *Bioorg. Med. Chem.*, **13**, 6513 (2005). |
| 2005CMC127 | R. Martinez and L. Chacon-Garcia, *Curr. Med. Chem.*, **12**, 127 (2005). |
| 2005CPB1540 | J.-B. Bongui, A. Elomri, D. Cahard, F. Tillequin, B. Pfeiffer, A. Pierré, and E. Seguin, *Chem. Pharm. Bull.*, **53**, 1540 (2005). |
| 2005IC6159 | B. Wilson, L. Gude, M.-J. Fernández, A. Lorente, and K.B. Grant, *Inorg. Chem.*, **44**, 6159 (2005). |
| 2005JA1160 | C.G. Barry, C.S. Day, and U. Bierbach, *J. Am. Chem. Soc.*, **127**, 1160 (2005). |
| 2005JMS699 | L. Pérez-Flores, A.J. Ruiz-Chica, J.G. Delcros, F. Sánchez-Jiménez, and F.J. Ramírez, *J. Mol. Struc.*, **744–747**, 699 (2005). |
| 2005JNP804 | A.R. Carroll, A. Ngo, R.J. Quinn, J. Redburn, and J.N.A. Hooper, *J. Nat. Prod.*, **68**, 804 (2005). |
| 2005JPC(B)21997 | E. Kuruvilla, J. Joseph, and D. Ramaiah, *J. Phys. Chem. B*, **109**, 21997 (2005). |
| 2005MI118 | K.-M. Chen, Y.-W. Sun, Y.-W. Tang, Z.-Y. Sun, and C.-H. Kwon, *Mol. Pharm.*, **2**, 118 (2005). |

2005MI1489 A.M. Burger, F. Dai, C.M. Schultes, A.P. Reszka, M.J. Moore, J.A. Double, and S. Neidle, *Cancer Res.*, **65**, 1489 (2005).

2005MI159 A. Scorilas, K. Agiamarnioti, and K. Papadopoulos, *Clin. Chim. Acta*, **357**, 159 (2005).

2005MI183 E.A. Lyakhova, S.A. Lyakhov, L.A. Litvinova, S.A. Andronati, M.N. Lebedyuk, V.P. Fedchuk, and G.A. Khorokhorina, *Pharm. Chem. J.*, **39**, 183 (2005).

2005MI215 K. Agiamarnioti, T. Triantis, D. Dimotikali, and K. Papadopoulos, *J. Photochem. Photobiol. A: Chem.*, **172**, 215 (2005).

2005MI2312 R. Narayanan, P. Tiwari, D. Inoa, and B.T. Ashok, *Life Sci.*, **77**, 2312 (2005).

2005OL1793 P. Belmont and T. Belhadj, *Org. Lett.*, **7**, 1793 (2005).

2005P1724 D. Naidoo, P.H. Coombes, D.A. Mulholland, N.R. Crouch, and A.J.J. van den Bergh, *Phytochemistry*, **66**, 1724 (2005).

2005SOS(15)987 R.H. Prager and C.M. Williams, In D.StC. Black, editor: *Science of Synthesis*, **Vol. 15**, Thieme Verlag: Stuttgart (2005), p 987.

2005T9075 A. Molinos-Gómez, X. Vidal, M. Maymó, D. Velasco, J. Martorell, and F. López-Calahorra, *Tetrahedron*, **61**, 9075 (2005).

2006ARK111 A. Fraleoni-Morgera and P. Zanirato, *ARKIVOC*, **xii**, 111 (2006).

2006ASC1862 S.K. Schneider, P. Roembke, G.R. Julius, H.G. Raubenheimer, and W.A. Herrmann, *Adv. Synth. Catal.*, **348**, 1862 (2006).

2006BMC334 M.O. Anderson, J. Sherrill, P.B. Madrid, A.P. Liou, J.L. Weisman, J.L. DeRisi, and R.K. Guy, *Bioorg. Med. Chem.*, **14**, 334 (2006).

2006BMCL4913 B.C.H. May, J. Witkop, J. Sherrill, M.O. Anderson, P.B. Madrid, J.A. Zorn, S.B. Prusiner, F.E. Cohen, and R.K. Guy, *Bioorg. Med. Chem. Lett.*, **16**, 4913 (2006).

2006CC4242 C. Wolf, S. Liu, and B.C. Reinhardt, *Chem. Commun.*, 4242 (2006).

2006CPB292 J.D. Wansi, J. Wandji, L.M. Meva'a, A.F.K. Waffo, R. Ranjit, S.N. Khan, A. Asma, C.M. Iqbal, M.-C. Lallemand, F. Tillequin, and Z. Fomum Tanee, *Chem. Pharm. Bull.*, **54**, 292 (2006).

2006EJMC1124 H. Cope, R. Mutter, W. Heal, C. Pascoe, P. Brown, S. Pratt, and B. Chen, *Eur. J. Med. Chem.*, **41**, 1124 (2006).

2006JMC3710 T.-L. Su, Y.-W. Lin, T.-C. Chou, X. Zhang, V.A. Bacherikov, C.-H. Chen, L.F. Liu, and T.-J. Tsai, *J. Med. Chem.*, **49**, 3710 (2006).

2006JMC582 M.J.B. Moore, C.M. Schultes, J. Cuesta, F. Cuanca, M. Gunaratnam, F.A. Tanious, W.D. Wilson, and S. Neidle, *J. Med. Chem.*, **49**, 582 (2006).

2006JMC6591 S. Dollinger, S. Löber, R. Klingenstein, C. Korth, and P. Gmeiner, *J. Med. Chem.*, **49**, 6591 (2006).

2006JPOC148 I. Correia, N. Ronzani, N. Platzer, B.-T. Doan, and J.-C. Beloeil, *J. Phys. Org. Chem.*, **19**, 148 (2006).

2006MI605 J. Šebestík, M. Šafařík, I. Stibor, and J. Hlaváček, *Biopolymers (Pept. Sci.)*, **84**, 605 (2006).

2006MI950 I. Géci, V.V. Filichev, and E.B. Pedersen, *Bioconjugate Chem.*, **17**, 950 (2006).

2006P475 J.D. Wansi, J. Wandji, A.F.K. Waffo, H.E. Ngeufa, J.C. Ndom, S. Fotso, R.P. Maskey, D. Njamen, T.Z. Fomum, and H. Laatsch, *Phytochemistry*, **67**, 475 (2006).

2006S1882 A. Schmidt, A. Beutler, T. Habeck, T. Mordhorst, and B. Snovydovych, *Synthesis*, 1882 (2006).

2006THL127 C. Uyeda, M. Biscoe, P. LePlae, and R. Breslow, *Tetrahedron Lett.*, **47**, 127 (2006).

2007BMC3278 L. Guetzoyan, F. Ramiandrasoa, H. Dorizon, C. Desprez, A. Bridoux, C. Rogier, B. Pradines, and M. Perrée-Fauvet, *Bioorg. Med. Chem.*, **15**, 3278 (2007).

2007BMCL1418 A.L. James, J.D. Perry, A. Rigby, and S.P. Stanforth, *Bioorg. Med. Chem. Lett.*, **17**, 1418 (2007).

2007CEJ3842 N. Agorastos, L. Borsig, A. Renard, P. Antoni, G. Viola, B. Spingler, P. Kurz, and R. Alberto, *Chem. Eur. J.*, **13**, 3842 (2007).

2007CMC3079 M. Kukowska-Kaszuba and K. Dzierzbicka, *Curr. Med. Chem.*, **14**, 3079 (2007).

2007EJMC752 M.-H. David-Cordonnier, M.-P. Hildebrand, B. Baldeyrou, A. Lansiaux, C. Keuser, K. Benzschawel, T. Lemster, and U. Pindur, *Eur. J. Med. Chem.*, **42**, 752 (2007).

2007JMC4096 Z. Yan, S. Sikri, D.L. Beveridge, and A.M. Baranger, *J. Med. Chem.*, **50**, 4096 (2007).

2007JOC8501 D.V. LaBarbera, T.S. Bugni, and C.M. Ireland, *J. Org. Chem.*, **72**, 8501 (2007).

2007JPOC484 I.-S.H. Lee, H.J. Kil, and Y. Ji, *J. Phys. Org. Chem.*, **20**, 484 (2007).

2007MI1067 I.-S.H. Lee and C.K. Lee, *Bull. Korean Chem. Soc.*, **28**, 1067 (2007).

2007MI139 P. Belmont, J. Bosson, T. Godet, and M. Tiano, *Anti-Cancer Agents Med. Chem.*, **7**, 139 (2007).

2007MI1863 S. Vispé, I. Vandenberghe, M. Robin, J.P. Annereau, L. Créancier, V. Pique, J.P. Galy, A. Kruczynski, J.M. Barret, and C. Bailly, *Biochem. Pharm.*, **73**, 1863 (2007).

2007MI205 K. Kusuzaki, H. Murata, T. Matsubara, H. Satonaka, T. Wakabayashi, A. Matsumine, and A. Uchida, *In Vivo*, **21**, 205 (2007).

2007MI233 M. Faisal, M. Abu Tariq, and M. Muneer, *Dyes Pigm.*, **72**, 233 (2007).

2007MI4169 D.W. King, W.J. Cooper, S.A. Rusak, B.M. Peake, J.J. Kiddle, D.W. O'Sullivan, M.L. Melamed, C.R. Morgan, and S.M. Theberge, *Anal. Chem.*, **79**, 4169 (2007).

2007MI65 F. Tillequin, *Phytochem. Rev.*, **7**, 65 (2007).

2007MI662 S.C. Bhakdi, P. Sratongno, P. Chimma, T. Rungruang, A. Chuncharunee, H.P.H. Neumann, P. Malasit, and K. Pattanapanyasat, *Cytometry Part A*, **71A**, 662 (2007).

2007MI679 M. Gunaratnam, O. Greciano, C. Martins, A.P. Reszka, C.M. Schultes, H. Morjani, J.F. Riou, and S. Neidle, *Biochem. Pharmacol.*, **74**, 679 (2007).

2007MI93 L. Janovec, D. Sabolová, M. Kožurková, H. Paulíkova, P. Kristian, J. Ungvarský, E. Moravčíková, M. Bajdichová, D. Podhradský, and J. Imrich, *Bioconjugate Chem.*, **18**, 93 (2007).

2007PLM41 B. Réthy, I. Zupkó, R. Minorics, J. Hohmann, I. Ocsovszki, and G. Falkay, *Planta Med.*, **73**, 41 (2007).

2007T11048 M. Bobin, A. Kwast, and Z. Wróbel, *Tetrahedron*, **63**, 11048 (2007).

2007TL1099 T. Shoji, R. Yokoyama, S. Ito, M. Watanabe, K. Toyota, M. Yasunami, and N. Morita, *Tetrahedron Lett.*, **48**, 1099 (2007).

2008AGE8661 C. Gunanathan and D. Milstein, *Angew. Chem. Int. Ed.*, **47**, 8661 (2008).

2008BMC10022 J.A. Clement, J. Kitagaki, Y. Yang, C.J. Saucedo, B.R. O'Keefe, A.M. Weissman, T.C. McKee, and J.B. McMahon, *Bioorg. Med. Chem.*, **16**, 10022 (2008).

2008BMC6560 V.S. Nukoolkarn, S. Saen-oon, T. Rungrotmongkol, S. Hannongbua, K. Ingkaninan, and K. Suwanborirux, *Bioorg. Med. Chem.*, **16**, 6560 (2008).

2008CPL378 A. Mohajeri and A. Ashrafi, *Chem. Phys. Lett.*, **458**, 378 (2008).

2008EJO4313 A. Schmidt, B. Snovydovych, and S. Hemmen, *Eur. J. Org. Chem.*, 4313 (2008).

2008H1943 T. Higashi, M. Sakamoto, and M. Mochizuki, *Heterocycles*, **75**(8), 1943 (2008).

2008JHC837 E.C. Creencia, K. Taguchi, and T. Horaguchi, *J. Heterocycl. Chem.*, **45**, 837 (2008).

2008JMC7574 Z. Ma, J.R. Choudhury, M.W. Wright, C.S. Day, G. Saluta, G.L. Kucera, and U. Bierbach, *J. Med. Chem.*, **51**, 7574 (2008).

2008JOC4267 S. Liu, J.P.C. Pestano, and C. Wolf, *J. Org. Chem.*, **73**, 4267 (2008).

2008JPC(B)9243 D. Sarkar, P. Das, S. Basak, and N. Chattopadhyay, *J. Phys. Chem. B*, **112**, 9243 (2008).

2008MI1211 P. Belmont and I. Dorange, *Expert Opin. Ther. Pat.*, **18**, 1211 (2008).

2008MI88 R.P. Verma and C. Hansch, *J. Pharm. Sci.*, **97**, 88 (2008).

2008NPR166 J.P. Michael, *Nat. Prod. Rep.*, **25**, 166 (2008).

2008P2616 A. Chukaew, C. Ponglimanont, C. Karalai, and S. Tewtrakul, *Phytochemistry*, **69**, 2616 (2008).

2008S2798 A. Schmidt and B. Snovydovych, *Synthesis*, 2798 (2008).

2008TL1261 H.N. Lee, H.N. Kim, K.M.K. Swamy, M.S. Park, J. Kim, H. Lee, K.-H. Lee, S. Park, and J. Yoon, *Tetrahedron Lett.*, **49**, 1261 (2008).

2008TL3663 S. Petit, R. Azzouz, C. Fruit, L. Bischoff, and F. Marsais, *Tetrahedron Lett.*, **49**, 3663 (2008).

2009ARK(XII)31 V. Srinivas and K.C.K. Swamy, *ARKIVOC*, **XII**, 31 (2009).

2009BMC4836 L. Bouffier, R. Dinica, J. Debray, P. Dumya, and M. Demeunynck, *Bioorg. Med. Chem. Lett.*, **19**, 4836 (2009).

2009EJMC3880 M. Laronze-Cochard, Y.-M. Kim, B. Brassart, J.-F. Riou, J.-Y. Laronze, and J. Sapi, *Eur. J. Med. Chem.*, **44**, 3880 (2009).

2009EJMC4758 P. Peixoto, W. Zeghida, D. Carrez, T.-D. Wu, N. Wattez, A. Croisy, M. Demeunynck, J.-L. Guerquin-Kern, and A. Lansiaux, *Eur. J. Med. Chem.*, **44**, 4758 (2009).

2009JCS(D)1223 S. Bartoli, C. Bazzicalupi, S. Biagini, L. Borsari, A. Bencini, E. Faggi, C. Giorgi, C. Sangregorio, and B. Valtancoli, *Dalton Trans.*, 1223 (2009).

2009JMC6519 L.C. Eiter, N.W. Hall, C.S. Day, G. Saluta, G.L. Kucera, and U. Bierbach, *J. Med. Chem.*, **52**, 6519 (2009).

2009MI1351 J. You, H. Zhao, Z. Sun, L. Xia, T. Yan, Y. Suo, and Y. Li, *J. Sep. Sci.*, **32**, 1351 (2009).

2009MOL1725 A.D.N. Brenno and A.A.M. Lapis, *Molecules*, **14**, 1725 (2009).

2009OBC386 R.C. Brown, Z. Li, A.J. Rutter, X. Mu, O.H. Weeks, K. Smith, and I. Weeks, *Org. Biomol. Chem.*, **7**, 386 (2009).

2009T(A)2795 J. Kertész, P. Huszthy, A. Kormos, F. Bertha, V. Horváth, and G. Horvai, *Tetrahedron: Asymmetry*, **20**, 2795 (2009).

2009TL6894 N. Desbois, A. Szollosi, A. Maisonial, V. Weber, E. Moreau, J.-C. Teulade, O. Chavignon, Y. Blache, and J.M. Chezal, *Tetrahedron Lett.*, **50**, 6894 (2009).

2010AG2851 A. Schmidt, N. Münster, and A. Dreger, *Angew. Chem.*, **122**, 2851 (2010).

2010AGE2790 A. Schmidt, N. Münster, and A. Dreger, *Angew. Chem. Int. Ed.*, **49**, 2790 (2010).

2010ARK338 A.A. Khalaf, A.M. El-Khawaga, I.M. Awad, and H.A.K. Abd El-Aal, *ARKIVOC*, **X**, 338 (2010).

2010BMC6324 M.M. Patel, M.D. Mali, and S.K. Patel, *Bioorg. Med. Chem. Lett.*, **20**, 6324 (2010).

2010BMCL3491 G.D. Cuny, M. Robin, N.P. Ulyanova, D. Patnaik, V. Pique, G. Casano, J.-F. Liu, X. Lin, J. Xian, M.A. Glicksman, R.L. Stein, and J.M.G. Higgins, *Bioorg. Med. Chem. Lett.*, **20**, 3491 (2010).
2010EJMC1772 T. Chandra, N. Garg, S. Lata, K.K. Saxena, and A. Kumar, *Eur. J. Med. Chem.*, **45**, 1772 (2010).
2010EJO4296 A. Dreger, R. Cisneros Camuña, N. Münster, T.A. Rokob, I. Pápai, and A. Schmidt, *Eur. J. Org. Chem.*, 4296 (2010).
2010JA12263 S. Sparapani, S.M. Haider, F. Doria, M. Gunaratnam, and S. Neidle, *J. Am. Chem. Soc.*, **132**, 12263 (2010).
2010JA14006 C. Wang, S. Rakshit, and F. Glorius, *J. Am. Chem. Soc.*, **132**, 14006 (2010).
2010JA14048 D. Tsvelikhovsky and S.L. Buchwald, *J. Am. Chem. Soc.*, **132**, 14048 (2010).
2010JNP1044 N. Bontemps, D. Bry, S. López-Legentil, A. Simon-Levert, C. Long, and B. Banaigs, *J. Nat. Prod.*, **73**, 1044 (2010).
2010JOC204 T. Mahmood, A. Paul, and S. Ladame, *J. Org. Chem.*, **75**, 204 (2010).
2010JOC2289 D.C. Rogness and R.C. Larock, *J. Org. Chem.*, **75**, 2289 (2010).
2010JPC(A)10550 K. Krzymiński, A.D. Roshal, B. Zadykowicz, A. Białk-Bielińska, and A. Sieradzan, *J. Phys. Chem. A*, **114**, 10550 (2010).
2010JPO382 A.A. Taherpour, D. Kvaskoffa, P.V. Bernhardta, and C. Wentrup, *J. Phys. Org. Chem.*, **23**, 382 (2010).
2010MI1243 R. Ghosh, S. Bhowmik, A. Bagchi, D. Das, and S. Ghosh, *Eur. Biophys. J.*, **39**, 1243 (2010).
2010MI1769 X. Wei, T.S. Bugni, M.K. Harper, I.T. Sandoval, E.J. Manos, J. Swift, R.M. Van Wagoner, D.A. Jones, and C.M. Ireland, *Mar. Drugs*, **8**, 1769 (2010).
2010MI614 D. Chen, Q. Xu, X. Xia, J. Ge, J. Lu, and N. Li, *Mat. Chem. Phys.*, **120**, 614 (2010).
2010MI74 N.J. Baumhover, K. Anderson, C.A. Fernandez, and K.G. Rice, *Bioconjugate Chem.*, **21**, 74 (2010).
2011ACR588 C. Gunanathan and D. Milstein, *Acc. Chem. Res.*, **44**, 588 (2011).
2011BMC1477 R.P. Severino, R.V.C. Guido, E.F. Marques, D. Brömme, M.F.D.G.F. da Silva, J.B. Fernandes, A.D. Andricopulo, and P.C. Vieira, *Bioorg. Med. Chem.*, **19**, 1477 (2001).
2011BMC1790 L. Janovec, M. Kožurková, D. Sabolová, J. Ungvarský, H. Paulíková, J. Plšíková, Z. Vantová, and J. Imrich, *Bioorg. Med. Chem.*, **19**, 1790 (2011).
2011CC5079 S.C. To and F.Y. Kwong, *Chem. Commun.*, **47**, 5079 (2011).
2011COC2897 A. Schmidt and A. Dreger, *Curr. Org. Chem.*, **15**, 2897 (2011).
2011EJO5311 T. Shoji, S. Ito, J. Higashi, and N. Morita, *Eur. J. Org. Chem.*, 5311 (2011).
2011JA3780 J. Kwak, M. Kim, and S. Chang, *J. Am. Chem. Soc.*, **133**, 3780 (2011).
2011JOC7005 Y. Kuninobu, T. Tatsuzaki, T. Matsuki, and K. Takai, *J. Org. Chem.*, **76**, 7005 (2011).
2011MI T.A. Wynn, In J.J. Li, editor: *Name Reactions for Heterocyclic Chemistry II*, Hoboken: New Jersey (2011), p 360.
2011MI11 A.F.-C. Valdés, *Open Med. Chem. J.*, **5**, 11 (2011).
2011MI126 Y. Wang, X.Y. Hu, L. Wang, Z.B. Shang, J.B. Chao, and W.J. Jin, *Sens. Act. B*, **156**, 126 (2011).
2011MI131 S. Das and A.J. Thakur, *Green Chem. Lett. Rev.*, **4**, 131 (2011).
2011MI193 N.P. Singh, R. Kumar, D.N. Prasad, S. Sharma, and O. Silakari, *Int. J. Biol. Chem.*, **5**, 193 (2011).

2011MI2945 B. Zomer, L. Collé, A. Jedyńska, G. Pasterkamp, I. Kooter, and H. Bloemen, *Anal. Bioanal. Chem.*, **401**, 2945 (2011).

2011MI305 G. Cholewiński, K. Dzierzbicka, and A.M. Kołodziejczyk, *Pharm. Rep.*, **63**, 305 (2011).

2011T3803 C.-L. Do-Thanh, M.M. Rowland, and M.D. Best, *Tetrahedron*, **67**, 3803 (2011).

2012CC308 I. Hyodo, M. Tobisu, and N. Chatani, *Chem. Commun.*, **48**, 308 (2012).

2012CHE301 N.E. Shchepina, V.V. Avrorin, G.A. Badun, S.B. Lewis, and S.E. Ukhanov, *Chem. Heterocycl. Compd. Engl. Trans.*, **48**, 301 (2012).

2012EJO6586 Z. Huang, Y. Yang, Q. Xiao, Y. Zhang, and J. Wang, *Eur. J. Org. Chem.*, 6586 (2012).

2012JA14099 R.P. Johnson, J.A. Richardson, T. Brown, and P.N. Bartlett, *J. Am. Chem. Soc.*, **134**, 14099 (2012).

2012JFC272 A.F. Shidlovskii, A.S. Golubev, D.V. Gusev, K.Y. Suponitsky, A.S. Peregudov, and N.D. Chkanikov, *J. Fluor. Chem.*, **143**, 272 (2012).

2012JMC10198 S. Ding, X. Qiao, G.L. Kucera, and U. Bierbach, *J. Med. Chem.*, **55**, 10198 (2012).

2012JMC5231 X. Chen, K. Zenger, A. Lupp, B. Kling, J. Heilmann, C. Fleck, B. Kraus, and M. Decker, *J. Med. Chem.*, **55**, 5231 (2012).

2012JOC11232 A.V. Dubrovskiy and R.C. Larock, *J. Org. Chem.*, **77**, 11232 (2012).

2012JOC5173 X.-D. Han, Y.-L. Zhao, J. Meng, C.-Q. Ren, and Q. Liu, *J. Org. Chem.*, **77**, 5173 (2012).

2012MI1109 L. Bessi, C. Bazzicalupi, C. Richter, H.R.A. Jonker, K. Saxena, C. Sissi, M. Chioccioli, S. Bianco, A.R. Bilia, H. Schwalbe, and P. Gratteri, *Chem. Biol.*, **7**, 1109 (2012).

2012MI2736 F. Karagöz, O. Güney, M. Kandaz, and A.T. Bilgiçli, *J. Luminescence*, **132**, 2736 (2012).

2012MI330 M.A.V. Ribeiro da Silva, T.L.P. Galvão, I.M. Rocha, and A.F.L.O.M. Santos, *J. Chem. Thermodyn.*, **54**, 330 (2012).

2012MI5966 A. Kupka, V. Vasylyeva, D.W.M. Hofmann, K.V. Yusenko, and K. Merz, *Cryst. Growth Des.*, **12**, 5966 (2012).

2012OL4250 T. Tu, Z. Sun, W. Fang, M. Xu, and Y. Zhou, *Org. Lett.*, **14**, 4250 (2012).

2013AGE14131 D. Srimani, Y. Diskin-Posner, Y. Ben-David, and D. Milstein, *Angew. Chem. Int. Ed.*, **52**, 14131 (2013).

2013AGE3949 Q. Zhou, A. Ruffoni, R. Gianatassio, Y. Fujiwara, E. Sella, D. Shabat, and P.S. Baran, *Angew. Chem. Int. Ed.*, **52**, 3949 (2013).

2013AGE6983 J. Wu, D. Talwar, S. Johnston, M. Yan, and J. Xiao, *Angew. Chem. Int. Ed.*, **52**, 6983 (2013).

2013BMC4170 X. Lang, L. Li, Y. Chen, Q. Sun, Q. Wu, F. Liu, C. Tan, H. Liu, C. Gao, and Y. Jiang, *Bioorg. Med. Chem.*, **21**, 4170 (2013).

2013CCC2183 D. Ge, L. Hu, J. Wang, X. Li, F. Qi, J. Lu, X. Cao, and H. Gu, *Chem. Cat. Chem.*, **5**, 2183 (2013).

2013CEJ4271 J. Yu, H. Yang, Y. Jiang, and H. Fu, *Chem. Eur. J.*, **19**, 4271 (2013).

2013EJMC469 T. Fuchigami, N. Kobashi, M. Haratake, M. Kawasaki, and M. Nakayama, *Eur. J. Med. Chem.*, **60**, 469 (2013).

2013JA12548 Y. Lian, J.R. Hummel, R.G. Bergman, and J.A. Ellman, *J. Am. Chem. Soc.*, **135**, 12548 (2013).

2013JMCC63 P. Mignon, M. Tiano, P. Belmont, A. Favre-Réguillon, H. Chermette, and F. Fache, *J. Mol. Catal., A: Chem.*, **371**, 63 (2013).

2013JNP1801 N. Bontemps, F. Gattacceca, C. Long, O.P. Thomas, and B. Banaigs, *J. Nat. Prod.*, **76**, 1801 (2013).

2013JOC7436 Z. Liu, N. Dong, M. Xu, Z. Sun, and T. Tu, *J. Org. Chem.*, **78**, 7436 (2013).

2013MI1220 R. Kumar, M. Kaur, and O. Silakari, *Mini-Rev. Med. Chem.*, **13**, 1220 (2013).

2013MI1921 M. Manikandana and H.-F. Wu, *Photochem. Photobiol. Sci.*, **12**, 1921 (2013).

2013MI298 W. Sun, Y. Wang, S. Gong, Y. Cheng, F. Shi, and Z. Sun, *Electrochim. Acta*, **109**, 298 (2013).

2013MI909 S.K.H. Nejad-Darzi, A. Samadi-Maybodi, and M. Ghobakhluo, *J. Porous Mater.*, **20**, 909 (2013).

2013MMCIAE1256 M.R. Galdino-Pitta, M.G.R. Pitta, M.C.A. Lima, S.L. Galdino, and I.R. Pitta, *Mini-Rev. Med. Chem.*, **13**, 1256 (2013).

2013OBC3375 D. Zhang, X. Jiang, H. Yang, A. Martinez, M. Feng, Z. Dong, and G. Gao, *Org. Biomol. Chem.*, **11**, 3375 (2013).

2013OBC3558 Z. Guan, S. Wiechmann, M. Drafz, E. Hübner, and A. Schmidt, *Org. Biomol. Chem.*, **11**, 3558 (2013).

2013OL5460 H.-M. Guo, R.Z. Mao, Q.-T. Wang, H.-Y. Niu, M.-S. Xie, and G.-R. Qu, *Org. Lett.*, **15**, 5460 (2013).

2013OM6712 E.R. Clark and M.J. Ingleson, *Organometallics*, **32**, 6712 (2013).

2013T10457 D. Zhang, X. Jiang, Z. Dong, H. Yang, A. Martinez, and G. Gao, *Tetrahedron*, **69**, 10457 (2013).

2014BJOC832 Z. Guan, J.C. Namyslo, M.H.H. Drafz, M. Nieger, and A. Schmidt, *Beilstein J. Org. Chem.*, **10**, 832 (2014).

2014BMC3014 A.R. Benoit, C. Schiaffo, C.E. Salomon, J.R. Goodell, H. Hiasa, and D.M. Ferguson, *Bioorg. Med. Chem. Lett.*, **24**, 3014 (2014).

2014BMC4684 L. Čižeková, A. Grolmusová, Z. Ipóthová, Z. Barbieriková, V. Brezová, L. Hunáková, J. Imrich, L. Janovec, I. Dovinová, and H. Paulíková, *Bioorg. Med. Chem.*, **22**, 4684 (2014).

2014EJO4170 X. Chen, Y. Zhang, Y. Chen, J. Zhang, J. Chen, M. Li, W. Cao, and J. Chen, *Eur. J. Org. Chem.*, 4170 (2014).

2014H(88)535 G.W. Gribble, M.D. Mosher, G.D. Jaycox, M. Cory, and T.A. Fairley, *Heterocycles*, **88**, 535 (2014).

2014H(89)1877 G. Bratulescu, *Heterocycles*, **89**, 1877 (2014).

2014H(89)2356 Z. Guan, M. Gjikaj, and A. Schmidt, *Heterocycles*, **89**, 2356 (2014).

2014IC11540 K.E. Siters, M.A. Fountain, and J.R. Morrow, *Inorg. Chem.*, **53**, 11540 (2014).

2014JA13194 R.-Y. Zhu, J. He, X.-C. Wang, and J.-Q. Yu, *J. Am. Chem. Soc.*, **136**, 13194 (2014).

2014JCP(A)11591 S. Radenković, J. Kojić, J. Petronijević, and M. Antić, *J. Phys. Chem. A*, **118**, 11591 (2014).

2014JOC7451 Z. Zheng, L. Dian, Y. Yuan, D. Zhang-Negrerie, Y. Du, and K. Zhao, *J. Org. Chem.*, **79**, 7451 (2014).

2014JPC(B)11090 K. Vivek, A. Sengupta, K. Gavvala, R.K. Koninti, and P. Hazra, *J. Phys. Chem. B*, **118**, 11090 (2014).

2014MI1 W. Zhang, G. Wang, J.R. Falconer, B.C. Baguley, J.P. Shaw, J. Liu, H. Xu, E. See, J. Sun, J. Aa, and Z. Wu, *Pharm. Res.* Online. http://dx.doi.org/10.1007/s11095-014-1551-8.

2014MI13 J. Ungvarsky, J. Plsikova, L. Janovec, J. Koval, J. Mikes, L. Mikesová, D. Harvanova, P. Fedorocko, P. Kristian, J. Kasparkova, V. Brabec, M. Vojtickova, D. Sabolova, Z. Stramova, J. Rosocha, J. Imrich, and M. Kozurkova, *Bioorg. Chem.*, **57**, 13 (2014).

2014MI150 C.-H. Liu, S.L. Sahoo, and M.-H. Tsao, *Colloids Surf. B: Biointerfaces*, **115**, 150 (2014).

2014MI16 J.P. Joubert, F.J. Smit, L. du Plessis, P.J. Smith, and D.D. N'Da, *Eur. J. Pharm. Sci.*, **56**, 16 (2014).

2014MI169 Q. Li, K. Xu, P. Song, Y. Dai, L. Yang, and X. Pang, *Dyes Pigm.*, **109**, 169 (2014).

2014MI2 S. Arya, A. Kumar, N. Kumar, P. Roy, and S.M. Sondhi, *Med. Chem. Res.* http://dx.doi.org/10.1007/s00044-014-1268-6.

2014MI2366 D. Srikanta, D. Prusty, D. Dhayal, M.K. Gupta, A. Dar, S. Sen, P. Mukhopadhyay, T. Adak, and S.K. Dhar, *Chem. Biol.*, **9**, 2366 (2014).

2014MI2903 H. Kaplan Can, G. Karakus, and N. Tuzcu, *Polym. Bull.*, **71**, 2903 (2014).

2014MI320 C.-C. Yu, S.-L. Pan, S.-W. Chao, S.-P. Liu, J.-L. Hsu, Y.-C. Yang, T.-K. Li, W.-J. Huang, and J.-H. Guh, *Biochem. Pharm.*, **90**, 320 (2014).

2014PEA320 D. Patra, F. Sleem, Colloids, and A. Surfaces, *Physicochem. Eng. Aspects*, **443**, 320 (2014).

2014MI388 X.-J. Jiang, Y. Fu, L.-H. Xu, H.-L. Lu, S.-Q. Zang, M.-S. Tang, and T.C.W. Mak, *Sens. Actuators B*, **202**, 388 (2014).

2014MI43 W. Ju, M. Zhang, M. Petrus, M. Maeda, C.A. Pise-Masison, and T.A. Waldmann, *Retrovirology*, **11**, 43 (2014).

2014MI528 E. See, W. Zhang, J. Liu, D. Svirskis, B.C. Baguley, J.P. Shawa, G. Wang, and Z. Wu, *Int. J. Pharm.*, **473**, 528 (2014).

2014MI636 P. Mitra, P.K. Chakraborty, P. Saha, P. Ray, and S. Basu, *Int. J. Pharm.*, **473**, 636 (2014).

2014MI647 B. Zhang, X. Li, B. Li, C. Gao, and Y. Jiang, *Expert Opin. Ther. Pat.*, **24**, 647 (2014).

2014MI695 V. Murray, J.K. Chen, and A.M. Galea, *Anti-Cancer Agents Med. Chem.*, **14**, 695 (2014).

2014MI8093 M. Kara, O. Yumrutas, R. Atilgan, M. Baspinar, E. Sapmaz, and T. Kuloglu, *Mol. Biol. Rep.*, **41**, 8093 (2014).

2014MI997 H.W. Kava, A.M. Galea, F.Md. Jamil, Y. Feng, and V. Murray, *J. Biol. Inorg. Chem.*, **19**, 997 (2014).

2014NJC1004 K.-X. Xu, H.-J. Kong, P. Li, L. Yang, J.-L. Zhang, and C.-J. Wang, *New J. Chem.*, **38**, 1004 (2014).

2014OBC6677 S. Basili, T. Del Giacco, F. Elisei, and R. Germani, *Org. Biomol. Chem.*, **12**, 6677 (2014).

2014OL18 Q. Su, P. Li, M. He, Q. Wu, L. Ye, and Y. Mu, *Org. Lett.*, **16**, 18 (2014).

2014OL2150 M. Yousuf, N. Ahmed, B. Shirinfar, V.M. Miriyala, I.S. Youn, and K.S. Kim, *Org. Lett.*, **16**, 2150 (2014).

2014T5820 J. Zhang, Y. Chen, X. Chen, X. Zheng, W. Cao, J. Chen, and M. Zhang, *Tetrahedron*, **70**, 5820 (2014).

2014T944 M. Vilková, M. Prokaiová, and J. Imrich, *Tetrahedron*, **70**, 944 (2014).

2014THL3308 S.C. Johns, L.L.E. Crouch, S. Grieve, H.L. Maloney, G.R. Peczkowski, A.E. Jones, D. Sharp, and R.B. Smith, *Tetrahedron Lett.*, **55**, 3308 (2014).

| | |
|---|---|
| 2015JPC(A)127 | R.G.A.R. Maclagan, S. Gronert, and M. Neot-Ner (Mautner), *J. Chem. Phys. A*, **119**, 127 (2015). |
| 2015MI33 | Y. Zhou, Q. Huang, Q. Zhang, Y. Min, and E. Wang, *Spectrochim. Acta Part A: Mol. Biomol. Spec.*, **137**, 33 (2015). |
| 2015MI118 | A. Wiseman, L.A. Sims, R. Snead, S. Gronert, R.G.A.R. Maclagan, and M. Meot-Ner (Mautner), *J. Phys. Chem. A*, **119**, 118 (2015). |
| 2015MI1292 | K.B. Ayaz Ahmed, E. Reshma, M. Mariappan, and V. Anbazhagan, *Spectrochim. Acta Part A: Mol. Biomol. Spectros.*, **137**, 1292 (2015). |
| 2015ZN(B)83 | Z. Guan, M. Gjikaj, and A. Schmidt, *Z. Naturforsch.*, **70b**, 83 (2015). |

# INDEX

## B

## C

## D

Kolibri 360 GmbH
Ferdinand-Jühlke-Straße 7
99095 Erfurt, Deutschland
produktsicherheit@kolibri360.de